DEEP-WATER CARBONATE ENVIRONMENTS

Based on a Symposium
Sponsored by The Society of
Economic Paleontologists and Mineralogists

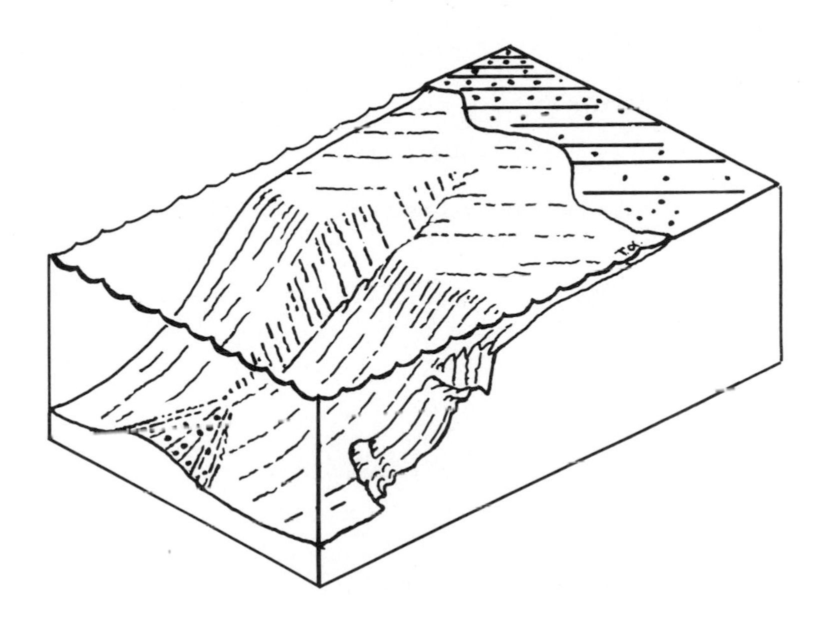

Edited by

Harry E. Cook, U.S. Geological Survey, Menlo Park, California
and
Paul Enos, State University of New York, Binghamton

SOCIETY OF ECONOMIC PALEONTOLOGISTS AND MINERALOGISTS
Special Publication No. 25
Tulsa, Oklahoma, U.S.A. November, 1977

A Publication of

The Society of Economic Paleontologists and Mineralogists

a division of

The American Association of Petroleum Geologists

PREFACE

A symposium on basinal carbonate facies was held in Dallas, Texas, on April 8, 1975, at the joint meeting of the American Association of Petroleum Geologists and the Society of Economic Paleontologists and Mineralogists. The technical program consisted of two half-day sessions during which fourteen papers were presented on carbonate facies in deep-water environments. The Dallas symposium was organized by Lloyd C. Pray and James Lee Wilson. After the symposium the authors of the papers accepted the offer of the Society of Economic Paleontologists and Mineralogists to publish the symposium as one of the Special Publication series of that society. Two of the contributors to the symposium, Harry E. Cook and Paul Enos, agreed to act as editors for Special Publication No. 25.

Prior to the symposium a three day field trip to study Mississippian and Permian carbonate basinal facies in southern New Mexico and west Texas was led by W. J. Meyers (State University of New York, Stony Brook), L. C. Pray (University of Wisconsin), J. L. Wilson (Rice University), and D. M. Yurewicz (University of Wisconsin). This excellent trip provided background and additional perspective for the basinal carbonate facies symposium that followed in Dallas.

The AAPG has given permission for publication of this volume. The editors are greatly appreciative for the cooperation and help of the authors of this volume and the many other people who worked hard to meet all deadlines so that this collection of papers could become a reality. We heartily thank Terry Coit, Jean Newcombe, Carol Hirozawa, Kay Kinoshita, Sandy Maroney and LaVernne Hutchison of the U.S. Geological Survey, Menlo Park, who typed much of the volume and helped in the preparation of the index and correction of the printer's proofs. Doris Woofter of the SEPM headquarters office helped us in many ways during the critical last stages of this volume. Raymond Ethington, Chairman of the SEPM Publications Committee offered valuable assistance throughout the evolution of this publication and was especially helpful during the final editorial phases.

Special thanks goes to Judy Cook who spent many days helping the editors index this book.

HARRY E. COOK AND PAUL ENOS
Editors

CONTENTS

SEPM Special Publication No. 25, p. 1-3, November 1977

DEEP-WATER CARBONATE ENVIRONMENTS—AN INTRODUCTION

HARRY E. COOK and PAUL ENOS
U.S. Geological Survey, Menlo Park, California 94025, and
State University of New York at Binghamton, Binghamton, New York 13901

The need to expand our search for energy resources into deeper marine environments makes a volume devoted entirely to deep-water carbonates most timely. It has become increasingly important to better understand deep marine environments and their transitions to shoal water settings. Of importance are the processes operating in deep-water settings and the magnitude and periodicity of these processes especially as they affect the geometry, textural character, and overall lithologic and biologic patterns of the sediment.

Research concerned with carbonates deposited in deep-water environments has tended to lag behind that in shallow water carbonates. This is inevitable because many of our modern concepts of carbonate sedimentology were developed by major oil company research centers. Research by these groups focused on modern and ancient shoal-water facies because of the large number of reservoirs formed in rocks of these facies. Furthermore, modern shoal-water environments are more accessible for direct examination, promoting comparative studies between ancient and modern carbonate sediment. However, within the past 5 to 10 years accelerated effort has been directed toward deeper water settings by industry, academia, and government. A manifestation of this was the flood of abstracts which greeted the original call for a symposium on "the deep, dark, and dirty" by Lloyd C. Pray and James Lee Wilson through the SEPM carbonate research group. The flood was of such magnitude that the planned half-day symposium was doubled in length and narrowed in focus by shunting many papers to other sessions.

This increased interest has been due in part to the JOIDES Deep Sea Drilling Project, the recent policy of the Department of the Interior to accelerate the leasing of continental margins of the United States, the increased world wide exploration effort directed toward energy sources on continental margins, and the necessity to conduct research on the stability and other environmental parameters of marine slopes involved with exploitation on continental margins and deep marine settings.

This volume represents contributions from researchers who have diverse perspectives. Their papers include new data on deep-water carbonate environments ranging in age from the Cambrian to the present with examples from Canada, the United States, Mexico, Africa, and from modern ocean basins. The geologic settings of these studies are also quite varied and include modern and ancient deep open oceans, continental slopes and shelves and continental-interior basins. The order of presentation of the papers is stratigraphic except for the first two which are more general.

Because modern deep-water environments are largely inaccessible to direct observation, with the notable exception of submersibles, the authors often use the approach that "the ancient can provide a key to the present." In the ancient environment we have three-dimensional control on stratigraphic features and can study the evolution of these features through time.

The authors of this volume not only point out criteria for making environmental interpretations and stratigraphic predictions but also attempt to elucidate processes operating in these environments and how these processes contribute to the overall development of the stratigraphic sequence. Their papers clearly demonstrate that deep-water carbonate environments are not passive; rather they are active areas where both depositional processes and significant amounts of mass wasting processes contribute to their development. The outstanding example of this is the recurring theme of small-to-large surfaces, normally concave up, which truncate underlying strata until they merge tangentially with underlying beds and lose their identity. The surfaces appear to be of submarine origin. The beds above the surface are usually lithologically like those which underlie it and parallel the surface. Such surfaces are termed "intraformational truncation surfaces" by most authors, at the suggestion of the editors, to establish a uniform non-genetic term.

In the first paper, Byers succinctly develops a general model for biofacies patterns in basins based on a zonation of oxygen content by water depth. The zonation goes from well-oxygenated surface water (characterized by shelly fauna, infauna bioturbation) to marginally oxygenated waters at greater depths (shelly fauna absent, bioturbation by infauna persists) to deeper water that completely lacks oxygen (shelly fauna absent, bioturbation lacking, sediments laminated). Byers points out that recognition of these facies can assist in establishing paleoslope directions, basin contours, and paleohydrography. The elegant simplicity of Byers' model makes it readily adaptable to testing in the field.

Fischer and Arthur follow with a paper on secular variations in pelagic environments. Using an abundance of data from JOIDES Deep Sea Drilling Project and field studies they develop a global model that includes a biotic and lithologic cycle with a rhythm of about 32 million years for the past several hundred million years. These cycles are characterized by times of maximum diversity of pelagic biotas that alternate with times of lowered diversity of pelagic life. Maximum diversity of pelagic biota coincides with higher and more even oceanic temperatures, widespread marine anaerobism and several other parameters. Periods of lowered diversity of pelagic biotas are accompanied by lower and less uniform marine temperatures, lack of anaerobic marine sediment and other variables. The authors suggest that these cycles may directly influence the sedimentation patterns of petroleum source rocks, major phosphate deposits, deep-sea carbonate and biogenic silica. This very thought-provoking paper suggests that changes in rates of oceanic circulation resulting from climatic variation cause the apparent cyclicity. The climatic variations may result from solar energy fluctuations or from changes that are internal to the earth.

Cook and Taylor show the value of an intergrated sedimentological and paleontological approach in the study of a coeval deep-water to shoal-water continental margin in the Cordillera of the western United States. Their study shows that a seaward progradation of the continental margin took place from the Late Cambrian through the earliest Ordovician. This seaward progradation of environments appears to be a mirror image of sedimentological patterns that occur during the same span of time on the continental margin of the eastern United States as described by Reinhardt (this volume). Cook and Taylor stress that massflow and slump deposits make up at least onefourth of the deep-water continental slope section. Because most of this allochthonous sediment is relatively thin bedded (less than 2 m thick), they speculate that the smooth undisturbed nature of some modern continental slopes as seen on seismic profiles may be in part a function of the limited resolving power of much surface seismic equipment. On the basis of stratigraphic and geographic reconstructions they estimate water depth on the slope was on the order of 1500 m.

Reinhardt's study of the Cambrian off-shelf sedimentation in the central Appalachians documents a mirror-image of what is found in the western Cordillera (see Cook and Taylor, this volume). The possible factors responsible for the development of seaward progradation on both sides of the continent is an intriguing question with regional significance. Reinhardt shows that both erosional and depositional processes were active during the development of a broad carbonate shelf and an adjacent slope and basin. Tidal channels and submarine canyons were local features at or near the continental shelf margin and water depths were as great as 500 m in the basin.

McIlreath discusses a debris apron of Cambrian deep-water limestone that developed adjacent to a rather unusual, nearly vertical submarine carbonate escarpment in the southern Rocky Mountains of Alberta, Canada. This apron was deposited in water depths of about 200 m. Lateral facies changes in this debris apron are quite sharp within the 3-km-wide apron exposed. Submarine sliding was important in transporting the debris apron further basinward. Good faunal control within individual slide masses shows that the youngest of a repetitive series of slides extends furthest into the basin.

In the next paper Hubert, Suchecki, and Callahan interpret the well-known Cow Head Breccia of Newfoundland as a continental-margin slope sequence within a Taconic-age klippe. They trace the history of the western side of the proto-Atlantic ocean from Middle Cambrian through Middle Ordovician time. This slope sequence contains spectacular allochthonous carbonate megabreccias containing clasts up to 60 by 150 m in size. As the proto-Atlantic narrowed, the continental margin rapidly subsided, and red and gray mud and volcanogenic sandstone were deposited. Through detailed analyses of paleocurrent and paleoslope data, the authors show that paleocurrents followed contours and that topography altered as the proto-Atlantic closed.

Hopkins has documented an important new type of basinal breccia in the Upper Devonian of Canada. He interprets breccia clasts as representing differentially cemented foreslope carbonate sand. This cementation is believed to be a synsedimentary process near the sediment-water interface in a submarine environment. Downslope displacement of uncemented carbonate sand intermixed with partially cemented sand resulted in the deposition of breccia beds.

Bissell and Barker have synthesized a great deal of stratigraphic data pertinent to our understanding of deep-water environments in the eastern part of the Cordilleran during the Mississippian. They trace the structural and stratigraphic evolution of this region and discuss numerous lines of evidence that indicate the rocks they studied formed under deep-water conditions.

The following paper by Smith focuses on the Mississippian of Montana. Limestone and dolomite sequences in this area of the northern Rocky

Mountains and adjacent plains document a fluctuation from shallow to deep to shallow marine water. This cyclicity was strongly influenced by both subsidence of an unstable shelf and marine inundation. As in several other papers of this volume intraformational truncation surfaces are present in the deeper water facies that may represent submarine mass movements.

Yurewicz concisely documents the deep-water origin for the Mississippian Rancheria Formation of southern New Mexico and west Texas through the study of stratigraphic sections over an area of 60 by 120 km. Coarse skeletal grainstone deposited by sediment gravity flows at the northern edge of the basin are gradually replaced southward by deeper water hemipelagic and low-energy turbidity-current deposits further basinward. Drawing on biofacies patterns presented in Byers' paper, Yurewicz shows by fluctuations in benthonic fossil abundances and bioturbation patterns that conditions ranged from aerobic to anaerobic during Mississippian time. The Rancheria includes intraformational truncation surfaces that Yurewicz interprets as being submarine in origin and representing broad, shallow channels produced by corrasion.

Similar but larger features are interpreted as a product of submarine sliding by Davies in his study of Pennsylvanian and Permian rocks in the Sverdrup Basin of the Arctic Archipelago. Pelagic carbonate, turbidites and debris sheets were deposited in trough and slope environments seaward of a shallow-water shelf, at water depths estimated to range from 300 to 1200 m. Depositional dips were locally as much as 40°. Davies documents the abundant contribution to this setting by a wide variety of sediment gravity flow deposits and beautifully illustrates spectacularly exposed truncation surfaces. These surfaces are submarine in origin and occur in a setting distal to the zone of debris sheets. Individual truncation surfaces can be traced for at least 1.5 km and have as much as 150 m of section removed. These features are similar both in size and geologic setting to structures in modern continental slopes and rises that are also interpreted as gravity slides and slumps.

During the Jurassic an east-west seaway existed in the present High Atlas Mountains of Morocco. Sediment in this seaway fills a rift zone that developed during the initial breakup of Africa, Europe, and North America and today represents a tectonic scar on the African plate. Evans and Kendall show that initial supratidal carbonate was followed by deep-water carbonate. This deep-water carbonate formed in two major settings within this tectonic seaway—a slope along both margins of the seaway and a basin along the axial part. Water depth in the basin is estimated to have been on the order of 300 to 700 m. Turbidites almost a kilometer thick are present along the slope and base of slope. Characteristic of the axial part of the seaway are deep-water lithoherms similar to those that occur today in the Straits of Florida.

The final two papers in this symposium deal with a facies of the Gulf-Coast Cretaceous limestone which has produced one of the world's giant oil fields in Poza Rica, Veracruz, Mexico. Carrasco's study of the outcropping deep-water carbonate of the Tamaulipas Basin in the Sierra Madre Oriental shows significant differences in the style of sedimentation from each of three areas. These areas all border the larger Valles-San Luis Potosi platform of central Mexico and received major contributions of shallow-water debris. In one area most of the debris is concentrated in two massive beds of coarse breccia, in the second large exotic blocks are of major importance, whereas the third area is characterized by thin beds of finer-grained allochthonous carbonate.

The actual reservoir rocks of Poza Rica and adjacent fields were studied by Enos from cores. Controversy has arisen as to whether the reservoir rock is shallow-water debris resedimented in deep water or an *in situ* reef. Detailed sedimentology and regional stratigraphy lead Enos to the conclusion that the shallow-water component of the rocks was deposited by various types of sediment gravity flow from the adjacent Golden Lane escarpment into water as much as 1000 m deep. This points to the petroleum potential of basinal carbonate.

The data and interpretations presented in this volume provide criteria for making stratigraphic and environmental interpretations in both the surface and subsurface and should serve as a stimulus for new investigations that will further our understanding of the processes operating in deeper water environments. Material presented in these papers also provides a variety of new ideas and interpretations which need future testing and refinement through research in modern and ancient carbonate deposits of this long-neglected realm.

SEPM Special Publication No. 25, p. 5-17, November 1977

BIOFACIES PATTERNS IN EUXINIC BASINS: A GENERAL MODEL

CHARLES W. BYERS
University of Wisconsin, Madison, 53706

ABSTRACT

Basinal sediments in the stratigraphic record can be recognized using a biofacies model based on the distribution of shelly fauna and bioturbation in modern basins. Stagnation in modern basins is produced by density stratification which insulates bottom water from atmospheric oxygen, and by sills which prevent lateral exchange at depth with oxygenated ocean water. Such basins exhibit a tripartite layering in their water columns: (1) a mixed surface layer, well-oxygenated, approximately 50 m deep, (2) a stratified layer (pycnocline) in which oxygen decreases rapidly with depth, approximately 100 m in thickness, (3) a stagnant zone in which oxygen is absent, extending from the 150 m depth down to the basin floor. The major changes in faunal composition take place within the pycnocline, as dissolved oxygen drops below 2 ml/l. Marginally oxygenated waters (dysaerobic) contain communities with lower diversity, generally smaller body sizes, a greater dominance by infauna, and fewer calcified species than communities in well-oxygenated (aerobic) conditions. A complete lack of oxygen (anaerobic) eliminates metazoan animals. Basinal water conditions are reflected by corresponding biofacies:

Aerobic—shelly fauna, infaunal bioturbation
Dysaerobic—shelly fauna lacking, bioturbation by infauna persists
Anaerobic—shelly fauna lacking, bioturbation also lacking; sediments laminated

Recognition of these biofacies in the stratigraphic record should permit the reconstruction of basin contours and hydrography; paleoslope direction and dip angle and absolute water depths should be calculable from the biofacies distribution. Cyclic alternations of normal marine and euxinic facies can be explained as repeated lateral shifts of the basinal environments into a shallow shelf.

An ancient basinal deposit, the Upper Devonian Middlesex Shale of New York, was sampled laterally in order to test the model; because the shale is thought to represent a time plane, facies changes should depict "instantaneous" paleoenvironments in the basin in Late Devonian time. A west-to-east trend was established from laminated unfossiliferous shale through bioturbate non-fossiliferous mudstone to bioturbate and sparsely fossiliferous mudstone, indicating a transect from anaerobic through dysaerobic to nearly aerobic conditions. Basin depth therefore increased westward; the width of outcrop of dysaerobic facies is a function of the bathymetric gradient, calculated as 1:400 (0°09′) for the Middlesex basin. The eastward limit of anaerobic facies marks the ancient pycnocline base at 150 m depth, and provides a bathymetric tie-point on the basin slope. West of this point the basin floor was continuously anoxic, while to the east dysaerobic and aerobic conditions alternated. Alternations are explained by transgression of the pycnocline up onto the eastern shelf, bringing basinal environments into areas which were normally oxygenated. Repeated shifts produced the alternating euxinic and fossiliferous sediments in the Upper Devonian marine sequence.

Comparative examples of euxinic facies from carbonate basins show similar patterns. The Mississippian Rancheria Formation of New Mexico and the Permian Cutoff Formation of West Texas both overstep erosion surfaces cut into fossiliferous shelf carbonates, indicating a shift of basinal conditions shelfward. From published descriptions, both the Rancheria and Cutoff are here interpreted as representing dysaerobic conditions in waters 50-150 m deep. A further example, the shelf carbonates of the Pennsylvanian Paradox Basin, differs from the present model as the euxinic rocks are associated with an evaporite basin. The euxinic carbonates (probably dysaerobic facies) may have resulted from salinity rather than oxygen fluctuations, but the concept of deep basinal water periodically invading the shelf to produce cyclicity is similar to the Middlesex Shale example. Transgressions of the pycnocline may have accompanied actual sea level shifts or have occurred because of a change in mixing depth or sill depth. The latter situation would produce euxinic conditions across a shelf without increasing water depth there. Regressions of the pycnocline would return the shelf to aerobic conditions while the majority of the basin remained anaerobic.

INTRODUCTION

A large part of the stratigraphic record was deposited under conditions rare on the earth today, in epeiric seas, interior basins, and marginal basins, whose connections with the open ocean may have been restricted. For many years, black shales have been interpreted as products of restricted, anoxic basins, and more recently some carbonates have also been recognized as euxinic basin sediments (Wilson, 1969). That an entire symposium can now be devoted to deep-basin carbonates suggests the prevalence of this environment in the geologic past; it also suggests that geologists are ready to accept and espouse the concept of deep-water sedimentation, even on the continents. One suspects that there is still more carbonate, shale, and sand waiting to be recognized as products of basin sedimentation. An anology might be made with the turbidite concept. Once it was understood what turbidites are and what their defining features should be, they were found in profusion; new turbidite sand was quickly located and well-known sand was reinterpreted. Perhaps if sediments in restricted basins are similarly understood and defined, more of them will be recognized in the stratigraphic record.

Criteria generally used in recognizing basinal carbonates are microfacies, sedimentary structures, and distinctive faunas (see Wilson, 1969; Tyrrell, 1969; other papers in this volume), all of which suggest downslope transport and pelagic sedimentation into a quiet water zone. In this paper I will focus on biogenic structures as *in situ* recorders of basin conditions; thus the basin is defined less by what is brought into it and more by its own benthic environments. The model proposed here should be applicable to any anoxic basin, regardless of sediment type.

The model will be built up from observations on the water stratification and benthic ecology of modern euxinic basins, and then tested against a suspected basinal deposit in the sedimentary record. It will be apparent that much of the terminology for the model is taken from the descriptive scheme developed by Rhoads in his discussion of modern basin ecology (Rhoads and Morse, 1971). In particular, Rhoads and Morse synthesized the data on species diversity in anoxic environments and showed that basins have the three zones of oxygenation discussed below, each of which has a characteristic fauna and sedimentary fabric. Rhoads and Morse found that a decrease in dissolved oxygen led to a diversity drop and a loss of calcification in benthic invertebrates; these trends were then compared with the evolutionary record of invertebrate fossils, the oxygen gradient in a modern basin being taken as an analogue for the rise of atmospheric oxygen over geologic time. The primary focus of the paper was on the evolution of metazoans. In the present paper, on the other hand, I have attempted to recognize the features of Rhoad's ideal anoxic basin in a particular set of rocks, and the primary focus is on paleoecology and environmental reconstruction.

STAGNANT BASINS

In the modern world, enclosed seas may be anoxic at depth due to a density stratification of the water column. Although the open ocean waters are well mixed by surface waves and thermal currents, semi-enclosed seas are easily stagnated by restrictions in mixing, both at the surface and at depth. Surface mixing by waves is limited by the smaller wave heights that can be built up in basins of limited fetch, and lateral movement of bottom water may be impeded by sills. Where semi-enclosed seas have free access to the open ocean at all their depths, bottom waters in the seas will mix horizontally with ocean water and stagnation will not occur. However, where a sill is present which blocks movement of bottom water, lateral exchange is stopped and the bottom water stagnates.

Oxygen is mixed in from the atmosphere at the surface of the sea, and all seas are oxygenated in their shallow surface zones. Oxygen will not diffuse down through the water column very rapidly, however, so oxygenation of deeper waters requires vertical currents which carry surface water downward. Although some vertical circulation is achieved by the small surface waves in enclosed seas, the most important vertical movements are thermal currents, produced by the cooling of surface water, which then sinks, causing an overturn of the shallow zone of the sea. In temperate latitudes, surface water is rarely cooled enough to sink and displace the very cold water on the basin floor; instead it sinks until it encounters water of equal density, which is at a relatively shallow depth (100–200 m). Vertical currents rarely reach the bottom water of the basin, and if the basin is silled then this water cannot exchange either horizontally or vertically.

Stagnation is most easily produced when there is a large freshwater influx into an enclosed sea. Where this occurs a salinity gradient is added to the normal temperature gradient, producing a strong density stratification; the lighter fresh or brackish water floats above heavier, more normal marine water. Thus in a silled basin, the light surface water flows out over the sill and traps cold salty bottom water in the basin, where no oxygen can be added. Examples of modern stratified seas are the Baltic and Black Seas, both of which have a large freshwater influx which forms a low-salinity surface layer. Connection with the ocean is by means of shallow sills, the Bosporous acting as a sill for the entire Black Sea, while a series of basins in the Baltic are silled at various depths, some of which are deep enough for stagnation to develop.

When water stagnates, dissolved oxygen originally present is removed by oxidation of organic matter, so the deep basins in the Baltic are poorly-oxygenated (Segerstrale, 1957) and the Black Sea bottom water lacks oxygen entirely (Caspers, 1957).

Many anoxic basins of smaller scale are found in coastal inlets, especially in fjords, but in all cases the mechanism for stagnation is the same, water stratification produced by a salinity gradient in combination with a sill at depth (see, for example, Strøm, 1939; Gucluer and Gross, 1964; Seibold, 1970).

WATER STRATIFICATION IN STAGNANT BASINS

The water column in an enclosed basin exhibits a tripartite layering. Figure 1 depicts an idealized stratified basin based on the Black Sea (Caspers, 1957). While the pattern of water stratification is similar in all anoxic basins, the depth and oxygenation values given here will vary with latitude, season of year, and sill depth. Because the Black Sea is the largest modern anoxic sea, the 50 m and 150 m reference points quoted below are probably maximal values. They are specified for the purpose of developing a general model, however, not as absolute or invariable limits. Within the surface layer (upper 50 m) conditions are nearly uniform, due to vertical mixing by waves and thermal overturn. The water throughout the surface layer is of nearly equal salinity and density, and atmospheric oxygen is continually mixed in at the surface (Richards, 1957). Although oxygen content in marine water varies because saturation depends on temperature, surface water in temperate climates normally contains about 7 ml/1 dissolved oxygen.

Deep-basin water also is nearly uniform in salinity and density, but it is not mixed vertically and the sill prohibits lateral exchange with the open ocean. The only mechanism for oxygenation is diffusion through the water column, which is too slow to reoxygenate the basin in the face of much higher rates of oxygen uptake by organisms and decay processes. As a result the oxygen content of deep-basin water will be at or near zero.

Between the surface and deep-basin layers, there is a zone (100 m thick) of rapid change in salinity and density (pycnocline) in which the water column is strongly stratified and vertical mixing is almost nil; in the pycnocline, oxygen drops rapidly from normal surface water values to near zero.

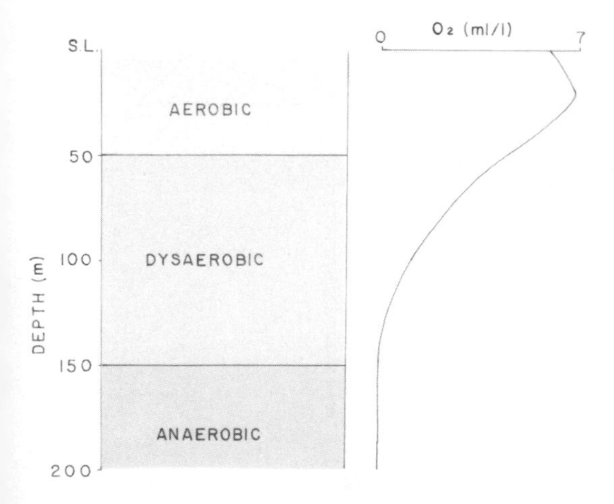

FIG. 1.—Diagram of water layering in an idealized enclosed basin. Curve at right shows dissolved oxygen content, and layer names refer to degree of oxygenation. Values are general estimates based on the Black Sea (Caspers, 1957).

It is important to note that the thicknesses of the three layers are dependent on a variety of parameters. The surface layer thickness depends on the depth of mixing, which is in turn controlled by wave base and climate variability. Larger waves may be generated in wider basins, and seasonal temperature change produces vertical thermal currents.

Further, the relation between oxygen decrease and depth is influenced by any lateral influx of oxygenated water across a deep sill and by the input of organic material from surface production or sediment gravity flows from the basin margin which consumes oxygen at depth.

Layer thicknesses shown in Figure 1 are for the Black Sea, the largest modern euxinic basin, the "Type Euxine" in fact. A smaller basin would have a thinner mixed layer, as wind fetch and therefore wave height would be less; hence oxygen would begin to decrease only a few meters below the surface.

In general, we should expect development of stratified basins to have been more common in the geologic past, and anoxic conditions to have been more widespread. The greater number of epeiric seas would have provided more enclosed basins than we find in the Holocene, not that every epeiric sea was totally anoxic, but isolated deep basins within epeiric seas probably developed anoxic conditions while the shallow zones remained normally oxygenated. This is the situation in the modern Baltic Sea (Segerstrale, 1957); much of the bottom of this small seaway is fully habitable, since oxygenated water is replenished by exchange with the ocean. The deeper areas, which have closed bathymetric contours, are isolated from exchange and become stagnant. The Black Sea is a more extreme case; there the entire basin's circulation is governed by a single sill.

Since a sill is crucial to the development of any stratified basin, now or in the past, sill depth becomes a complicating unknown factor in the estimation of ancient bathymetry. Great depth alone will not insure that a basin will become stagnant, as has often been implicitly assumed in the geologic literature. Lateral exchange of water with the open ocean must have been restricted by a sill or sills of unknown depth. Thus a very deep sill would produce anoxic conditions only far below the 150-m depth cited above. On the other hand, a very shallow sill will not restrict oxygenation completely, because vertical mixing from the surface may extend below the sill depth. For epeiric seas in the past it seems reasonable to assume that sill depths would have been relatively shallow, and the oxygenation depth was controlled by mixing from the surface rather than lateral influx alone.

EUXINIC BIOFACIES

The effect of low oxygen concentrations on marine invertebrates has been studied both in laboratory experiments and field surveys (see, for example, Hartman and Barnard, 1958, and 1960; Parker, 1964; Theede and others, 1969; Fenschel and Riedl, 1970; Raff and Raff, 1970). In all cases, similar trends are observed: as dissolved oxygen is lowered the benthos becomes less diverse, less abundant, smaller in body size, less heavily calcified, and dominated by infauna.

Hartman and Barnard (1958, 1960) documented these trends in their survey of benthic life in poorly oxygenated basins off southern California. Comparing basin and shelf, biomass values are 4 to 5 g/cm^2 versus 4000 g/m^2 respectively, the low values in the basin being due to both impoverishment of populations and smaller animal sizes. In particular, the large shelf species, especially shelled molluscs, are rare in the basins, and where they do occur *in situ* are typically smaller than normal or incompletely developed. Basin animals average less than 2 mm long and 1 g mass.

Basin faunas show a comparative increase in dominance of wormlike animals; for example, at shelf depth the polychete-crustacean ratio is 1:1, while in the basins it is 8:1. Hartman and Barnard concluded that the low-oxygen basinal environment strongly favors wormlike infauna over epifauna, especially shelled epifauna. Parker's (1964) data (graphed in Figure 2) from a survey of anoxic environments in the Gulf of California also show the diversity decrease which appears when oxygen drops below 2 ml/l. Species diversity is reduced in each taxonomic group, with the strongest reductions in heavily calcified epifauna. In the oxygen range 0.01 to 1.0 ml/l, termed "dysaerobic" by Rhoads and Morse (1971), the soft-bodied, generally infaunal polychaetes and nematodes and the poorly calcified crustaceans become the dominant components of the benthos, in sharp contrast to the calcified epifauna which dominate in normal aerobic ranges (2.0–7.0 ml/l). It is significant that within the calcified taxa, those forms which do survive in dysaerobic conditions are the infauna, protobranch bivalves and holothurian echinoderms, again suggesting a selection for the infaunal habit in low-oxygen biotopes.

Similar trends are apparent in the Black Sea, where molluscs and echinoderms are restricted to the upper levels of the water column, at oxygen values greater than 1.0 ml/l. Below, as the oxygen content continues to fall, the faunas are dominated

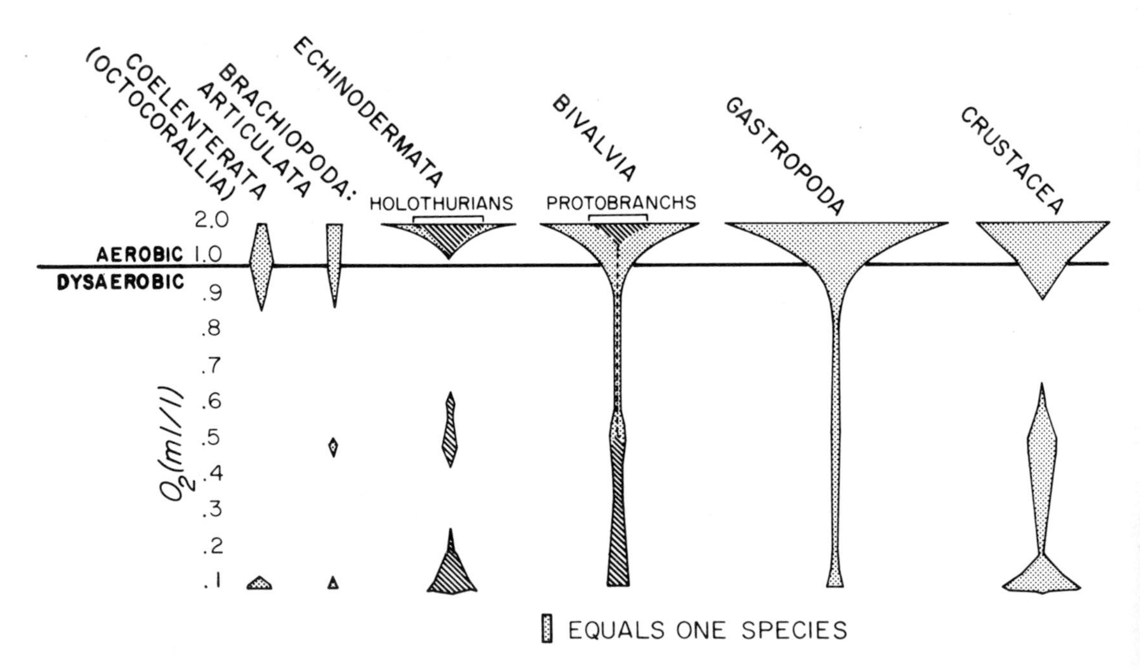

FIG. 2.—Species diversity in low-oxygen environments, Gulf of California. Major groups of shelled epifauna are strongly reduced below 1.0 m/l O_2, the aerobic-dysaerobic boundary. Note persistence of weakly calcified infaunal groups (holothurians and protobranchs, cross-hatched) into the dysaerobic environment. (Redrawn from Rhoads and Morse, 1971; data from Parker, 1964).

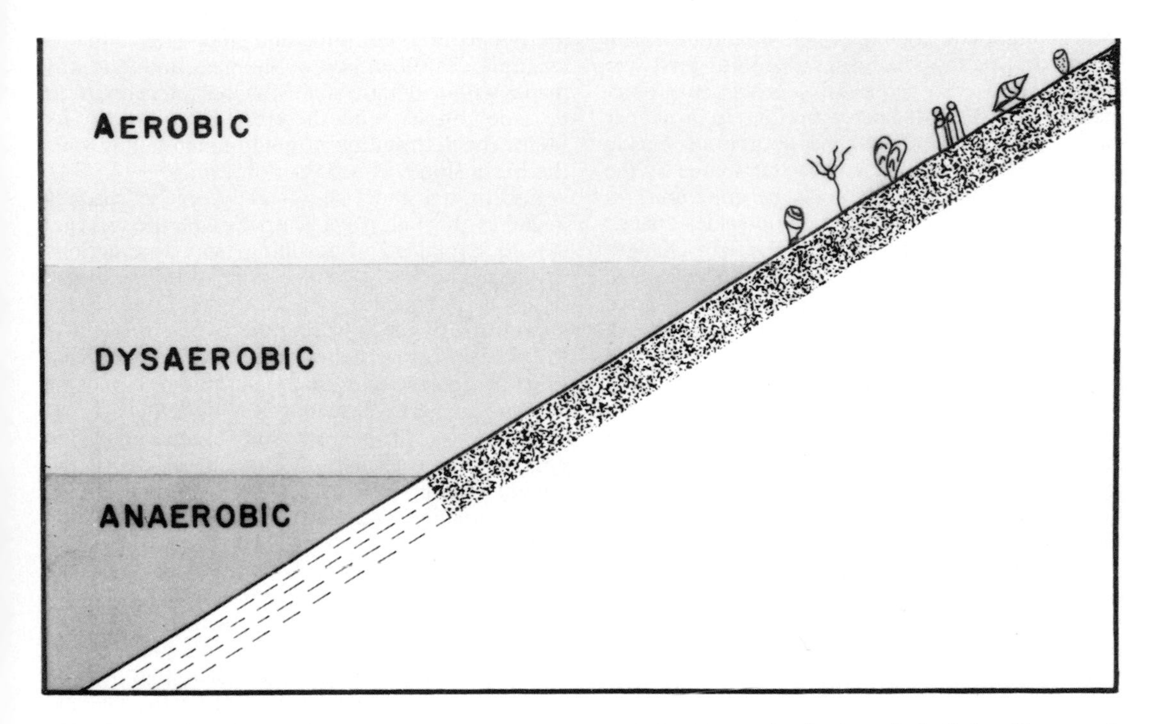

FIG. 3.—Model of biofacies change in a stagnant basin, derived from distribution of shelled epifauna and bioturbation in modern basins. Stippled pattern indicates bioturbate sediment, dashed pattern laminated sediment. (Modified from Rhoads and Morse, 1971).

by sponges, non-calcareous coelenterates, polychaetes, crustaceans, and aschelminths, the last extending deepest (data from Bacescu, 1963, replotted in Rhoads and Morse, 1971).

When dissolved oxygen drops to zero, metazoan life is excluded (but see Fenschel and Riedl, 1970, for a discussion of facultative anaerobes among the Protozoa and some very tiny metazoans which seem to "visit" the anoxic zone).

Combining the trends in animal distribution with the three-layer basin model, the following biotopes can be predicted (Fig. 3):

1. Aerobic zone—in shallow water; characterized by calcareous epifauna and strongly bioturbate sediment (reworked biologically by abundant infauna). "Shallow water" in this sense denotes subtidal shelf depths, not the extremely shallow nearshore surf or tidal flat zones, where oxygen is never lacking but bioturbation is rare because of continuous physical reworking.

2. Dysaerobic zone—within the pycnocline; calcareous epifauna lacking, but sediment bioturbate due to the activities of resistant infauna.

3. Anaerobic zone—below the pycnocline; lack of all benthos, shells absent and sediment undisturbed. This zone is expressed in the fine muds which settle from suspension; sediments which move into the basin by sediment gravity flow will be different, containing distinctly different sedimentary structures and textures. In addition, sediment gravity flows may bring in fauna and shells from shallower regions, and perhaps even accompanying pulses of oxygenated water. The result of such a flow might be a short-lived benthic community, complete with a bioturbation horizon, in the midst of an anaerobic facies (H. Cook, Pers. Comm.). If mass flows were common, a sequence of such bioturbated horizons could be expected, separated by zones of laminated, nonfossiliferous sediment.

It is important to note the Figure 3 is only diagrammatic; in particular the scale is not consistent. In a deep basin the pycnocline is very thin vertically compared to the total depth, as in the Black Sea, where the 100 m pycnocline covers a 2000 m water column; the anaerobic zone is thus overwhelmingly predominant. Only a thin surface film is oxygenated at all, and this film intersects only a narrow interval of the basin slope. The result should be a very rapid facies change in the slope sediments proceeding downslope from fossiliferous bioturbate mud to shell free bioturbate mud to laminated mud. The steeper the slope, the more rapid the lateral change in the sediments, and the narrower the band of dysaerobic facies.

From the general model shown in Figure 3,

four implications follow, which should be useful in interpreting the stratigraphic record. First, the facies changes, aerobic-dysaerobic-anaerobic, indicate the basin slope direction. In a deeper marine basin, far from the source of clastic material, sedimentation will be dominated by the pelagic component, which may be quite uniform over a wide area, with little or no facies change of a purely lithologic nature to reflect changes in water depth. However, if a lateral change in sedimentary fabric can be recognized, say from dysaerobic to anaerobic, then increasing depth is indicated and the direction to shoreline can be determined, even far out in the basin, and in the absence of coastal sediments.

Second, if the bases of ancient pycnoclines were at about 150 m depth, then recognition of the pycnocline in sedimentary rocks provides a direct estimate of ancient water depths. This is potentially a useful tool, since there are very few methods of estimating actual bathymetry in ancient seas. The estimates given here, based on biofacies, must be regarded as approximations, due to the large number of unknowns which control pycnocline depth, but the order of magnitude is almost certainly correct; even in the open ocean, surface mixing extends only to about 200 m, and at the other extreme, even small fjords are mixed through several tens of meters.

Third, if ancient pycnoclines were 100 m thick, as modern ones are (Fig. 1), then the lateral extent of the contemporaneous dysaerobic facies is dependent on the angle of slope in the ancient basin. Consider a basin sloping very gently: the sediments on such a slope will be within the pycnocline layer for a considerable lateral distance, and when the sediment has become rock, the dysaerobic facies will be traceable through that distance. On the other hand, a steeply sloping basin floor will pass through the 100 m pycnocline quickly, and the resulting facies band will be correspondingly narrower. The facies band width can be measured on outcrop; if we assume a 100 m pycnocline, then the slope angle is easily calculated. Ancient slopes, like depth, are extremely difficult to determine; the method of biofacies change seems to have great potential for defining the characters of basins, but the method is subject to an obvious pitfall. In measuring the width of a facies band we are recording a rise or fall of the pycnocline through time rather than its impingement on the slope at any one instant. A gradually moving pycnocline might produce a wide dysaerobic facies band as the zone moved across the basin slope, leading to an interpretation of a gentle slope gradient. In fact, unless the measurement band can be shown not to be time-transgressive, there is no way to distinguish between the effects of movement of the pycnocline with time and slope angle. In the example described below, the measurement was made within a thin shale tongue interpreted to be isochronous, and therefore recording in its facies the distribution of environments (and thus the basin slope) at a single "instant."

Fourth, and most significantly, cyclic changes in facies through a vertical stratigraphic interval can be explained as resulting from fluctuations in the pycnocline depth relative to the seafloor. A rising pycnocline will shift the three zones shoreward, while lowering the pycnocline shifts them basinward. In this way, rhythmic alternations of aerobic and anaerobic biotopes may be stacked in vertical succession, and there is no necessity to change the water character of the entire basin for each alternation; at depth the anaerobic zone persists continuously, but its top, the pycnocline, varies in position on the basin slope.

AN ANCIENT EXAMPLE—MIDDLESEX SHALE

The model outlined above was tested by comparison with some alternations of fossiliferous and euxinic rocks in the Upper Devonian of New York State. The sediments, part of the Catskill wedge, are entirely terrigenous clastics, but the relations they illustrate should be applicable to carbonate basins as well. The marine Upper Devonian of New York consists of thousands of meters of sandstone, siltstone, and shale, mostly fossiliferous, in which the principal marker horizons are a series of thin (a few meters thick), laterally persistent, nonfossiliferous black shale tongues (Sutton and others, 1970). The tongues serve both to define lithostratigraphic groups and local chronostratigraphic units, since they are interpreted as being isochronous lines in a classic area of facies change (Rickard, 1964). Because each shale tongue represents a time plane, facies change within the tongue itself should reflect various paleonenvironments at a single "instant" in the Late Devonian. One black shale tongue, the Middlesex Shale, was examined for lithologic and faunal facies changes. As shown in Figure 4, the single shale unit in western New York splits into two tongues eastward, near Ithaca, to enclose a thick fossiliferous sequence, the Triangle Formation, interpreted by Sutton and others (1970) as a deltaic lobe. The eastern tongues, the Montour and Sawmill Creek Shales, are superficially similar to the Middlesex: thin, black, and mostly nonfossiliferous. However, lateral sampling of the Middlesex and these eastern extensions revealed significant changes in sedimentary structure (Fig. 5).

In the area west of Ithaca, the entire outcrop of the Middlesex and the western parts of the Montour and Sawmill Creek are all devoid of

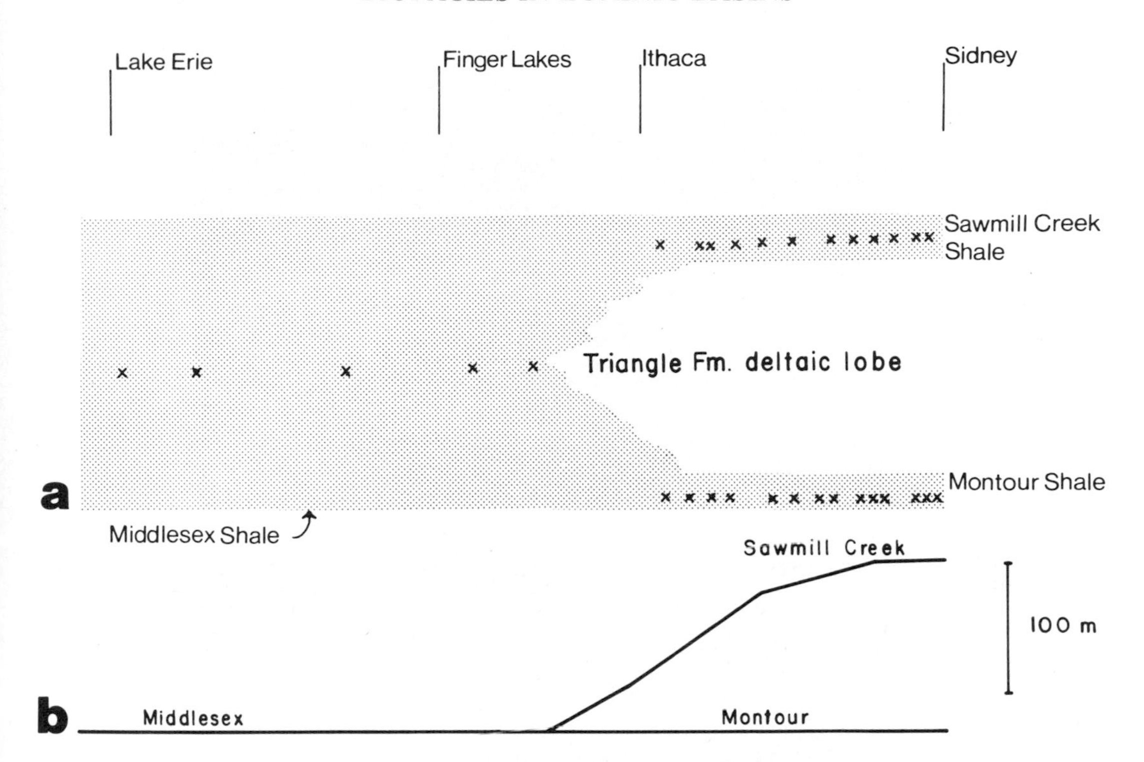

FIG. 4.—Stratigraphy of the Upper Devonian Middlesex Shale and its extensions. *a.* Correlation diagram showing lateral equivalence of Middlesex to the two shale tongues and the enclosed Triangle Formation to the east. Width of outcrop belt is 320 km. Vertical scale is time; i.e., thick Triangle Formation was deposited during same time span as much thinner Middlesex. Sampling localities shown by x's. *b.* Stratigraphic section diagram showing thickness relations of the shales (<5 m thick) and deltaic lobe (>100 m). Thin persistent shales are major marker horizons; bases of shales are thought to be isochronous lines.

benthic fossils, and the rock is perfectly laminated on a scale of 5-10 laminae per millimeter. Biogenic disturbance of the sedimentary fabric is totally absent (Fig. 5A).

Eastward, both the Montour and Sawmill Creek tongues are bioturbated. Small burrows produced by deposit-feeding infauna mix clay and silt layers to form a mottled appearance in polished vertical sections (Fig. 5B). This bioturbate fabric persists as the shale is traced eastward, and in addition there is the appearance of occasional body fossils, mainly protobranch bivalves, near the eastern limit of the tongues at Sidney, New York. (Fig. 5C). In terms of the zones of oxygenation discussed previously, the Middlesex and its tongues can be subdivided laterally (see Fig. 6):

1. Western area—anaerobic
2. Middle area—dysaerobic
3. Eastern area—marginally aerobic

The Triangle lobe between the shale tongues is highly fossiliferous, and clearly represents the fully aerobic zone.

The above-mentioned four corollaries can be applied to the Middlesex as well. The basin slope was to the west, because the facies became less well oxygenated in that direction. In Figure 6, the central basin is to the left, the Catskill shoreline to the right. In this particular instance we can confirm the paleogeography by the presence of redbeds and alluvial plain sequences in the coeval rocks east of Sidney; however, the basin paleoslope can be deduced from facies changes in the shale alone. This method may be of significance in other areas where shoreline sediments are covered or eroded.

In the area west of Ithaca the entire Middlesex and both tongues record anaerobic conditions; that is, the basin was continuously anoxic west of Ithaca, even during the deposition of the deltaic lobe. The Ithaca area marks the top of anoxic water (the base of the pycnocline), and the Devonian sea was probably about 150 m deep at this point.

The actual basin gradient can be calculated from the width of the dysaerobic facies belt on outcrop. In the Montour and Sawmill Creek tongues the average width of outcrop of bioturbate, nonfossiliferous shale is about 40 km; if the ancient

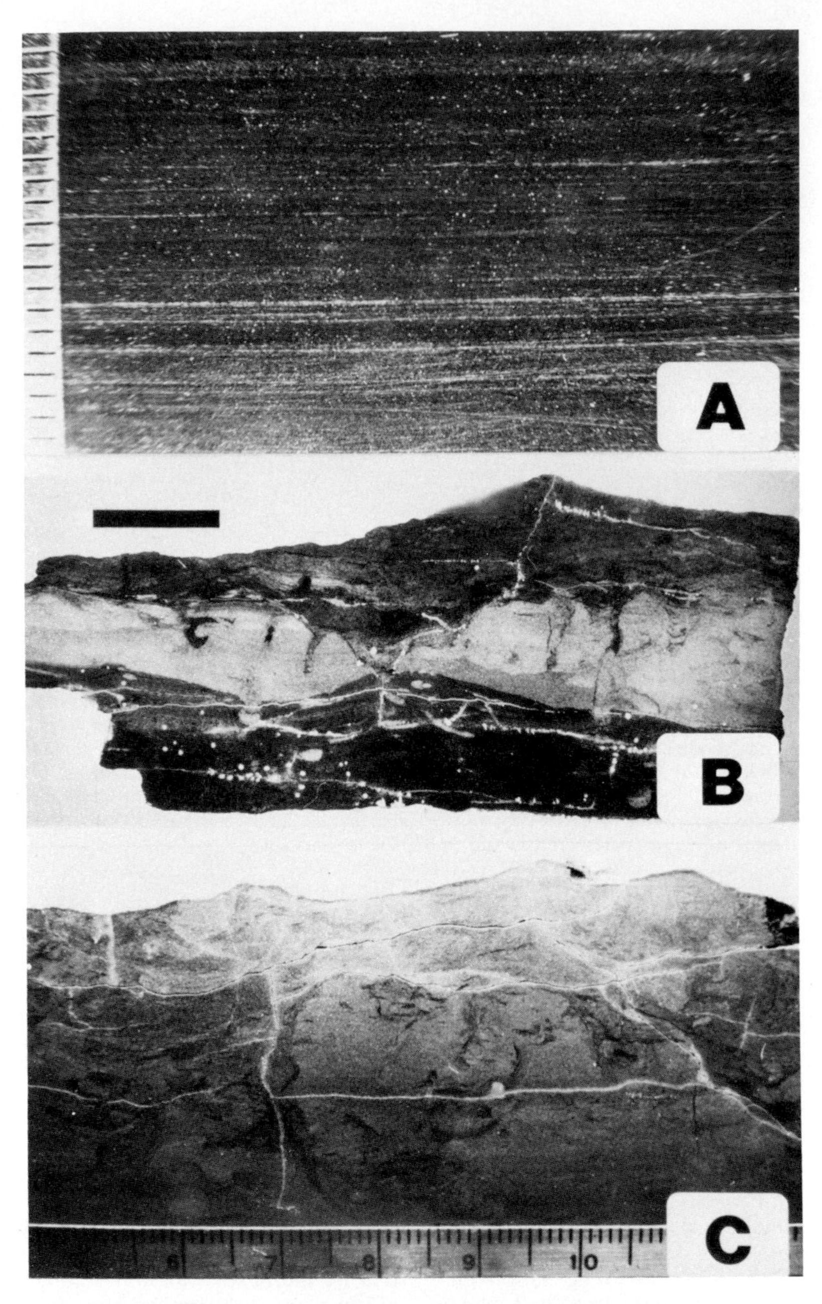

FIG. 5.—Lithologies in the Middlesex Shale and Sawmill Creek Shale. *A.* Laminated Middlesex Shale from westernmost outcrop at Lake Erie. Clay-silt laminae couplets are undisturbed by bioturbation. Scale divisions are millimeters. *B.* Bioturbate Sawmill Creek Shale from area east of Ithaca. Dark mudstone has been completely reworked; lighter silt layer is broken and mottled by burrowers. Bar scale equals 1 cm. *C.* Totally bioturbate Sawmill Creek Shale from easternmost area of outcrop, near Sidney. Both silt and clay layers have been obliterated by reworking; almost no depositional structure remains. Outcrops in this facies are sparsely fossiliferous. Centimeter scale.

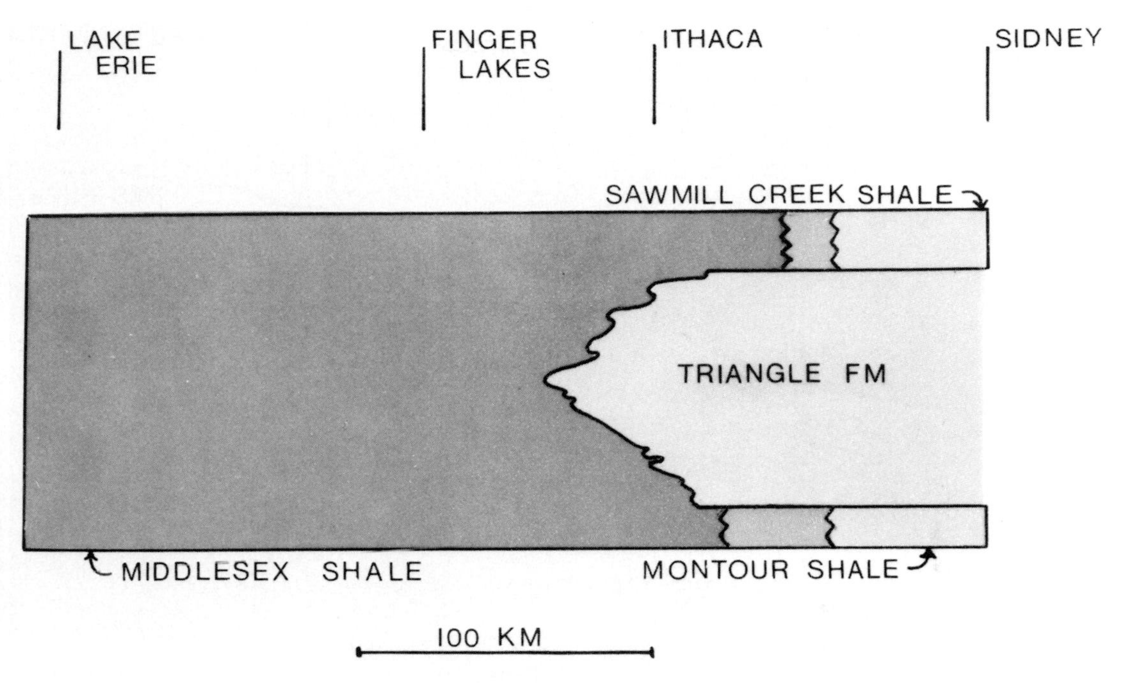

FIG. 6.—Facies change diagram for the Middlesex Shale and its equivalents, gray tones as in Figs. 1 and 3: Dark—anaerobic facies; intermediate—dysaerobic facies; light—aerobic facies.

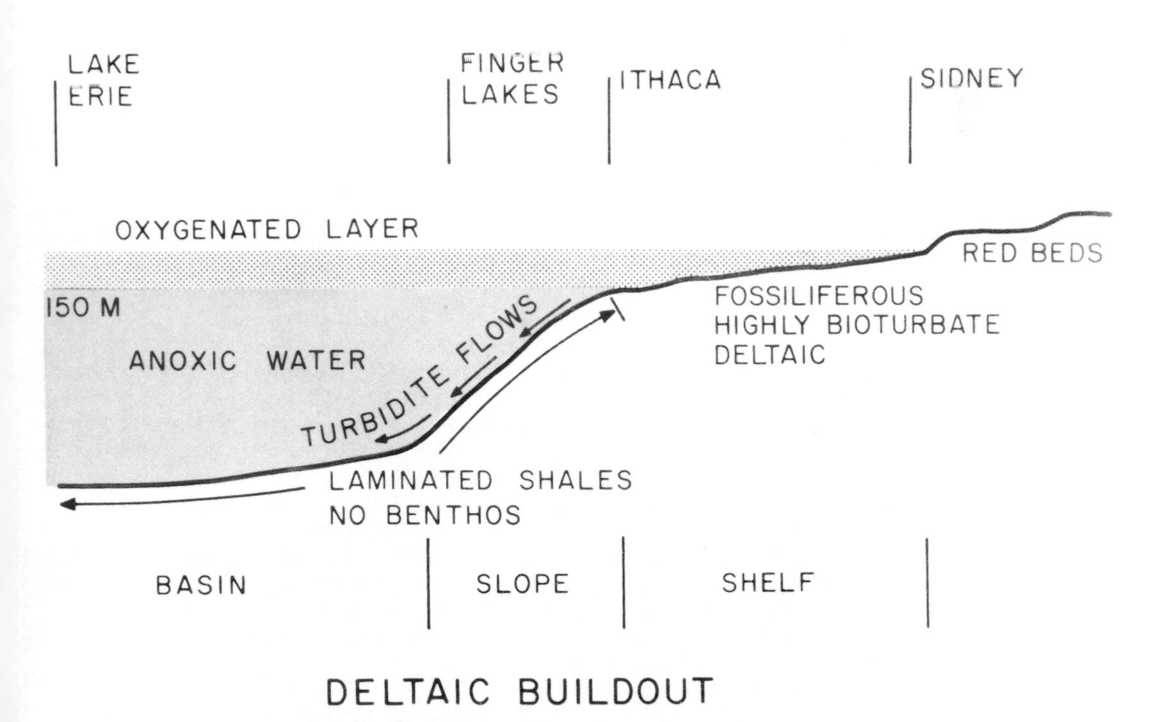

FIG. 7.—Upper Devonian basin environments: deltaic progradation. Shelf mostly aerobic, deep basin anaerobic. Dysaerobic facies which should be on outer shelf and slope edge is obscured by turbidite sedimentation. Stippled "oxygenated layer" includes both aerobic and dysaerobic zones.

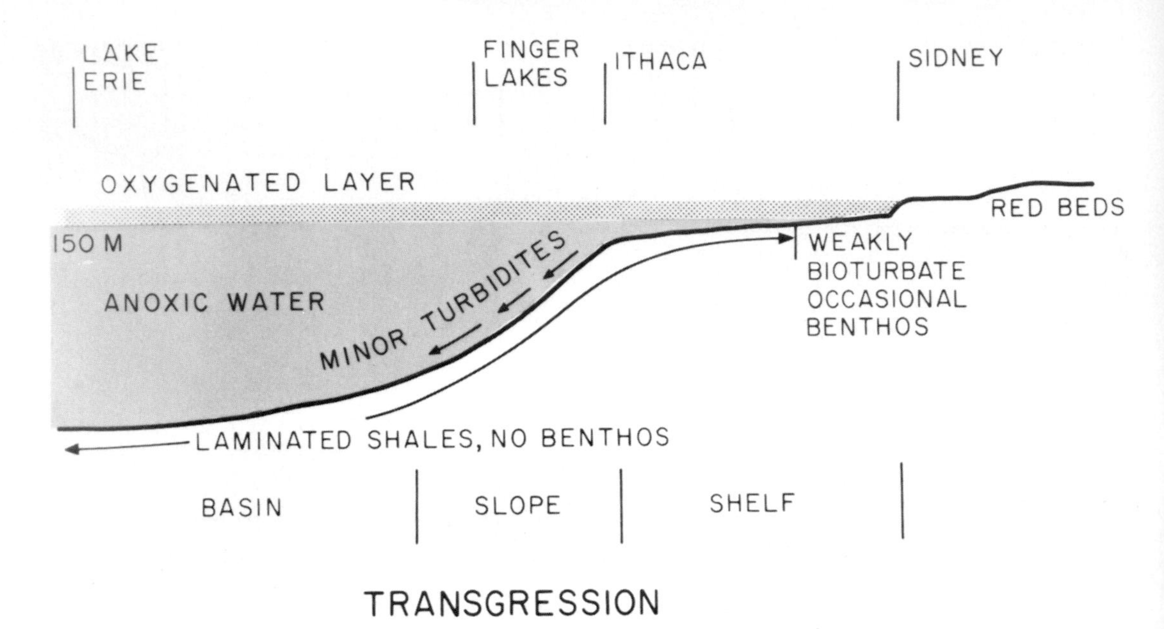

FIG. 8.—Upper Devonian basin environments: transgression. Shelf dysaerobic and marginally aerobic (in east). Shelf edge, slope, and basin all anaerobic. Stippled "oxygenated layer" includes both aerobic and dysaerobic zones.

pycnocline were 100 m thick, then the inclination was 1:400 (0°09′). Thus, the Late Devonian shelf probably sloped westward at a gradient similar to that of modern continental shelves (0°07′; Shepard, 1963).

Looking beyond the single Middlesex tongue, we can explain the recurrence of black shale units in the Upper Devonian section in terms of fluctuations in the pycnocline depth. Two phases of deposition alternated throughout the Late Devonian, deltaic progradation and transgression. During the deltaic phases, units such as the Triangle Formation built westward across the marine shelf. Deposition took place within the shallow aerobic zone (Fig. 7). During transgressions the pycnocline was raised, and basinal environments were shifted shoreward. The formerly aerobic shelf became the site of dysaerobic and marginally aerobic conditions, and totally anaerobic water lapped up onto the shelf edge in the Ithaca area (Fig. 8). While the main basin is western New York (and possibly extending far out into the Midcontinent as the Chattanooga Sea) remained anaerobic throughout the course of the fluctuations which affected the eastern shelf, each transgression was recorded by a shale tongue, as dysaerobic conditions were brought up onto the shelf.

During the intervening deltaic episodes the entire shelf was aerobic and the dysaerobic biotope should have been located on the basin slope, to be represented today by rocks in the Middlesex Shale directly west of the Triangle lobe. Unfortunately, the actual stratigraphy is more complex than diagrammed, and the Triangle deltaic rocks are replaced by turbidity current deposits westward (see Sutton and others, 1970), which obscure any facies changes expected in the shales of the shelf edge. These turbidites pinch out westward into typical laminated Middlesex, so the turbidity currents must have reached below the pycnocline and into the anaerobic environment. Although the turbidity flows probably originated in oxygenated waters, any oxygen brought down into the basin was insufficient to support the establishment of benthic life. If it had been, the tops of turbidities should show thin horizons of bioturbation produced by a short-lived community. However, turbidities interfingering with the Middlesex grade from Bouma divisions into laminated shale with no trace of infaunal disturbance. It seems probable that the Late Devonian basin remained anoxic for long periods of time, and that the fluctuations in the pycnocline which affected the basin margin were small in comparison to the total depth.

DISCUSSION

Two points require additional discussion: first, that pycnocline depth changes might shift the anoxic environment laterally, and second, that repeated shifts of this nature could explain cyclic successions. In the Middlesex Shale example described above, it is clear that the main basin remained in an anaerobic state, even while aerobic conditions prevailed on the shelf, and that black mud deposition was extended onto and across the shelf by a rise in the pycnocline relative to the sea floor. It is unclear whether this rise was simply part of an overall marine transgression and deepening, or whether it was a shift in the pycnocline alone, perhaps produced by a change in sill depth or mixing depth. A true transgression is suggested by the change in sedimentation pattern from deltaic sand to open shelf lutite, as if the river were failing to reach the sea (drowned coastline). It is perfectly conceivable, however, that pycnoclines in other ancient basins may have fluctuated independently of any change in sea level. In that case one would expect a change from aerobic to dysaerobic or anaerobic conditions with little concomitant change in grain size.

Whatever the ultimate cause, a rise in the pycnocline is manifested by a change from aerobic to dysaerobic conditions as the basinal environment spreads up and across the adjacent shelf; this process has been illustrated for a terrigenous clastic basin, and three further examples, all from carbonate basins, will be briefly discussed for comparison.

The Meramecian Rancheria Formation in the Sacramento Mountains of New Mexico has been shown by Yurewicz (1975; this volume) to consist of typical deep-water carbonates. He cited the prevalence of dark lime mudstone, and presence of fine lamination as indicators of a quiet, anoxic environment. This conclusion is supported by the sparseness of body fossils in the section, and by the absence of sedimentary structures indicative of shallow-water or emergent conditions.

The Rancheria onlaps an unconformity cutting the Lake Valley Formation (Osagean), which Meyers (1975) has shown to be subtidal marine shelf limestone, and the Rancheria thins progressively up the unconformity surface, indicating an actual shift of the basinal environment up onto the shelf. In the terminology of this paper, the Rancheria lime mudstones would be classed as dysaerobic facies; the rocks are often bioturbate, but body fossils are rare. Water depths for the Rancheria would be 50–150 m, according to the basin model (Fig. 1); this compares favorably with Yurewicz's estimate of 100 m minimum depth over the drowned shelf.

A similar situation has been described from the Permian rocks of the Guadalupe Mountains of West Texas, in which a basinal carbonate unit, the Cutoff Formation, blankets an unconformity cutting bank-margin carbonates of the Victorio Peak Formation. Pray (1968, 1971) and Harms and Pray (1974) interpreted the unconformity as submarine in origin, and suggested density currents from the shelf were the erosive agents.

During Leonardian time, bank sediments (Victorio Peak) graded laterally into dark lime mudstone interpreted as anoxic basin sediments (Bone Spring). Harms and Pray (1974) noted the abruptness of the facies change and pointed out that an euxinic interface must have persisted low on the bank-to-basin slope throughout Leonardian time. In the terminology of the present model, this interface is the pycnocline. If the Bone Spring is truly anaerobic, below the pycnocline, then basin deposition occurred at depths greater than 150 m.

Submarine erosion in latest Leonardian time truncated and steepened the bank margin, and the Cutoff Formation, a dark carbonate unit, was deposited over the unconformity surface. Pray (1968) stated that there is little apparent lateral change in the Cutoff carbonate blanket, and he interpreted the unit as a deep-water facies throughout its extent. Apparently there was a rise in the pycnocline in early Guadalupian time, perhaps accompanying a transgression, and euxinic conditions spread out of the basin, up the unconformity slope and across the old Leonard shelf. My own preliminary observations of the Cutoff lithology show bioturbate fabric, indicative of dysaerobic conditions, but there may be facies changes in the Cutoff that would only be evident with systematic sampling. A transect up the unconformity might be expected to show the anaerobic-dysaerobic transition, as in the Middlesex Shale discussed earlier.

A final example on the shift of euxinic conditions from basin to shelf is seen in Pennsylvanian carbonates of the Paradox and Honaker Trail Formations, Paradox Basin, southern Utah (Pray and Wray, 1963). These rocks also exhibit a repeated alternation of normally oxygenated and euxinic facies as described above for the Upper Devonian of New York. The rocks are all shelf carbonates, but their lateral equivalents to the north and east, in the Paradox Basin proper, are carbonate-evaporite mixtures, which also display cyclic alternations. Hite (1966) related the cyclicity of the shelf carbonates and basin evaporites using a transgressive-regressive model. He postulated that sea level rise would increase reflux from the evaporite basin, and the refluxed water, highly saline and anoxic, would flow across the shelf to the open sea, creating unfavorable conditions for life and producing euxinic carbonates. Reflux was minimized during low-stand times, and only normal seawater flowed across the shelf on its way into the basin; the result was fossiliferous carbonate deposition on the shelf, especially algal buildups.

It should be noted that Hite's model is different

from that proposed in this paper since an evaporite basin develops a density gradient by surface evaporation and sinking of saline water. Where evaporation is not extreme, in the modern Mediterranean or Persian Gulf for example, the bottom water will be kept continuously oxygenated by descending currents of surface water (see Seibold, 1970). Only when strong surface evaporation produces hypersaline brines is oxygen content diminished due to the reduced solubility of oxygen in very saline solutions; Hite's (1966) graph shows that NaCl solutions of 200 percent salinity can contain less than 2 ml/l dissolved oxygen, nearly the lower limit of aerobic conditions. However, invertebrate animals would be excluded by much smaller increases in salinity, long before oxygen became limiting.

Nevertheless, Hite's model for cyclicity is similar to the one described here in that it involves shifting basinal conditions up onto the shelf without necessarily changing water depth on the shelf itself. This point was noted by Pray and Wray (1963) who delineated cycles in deposition on the shelf by recurrence of facies, particularly the alternation of dark spiculitic mudstone (euxinic) with fossiliferous wackestone and algal grainstone (normal marine). They (Pray and Wray, 1963, p. 216) specifically discussed the problem of abrupt transitions from euxinic to normal facies up through the section, with no indication of change in water agitation: "In this stratigraphic sequence, which contains many rocks showing low degrees of water turbulence, we believe that a fluctuation of the interface between heavier, deeper, stagnant waters and overlying more normal waters could exert a major control." Pray and Wray concluded that absolute depth changes need not have occurred, only transgressions of basin-type water. It should be noted that the depth estimates for pycnocline thickness given in this paper may not be applicable in evaporite settings such as the Paradox Basin. In these types of settings brine formation greatly increases the density of the bottom waters which results in a rapid density increase with depth and a thin pycnocline. In such a basin, azoic conditions, due to either high salinity or low oxygen, may have been achieved at depths much shallower than the 150 m shown in Figure 1.

The euxinic facies in the Honaker Trail section was probably deposited in still shallower waters, as the muds were clearly not anaerobic. Sponge spicules are abundant in the mudstone and a few other thin shell fragments are found. The published photographs of slabs and thin sections of this lithology appear to be bioturbate (Pray and Wray, 1963; Pl. 1-1, 3-1), placing it in the dysaerobic facies. Maximum water depth in the paleoenvironment would have been 150 m, but was probably shallower (tens of meters?).

CONCLUSIONS

Biofacies changes based on the presence or absence of body fossils and bioturbation present a coherent pattern in modern basins. These changes result in response to oxygen content, which correlates with depth. Similar patterns can be discerned in sediments from ancient basins, permitting the reconstruction of a basin's physical features and hydrography, including paleoslope direction and dip angle and depths of deposition of the various facies.

Stratigraphic sections in which normal marine rocks alternate with euxinic rocks can be explained as transgressions of basinal conditions up onto a shelf, either as a part of an actual transgression or simply a shift in the pycnocline depth relative to the seafloor. Repeated cycles of normal and euxinic lithologies can be produced by a series of shifts, which affect only the upper reaches of the water column. Conditions may vary repeatedly on the shelf while the basin floor remains continuously anoxic.

ACKNOWLEDGEMENTS

This article derives from a Ph.D. dissertation completed at Yale University under the direction of Dr. Karl M. Waage. Dr. Donald C. Rhoads at Yale introduced me to the literature on anoxic basins and suggested techniques for collection and examination of shale samples. Several of the basic ideas in this paper can be traced directly to Dr. Rhoads, in particular to his 1971 paper on modern anoxic basins. Beyond the normal citations, I want to acknowledge him as the source of concepts which I have pursued and extended to the stratigraphic record.

Dr. Lloyd C. Pray suggested the applicability of this model to the problem of basinal carbonates. He has supplied me with several carbonate sequence examples as analogues to my own shale units, by drawing on his own field experience and guiding me through the carbonate literature. Dr. Pray and Dr. David L. Clark both read and critically evaluated the first draft of this paper; their efforts and improvements are greatly appreciated.

REFERENCES

BACESCU, M., 1963, Contribution à la biocéanologie de la Mer Noire. L'étage Périozoique et le faciès paléodreissénifère, leurs charactéristiques: Rapport et procès-verbaux des réunions: Comm. Internat. pour l'exploıation scientifique de la Mer Méditerranee, Paris, France, v. 17, p. 107-122.

Caspers, H., 1957, Black Sea and Sea of Azov: *In* J. W. Hedgepeth (ed.), Treatise on marine ecology and paleoecology: Geol. Soc. America, Mem. 67, v. 1, p. 801-889.
Fenschel, T. M., and Riedl, R. J., 1970, The sulfide system: A new biotic community underneath the oxidized layer of marine sand bottoms: Marine Biology, v. 7, p. 255-268.
Gucluer, S. M., and Gross, M. G., 1964, Recent marine sediments in Saanich Inlet, a stagnant marine basin: Limnology and Oceanography, v. 9, p. 359-376.
Harms, J. C., and Pray, L. C., 1974, Erosion and deposition along the mid-Permian intracratonic basin margin, Guadalupe Mountains, Texas [abs.]: *In* R. J. Dott, Jr., and R. H. Shaver (eds.), Modern and ancient geosynclinal sedimentation: Soc. Econ. Paleontologists and Mineralogists, Spec. Pub. 19, p. 37.
Hartman, O., and Barnard, J. L., 1958, The benthic fauna of the deep basins off southern California: Allan Hancock Pacific Expeditions, v. 22, no. 1, 67 p.
—— and ——, 1960, The benthic fauna of the deep basins off southern California: Allan Hancock Pacific Expeditions, v. 22, no. 2, p. 69-297.
Hite, R. J., 1966, Shelf carbonate sedimentation controlled by salinity in the Paradox Basin, southeast Utah: *In* J. L. Rau and L. F. Dellwig and others (eds.), Third symposium on salt: Northern Ohio Geol. Soc., Cleveland, Ohio, p. 48-66.
Meyers, W. J., 1975, Stratigraphy and diagenesis of the Lake Valley Formation, Sacramento Mountains, New Mexico: *In* Guidebook to the Mississippian shelf-edge and basin facies carbonates, Sacramento Mountains and southern New Mexico region: Dallas Geol. Soc., Dallas, Texas, p. 45-65.
Parker, R. H., 1964, Zoogeography and ecology of some macroinvertebrates, particularly molluscs, in the Gulf of California and the continental slope off Mexico: Dansk Naturhistorisk Foren. Videnskabelige Medd., Bd. 126, 178 p.
Pray, L. C., 1968, Basin-sloping submarine (?) unconformities at margins of Paleozoic banks, west Texas and Alberta [abs.]: Geol. Soc. America Programs with Abs., 1968 Ann. Meetings, p. 243.
——, 1971, Submarine slope erosion along the Permian bank margin, west Texas [abs.]: Am. Assoc. Petroleum Geologists Bull., v. 55, p. 358.
—— and Wray, J. L., 1963, Porous algal facies (Pennsylvanian) Honaker Trail, San Juan Canyon, Utah: *In* M. O. Bass (ed.), Shelf carbonates of the Paradox Basin: Four Corners Geol. Soc., Guidebook 4th Field Conf., p. 204-234.
Raff, R. A., and Raff, E. C., 1970, Respiratory mechanisms and the metazoan fossil record: Nature, v. 228, p. 1003-1005.
Rhoads, D. C., and Morse, J. W., 1971, Evolutionary and ecologic significance of oxygen-deficient marine basins: Lethaia, v. 4, p. 413-428.
Richards, F. A., 1957, Oxygen in the ocean: *In* J. W. Hedgpeth (ed.), Treatise on marine ecology and paleoecology: Geol. Soc. America, Mem. 67, v. 1, p. 185-238.
Rickard, L. V., 1964, Correlation of the Devonian rocks in New York State: New York State Mus. and Sci. Service, Geol. Survey Map and Chart Ser., no. 4.
Segerstrale, S. G., 1957, Baltic Sea: *In* J. W. Hedgepeth (ed.), Treatise on marine ecology and paleoecology: Geol. Soc. America, Mem. 67, v. 1, p. 751-802.
Seibold, E., 1970, Nebenmeere in humiden und ariden Klimabereich: Geol. Rundschau, v. 60, p. 73-105.
Shepard, F. P., 1963, Submarine geology, 2nd ed.: Harper and Row, New York, New York, 557 p.
Ström, K. M., 1939, Land-locked waters and the deposition of black muds: *In* P. D. Trask (ed.), Recent marine sediments: Am. Assoc. Petroleum Geologists, Tulsa, Oklahoma, p. 256-372.
Sutton, R. G., Bowen, Z. P., and McAlester, A. L., 1970, Marine shelf environments of the Upper Devonian Sonyea Group of New York: Geol. Soc. America Bull., v. 81, p. 2975-2992.
Theede, H., Ponat, A., Hiroki, K., and Schlieper, C., 1969, Studies on the resistance of marine bottom invertebrates to oxygen deficiency and hydrogen sulfide: Marine Biology, v. 2, p. 325-337.
Tyrrell, W. W., Jr., 1969, Criteria useful in interpreting environments of unlike but time-equivalent carbonate units (Tansill-Capitan-Lamar), Capitan reef complex, west Texas and New Mexico: *In* G. M. Friedman (ed.), Depositional environments in carbonate rocks: Soc. Econ. Paleontologists and Mineralogists, Spec. Pub. 14, p. 80-97.
Wilson, J. L., 1969, Microfacies and sedimentary structures in "deeper water" lime mudstones: *In* G. M. Friedman (ed.), Depositional environments in carbonate rocks: Soc. Econ. Paleontologists and Mineralogists, Spec. Pub. 14, p. 4-19.
Yurewicz, D. A., 1975, Basin margin sedimentation, Rancheria Formation, Sacramento Mountains, New Mexico: *In* Guidebook to the Mississippian shelf-edge and basin facies carbonates, Sacramento Mountains and southern New Mexico region: Dallas Geol. Soc., Dallas, Texas, p. 67-86.

SEPM SPECIAL PUBLICATION NO. 25, P. 19–50, NOVEMBER 1977

SECULAR VARIATIONS IN THE PELAGIC REALM

ALFRED G. FISCHER AND MICHAEL A. ARTHUR
Department of Geological and Geophysical Sciences
Princeton University, Princeton, New Jersey 08540

ABSTRACT

Diversity of pelagic biotas varies with a rhythm of about 32 million years. "Polytaxic" times of maximal diversity coincide with higher and more uniform oceanic temperatures, with continuous pelagic deposition and with widespread marine anaerobism, eustatic sea-level rises, and heavier carbon isotope values in marine calcareous organisms and organic matter. Pelagic communities reach maximal complexity, expressed in numbers of taxa and in predator size.

"Oligotaxic" episodes are characterized by lower marine temperatures and sharper latitudinal and vertical temperature gradients, by interruptions of submarine sedimentation caused by intensified current systems, by marine regression, by a lack of anaerobic marine sedimentation, and by lighter carbon isotope values in marine carbonate skeletons and organic compounds. Degradation of pelagic communities is reflected by loss of large predators and lowered diversity; blooms of opportunistic species occur during these intervals.

The fluctuation from oligotaxic to polytaxic conditions is attributed to changes in rates of oceanic circulation as a direct result of climatic variation: during polytaxic times warm, globally equable climates result in reduced oceanic convection rates, causing expansion and intensification of the oxygen minimum layer. Colder climatic intervals lead to increased circulation rates, more efficient oxygenation of ocean waters, and oligotaxy. The climate changes in response to either fluctuations in receipt of solar energy, or to internal causes such as rhythms in mantle convection and associated processes: rates of plate motions, volcanism and orogeny. These cycles directly influence the accumulation patterns of petroleum source rocks, major phosphorite deposits, deep-sea carbonates and biogenic silica.

INTRODUCTION

Most studies of sediment are focused appropriately on the nature and interpretation of sedimentary and biotic features related to local processes. Sources of sediment, distance from shore, depth of water, current and wave patterns, local chemistry—these are the data from which local and regional reconstructions are made, and with which the search for minerals is conducted. Hutton's actualistic philosophy, so brilliantly employed in sedimentology by Johannes Walther and his followers, has been eminently successful at this level.

Rearrangement of lands and seas by plate tectonics involves changes on a larger scale. But even beyond this the overall state of the outer earth appears to have changed through geological time. The atmosphere must have evolved from anoxia to its present oxidizing state, the lands and shoal water areas have expanded and contracted reciprocally, and the episodic development of great glacial ages suggests that patterns of energy distribution and perhaps the energy budget as a whole have undergone fluctuations. The evolution of organisms has also influenced sedimentary processes and products. One might expect, then, that variations in sedimentation through time reflect not only locally generated changes in the environment but also carry an overprint produced by shifts in the state of the oceans, of the biosphere, and perhaps of the earth as a whole—changes of a sort not considered by Hutton, Lyell, and other classical uniformitarianists.

This overprint is likely to be a subtle one, readily masked in shoal water sediment by the multitude of local tectonic and geomorphic events. It is much more likely to be apparent in the pelagic deposits of oceans and starved epicontinental basins, remote from tectonic and geomorphic "noise." We have therefore turned primarily to pelagic sediment in a search for global patterns of secular variation. Our conclusion is that such a variation is demonstrable by various independent lines of evidence, and further, that it appears to include a cyclic fluctuation with a period of about 32 million years.

The pelagic realm as a whole was a stepchild of sedimentology and historical geology until the relatively recent emphasis on ocean coring, drilling, and chemical oceanography. This is producing a flood of information on the functioning and history of the pelagic realm from which most of our data derive.

Our interest was initially aroused by the well-known fact that black, laminated marine sediment (sapropel) devoid of benthic fauna is very restricted in Recent marine deposits, but was widespread round the globe in certain ages of the geologic past, such as in parts of Ordovician, Silurian, and Devonian times, in the early Jurassic, and in the middle Cretaceous (Aptian-Albian). Strøm (1939), in a classical paper on deposition of black sediment in locally restricted basins, suggested that the "local" hypothesis might not adequately explain these more general "black" episodes, and appealed to the possibility of an ocean more prone to anoxia, due to inhibited circulation at times of greater climatic uniformity. Although the "local" model of barred basins has been the generally accepted one (Trask, 1932; Krumbein and Sloss, 1963), Hallam (1975) is disenchanted with its applicability to Jurassic bituminous facies, and we find it inadequate to explain the distribution of Early Cretaceous ones.

This prompted us to investigate the temporal distribution of various other phenomena of the pelagic realm, ranging from organisms to paleotemperatures, the calcite compensation depth, the relative constancy of sedimentation versus incidence of stratigraphic gaps (erosion intervals) in deep-water settings, and the distribution of carbon isotopes in skeletal matter and organic compounds. The data on these various characteristics of the pelagic realm show secular fluctuations, plotted in Figures 1, 4, 6 and 7.

The chronologic base for these plots is essentially the "London time scale" of Harland and others (1964) for the Mesozoic, and that of Berggren (1972) for the Cenozoic. The reliability of points on this scale is probably within 10 percent of their assigned age. In view of uncertainties in the absolute time scale, the figures in this paper have been plotted relative to stages or epochs. The Cenozoic scale of Berggren (1972) has undergone little revision, but a number of different absolute time scales have been proposed for the Cretaceous (Kauffman, 1970; Lambert, 1971; Obradovich and Cobban, 1975; van Hinte, 1976) since that of Harland and others, (1964); some of the newest compilations are, as yet, unpublished (e.g. Thierstein, in press; Douglas and Bukry, personal comm.). The most significant change between the scale used in this paper and most of the newly proposed time scales is the age assigned to the Albian-Cenomanian boundary. Van Hinte (1976) retains a 100 m.y.b.p. age for this boundary, while Kauffman (1970), Obradovich and Cobban (1975) and most others suggest a 94 m.y.b.p. age. This is the only revision having a major effect on the calibration of our proposed cycle to absolute time intervals, and the 94 m.y.b.p. date, in fact, provides a better fit with our observations (32 m.y. cycle). Although our choice of absolute time scales is arbitrary, we have chosen not to incorporate any of the newer scales since acceptance of any one remains uncertain.

Some of the fluctuations shown are surely induced by unrepresentative sampling, but others are "real." In varying degrees, all of these curves show a common element—a response to a cycle on the order of 30 million years long (we suggest 32 million for reasons explained below). This cycle appears to reflect an oscillation between two oceanic states or modes, which we term polytaxic and oligotaxic.

The *oligotaxic mode* is that of the Pleistocene and present seas. It brings a notable increase of extinction rates in nekto-planktic organisms, reducing the complexity of biotic communities and diminishing biotic diversity on a global scale. A major regression seems to be normally associated.

The *polytaxic mode* is characterized by seas warmer than the present and Pleistocene, and by gentler latitudinal and vertical temperature gradients. Thus rate of oceanic circulation as well as the vigor of deep currents is decreased. The oxygen minimum zone, located below the oceanic thermocline is intensified and expanded, leading to widespread sedimentation of organic-rich sedi-

FIG. 1.—Pelagic diversity, superpredators and blooms of opportunists over last 220 million years.

On left, changes in global diversity. Genera of ammonities (A) and species of planktic (globigerinacean) foraminifera (G), plotted logarithmically. Episodes of increasing diversity are separated by biotic crises of varying magnitude. Crises of moderate and high intensity recur at intervals of approximately 32 million years (shaded bands, defining seven cyclic episodes or pulses of diversification: polytaxy). These essentially coincide with transgressive pulses of Grabau (right).

Each polytaxic pulse brought superpredators exceeding 10 m in length, a role which has been successively filled by ichthyosaurs, pliosaurian plesiosaurs, mosasaurs, whales and sharks, as shown in middle. Superpredators are known only from stages opposite the names. Mid-Triassic ichthyosaurs: *Cymbospondylus* and *Shastasaurus*; Toarcian ichthyosaur: *Stenopterygius*; Oxfordian pliosaur; *Stretosaurus*; Albian pliosaur: *Kronosaurus*; Campanian-Maastrichtian mosasaurs: *Hainosaurus* and others. Eocene whale: *Basilosaurus*; Mio-Piocene shark: *Carcharodon megalodon*.

Biotic crises are accompanied by local mass-occurrences of single pelagic species, rare in normal biotas. These are interpreted as blooms of opportunists, and have been plotted in black circles. B, *Braarudosphaera*, a coccolithophorid; P, *Pithonella*, a problematicum; E, *Ethmodiscus rex*, a giant diatom.

Sources of data: Foraminifera, Postuma (1971); ammonities, Moore (1957). Opportunists, Initial Reports of the Deep Sea Drilling Project, Vols. 1-30. Superpredators, Romer (1966), Ida Thompson, pers. comm., and various colleagues. Transgressive pulses, Grabau (1940).

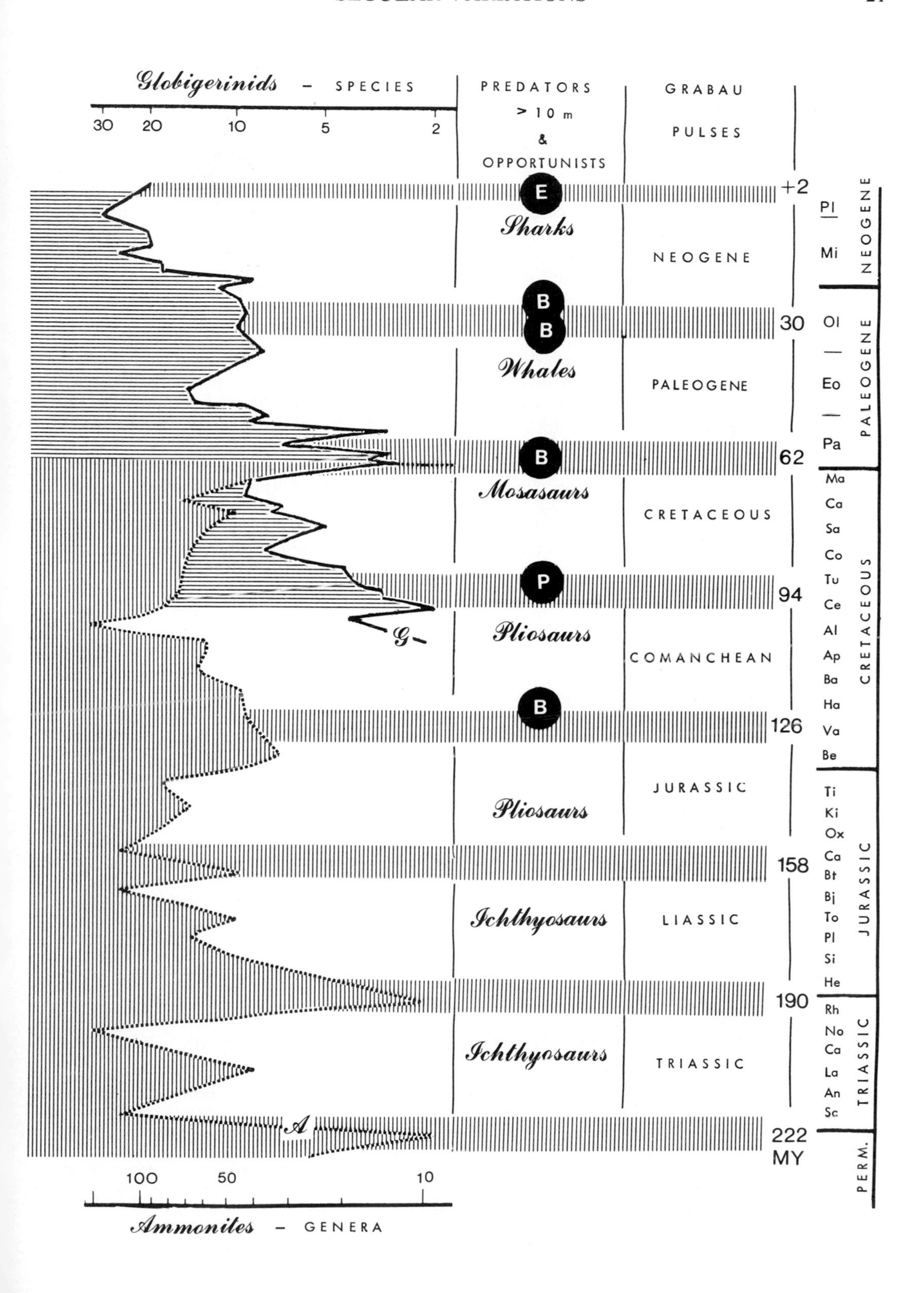
Globigerinids - SPECIES
30
20
10
5
2
PREDATORS
> 10 m
&
OPPORTUNISTS
GRABAU
PULSES
E
Sharks
NEOGENE
B
B
Whales
PALEOGENE
B
Mosasaurs
CRETACEOUS
P
Pliosaurs
G
COMANCHEAN
B
JURASSIC
Pliosaurs
Ichthyosaurs
LIASSIC
Ichthyosaurs
TRIASSIC
A
+2
30
62
94
126
158
190
222
MY
Pl
Mi
NEOGENE
Ol
Eo
Pa
PALEOGENE
Ma
Ca
Sa
Co
Tu
Ce
Al
Ap
Ba
Ha
Va
Be
CRETACEOUS
Ti
Ki
Ox
Ca
Bt
Bj
To
Pl
Si
He
JURASSIC
Rh
No
Ca
La
An
Sc
TRIASSIC
PERM.
100
50
10
Ammonites - GENERA

ment and to net loss of carbon from the atmosphere-hydrosphere system to the lithosphere. The carbonate compensation depth rises to shoaler levels. Rates of speciation exceed rates of extinction, and the complexity of communities as well as global diversity increases accordingly. Marine transgressions seem to be associated.

The successive occurrence of these modes bears no obvious relation to the plate tectonic dispersal patterns of continents (Smith and others, 1973), nor directly to magnetic reversal history (Larson and Pitman, 1972) although each of these factors may influence the expression of such modes. The cyclic pattern of the two modes discussed above shows some resemblance to the history of terrestrial climates (Dorf, 1970), suggesting that the oceanic cycle is tied to oscillations of global climate. These in turn may result from variations in receipt of solar energy or from changes in the manner in which energy is distributed over the earth's surface. It also appears to be in phase with the record of marine eustatic oscillations visualized by Grabau (1940), and currently being elucidated by Vail and others (P. Vail, and others, 1974 and pers. comm.): Oceanic and atmospheric behavior appear linked with eustasy, in ways which remain obscure. Frerichs (1970), Moore (1972), Hays and Pitman (1973), Berger and Roth (1975), and others have also been struck by the apparent link of paleoclimates to eustasy, to organic diversity, and to sedimentation.

The quantitative data base for a rhythmic or cyclic alternation of these two modes is as yet marginal, but is growing rapidly. Isotope studies in particular seem promising for testing and refining the pattern, and for tracking it over longer time spans. The emergence of this cycle is important not only to academic historical geology, but for an understanding of the functioning of hydrosphere, atmosphere and biosphere, and to the search for minerals: incidence of petroleum source beds and of rock phosphate seems directly related to the cycle, and precipitation patterns of naturally oxidizable elements (e.g. Mn, Fe, Cu, U, S) are likely to have been influenced.

In the following pages we discuss each of the various lines of evidence examined, commencing with biotic patterns because fluctuations in taxic diversity have by far the largest data base. From these we proceed to sedimentological-stratigraphical consideration, not easily quantified, and to geochemical data which, while quantitative, are available in sufficient detail for only a few areas. Most of the data are summarized in Figures 1 and 4; the intervening figures provide additional detail.

CHANGES IN BIOTIC DIVERSITY

Of the numerous studies which have been made of fluctuation of biotic diversity through time (Raup, 1972) only few, such as those by Tappan and Loeblich (1971) and Lipps (1970a), have dealt with the pelagic realm, or have focused on ecological rather than strictly taxonomic compilations. Changes in diversity, at the local (community) level, the regional (province) level, or on a global scale, result from the interaction of two opposed processes: speciation which adds taxa, and extinction which subtracts them.

In this paper we present global taxonomic diversity fluctuations for several groups of organisms: for phytoplankton as a whole, (Fig. 2A), for dinoflagellates (also Fig. 2A), for planktic (globigerinacean) foraminifera (Figs. 1, 2B), for certain families of bony (teleost) fishes (Fig. 2C), for sharks, reptiles and whales (Fig. 2D), and for ammonites (Fig. 1). These curves are remarkably similar to each other. They suggest that diversity at present is relatively low, and that, through time, diversity in species and genera has fluctuated repeatedly.

The question of whether our curves adequately depict the actual flux requires discussion. At stake are (1) the integrity of the units counted; (2) the adequacy of sampling; and (3) the time-resolution achievable with the data at hand. Because species,

→

FIG. 2.—Pelagic diversity changes over past 110 million years.

A—Specific diversity of total phytoplankton (scale at top) and of dinoflagellates only (scale at bottom).

B—Specific diversity in globigerinacean foraminifera (excluding heterohelicids). For logarithmic plot, see Fig. 1.

C—Generic diversity in several groups of bony fishes, formerly or presently important in pelagic faunas: (1) archaic teleosts (leptolepomorphs and elapomorphs incl. tarpons, ladyfishes, etc.); (2) scombroids, the mackerels, tunas, swordfishes, etc.; (3) clupeiformes, the sardines and herrings.

D—Generic diversity in other chordate groups: (1) galeoid sharks; (2) marine reptiles excluding crocodilians and turtles; (3) cetaceans (whales).

Intense blooms of single protistan species here interpreted as opportunists are plotted between B and C: *Ethmodiscus rex* (E); *Braarudosphaera* sp. (BB); and *Pithonella* (PP). In conjunction, these groups support the conclusion of Fig. 1, that biotic diversity in the pelagic realm fluctuates in complex ways, but that major lows (oligoaxic times: Cenomanian-Turonian, Paleocene, Oligocene) recur with a period of 32 million years. Sources of oligotaxic data: A, Tappan and Loeblich (1971); Sarjeant (1967); B, Postuma (1971); C, D, Romer (1966).

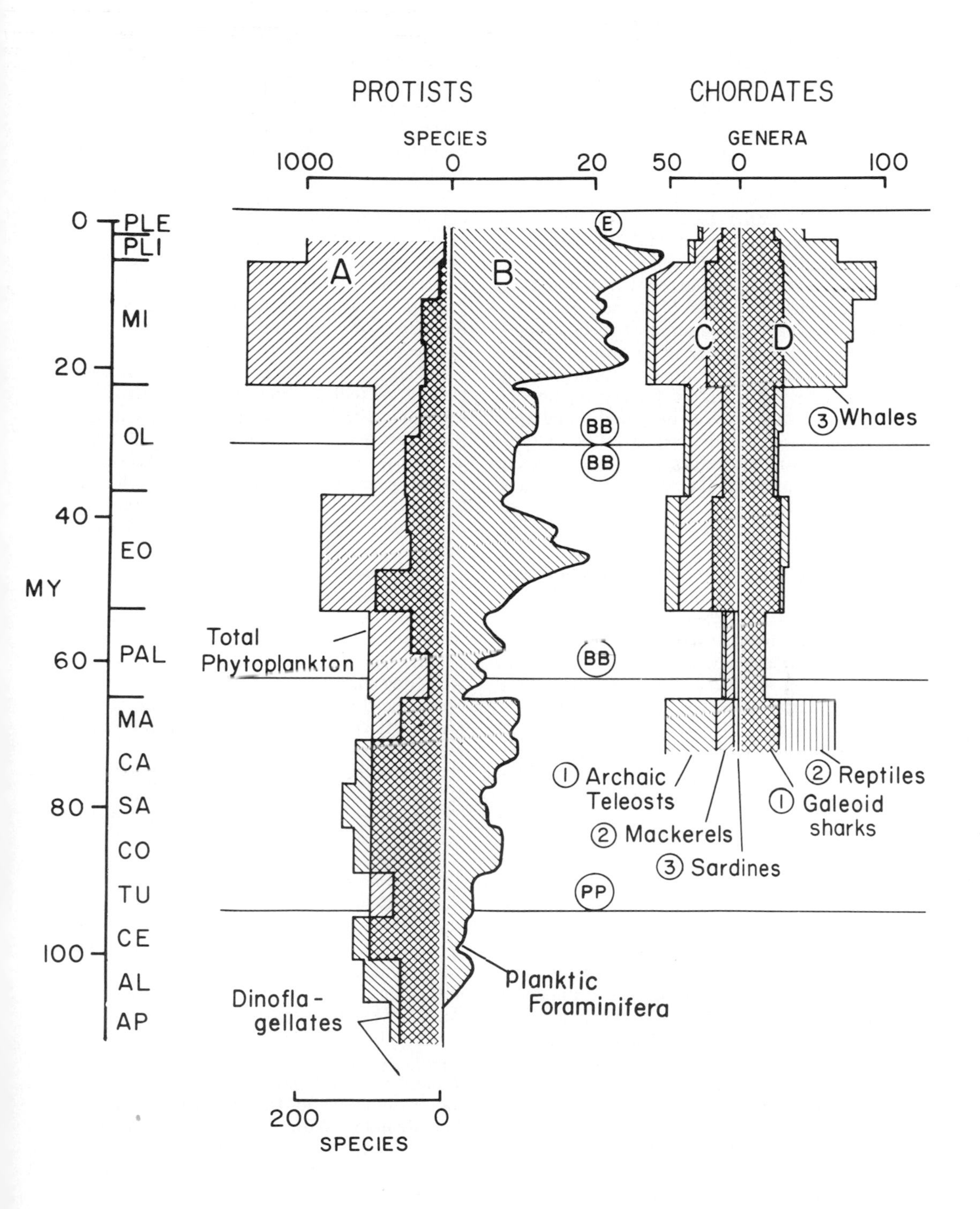
PROTISTS
CHORDATES
SPECIES
1000
0
20
GENERA
50
0
100
0
PLE
PLI
MI
20
OL
40
EO
MY
60
PAL
MA
CA
80
SA
CO
TU
CE
100
AL
AP
A
B
E
BB
BB
BB
PP
C
D
③ Whales
Total
Phytoplankton
① Archaic
Teleosts
② Mackerels
③ Sardines
② Reptiles
① Galeoid
sharks
Planktic
Foraminifera
Dinofla-
gellates
200
0
SPECIES

genera, or higher ranks are based on human judgements, taxa as units are inevitably somewhat variable, and their credibility essentially varies in proportion to the number of *cognoscenti* who have found them justified. In this context, the great peak of ammonite diversity in Albian time (Fig. 1) is probably exaggerated, due largely to the excessive splitting of genera in the British Gault by several British authors, in particular Spath. But the Albian remains a peak even should the number of genera be reduced by a factor of two. On the other hand, the foraminiferal species recognized by Postuma (1971) and graphed in Figs. 1 and 2-B are the daily stock-in-trade of scores of specialists round the world, and have been thoroughly established. I. Premoli-Silva, W. Hay and other specialists assure us that Postuma's compilation is conservative, listing only species accepted by most of the profession, and omitting numerous species recognized locally by one or more workers. Inclusion of these additional species would strengthen the Albian, Late Cretaceous, and Eocene peaks, but would add nothing to the troughs, in which the scarcity of taxa and the difficulty of recognizing zones have led various workers to labor at discriminating species, to little avail.

Sampling represents another complication, and one that changes from group to group. Planktic foraminifera are almost ubiquitous in marine sediment above the calcite compensation depth. Many people identify them daily around the world, from outcrops, wells, and deep sea cores and bore holes. The data at hand, for any one epoch, are based on thousands of fossil assemblages. Fossil vertebrates on the other hand are comparatively rare, and much of what we know about them for any one epoch comes from a few areas and sites, studied by small number of people, at intervals of decades. In terms of sampling adequacy, the foraminifera are at one end of the scale, the chordates at the other.

The record of the calcareous microfossils has been biased by the selective dissolution of the more soluble species from the deeper marine sediments (Berger and Roth, 1975), and temporal fluctuations in the carbonate compensation depth may have introduced some systematic bias into the apparent diversity through time. We are inclined, however, to discount the suggestion that at times such as the early Paleocene the whole ocean was undersaturated with respect to calcium carbonate, and that the faunal paucity observed there has resulted from the post-mortem elimination of many taxa. On the contrary, we believe that for all epochs and fractions of epochs numerous samples have provided a faunal spectrum from depths essentially unaffected by sea-floor solution. Furthermore, as shown below, our data suggest that the carbonate compensation depth was shoalest at times of greatest biotic diversity.

Another bias is introduced into some of our data by the marine transgressions and regressions, a matter discussed at greater length by Raup (1972). Times of transgressions such as the Late Cretaceous, Eocene and Miocene have provided broader exposures of marine sediments and thereby permitted better sampling of the fossil record in outcrop than have regressive times. This may introduce a serious bias into the chordate data, because of the scarcity of chordate remains, and of our dependence on surface outcrops. It may have also affected the ammonite compilation. It is probably of little significance in the case of protists, which are derived not only from outcrops but also in myriads from the subsurface and off the deep sea floor.

Regarding the time resolution of the data presented, we feel justified in drawing the foraminiferal diversities as a simple curve, inasmuch as the basic data are given in terms of nearly 50 successive zones, i.e. in episodes averaging two million years long. Other diversity changes are more crudely dated in published sources and are presented in Figure 2 as histograms. Refinement, in progress, will require time.

Ammonites.—Ammonites, a large and predominantly pelagic group of organisms, have a long and fairly well known fossil record. Due to lack of stratigraphic precision and problems of dating, we have ignored here their Paleozoic history, to focus on changes in their generic diversity through Mesozoic time (Fig. 1). The data are taken from Moore (1957), and the plot is a logarithmic one, inasmuch as we are more interested in proportional changes than in absolute numbers. The addition or subtraction of two genera is vital when the total number is four, but is trivial when the total is a hundred.

The curve shows that over this timespan of 220 million years ammonite diversity has ranged from one or a few genera to several hundred, and that the group approached extinction on three separate occasions: Once in the Permo-Skythian crisis, once in the Rhaeto-Liassic one, and finally in the actual extinction of the Maastrichtian-Danian. These may be called first-order crises.

Second-order crises in which diversity was rapidly reduced to about half of the preexisting number of genera occurred in Ladinian time, in the Bathonian, in the Berriasian, and in the Cenomanian. With exception of the Ladinian crisis, these first- and second-order crises fit into a 32 million year pattern, matching the shaded bars drawn across the middle of Figure 1.

In addition, one may note a series of third-order crises, which include one in the Toarcian, one in the Kimmeridgian, a possible one in the Aptian, and one in the Santonian, during which diversity decreased by a lesser amount. Together with the second order crisis of the Ladinian these appear to define a smaller cycle, half the length of the main one.

Planktic foraminifera.—We turn next to planktic foraminifers—specifically, Globigerinacea exclusive of heterohelicids. This small group of organisms has now the best documented geological record of any. The plots are numbers of species (which are mainly pan-tropical). The plot in Figure 1 is a logarithmic one, whereas the same data are shown arithmetically in Figure 2B. Unlike the ammonites, the globigerinaceans are a young group whose history to date has been one of an overall expansion; the curve therefore shows a strong overall trend toward greater diversity.

A first-order crisis that almost brought extinction occurred in the Maastrichtian-Danian event. Second-order crises that reduced the diversity to about half occurred in the Cenomanian and in the Oligocene. Third-order crises, of lesser magnitude, occurred in the Santonian, in the middle Miocene, and in the present Pleistocene-Recent episode. The first- and second-order crises, taken together, overlap and continue the 32 million year cycle shown by the ammonites, and make it appear likely that the present third-order crisis will grow into a second-order or even a first-order one. The third-order crisis of the Santonian and middle Miocene match the subsidary 16 million year periodicity suggested by the ammonites, but no such minor crisis is shown where expected in the Eocene.

A periodic reduction of diversity is demonstrated not only by species tabulation, but also by the occurrence of two structural types: each of the four major episodes of diversity contains species with simple chambers as well as species whose chambers bear equatorial flanges (keels). During three of the four main crises, only one or the other of these morphotypes survives: in the early Cenomanian crisis only keeled species, during the Paleocene and Oligocene crises only unkeeled ones. Both keeled and unkeeled species are currently in existence, but the worst of the present crisis may be yet to come. Taken together, the ammonites and globigerinacean foraminifera demonstrate a periodic recurrence of biotic crises, in which a 32 million year rhythm emerges as a consistent one of first- and second-order crises. There is a suggestion of a rhythm of double that length; alternate biotic crises tend to be more severe. There is also some indication that a lesser (third-order) fluctuation splits the 32 million year rhythm in half.

Nannoplankton.—The recent effort expended on biostratigraphic studies of calcareous nannoplankton (coccolithophycean algae) makes this the best known group of pelagic organisms next to planktic foraminifera. A general compilation (Tappan and Loeblich, 1971; Lipps, 1970a) shows a peak in the Late Cretaceous, a severe reduction in the Paleocene, high diversity in the Eocene, a crisis in the Oligocene, a further peak in the Miocene, and a reduction during the Pliocene leading to extreme crisis in the Pleistocene. More detailed data on calcareous nannoplankton are provided by Haq (1973), see also Berger and Roth, 1975.

Dinoflagellates.—The specific diversity of dinoflagellates is plotted in Figure 2A, from a compilation by Sarjeant (1967), as figured in Tappan and Loeblich (1971). These fossils are not remains of the active life stage, but of resting cysts which are formed by only some of the species. A Cenomanian to Turonian crisis and a major crisis in the Paleocene are the most outstanding features in this record, diversity having declined greatly during the Cenozoic. Apparently, the very abundant recent species are not leaving an adequate fossil record.

Silicoflagellates.—The history of silicoflagellates (Lipps, 1970b) shows a diversity peak in the Late Cretaceous, and a decline in the Paleocene. Diversity increased in the Eocene, fell in the Oligocene, reached an all-time high in the Miocene, and declined through the Pliocene to a Quaternary low (58 living species contrast with 97 Miocene ones). No separate plot of these data is given here.

Total phytoplankton.—Species diversity of phytoplankton as a whole (from Tappan and Loeblich, 1971) is shown in Fig. 2A, along with the dinoflagellate plot. Although the data are not well refined chronologically, they do indicate a major diversity peak in the Eocene, an Oligocene crisis, a Miocene peak, and a more recent decline. Taken together with the more refined dinoflagellate data, it suggests that the phytoplanktic changes in diversity through time closely match the foraminiferal ones.

Chordates.—Variations in the generic diversity of selected groups of chordates represented in the epipelagic biota, after Romer (1966), are shown in Figures 2C and 2D. The groups of teleost fishes shown (Fig. 2C) include the ancestral root stock of teleosts, the leptolepimorphs, which comprise the nonpelagic tarpon and the lady fishes of present seas; the Clupeiformes (herrings and sardines) which are important filter-feeding members of present pelagic fish faunas; and the Scombreidae (mackerels, tunas, swordfishes) which are the "top" teleost predators of the high seas. Diversity curves for genera of galeoid sharks (the main group of predatory sharks in the Cenozoic), for marine reptiles (excluding sea snakes, sea turtles, and crocodiles), and for whales and porpoises (cetaceans) are plotted in Figure 2D. Collectively these groups show diversity fluctuations which closely parallel pelagic protist patterns.

As pointed out above, the sample base for chordates is much less adequate than is that for protists; nevertheless, no sampling errors can explain the extinction of the great sea reptiles at the end of the Cretaceous, nor the spectacular diversification of cetaceans in the Miocene and their subsequent decline.

COMMUNITY STRUCTURE

Superpredators.—The variations in diversity, discussed above, suggest significant changes in the ocean as a habitat. Such fluctuation should be reflected not only in global diversity, but also in community structure, and in the proliferation of certain species. Accordingly, two other kinds of observations have been plotted on Figure 1: the time-occurrence of superpredators, and the mass-appearance of opportunistic "disaster forms."

The largest near surface predators of the present ocean, the great white shark *Carcharodon carcharias* and related species, reach lengths of about 6 m, while some of the great marine predators of the past exceeded 10 and approached 20 m. Such superpredators include the ichthyosaurs *Cymbospondylus* and *Shastasaurus* (Mid-Triassic), and *Stenopterygius* or *Temnodontosaurus* of the Liassic; the pliosaurian plesiosaurs *Stretosaurus* (Oxfordian) and *Kronosaurus* (Albian); the large Campanian and Maastrichtian mosasaurs such as *Hainosaurus*; the Eocene whale *Basilosaurus*; and the Mio-Pliocene shark *Carcharodon megalodon*. The only present-day predator which achieves that size range is the sperm whale, *Physeter*, feeding on squids of the midwater masses by extreme specialization for deep diving. In the geologic record superpredators in the 10–20 m range occur roughly at the diversity peaks, and no such large predators are known from the intervening episodes of lower diversity. Although the evolutionary pressures that lead to the development of superpredators are not clear, such organisms probably were an expression of community complexity. Their history suggests that in the pelagic realm the development and the collapse of global diversity and of community complexity coincide.

Disaster forms.—If superpredators are indices to times of diversity, one may ask whether opportunistic species in the sense of MacArthur (1955) and Levinton (1970) proliferated at times of biotic crises. In the pelagic settings, we suggest that the coccolithophycean *Braarudosphaera*, the problematicium *Pithonella*, and the diatom *Ethmodiscus rex* were disaster forms.

Braarudosphaera is the name applied to a highly distinctive group of coccospheres, 12-side (pyritohedron-shaped), thick-walled boxes of calcite, lacking pores or apertures (Fischer and others, 1967). These most likely represent resting stages or cysts, designed to sink to the bottom and to remain dormant during a period of unfavorable environmental conditions. They range from the Late Jurassic to the present, and have been found living in bays in Maine, Panama and Japan, i.e. under conditions unfavorable to pelagic organisms. *Braarudosphaera* is represented, throughout its time range and throughout the world, as a rare member of the coccolith associations in "globigerina oozes" and chalks, but in some instances it occurs in large numbers, making up an appreciable part of the sediment. Such mass occurrences which have become known to us are plotted in Figures 1 and 2. These are found in Neocomian chalks of the western North Atlantic (Ewing, Worzel, and others, 1969, Hollister, Ewing, and others, 1972); in the Paleocene (Danian) of northern Spain (Fischer and others, 1967; Percival, 1972) and central Italy (Monechi and Pirini-Radrizzani, 1975) and in the Oligocene of the South Atlantic (Maxwell, von Herzen, and others, 1970). *Braarudosphaera* is abundant, though not dominant, in the Pliocene of the northwestern Indian Ocean. It is prominent in all post-Jurassic oligotaxic episodes save the Cenomanian one. In our interpretation, *Braarudosphaera* is normally confined to marginal and environmentally unstable settings, but spreads and blooms over parts of the oceans when widespread environmental upsets have damaged the normally more efficient and competitive pelagic species assemblages.

In Cenomanian time this role seems to have been played in the Mediterranean region (Premoli Silva, pers. comm.) and in England (Banner, 1974) by *Pithonella*, a microscopic calcareous capsule of uncertain relationships. In Pleistocene and Holocene times, the main opportunist, at least in the tropics, is the giant diatom *Ethmodiscus rex*, which is a rare constituent of normal marine floras, but whose episode blooms in the past are documented by beds of pure *E. rex* ooze such as those cored in the Pleistocene of the western Pacific (Fischer, Heezen, and others, 1971), and of the eastern Equatorial Atlantic (Gardner and Burkle, 1975).

POLYTAXIC AND OLIGOTAXIC EPISODES

These data suggest that the upper ocean as a habitat for nekto-planktic organisms undergoes rhythmic fluctuations, reflected in global diversity, in community structure, and in the episodic spread of "disaster forms."

The episodes of taxonomic proliferation we term polytaxic, and those of taxonomic decline, oligotaxic. Oligotaxy recurs primiarily at 32 million year intervals. The last eight such episodes (Fig. 1) were: the Permo-Triassic crisis, the Triassic-Liassic crisis, a Bathonian-Callovian impoverishment, one in the Neocomian (Valenginian), one in the Cenomanian-Turonian, the Maastrichtian-Danian crisis, the Oligocene improverishment, and the Plio-Pleistocene crisis. There is some suggestion that alternate crises are stronger (but the Permian one was very severe and the present one is at least not yet so severe as might be expected). There is also a strong suggestion for a recurrence of minor oligotaxy at 16 million year intervals.

CHANGES IN OCEANIC TEMPERATURE RÈGIMES

Investigations of ancient marine temperatures by means of oxygen isotope ratios in calcareous fossils (Epstein and others, 1951; Bowen, 1966) are beginning to contribute substantially to historical geology. Errors may be introduced by vital effects upon carbonate precipitation (Keith and Weber, 1965; Duplessy and others, 1970; Vinot-Bertouille and Duplessy, 1973), by local changes in marine isotope ratios due to net evaporation or fresh-water contamination (Craig and Gordon, 1965), and by permineralization and recrystallization during diagenesis (Anderson and Schneidermann, 1973; Douglas and Savin 1973; Savin and Douglas, 1973; Spaeth and others, 1971). Despite these problems two sets of isotopic paleotemperature determinations seem to be so consistent and coherent that they are of great geological significance. They are the values obtained from belemnites, mainly from the deposits of European shelf seas, and those yielded by foraminifera and by bulk analysis of calcareous ooze retrieved from the deep ocean floor, untouched by fresh-water. A general compilation of available data is shown in Figure 4, and, in more detail, in Figure 3.

The belemnite values for the Jurassic and Cretaceous of northwestern Europe (Bowen, 1966; Lowenstam and Epstein, 1954; Fritz, 1965) supplemented by a few temperature determinations from ammonite aptychi, are plotted in Figure 4, and the Cretaceous values also in Figure 3 and 4. Unfortunately no paleotemperature determinations have been published for the late Triassic and earliest Jurassic. Numerous temperature fluctuations are evident throughout the Jurassic and Cretaceous and a decline precedes the extinction of nearly all belemnites at the end of Cretaceous time. On basis of form and occurrence the squid-like belemnites are thought to be nektic or semi-nektic members of the upper water masses. Their isotopic variation suggests that the mean temperature of these water masses, in the middle latitudes, varied through a relative range of about 10°C, the warm times coinciding in general with times of biotic diversity, the cool times with biotic crises.

This temperature history of the upper waters is overlapped in the Cretaceous and continued through the Cenozoic by values obtained from planktic foraminifera from deep-sea cores (Figs. 3 and 4A). Douglas and Savin (1975) have added to the knowledge of mid-latitude Pacific Ocean temperature trends for the Cretaceous through analyses of both planktic foraminifera and calcareous nannoplankton. Temperature highs occurred during the Albian, and the Coniacian-Santonian; lows in the Cenomanian and the late Maastrichtian-earliest Paleocene. Saito and van Donk (1974) show, for the South Atlantic, a marked cooling in the Maastrichtian, followed by a rise in the Paleocene (Figs. 3, 4A). The analyses of Douglas and Savin (1971, 1973) and the summary of Savin and others (1975) for the tropical Pacific (same figures) carry this trend on to a peak in the middle Eocene, followed by a decline to a low in mid-Oligocene time. A minor peak in the late Oligocene, a major one in the early Miocene, and a minor one in the Pliocene precede the present decline. Yet another, even more detailed set of determinations by Shackleton and Kennett (1974) for higher South Pacific latitudes illustrates the same general fluctuations, while showing a much steeper historical trend to cooler temperatures, and differing in details (Figs. 3, 4A).

Paleotemperatures from the deep ocean floor, recorded by benthic foraminifera, provide an interesting comparison (Fig. 3). Fluctuations which parallel those of their epipelagic counterparts are superimposed over a strong cooling trend that persisted through the entire Cenozoic. As noted above, this trend is also strongly expressed in the higher-latitude plankton. Discovered by Emiliani (1954, 1961), it has now been documented in detail by Douglas and Savin (1973) Savin and others (1975) and Shackleton and Kennett (1974).

These studies, along with other scattered results not reviewed here, represent the first attempts at a thermal history of the oceans. Admittedly there is an alternative interpretation: a secular variation in the isotopic composition of the ocean waters, by some such effect as the growth of ice caps in the Pleistocene or by changes in patterns of evaporation and precipitation over the sea-surface. At the present state of knowledge we hold a temperature fluctuation to be the more likely predominant cause of the phenomenon over these largely non-glacial parts of earth history.

As shown in Figures 3 and 4A, warm periods largely correspond to the polytaxic ones, while colder episodes coincide with states of oligotaxy or biotic crisis. This same observation has been

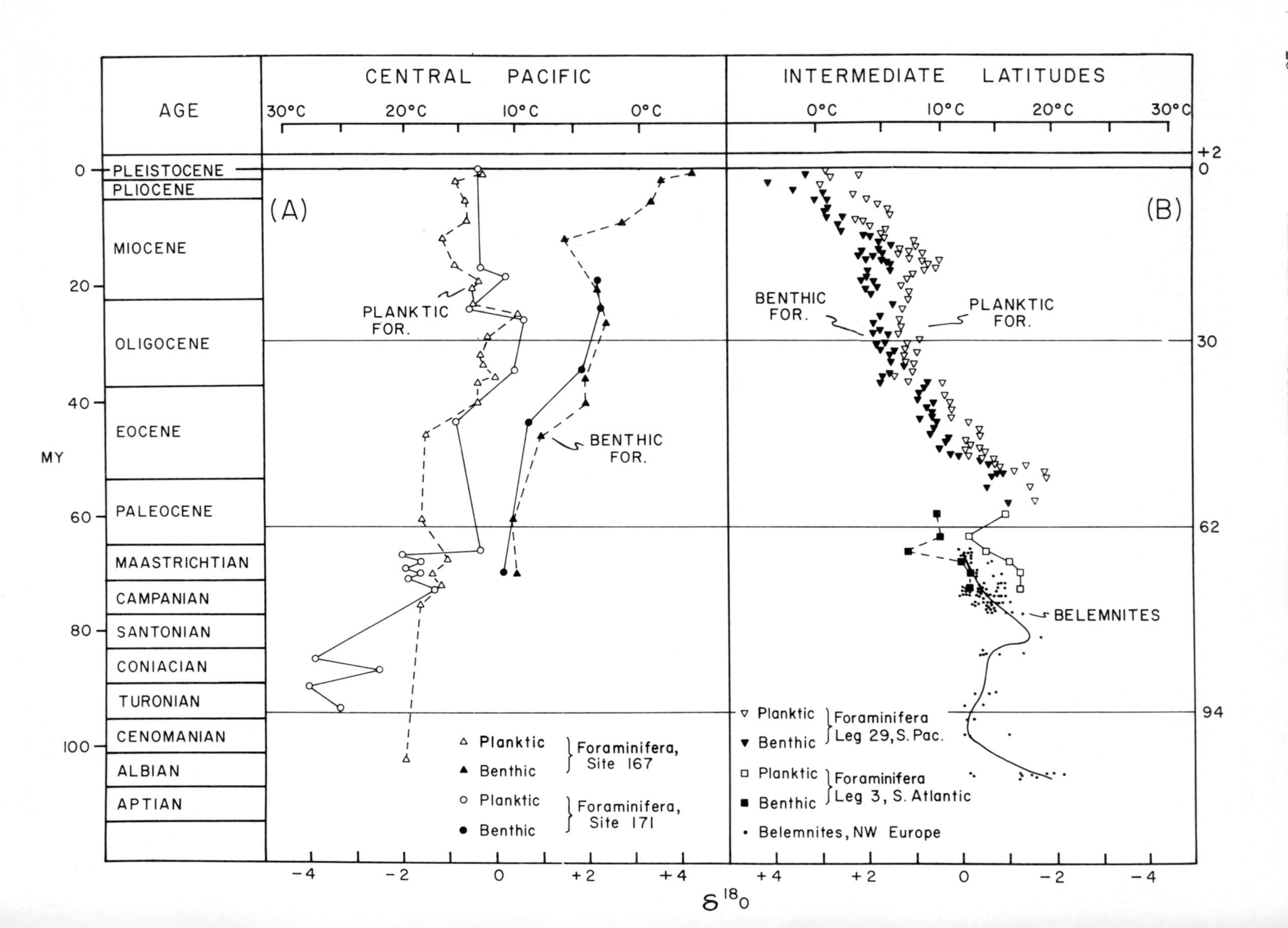
CENTRAL PACIFIC
INTERMEDIATE LATITUDES
AGE
PLEISTOCENE
PLIOCENE
MIOCENE
OLIGOCENE
EOCENE
PALEOCENE
MAASTRICHTIAN
CAMPANIAN
SANTONIAN
CONIACIAN
TURONIAN
CENOMANIAN
ALBIAN
APTIAN
MY
(A)
(B)
PLANKTIC FOR.
BENTHIC FOR.
BELEMNITES
Planktic Benthic } Foraminifera, Site 167
Planktic Benthic } Foraminifera, Site 171
Planktic Benthic } Foraminifera Leg 29, S. Pac.
Planktic Benthic } Foraminifera Leg 3, S. Atlantic
Belemnites, NW Europe
$\delta^{18}O$

made by earlier authors (e.g. Devereaux, 1967; Valentine, 1968; Cifelli, 1969; Funnell, 1971; Haq, 1973) on the basis of less information.

OXYGEN ON THE SEA FLOOR

Sediment-organism relations on pelagic bottoms.—The extensive dredging, gravity coring, piston coring and drilling which have been carried out since the Challenger expedition provide a reasonably good picture of the nature and distribution of Quaternary pelagic sediment on the ocean floor. Except for areas of rapid diatom sedimentation in high latitudes, this sediment is deposited at mean rates ranging from 2-100 mm/thousand years (before compaction). The bottoms on which it accumulates are, with very few exceptions, supplied with oxygen contents of 1 to 8 ml/l, sufficient for maintenance of aerobic metazoans. For these organisms the sediment provides both a means of shelter and a source of food. Although some of them burrow to depths of a meter or more, the majority confine their activities to the upper decimeter of the sediment (Berger and Heath, 1968; Frey, 1971). Thus some 1,000 to 50,000 years of stratigraphic record remains within their easy reach, and is homogenized by their stirring. At the same time, most of the organic matter within the sediment is extracted (oxidized) so that the mean carbon content of these pelagic sediments is reduced to the order of 0.1 percent.

This pattern dominates present deep-sea sedimentation as well as that on rises, sea mounts, and on some shelves. There are local exceptions of great significance, where free molecular oxygen is lacking on the sea floor, and where metazoans and aerobic decomposing organisms are absent (Rhoads and Morse, 1971). Such conditions occur in topographically restricted basins such as the Cariaco trench or the Santa Barbara Basin, but also locally where the "oxygen minimum" layer of the oceans (Fig. 8) is especially well-developed, as on the western side of India (von Stackelberg, 1972), Gulf of California (van Andel, 1964) and Peru (Veeh and others, 1973). Under such conditions the sediment is not bioturbated; instead it is generally finely laminated in response to short-term variations in sedimentary supply. Here it also normally retains organic carbon in amounts ranging from a fraction of 1 percent to several per cent (van Andel, 1964; Richards, 1965; Ronov, 1958). Bitterli (1963) presents a thorough discussion of structures and mineralogy of ancient analogues in Europe. Associated with this variation of carbon content is the general oxidation state, acquired during diagenesis. This condition, in need of more study, finds a general expression in the color of the sediment, (see Revelle, 1944).

Color and structure of marine sediments.—Black and dark gray colors (those not due to manganese oxides) are characteristic of marine sediment in which the organic carbon content is about 1 percent or greater and in which iron occurs in the reduced form in sulfide minerals. Such sediment, upon deposition, conceptually has a large unsatisfied capacity for taking up oxygen, which we refer to as its "oxygen debt." In shoal water settings, where sedimentation rates and organic supply can both be high, as in tidal marshes, lagoons, and estuaries, organic-rich sediment can develop under aerobic bottoms, can be thoroughly bioturbated and contain a benthic fauna. In the slowly accumulated pelagic settings, where every particle has reacted for hundreds if not thousands of years with the supernatant water, organic-rich sediment is an index to anoxic bottoms, and is generally devoid of bioturbation or benthic fossils. Such black pelagic marls, clays, and cherts are known.

In summary, the slowly accumulated sediment of the pelagic realm, deposited under anaerobic conditions, is laminated and ranges from dark gray to black and dark green. Sediment deposited on aerobic bottoms is normally homogenized by burrowers, and its colors range from tan-pink or brown-red to white, pale drab or greenish.

Secular variation.—Some 38 percent of the present sea-floor is covered by the highly oxidized "red clay"; 48 percent is covered by the pale carbonate oozes, but some of these are tan col-

FIG. 3.—Paleotemperatures derived from oxygen isotope ratios in calcitic fossil skeletons, assuming constant $\delta^{18}O$ values for sea water.

A—Tropical Pacific bottom temperatures from benthic foraminiferal assemblages; upper water temperatures from planktic foraminiferal assemblages.

B—Intermediate latitudes of Pacific and Atlantic regions; bottom temperatures from benthic foraminiferal assemblages; upper water temperatures from planktic foraminiferal assemblages, and belemnites.

During the Cenozoic, bottom values and high-latitude surface values trend toward lower temperatures. Superimposed over this trend are fluctuations corresponding to a warming in polytaxic and a cooling in oligotaxic times. Sources of data: Tropical Pacific, DSDP Leg 17, uncorrected latitude 7-19°N (Douglas and Savin, 1973); South Pacific, Leg 29, uncorrected latitude 47-52°S (Shackleton and Kennett, 1974); South Atlantic, Leg 3, uncorrected latitude 30-32°S (Saito and Van Donk, 1974); belemnites northwest Europe, uncorrected latitude 45-55° (Lowenstam and Epstein, 1954; Bowen, 1966; Fritz, 1965).

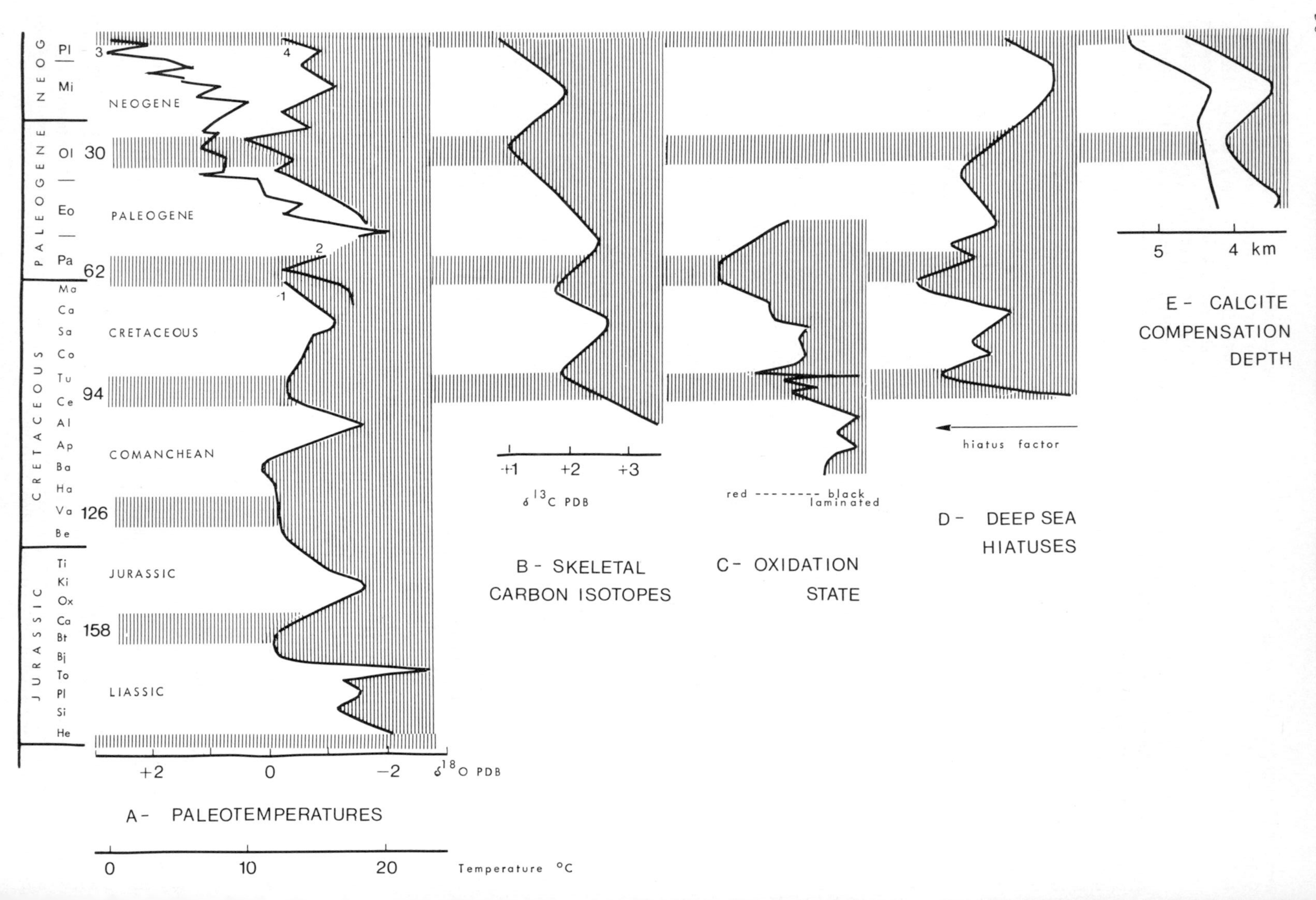
NEOGENE
PALEOGENE
CRETACEOUS
JURASSIC
Pl Mi
Ol Eo Pa
Ma Ca Sa Co Tu Ce Al Ap Ba Ha Va Be
Ti Ki Ox Ca Bt Bj To Pl Si He
30
62
94
126
158
NEOGENE
PALEOGENE
CRETACEOUS
COMANCHEAN
JURASSIC
LIASSIC
3
4
2
-1
+2
0
-2
δ18O PDB
A - PALEOTEMPERATURES
0
10
20
Temperature °C
+1
+2
+3
δ13C PDB
B - SKELETAL CARBON ISOTOPES
red -------- black laminated
C - OXIDATION STATE
hiatus factor
D - DEEP SEA HIATUSES
5
4 km
E - CALCITE COMPENSATION DEPTH

TABLE 1.—OXIDATION STATES IN PELAGIC SEQUENCE AT GUBBIO (ITALY)

					Structure		
Age	Lithology	Color	% Organic C	Fe-mineral	laminated	burrow mottled	homoge-nized
Late Eocene	shale	dark brown		pyrite	X		
Middle Eocene	cherty limestone, marlstone	cream to light gray, some red	<0.5	pyrite? hematite		X	
Campanian-early Eocene	limestone and marlstone	pink to red	<0.1	hematite		X	XX
Coniacian-Santonian	cherty limestone	cream to gray, some pink	<0.2	hematite some pyrite	X	XX	
Late Turonian	cherty limestone	pink to red	<0.1	hematite		X	XX
Early to mid-Turonian	cherty limestone and black shale	dark gray to black	0.02–13.6	pyrite	X	X	
Cenomanian	cherty marlstone and limestone	light gray to cream and pink	0.02–0.6	pyrite			XX
Aptian-Albian	claystone and marlstone	green, gray and black	0.02–2.5	pyrite	XX	X	
Neocomian	cherty limestone	white to light gray	<0.02	pyrite			XX

ored, and will probably turn pink with time and burial. Only a minute fraction of shelf and deep sea floors is anaerobic and accumulates dark gray to black, laminated sediment (e.g. Black Sea, Cariaco Trench, some fiords and areas of impingement of pronounced oxygen minimum on sea floor: Richards, 1965; Calvert, 1964). We suggest that the present ratio of highly oxidized red sediment to laminated black sediment is not representative of geological time as a whole; red muds have always been deposited somewhere, black muds at other sites, but the proportions have fluctuated. The mid-Ordovician, the Late Devonian, the Early Jurassic and the Late Jurassic, for example, show a wide spread of black anaerobic facies in geosynclines and epicontinental basins. The Jurassic occurrences in Europe are discussed by Hallam (1975). On a shorter time scale, a variation can be traced with some detail through the last 110 million years.

Table 1 and Figure 4-C present data on the pelagic sequence at Gubbio, in the Umbrian Appennines of Italy. This continuously deposited, 600-meter section ranges from the base of the

FIG. 4.—Synopsis of various chemical and geological parameters in relation to the 32 million year cycle.

A—Paleotemperatures since the early Jurassic, inferred from oxygen isotope ratios in fossils: (1) belemnites, northwestern Europe; (2) globigerinids, southern Atlantic; (3) globigerinids, South Pacific, showing the progressive cooling of the high-latitude waters; (4) globigerinids, tropical Pacific. For details and sources, see Fig. 3.

B—Carbon isotope ratios in pelagic skeletal carbonates. For details and sources, see Fig. 6.

C—Relative oxidation state of sediments in pelagic sequence (Scisti a Fucoidi and Scaglia Formations) at Gubbio, Italy, based on color, organic carbon content and lamination (unpublished study by Arthur). We infer that much of the fluctuation shown here is widespread in the upper pelagic realm of the world's oceans. (see, for example, the spread of black sediments in Aptian-Albian, Fig. 5).

D—Deep Sea hiatuses. Deep sea drilling has shown many hiatuses in the pelagic record; their incidence and magnitude vary through time, as shown semiquantitatively in this graph. Incidence of hiatuses increases toward left. For details, see Fig. 7.

E—Fluctuations in level of calcite compensation depth, as reconstructed from Deep Sea Drilling data. The two curves bracket the range in values calculated for the different oceans. After Van Andel (1975).

Cretaceous up through the Eocene. We have been studying it in detail (Arthur and Fischer, 1977. It seems clear that a history of oxidation here led from the Neocomian with moderately oxidized sea floor and sediment, to a time of drastic oxygen depletion (with short aerobic intervals) in the Aptian-Albian, a better oxidized episode in the Cenomanian, another (short) episode of anoxia in the early or middle Turonian, good oxygenation in the late Turonian, only moderately good oxidation in the Coniacian-Santonian, strong aeration in the Camponian to early Eocene, and progressively poorer oxygenation into the late Eocene.

Comparison of this local sequence with the nature of Cretaceous and Eocene pelagic sediment in general, as it is known from the surface and especially from Deep Sea Drilling, leads to the following generalities:

1. The entire Gubbio section is relatively more oxidized than is the average sediment of these ages.

2. The anoxic Aptian-Albian interval is not a local anomaly; it is widespread round the world (Fig. 5). In the mountain belts the geosynclinal Aptian-Albian is distinctively characterized by the occurrence of black, laminated sediment, contrasting with aerobically deposited sequences above and below. Well known examples include the black Pariatambo and Muerto Limestone of the Peruvian Andes, the Albian of Venezuela, the Mowry Formation of the Middle Rocky Mountain region and the "Flysch negro" of northern Spain. Even more impressive is the record of in Deep Sea Drilling Project: with completion of Leg 42, Albian Beds had been encountered at 39 sites, scattered through the Atlantic, Pacific and Indian Oceans. At 25 of these sites, the Aptian-Albian section yielded evidence of anoxic bottoms: dark or black, often laminated beds in chalk, shale, or radiolarite facies. Because of spotty coring, presence or absence of such beds at the remaining sites is not certain. Most of the anaerobic sediment cored to date in the deep sea has come from beds of Aptian-Albian age. So long as such discoveries were limited to the Atlantic basin, a possible explanation could be sought in the physical restrictions of the early Atlantic Ocean. However, this hypothesis fails to explain why the underlying Neocomian and especially Jurassic beds do not show equal signs of stagnation. Now that Aptian-Albian deposits of parts of the Pacific and Indian oceans are known to share this character, it is reasonable to conclude that the global oceans at that time were susceptible to anoxia. Whether the rapid reversals in aeration revealed in the Cenomanian-Turonian sequence of the Italian Scaglia are generally reflected round the world remains to be seen.

3. Susceptibility of Santonian-Coniacian seas to stagnation was not as great as that of Aptian-

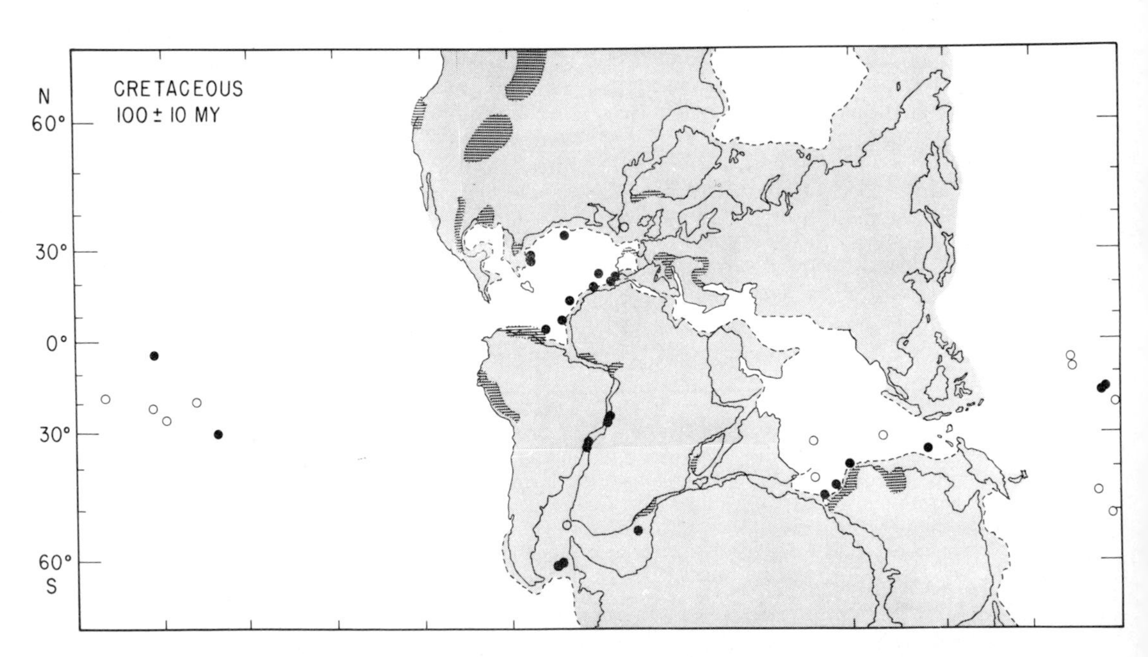

FIG. 5.—Occurrence of black Aptial-Albian marine sediments. Deep Sea Drilling sites which encountered pelagic Aptial-Albian sediments shown by circles; sites which recovered black sediment shown in black, others left open. Occurrences of black organic-rich sediment on continents shaded. Palinspastic base from Smith and others (1973). Compiled from various sources.

Albian ones, but is suggested by the occurrence of laminated chalk in the western Atlantic (DSDP Site 4, Ewing, Worzel, and others, 1969) carbonaceous chalk in the Caribbean Sea (Edgar, Saunders, and others, 1973), black shale in the South Atlantic (DSDP leg 40) and by the widespread occurrence of laminated, non-bioturbated "speckled" shale and chalk in the Austin and Niobrara Formations, from Florida to western Canada (Frey, 1972; Eicher and Worstell, 1970).

4. Nothing in the Cenozoic sequence rivals the Albian record of widespread marine anoxia, but organic-rich, laminated sediment appears to be more common in Eocene and mid-Miocene rocks than in the record of other times in the Cenozoic. Eocene examples are black sediment cored on the Ninety-East ridge (Davies, Luyendyk and others, 1974, site 253, Indian Ocean), black shales in the Cambay Basin of India (Bhandari and Chowdhary, 1975), and the black flysch of the inner Alps (e.g. Gignoux, 1955, p. 517). Well known Miocene occurrences are black, commonly diatomaceous shale of western California (Bramlette, 1946) and equivalents in Japan.

The widespread development of pink or tan chalk, documenting times of unusually vigorous oxidation, is notable. Such sediment apparently characterizes the lower Paleocene or Danian especially, for example, the "red Danian" limestone of northeastern Spain and Biarritz, Scaglia Rossa of the Appennines, and the red Velasco marl of the subsurface round the northern Gulf of Mexico, surfacing in the Tampico embayment. In the pelagic sequences of Cyprus, the Oligocene chalk differs from the Eocene and Miocene ones by virtue of its pinkish-tan color (Fischer, pers. obs.). Much of the Recent and Pleistocene globigerinid ooze exposed on the deep sea floor is stained buff by virtue of ferric iron; it may also eventually appear in the stratigraphic record as pink and red chalk.

In summary, the pelagic record shows that the Albian ocean floor was subject to widespread anoxia. From Albian time on the spread of anoxic pelagic bottoms has tended to decrease. A fluctuation is superimposed over this trend: polytaxic episodes show relatively widespread anoxia; oligotaxic times, widespread aeration.

CARBON ISOTOPE RATIOS

The earth's carbon exists in three main reservoirs: the ocean-atmosphere system, the bodies of living organisms, and the lithosphere. These are constantly interacting with each other. Most organic carbon in organisms and sediments is the result of photosynthetic fixation, which selectively favors the light isotope C^{12} over C^{13}. Organic matter in live photosynthetic organisms and in sediments is therefore enriched in the light isotope, and the atmosphere and hydrosphere are correspondingly depleted. This provides a means of tracking the carbon cycle through earth history (Keith and Weber, 1964; Weber, 1967, Broecker, 1970; Tappan and Loeblich, 1971; Perry and Tan, 1972; Garrels and Perry, 1975; Junge and others, 1975; Mackenzie, 1975).

As more analyses of carbon isotopes—both in organic matter and in limestones—have accumulated, and as the stratigraphic derivation of samples has become refined, a pattern of minor but significant variation in carbon isotope ratios[1] is beginning to emerge.

For our purposes some of this minor variation is "noise": (1) Carbon isotopes are not homogeneously distributed through the oceans, being locally enriched with C^{12} by photosynthesis. The uppermost waters are therefore especially liable to deviate from the oceanic mean. (2) Temperature may affect the isotope ratios of carbon withdrawn by organisms for both organic compounds and calcareous skeleton (e.g. Rogers and Koons, 1969; Rogers and others 1972; Sackett and others, 1965). Surface plankton, deeper plankton, and benthos, living in different water masses and different temperatures, will therefore yield somewhat different isotope ratios. Sediments of the same age may show geographic changes in isotopes reflecting climate. (3) Organisms may exert some vital effects on the ratios at which they withdraw isotopes from the medium. (Keith and Weber, 1965; Weber and Raup, 1968). (4) Not all of the organic matter deposited in a given sediment is autochthonous; thus, a fraction of that found in a given marine sediment is probably of terrestrial derivation, isotopically lighter than that produced by the marine organisms (Calder and others, 1974).

Many carbon isotope analyses of stratigraphically well-controlled and closely spaced deep sea samples have recently become available. These show the existence of systematic variations. Data on skeletal (carbonate) carbon (Figs. 4B, 6) are provided by Douglas and Savin (1971, 1973), by Shackleton and Kennett (1974), by Coplen and Schlanger (1973), and by Saito and Van Donk (1974). Analyses of organic carbon (Fig. 6) are provided by Rogers and others (1972) and by Calder and others (1974). As shown in Figure 6 the total range in the carbonates is one of about 4 per mil, and in organic matter about 7 per mil. At least half of this amplitude results from systematic excursions on a scale of tens of millions of years, which fits rather well the 32 million year cycle suggested by our other data, and is larger than that expected from coeval temperature variations, and their influence on carbon isotopic ratios.

[1] $\delta\,^{13}C = (^{13}C/^{12}C \text{ sample} - ^{13}C/^{12}C \text{ standard})\ 1000$

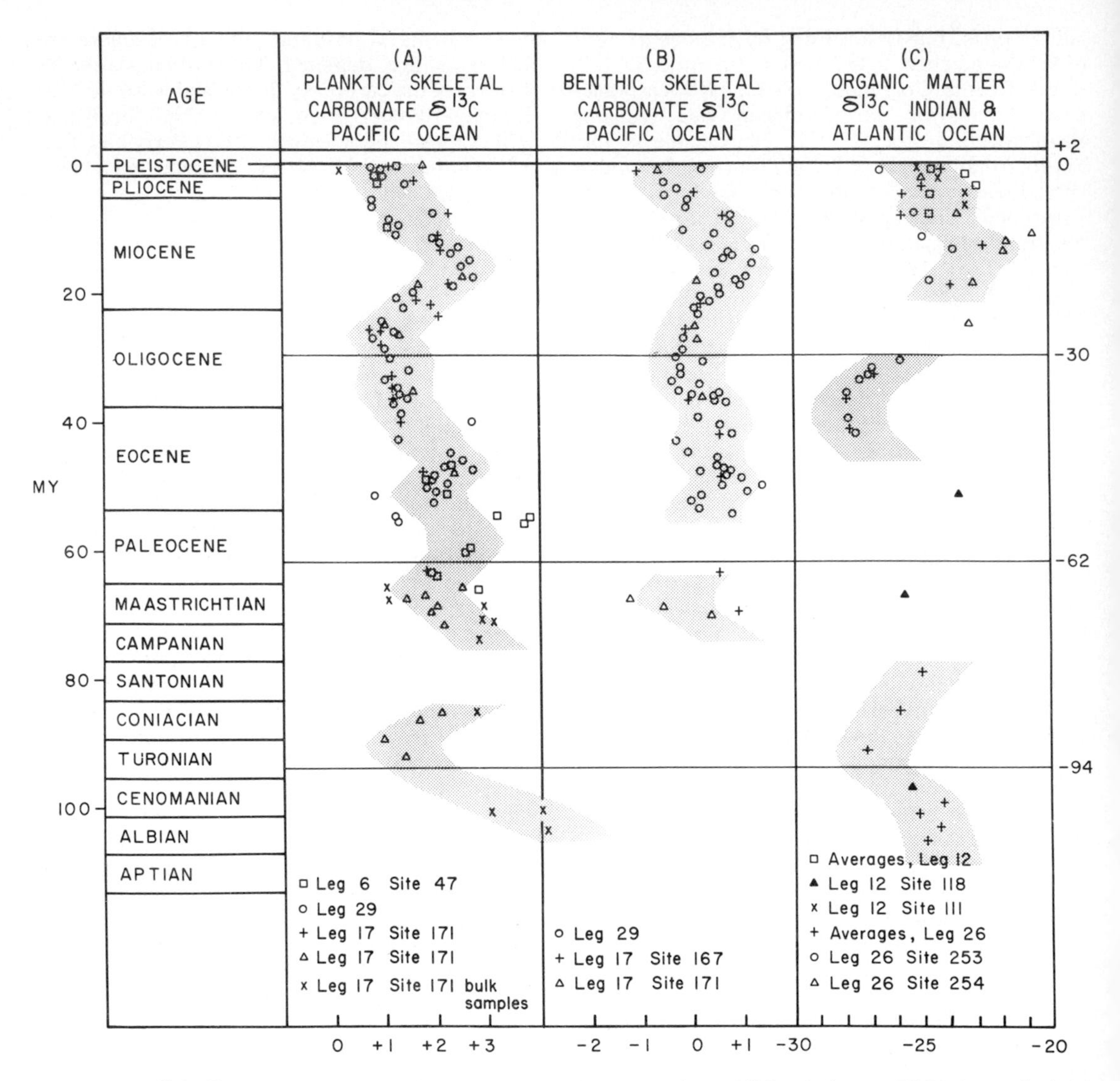

FIG. 6.—C^{13}/C^{12} ratios in pelagic sediments. Carbon isotope values ($\delta^{13}C$ relative to PDB) from planktic and benthic foraminifera, and from the organic matter in pelagic sediments recovered by the Deep Sea Drilling Project. Values for carbonate at any given time fall generally within a 1-2 per mil band (shaded), which shows braod fluctuations with a period of about 30 million years. This fluctuation is shared by individual sites, in the combined data shown in each column, and by planktic skeletons, benthic skeletons, and organic matter: We conclude that it depicts variations in the carbon isotope composition of the oceanic-atmospheric reservoir. Sources of data: DSDP Leg 12, North Atlantic (Rogers and others, 1972); Leg 17, equatoria Pacific (Douglas and Savin, 1973; Coplen and Schlanger, 1973); Leg 26, Indian Ocean (Calder and others, 1974); Leg 29, South Pacific (Shackleton and Kennett, 1974).

Such a secular variation in marine carbon-isotope ratios can be explained by any of at least three factors:

1. Supply of carbon isotopes to the ocean may vary significantly, both in quantity and in proportion of isotopes, in response to a number of variables. Amongst these are changing size of land areas, correlated with fluctuation of sea level; variation in rates of erosion correlated with relief, in turn dependent on tectonic activity; and changes in the sources of carbon (e.g. limestones, carbonaceous terrestrial sediments, and carbonaceous marine sediments).

2. Withdrawal of photosynthetically fixed carbon into sediments may fluctuate. Since photosynthetic fixation favors the lighter isotope, with-

drawal of carbonaceous matter drives the oceanic reservoir to the heavier side. This withdrawal of carbon into sediments results from the interaction of two factors: The rate at which organic carbon is being fixed by plants, and the rate at which it is redissolved by oxidative processes. Tappan and Loeblich (1971) appeal to the former, i.e. to changes in net oceanic productivity, to account for the fluctuation in carbon isotopes observed; they would equate our polytaxic periods with episodes of maximal organic productivity. We are doubtful about this correlation, and are inclined to seek a major cause of isotope fluctuation in rates of carbon burial related to changes in the oxidizing potential of the oceanic system, mainly through expansions and contractions of the oxygen minimum zone of the oceans, as explained below.

3. Temporal variations in the rates of carbonate deposition are also likely to change the isotope ratios. Since chemical or organic deposition of carbonates favors the heavier isotope, periods of maximal carbonate withdrawal might be expected to drive the oceanic reservoir to the light side.

The relative importances of these factors can only be evaluated qualitatively with the data at hand. In pelagic sedimentary sequences strong polytaxic episodes such as the Aptian-Albian commonly show a high organic carbon content combined with low carbonate values, and suggest that factors 2 and 3 may act in conjunction to drive the reservoir to the heavy side in polytaxic times.

CALCITE COMPENSATION DEPTH

Another index to the changing state of the oceans is the calcite compensation depth (CCD). This separates the shoaler, carbonate-rich pelagic sediment from the deeper carbonate-poor ones, and is interpreted as that level of the sea floor below which dissolution of calcite exceeds supply (Bramlette, 1961). As a dynamic equilibrium between dissolution and supply, the CCD is certain to be a sensitive indicator of changes in the physics, chemistry, and biology of the oceans (Berger and Winterer, 1974), although the complex interactions of these factors have so far defied rigorous analysis and prediction.

Attempts to chart the secular fluctuations of the CCD, for different parts of the oceans, using data of the Deep Sea Drilling Project with due attention to the progressive subsidence of the sea floor are summarized by van Andel (1975), and yield a well-defined pattern since late Paleocene time (Fig. 4-E, Legs 1-27). This reveals a shoaling of the CCD during polytaxic times, and a deepening during oligotaxic ones. Prior to the Eocene Epoch the CCD variation is less well defined due to the smaller number of observations, to problems of age determination, and to uncertainties in restoring sea floor elevations. Nevertheless, one particular observation is in direct conflict with the Eocene and later patterns. A reduction of carbonate content occurs in basal Danian beds, at the height of an oligotaxic episode (Tappan, 1968; Hay, 1970; Worsley, 1974), and has been interpreted as the result of a brief extreme shoaling of the CCD. Observations by Herm (1965) and coring of a complete carbonate sequence across the Cretaceous-Tertiary boundary in the deep Atlantic on Deep Sea Drilling Leg 43 throws doubt on this explanation.

SUBMARINE UNCONFORMITIES

The Deep Sea Drilling Project has repeatedly confirmed the presence of extensive submarine hiatuses which interrupt the pelagic depositional record on most submarine rises and ridges, and in some places on the abyssal floor (Legs 1-27; Rona, 1973a). These cannot be attributed to emergence. Incidence of such hiatuses in the Deep Sea Drilling Project through Leg 41 is graphically summarized in Figures 4D and 7. The diagram was constructed from data in Initial Reports volumes 1 to 29, and Geotimes reports for Legs 30 through 41. Only gaps reasonably well established through absence of one or more biostratigraphic zones (e.g. greater than one or two million years) were plotted. The data are incomplete and in part preliminary, but a pattern of synchroneity emerges. An even more striking case for concentration of hiatuses at certain stratigraphic levels (e.g. Oligocene of Indian Ocean) emerges from a more inferential approach (Veevers, Heirtzler, and others, 1974). The documentation of these hiatuses is, of course, confined largely to pelagic sequences which are fossiliferous, such as the carbonate ooze deposited on rises and ridges (though such settings now commonly lie at abyssal depths, as a result of subsidence). Erosion of abyssal bottoms occurs on the present sea floor off the Bahamas in the western Atlantic, where Cretaceous sediment is exposed, but recognition of such events at very great depths is hampered by the relatively unfossiliferous nature of the "red clay," and the minimal effort made so far to obtain continuous cores through that facies. Whereas some writers assumed that these unconformities in pelagic sediment are the result of chemical changes (i.e. a shoaling of the carbonate compensation depth, preventing carbonate sedimentation on the rises), many of them are shown on seismic reflection profiles as surfaces of physical truncation, and in such cases such as Shatsky rise (Fischer, Heezen, and others, 1971) hiatuses on top of the rise correspond to carbonate sediment on its flanks. The significant point is that these surfaces

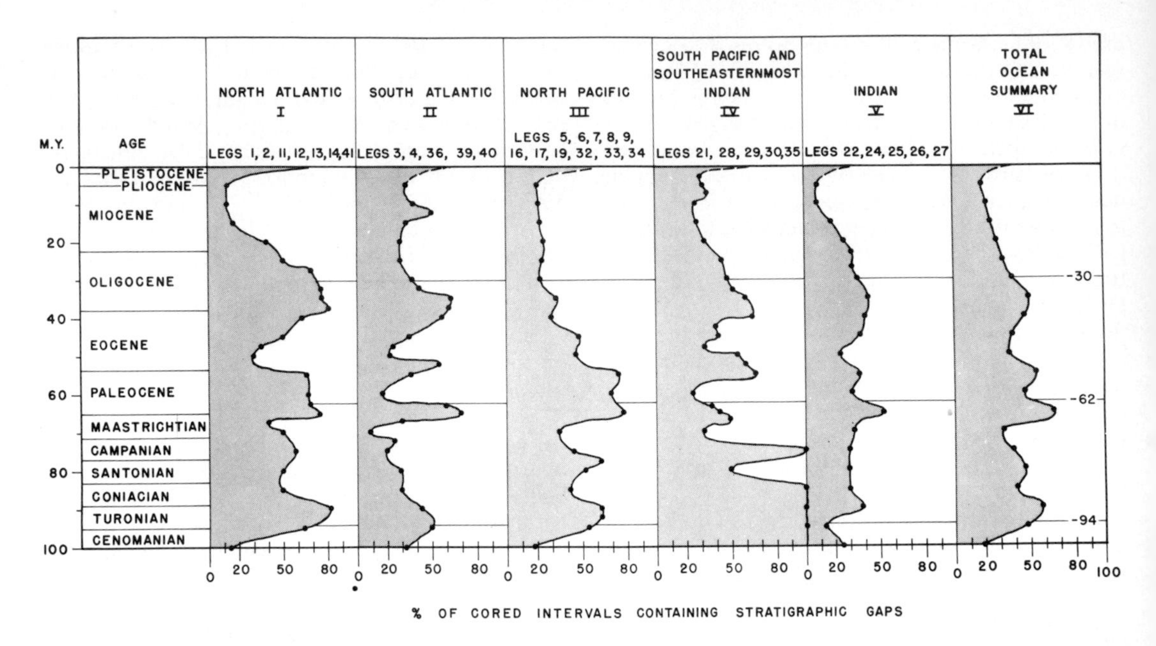

FIG. 7.—Stratigraphic completeness of pelagic sequences. Cores obtained by the Deep Sea Drilling Project commonly contain stratigraphic hiatuses. Values plotted are the percentage of cored intervals, of any given age, in which gaps of the magnitude of one or more zones (roughly 1.5 million years or more) have been recognized, plotted at 2.5 m.y. intervals. Gaps are particularly common, in all oceans, in the Turonian, late Maastrichtian-early Paleocene, late Paleocene-early Eocene and Oligocene. Sources of data: Legs 1-30, Initial Reports of Deep Sea Drilling Project. Legs 31-41 from various summaries in *Geotimes*.

are not randomly distributed through time, but fall mainly into three episodes: the Cenomanian-early Turonian, the early Paleocene, and the late Eocene-early Oligocene. The great incompleteness of Pleistocene sequences in pelagic carbonate sediment (Ericson and Wollin, 1964) and evidence of active current scour in the deep-sea (Johnson, 1972; Watkins and Kennett, 1972, 1973) indicates that Quaternary time is another such episode. The record suggests that whereas the ridges and rises commonly accumulate pelagic sediment at depths of one to several kilometers, they are at times scoured by currents. Details of the distribution of these submarine hiatuses in space and time are too complicated to present. Rona (1973a) points out, however, that the duration of each hiatus is longest on the western sides of ocean basins, suggesting an origin due to periodic intensification of deep circulation, especially the Western boundary currents (Stommel, 1958). Excellent discussions of the Oligocene and Recent submarine unconformities in the southeastern Indian Ocean and South Pacific have been given by Watkins and Kennett (1972, 1973), Kennett, Houtz, and others (1974), Johnson (1972) and Hollister, and others (1974). Minor hiatus intervals, including the Santonian and the latest Paleocene-early Eocene, may be due to upward excursions of the CCD which led to low rates of sedimentation in the deep-sea. Thus not all hiatuses may be due to current scour.

In summary, the incidence of interruptions in pelagic sedimentation corresponds in general to the oligotaxic episodes.

SUMMARY OF OBSERVATIONS

Observed fluctuations of pelagic phenomena and global diversity changes in nektoplanktic organisms (Figs. 1 and 4) show alternate episodes of oceanic history. Times of higher global diversity (polytaxia) alternate with times of reduced diversity or biotic crisis (oligotaxia). Nektoplanktic communities tend to be more elaborate during polytaxic times (superpredators), and simpler during oligotaxic ones. Opportunistic "disaster forms" may locally swamp the oligotaxic biotas.

Polytaxic times such as the Albian and Eocene are characterized by relatively warm oceans, having low latitudinal temperature gradients, and less contrast between surface and bottom temperatures than do oligotaxic times of which the Quaternary is an example. Carbonaceous sediment bearing evidence of anoxia is more widespread during polytaxic times, whereas red pelagic sediment, indicating a high degree of seafloor oxidation, reaches its widest distribution in oligotaxic ones. This trend, reflecting rates of carbon loss to the lithosphere, is also seen in carbon isotopes of skeletal calcite as well as of organic matter: isotope ratios shift toward light values during oligotaxic times. The carbonate compensation depth shoals in general during polytaxic times and deepens during oligotaxic ones. Finally, pelagic sedimentation is relatively constant during polytaxic times, but is interrupted during oligotaxic episodes by hiatuses.

Correlation of the various curves depicting these fluctuations is by no means exact, but is sufficiently good to suggest that these phenomena are linked in various ways, and express, in aggregate, a secular oscillation of the nektoplanktic habitat, and of the ocean as a whole.

Furthermore, this oscillation appears to be rhythmic through time: the record of the polytaxic states is somewhat diffuse, the more sharply defined first- and second-order oligotaxic states recur at intervals of about 30 million years. The best fit to the available chronology suggests a 32 million year cycle, with oligotaxic troughs centered at the following dates (Fig. 1): 222 m.y. (Permo-Triassic boundary); 190 m.y. (Triassic-Jurassic boundary); 158 m.y. (Bathonian-Callovian boundary) 126 m.y. (early Neocomian); 94 m.y. (Cenomanian); 62 m.y. (early Paleocene); 30 m.y. (mid-Oligocene); and +2 m.y. (Holocene).

A 30 million year temperature cycle has been noted by Dorman (1968) on the basis of oxygen isotopes in molluscan shells. In addition to this cyclicity, there appears to be a longer trend, from earlier times in which the physical phenomena of the polytaxic state were more pronounced (Cretaceous) to later ones in which the physical features of the oligotaxic state became more dominant, culminating in the Quaternary spread of glaciers. This trend may express a 300 million year cycle discussed, amongst others, by Umbgrove (1947), and supported by studies of isotopes in fresh-water carbonate (Keith and Weber, 1964).

This is not to deny the existence of shorter cycles: Our diversity data suggest the existence of a less drastic cycle about half as long; eustatic cycles on the order of 300, 000 years are suggested by the marine cyclothems of the of the Carboniferous; and the 50,000–100,000 year cycle in sea level changes, driven in the Pleistocene by glacial advances, also finds expressions in older sedimentary sequences (Fischer, 1969; Arthur and Fischer, 1977).

Failures of curves to correspond perfectly with each other and with the timing result from a variety of factors. These include imperfections in the nature of the data, (e.g. the dates assigned to stratigraphic intervals); reactions of the phenomena plotted to special factors, such as response of isotope ratios to local influences; and varying lags in the response of these phenomena to the driving forces, whatever they may have been. For example, diversity apparently lagged behind increases in temperature, and evolution of superpredators lagged behind diversity in protists. These are details that lie as yet beyond the resolution of the data.

ALTERNATIONS IN OCEANIC BEHAVIOR

These patterns of secular variation suggest a common cause: a significant variation in the general nature of the ocean as a physical-chemical-biological system. In spite of the difficulty of isolating cause and effect relationships, the variation can be reasonably explained by changes in the vigor and patterns of global oceanic convection. The polytaxic state is one of sluggish oceans in a world of low latitudinal temperature gradients; the oligotaxic state is characterized by highly convective oceans in a world having great geographic temperature contrasts, culminating in the glacial regime. This interpretation rests largely on two lines of evidence: secular changes in the oxidation of marine sediment, and long-term variations in the ocean's power to sweep sediment off rises, ridges and seamounts.

As shown above, the polytaxic oceanic state in its more extreme form (Albian) permits widespread accumulation of organic-rich sediment, whereas maximal sedimentary oxidation is accomplished in oligotaxic times. Since this oxidation state is the resultant of two opposing forces—the input of organic matter, and its removal by oxidative processes—the patterns can be explained by variations in organic supply (overall productivity), by variations in the oxidizing efficiency of the oceanic system, or by some combination of both.

If this pattern were due to fluctuations in organic (phytoplanktic) productivity, we should have to assume maximum productivity of the oceans as a whole during polytaxic times (Tappan and Loeblich, 1971) and productivity minima in the oligotaxic state such as that of today. We have no direct evidence that overall organic productivity has fluctuated in this manner. Indeed, long trophic chains and slow recycling of nutrients in highly complex biotas essentially preclude maximal productivity (Margalef, 1968; Ryther, 1969; Parsons and Lebrasseur, 1970; Valentine, 1971, 1974; Lipps, 1970a). Furthermore, the seas at any one time are a great mosaic of communities whose fertility base and productivity must be thought of as a statistical summation, which may be quite independent of the history of any specific community or group of communities. The greater vigor of oceanic circulation associated with oligotaxic times suggests that these may be the more productive periods, but until we have some direct criteria of paleoproductivity, at the community level and for the seas as a whole, proposed global fluctuations in this parameter remain speculative.

On the other hand, a good case can be made for secular variation in the ocean's oxidizing capacity. In the first place, the solubility of oxygen in water is inversely proportional to temperature: a drop from 12° to 2°C in normal sea water changes the solubility of oxygen from 7.1 to 8.5 ml/l. Because oligotaxic times are characterized by low temperature in the circumpolar regions, the oligotaxic ocean absorbs relatively more oxygen than a polytaxic one.

Secondly, the ocean's oxidizing capacity depends on the rate at which oxygen is advected to different parts of the system, and here, too, the oligotaxic times, having a circulation driven by higher temperature gradients, seem to have the advantage over the polytaxic ones.

Waxing and waning of oxygen minimum.—A secular variation in dissolved oxygen would affect primarily that part of the oceans known as the oxygen-minimum layer. In the present ocean the wind-stirred and current-churned surface waters exchange gases with the atmosphere and are internally supplied with oxygen by photosynthesis. In contrast, the intermediate and deeper water masses obtained almost all of their free oxygen at the time when they were surface waters in the high latitudes. In these waters, the free molecular oxygen content is subject to progressive losses due mainly to the metabolism of aerobic organisms, losses being proportional to the size of the living biomass and to the rate of metabolism. Since the biomass is largest near the source of food (the surface), and metabolism bears a direct relationship to temperature and perhaps an inverse one to pressure, depletion of oxygen develops most notably in the upper part of these lower water masses near the base of the thermocline (Fig. 8A). This oxygen-minimum layer is of variable thickness and intensity, its growth being checked by oxygen advection and diffusion from above and below (Wyrtki, 1962; Munk, 1966), and is best developed at the eastern edges of tropical oceans, where upwelling promotes high fertility and deep waters are relatively old. In the east tropical Pacific, for example, the zone normally extends from about 100 to about 1500 meters, and oxygen drops locally to levels about 0.1 ml/l (Reid, 1965; Richards, 1965; Menzel and Ryther, 1968; Goering, 1968). These are values at the limits of animal survival. Where waters showing such oxygen depletion impinge upon the seafloor (Fig. 8A), sediment is enriched in organic matter as compared to that at shoaler or greater depths, and usually contains more opaline silica; where oxygen values are too low for burrowing metazoans, the sediment is dark and laminated (Calvert, 1964; Baturin, 1969; von Stackelberg, 1972). In other parts of the oceans, such as the northwest Atlantic margin, the oxygen-minimum layer is less well developed.

The present patterns displayed by the oxygen-minimum layer are typical only of oligotaxic intervals. In polytaxic times the lowered initial oxygen supply, the diminished rate of advection and increased rates of metabolism due to higher temperatures tend to intensify the oxygen minimum, and to expand the zone characterized by it, both vertically and laterally (Fig. 8B). In extreme cases this might have had the effect of interposing a totally anaerobic layer between the surface waters and the deep-water masses. Presumably such changes had a considerable effect on the mid-water biota. Furthermore, they led to changes in areal extent of anoxic bottoms, minimal today and widespread in the past. When viewed in this light, the Aptian-Albian interval emerges as the most intense polytaxic episode in the last 100 million years, and the following ones were progressively less so. The general concept of the development of anaerobic conditions in the open oceans, due to slowed convection rates, was earlier proposed by Strøm (1939) to account for the black shale of the Paleozoic and by Hallam (1975) for the Jurassic ones. Piper and Codispoti (1975) have attributed episodes of phosphorite genesis at continental margins to expansion of the oxygen-minimum layer, and Eicher and Worstell (1970) have attributed the poverty of Upper Cretaceous foraminiferal faunas in the North American interior to reduced oxygen supply. Berger and von Rad (1972), after reconstructing paleodepths of the North Atlantic sea floor for sea-floor subsidence, proposed that an oxygen-minimum layer may have impinged on an Early Cretaceous ridge leading to development of organic-rich sediment.

The oxygen-minimum expansion model also

FIG. 8.—Expansion of oxygen-minimum layer.

A—Present-day oxygen-minimum zone in oceans, its impingement on upper continental slope, and its effect on oxidation state of sediments (generalized).

B—Model of expanded oxygen-minimum zone and raised sea level, as visualized by us for the more extreme polytaxic episodes. Expansion and intensification of oxygen-minimum layer leads to widely spread reducing conditions on ocean floor, and a corresponding expansion of organic-rich black sediments.

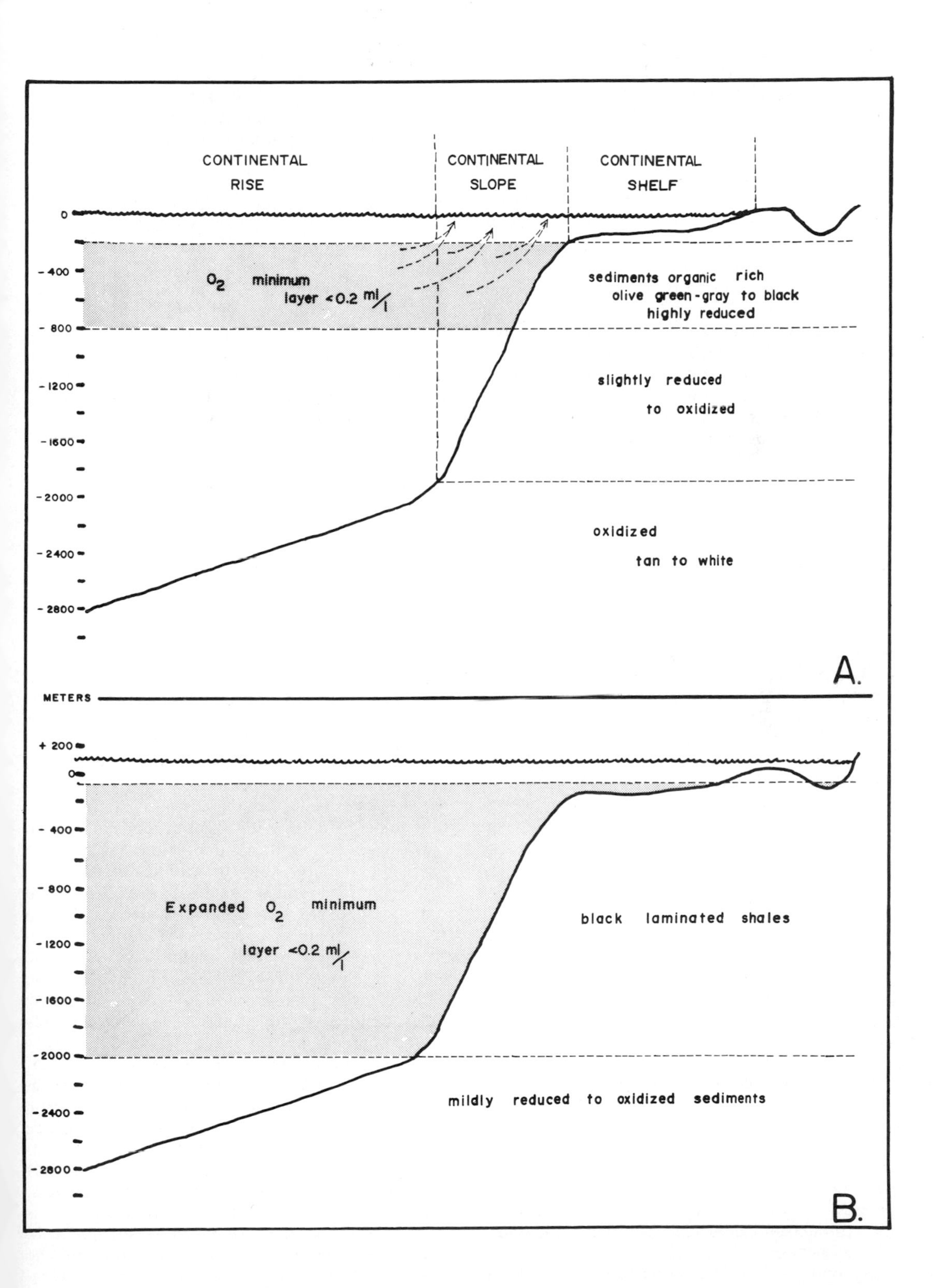
CONTINENTAL RISE
CONTINENTAL SLOPE
CONTINENTAL SHELF
0
-400
-800
-1200
-1600
-2000
-2400
-2800
O_2 minimum layer <0.2 ml/l
sediments organic rich olive green-gray to black highly reduced
slightly reduced to oxidized
oxidized tan to white
A.
METERS
+200
0
-400
-800
-1200
-1600
-2000
-2400
-2800
Expanded O_2 minimum layer <0.2 ml/l
black laminated shales
mildly reduced to oxidized sediments
B.

explains why during Aptian-Albian time (Fig. 5) red to olive green, generally organic-poor marl or clay accumulated in the abyssal Indian Ocean while thick sequences of black, organic-rich clay accumulated above about 3000 m on the western continental margin and inland seas of Australia (Veevers, Heirtzler and others, 1974). In this ocean, intermediate waters were essentially anoxic whereas bottom waters were aerobic. The same situation occurred in the Pacific Ocean where during Aptian-Albian times oxygenated sediment was deposited on deeper floors while anoxic environments persisted along continental margins (Fig. 5) and occasionally on oceanic rises (e.g. Shatsky Rise, Manahiki Plateau; Larson, Moberly, and others, 1975; Jackson and Schlanger, 1976).

Young ocean basins such as the North and South Atlantic, both smaller and shallower, were more effectively isolated from exchange with waters from high latitude. In these basins, mainly located in low latitudes, the entire water mass may have been anoxic during the Albian (Deep Sea Drilling Project Legs 11, 12, 14, 36, 39, 40) and, in the South Atlantic, again in the Santonian-Coniacian (Legs 39, 40). Another complicating factor in the nearly isolated South Atlantic basin is the possibility that a strong salinity stratification occurred periodically from Aptian through Santonian times, contributing to a stagnant, anoxic condition (Bolli, Ryan, and others, 1975).

During the polytaxic episodes, more isolated basins and shallower, low-latitude seas can become entirely euxinic. A rise in sea level also contributed materially to the spread of anoxic bottoms and, as explained below, seems to be characteristic of polytaxic times. This extended the oxygen minimum far into the continental interiors (Fig. 8B) and produced such deposits as the Albian Mowry Formation of the Rocky Mountain region.

We conclude that even though black shale can be deposited during any part of geologic time under the influence of the oxygen-minimum layer and local anoxic basins, its widespread distribution is achieved during polytaxic episodes, especially on ocean bottoms of intermediate depths.

ASSOCIATED PHENOMENA, AND SPECULATIONS

Beyond the pelagic setting and extending to the entire biosphere, several other fluctuations appear to correlate with the marine cycle; they are terrestrial climates, as revealed by latitudinal shifting of floras; the diversity of terrestrial organisms; and eustatic sea-level fluctuations expressed in the submergence and emergence of the continental platforms, recorded in sedimentary sequences.

Isotopic evidence for climatic changes has already been presented and we have appealed to climatic variations as an intermediate driving mechanism for the oceanic cycle. Independent evidence for changes in terrestrial climates is provided by the record of fossil plants. Dorf (1970) has presented a synthesis of his studies of plant communities in the north-central United States from Upper Cretaceous through Cenozoic strata. Subtropical and temperate floras replace one another in a manner basically parallel to the major cooling and warming episodes outlined in this paper.

Flessa and Imbrie (1973), in a study of patterns of diversity of marine and terrestrial organisms during the Phanerozoic, present diversity tabulations for terrestrial groups at the family level. During Cretaceous and Cenozoic time terrestrial oligotaxy matched the marine pattern, occurring in the early Cenomanian, at the Cretaceous-Paleocene boundary, in the Oligocene, and in the Pleistocene.

Secular variations in sea level have long intrigued geologists, but the complications by local tectonics have been an obstacle, and no standard curve based on well documented global data exists. However, a number of approaches to the problem may here be compared to our secular pattern.

In 1940, Grabau, who had dealt with global stratigraphy on a grand scale, proposed a reclassification of the Phanerozoic stratigraphic column on what he held to be an objective basis: The occurrence of major regressions, separating transgressive pulses. Grabau made no reference to radiometric time data, and seems to have been unaware that his Mesozoic and Cenozoic pulses (Fig. 1) are cyclic, his regressions matching our oligotaxic episodes with exception of one, which Grabau placed at the end of the Lias and we place slightly later, into the Middle Jurassic.

Damon (1971) and Wise (1974) have constructed sea level curves for the Phanerozoic. Damon suggested a 36 million year periodicity for regressions, and they roughly match our oligotaxic episodes. Rona (1973b) has shown that Cenozoic drops in sea level, corresponding to our oligotaxic times, resulted in reduction of sedimentation rates on the shelves.

A very promising study, as yet unpublished, is that of Vail and others (1974 and pers. comm.). Seismic profiles across the world's continental margins, stratigraphically calibrated by surface geology and drilling, show strikingly similar patterns of sedimentary onlaps and offlaps, and offer the most direct approach to a history of sea levels. A preliminary match of our data with theirs shows good correspondence. All of our Mesozoic-Cenozoic oligotaxic episodes save the one in the mid-Jurassic correspond to one of their regressions; not all of their regressions, however, are matched by one of our 32 million year breaks. A more detailed comparison will have to await publication of their data.

We previously concluded that the cyclic behavior of the oceans must be linked to the total global climate. In particular, the polytaxic times must have had more effective and uniform heat distribution over the earth's surface than we experience today. Is this simply the result of higher sea levels?

That higher sea levels would tend to moderate the climate seems certain. The higher ratio of water to land surface ameliorated climates because of the water's high relative thermal inertia; the development of intracontinental seaways provided additional avenues for marine circulation, and reduced the "continentality" of the remaining lands. Furthermore the increase in water surface probably increased the moisture content of the atmosphere as a whole, and the amount of heat transported thereby. On the other hand, the marine record suggests that oceanic circulation was sluggish, and probably less effective in heat distribution, relative to the atmosphere, than it is in the oligotaxic mode. Thus the adequacy of eustasy as a generating cause of the cycle observed remains in doubt, and the search is directed toward events in which both physical-chemical changes in the atmosphere and changes in sea level might find a common cause. We suggest that processes within the earth's interior influence sea levels by changing the earth's surface configuration, and may simultaneously affect the atmosphere and therefore climates through vulcanism.

Periodicity in tectonic activity and related plutonism and vulcanism has been suggested by various workers, notably by Dearnley (1966), Kistler and others (1971), Shaw and others (1971), and Damon (1971). Damon found a periodicity of about 36 m.y., which is close to the periodicity described here. Kistler and others and Damon related times of marine regression (our oligotaxic episodes) to increased tectonism, emplacement of plutons, and eruptive silicic vulcanism in western North America. The models introduced to account for this periodicity (see also Shaw and others, 1971) are highly speculative and have not been tested on a global scale, but the data offer support to Chamberlain's concept that "diastrophism is the ultimate basis for correlation," and for Umbgrove's (1947) theme that diastrophism and associated silicic magmatism result in mountain building, regression, climatic deterioration, and accelerated evolution.

Recently a number of workers (Larson and Pitman, 1972; Hays and Pitman, 1973; Rona, 1973a, 1973b) have suggested that sea-level fluctuations other than glacio-eustatic ones are related to changes in volume of the mid-oceanic ridges, engendered by variation in rates of ocean floor spreading. Hays and Pitman deduced from magnetic seafloor anomalies that global seafloor spreading rates were markedly accelerated in the Late Cretaceous, and are the cause of the great Late Cretaceous transgression; Rona suggested a similar cause for the Eocene transgression. However, until the magnetic anomalies on which these suggestions depend are more firmly and critically dated (Berggren and others, 1975), the matter remains an intriguing and attractive but as yet unsubstantiated hypothesis.

We are tempted to visualize a model in which intervals of slower sea floor spreading are correlated with marine regressions. A simultaneous slowing of subduction in the orogenic belts allows crustal temperatures to rise in these areas, and induces them to seek isostatic equilibrium, resulting in vulcanism and surface deformation-orogeny. The combination of volcanic dust in the atmosphere and of a global increase in continentality leads to a deterioration of climates, until resumption of faster spreading reverses the pattern.

The possibility that the 32 million year cycle is initiated by organisms seems unlikely, inasmuch as organisms are hardly in a position to influence sea-level changes. They may, however, have played an important role in modifying or "buffering" atmospheric composition (Tappan and Loeblich, 1971).

A secular variation in the solar constant remains another possible driving mechanism for the 32 million year cycle, but, as in the case of an organic cause, the correlated changes in sea level are left unexplained thereby.

We have yet to discuss the effect of plate tectonics in rearanging the patterns of continental distribution round the globe. These patterns of change seem to show no obvious relationship to the 32 million year cycle, which appears to operate whether continents are lumped into a Pangea, or are widely dispersed. However, these patterns probably modify the expression of the 32 million year cycle, by reinforcing or damping some of its phases, and may be partly responsible for certain climatic and oceanic events and regimes (Berggren and Hollister, 1974; Douglas and others, 1973; Frakes and Kemp, 1973; Robinson, 1973): Oligocene and Recent oligotaxic conditions were re-enforced by the location of Antarctica over the South Pole, and by the development of a circum-Antarctic seaway. By the same token, the predominance of the polytaxic mode over the oligotaxic one in the Jurassic and Cretaceous periods may be largely related to lack of polar land masses, and to the existence of a globe-girdling tropical Tethys. The role of plate tectonic patterns is thus viewed as a modifying rather than an initiating one.

SEDIMENTARY CONSEQUENCES—A PROJECTION

If the earth's climates and its atmosphere and hydrosphere have behaved cyclically throughout earth history, as here suggested, then a secular control of various other marine geological and geochemical processes is to be expected. Precise geochemical balances cannot yet be provided for different points on the cycle; however, a number of simple, qualitative predictions of sedimentological and biological effects can be made. The increased residence time of deep waters, widespread anoxia, and high sea levels associated with the polytaxic state would have produced patterns of productivity, of potential petroleum source-bed accumulation, of silica distribution, of phosphate sedimentation, and of carbonate dissolution, among others, which differ from those of the present (oligotaxic) ocean in qualitatively predictable ways. Thus, recognition of the oceanic cycle and its effects directly influence mineral resource evaluation.

Nutrients.—The lower turnover rates associated with the polytaxic state imply a slower recycling of nutrients (phosphates, nitrates, silica) from deep water back to the surface where primary production occurs. At the same time, the influx of nutrients from the shrunken land area was probably lowered, and much of it may have been taken up in the expanded and fertile epicontinental seas. The net result can only have been a decreased fertility and productivity of the open ocean. This effect may have been intensified by the loss of nitrate through the action of nitrate-reducing bacteria: a process which occurs in nearly anoxic parts of the present oxygen minimum of the tropical eastern Pacific Ocean (Goering, 1968; Goering and others, 1973; Dugdale, 1972), and which would have reached greater proportions in the polytaxic ocean. The extent to which regeneration of nitrates by bluegreen algae (Berger and Roth, 1975; Stewart, 1971; Mague and others, 1974) would have restored the nitrate-phosphate balance is an open question, but in any case the total nutrient supply in the upper oceanic waters is reduced,

At the same time, local areas of upwelling, delivering deep waters that had been accumulating phosphate and silica over longer periods than any water masses in the present ocean, may well have been exceedingly fertile. This may be expressed in local accumulations of phosphorites and siliceous sediments, discussed below.

Petroleum source beds.—The extraordinary abundance of carbonaceous (sapropelic) sediment of certain ages has already been used as evidence for the inferred cyclicity of changes in the character of oceanic circulation and chemistry, but this has further implications to resource evaluation. Such sediment occurs both on the continental margins and in the deeper epicontinental basins, to which the oxygen-minimum layer extended as sea level rose. We suggest that this sediment is responsible for the generation of most of the world's petroleum; more specifically, that in the last 200 million years, there have been six episodes (polytaxic) favorable to the accumulation of source beds: these were during the Early Jurassic (middle to Late Lias), the Late Jurassic (Oxfordian), mid-Cretaceous (Aptian-Albian), the Late Cretaceous (Coniacian-Santonian), the Eocene, and the Miocene. A peak in source bed generation was reached in the Late Jurassic and mid-Cretaceous episodes (which is one reason for the concentration of oil in the region of the Persian Gulf), and the succeeding polytaxic episodes were progressively less important. Such considerations bear on the evaluation of the petroleum potential of the world's continental margins and other basins.

Silica.—Deposition of authigenic silica in the ocean basins occurs through the agency of siliceous protists and sponges. Their demand on dissolved silica in upper waters is great enough to reduce the concentration to a small fraction of its saturation level (Calvert, 1966; Heath, 1974; Edmond, 1974). Skeletal silica is redissolved at depth, so that bottom waters are progressively enriched (Berger, 1970; Wollast, 1975; Hurd, 1973); but nowhere on the present sea floor does the concentration of dissolved silica reach the saturation point for opaline silica.

As a result, siliceous tests accumulate only where trapped by relatively high rates of sedimentation. These may be of terrigenous material, as in many shoal-water settings and continental margins, or of carbonate, as in other shoal-water areas or on oceanic rises, or of biogenic silica itself, as in the circum-polar belts. Over vast portions of the oceans, such as the abyssal Atlantic (Berger 1970, 1974), virtually no biogenic silica is being retained in present day sediment.

During polytaxic times, patterns of silica deposition and retention differed for two main reasons: a less intensive upwelling in the circum-polar belts, and a general slowing of oceanic convection.

The great silica-depositing role played by the present circum-polar belts of seasonal upwelling and circum-Antarctic circulation was presumably diminished at times of lower temperature gradients (and, incidentally, of different continental distributions). A decrease of circum-polar silica withdrawal from seawater would have left more of this substance available to organisms elsewhere.

The increased residence time of deep water

masses probably allowed silica concentrations in bottom waters to reach saturation with respect to opal. This would have permitted retention of siliceous tests in sediment over broader parts of the deep ocean floors, and would have facilitated the precipitation of authigenic silicate minerals such as zeolites and glauconite (see also Berger and Roth, 1975).

Silica is most apparent in sediment when diagenetic processes have caused conversion to chert; hence, polytaxic deep-water sediment should be characterized by more widespread cherts than oligotaxic ones. Most Miocene deep-sea sediment has not reached the chert stage, although Miocene siliceous sediment is common. The most widely known deep-sea chert, the middle Eocene Horizon A of the Atlantic (Ewing, Worzel, and others, 1969) corresponds to a polytaxic time and has correlatives in other oceans. In the Aptian-Albian and throughout much of the Upper Cretaceous cherty pelagic limestone is also widespread. We have not as yet attempted to determine the relative abundance of chert in the various stages of the deep-sea Cretaceous.

Although upwelling presumably diminished in polytaxic times, it must nevertheless have occurred, thereby delivering silica-enriched deep water to the surface in local areas. While most shoal-water sediment of polytaxic times may have been starved of silica, extraordinary quantities of this substance may have been dumped at the sites of upwelling along continental margins. Perhaps the great chert accumulations of the past such as the Devonian Caballos Chert of Texas (McBride and Thompson, 1970) and the Permian Rex Chert of Idaho (Mansfield, 1927), with depositional rates far in excess of radiolarian ooze, find an explanation in this phenomenon, along with the Miocene diatomites of California, Peru, and Japan.

We predict, then, that biogenic silica (and possibly authigenic silicates) would show widest distribution in deep-water sediment in the polytaxic episodes.

The accumulation of glauconite is probably influenced by the oceanic cycle. Deposition of glauconitic marine sediments seems to have been particularly prevalent during polytaxic times, especially during the Cretaceous. They may well be an index to moderately high dissolved silica and iron levels in sea water in association with lower dissolved-oxygen values, hence the curious combination of reduced and oxidized iron in the glauconite lattice (McCrae, 1972). Ancient glauconitic deposits may be closely associated with the oxygen-minimum layer, as they seem to be today, possibly indicating low dissolved-oxygen contents but not anoxia. The distribution of these deposits is also probably a function of lowered sedimentation rates on slopes and outer shelves during marine transgression typical of polytaxic intervals. Examples are the British Gault and Swiss Albian, the Upper Cretaceous and Eocene of the North American Atlantic coast, and the Eocene of the Gulf Coast. Glauconitic sediments of oligotaxic age, such as the Paleocene ones of New Jersey, demonstrates that such deposits are not unique to polytaxic episodes.

Phosphates.—Phosphate, like silica, is mainly fixed by organisms in the photic zone. It is contributed to the lower water masses in part by the bacterial disintegration of organic matter, and in part by the dissolution of phosphatic skeletal material. The present ocean appears to be slightly undersaturated in respect to apatite (Kramer, 1964), for the preservation of phosphatic skeletal matter is the exception rather than the rule, and evidence for chemical precipitation from solution is minimal (Kolodny and Kaplan, 1970). Exceptions are the areas where a well-developed oxygen-minimum zone impinges on the sea floor (Veeh and others, 1973; Baturin, 1971). We suggest that in polytaxic times the increased residence time of deeper waters leads to an increase in their phosphate content, possibly to apatite saturation (e.g. Berger and Roth, 1975). This should find expression in a more general retention of phosphatic skeletal debris in deep water sediment. This question can now be tested by systematic investigation of deep sea cores; for example, phosphatic intervals occur in Aptian through Cenomanian pelagic sediments (Legs 12 and 27; Kennedy and Garrison, 1975). The widespread phosphatization of sediment on seamounts (Heezen and others, 1973) may have occurred during polytaxic peaks—in the Albian, Coniacian, mid-Eocene, and late Miocene/early Pliocene.

We visualize a reduction in the general phosphate content of photic waters, but an abundance of phosphate and high productivity in the limited areas of upwelling, associated with silica deposition and sapropels. Piper and Codispoti (1975) have already suggested that the deposition of phosphorite is associated with an expansion of the oxygen-minimum layer. To us, the association of these phenomena is a general expression of the polytaxic state. The mid-Permian phosphate of the Phosphoria Formation (Idaho), the largest known phosphatic deposit, was probably deposited at the peak of a polytaxic episode, and the Late Cretaceous phosphate of Israel and North Africa, as well as the Miocene phosphate of Florida and Peru fit the pattern; however the partly Paleocene age of North African phosphate and the Oligocene Carolina phosphate show that some phosphorite, like some glauconite, was deposited during Oligotaxic times.

Calcite compensation depth.—Recurrent

changes in the calcite compensation depth (CCD) have been listed above as part of the evidence for a rhythmic behavior of the oceans, polytaxic states being characterized by a shoaling of the CCD relative to most oligotaxic times. Shallow epicontinental seas spread with eustatic sea-level rise during polytaxic episodes and led to large-scale extraction of carbonate due to buildup of reef complexes and increased area of chalk deposition above the lysocline, as well as decreased bicarbonate influx from correspondingly smaller land areas. This competition for available calcium carbonate probably led to a sympathetic CCD rise in the deep oceans (e.g. Berger and Winterer, 1974; van Andel, 1975). However, other possible contributing factors include decreased productivity of calcareous plankton because of a decline in marine fertility in the open oceans, and increased calcium carbonate solubility due to higher levels of dissolved Co_2 in deep waters associated with oxygen deficiency.

In theory, it seems possible that the old bottom waters of the polytaxial ocean approached saturation with respect to calcite while intense dissolution occurred at intermediate depths in the intensified oxygen minimum zone. This possibility may be examined in deep sea drilling cores.

CONCLUSIONS

Our conclusions are summarized in the following points:

1. The world oceans alternate rhythmically between two modes of behavior, which we have termed polytaxic and oligotaxic. In the polytaxic mode latitudinal and vertical temperature gradients in the oceans are low and the oceans convect sluggishly, in contrast to the oligotaxic mode in which we live.
2. Sea levels stand high during polytaxic times, while the oligotaxic mode is associated with relative regressions.
3. This alternation is not directly correlated with plate tectonics and patterns of continental distribution.
4. The alternation has had a strong influence on the evolution of marine biotas. Pelagic biotas expand in numbers of taxa and in complexity of communities during polytaxic times, while they decline in oligotaxic ones. The major biotic crises in the marine realm occur during particularly severe oligotaxic episodes.
5. The unit-area productivity of the open seas is probably highest during oligotaxic times, such as today. During polytaxic times, a larger share of the total oceanic organic production is carried by the expanded epicontinental seas. In the oceans the contrast between the highly fertile upwelling areas and the relatively less fertile open ocean probably was greater than is the present case.
6. The alternation between opposed states is rhythmic: oligotaxic episodes of varying severity recur at 32 million year intervals. The polytaxic intervals of the last 200 million years coincide with Grabau's Triassic, Liassic, Jurassic, Comanchean, Late Cretaceous or Gulfian, Paleogene, and Neogene transgressive pulses (Fig. 1).
7. The cycle is superposed over a broader climatic trend (of 200 to 300 m.y. duration) which tends to emphasize one state over the other, such as pronounced polytaxic episodes during Jurassic-Cretaceous time and oligotaxic periods during the Cenozoic.
8. The oxygen-minimum layer in the oceans responds to changes in circulation rate and oxygen solubility. It expands and intensifies during polytaxic periods, greatly increasing the area of anoxic sedimentation. During oligotaxic episodes, areas of anoxic conditions on ocean bottoms are minimal (Fig. 8).
9. This apparently cyclic alternation is responsible for secular variations in patterns of stratigraphy. Interruptions of sedimentation during oligotaxic times are noteworthy on the continents (where they correspond to base levels lowered by eustatic sea-level drop) and in the deep sea (where we attribute them to intensified deeper current activity).
10. Likewise, these rhythmic changes influence patterns of sedimentation. The widespread formation of petroleum source beds is associated with polytaxic episodes, as is the development of widespread deep-sea chert, the deposition of phosphorite, and upward migration of the calcite compensation depth.
11. This oceanic cycle appears to be linked to continental climates expressed in the paleobotanical record.
12. The ultimate cause remains unknown, but the link with sea-level oscillations and possible correlations with periodicity in magmatism suggest that it is internal to the earth, rather than tied to variations in solar or cosmic processes.

ACKNOWLEDGEMENTS

Thanks are due to the John Simon Guggenheim Memorial Foundation, for a fellowship which enabled the senior author to visit pelagic sediment around the world. Many of the problems dealt with in this paper first came into view during that trip. Princeton University's Tuttle Bequest enabled us to begin fieldwork in the pelagic sequence of central Italy. The National Science Foundation (grant DES 74-22214) has supported further work there, and the compilations on which the present paper is based. From amongst the many colleagues who have aided with data and discussions we can only name a few: Kirk Bryan, Robert Douglas, William Hay, Christopher Kendall, Fred Mackenzie, Suki Manabe, Isabella Premoli Silva, W. B. F. Ryan, Seymour Schlanger, Peter Scholle, Peter Supko, Ida Thompson, and Peter Vail. Wolfgang Berger and F. B. Van Houten provided thoughtful reviews and criticisms. We also thank the editors, Harry Cook and Paul Enos for many valuable suggestions which have measureably improved the manuscript.

REFERENCES

ANDERSON, T. F., AND SCHNEIDERMANN, NAHUM, 1973, Stable isotope relationships in pelagic limestones from the central Caribbean: *In* N. R. Edgar, J. B. Saunders and others, Initial reports of the Deep Sea Drilling Project, v. 15: U.S. Government Printing Office, Washington, D.C., p. 795-803.

ARTHUR, M. A., AND FISCHER, A. G., 1977, Upper Cretaceous-Paleocene magnetic stratigraphy at Gubbio, Italy, I. Lithostratigraphy and Sedimentology: Geol. Soc. Amer. Bull. v. 88, p. 367-371.

BANNER, F. T., 1974, *Pithonella ovalis* from the early Cenomanian of England: Micropaleontology, v. 18, p. 278-284.

BATURIN, G. N., 1969, Authigenic phosphate concretions in Recent sediments of the southwest African Shelf: Acad. Sci. USSR, Dokl. Earth Sci. Sect., v. 189, no. 1-6, p. 227-230.

——, 1971, Stages of phosphorite formation on the ocean floor: Nature, v. 232, p. 61-62.

BERGER, W. H., 1970, Biogenous deep-sea sediments; Fractionation by deep-sea circulation: Geol. Soc. America Bull., v. 81, p. 1385-1402.

——, 1974, Deep-sea sedimentation: *In* C. A. Burk and C. L. Drake (eds.), The geology of continental margins. Springer-Verlag, Berlin, p. 213-242.

—— AND HEATH, G. R., 1968, Vertical mixing in pelagic sediments: Jour. Mar. Research, v. 26, p. 134-243.

—— AND ROTH, P. H., 1975, Oceanic micropaleontology: Progress and prospect: Rev. Geophys. and Space Phys, v. 13, p. 561-585.

—— AND VON RAD, ULRICH, 1972, Cretaceous and Cenozoic sediments from the Atlantic Ocean: *In* D. E. Hayes, A. C. Pimm and others, Initial reports of the Deep Sea Drilling Project, v. 14: U.S. Government Printing Office, Washington, D.C., p. 787-954.

—— AND WINTERER, E. L., 1974, Plate stratigraphy and the fluctuating carbonate line: *In* K. J. Hsü and H. C. Jenkyns (eds.), Pelagic sediments on land and under the sea: Internat. Assoc. Sedimentol., Spec. Pub. 1, p. 11-48.

BERGGREN, W. A., 1972, A Cenozoic time scale—Some implications for regional geology and paleobiogeography: Lethaia, v. 5, p. 195-215.

—— AND HOLLISTER, C. D., 1974, Paleogeography, paleobiogeography, and the history of circulation in the Atlantic Ocean: *In* W. W. Hay (ed.), Studies in paleo-oceanography: Soc. Econ. Paleontologists and Mineralogists, Spec. Pub. 20, p. 126-186.

——, MCKENZIE, D. P., SCLATER, J. G., AND VAN HINTE, J. E., 1975, Worldwide correlation of Mesozoic magnetic anomalies and its implications, discussion: Geol. Soc. America Bull., v. 86, p. 267-269.

BHANDARI, L. L., AND CHOWDHARY, C. R., 1975, Stratigraphic analysis of Kadi and Kohol Formations, Cambay Basin, India: Am. Assoc. Petroleum Geologists Bull., v. 59, p. 856-871.

BITTERLI, P., 1963, Aspects of the genesis of bituminous rock sequences: Geologie en Mijnbouw, N. S., v. 42, p. 183-201.

BOLLI, H. M., RYAN, W. B. F., AND OTHERS, 1975, Deep sea drilling in the South Atlantic; Leg 40: Geotimes, v. 20(6), p. 22-24.

BOWEN, R., 1966, Paleotemperature analysis: Elsevier, Amsterdam, 265 p.

BRAMLETTE, M. N., 1946, The Monterey Formation of California and the origin of its siliceous rocks: U.S. Geol. Survey, Prof. Paper 212, 57 p.

——, 1961, Pelagic sediments: *In* Mary Sears (ed.), Oceanography: Am. Assoc. Adv. Sci. Pub., v. 67, p. 345-366.

BROECKER, W. S., 1970, A boundary condition on the evolution of atmospheric oxygen: Jour. Geophys. Research, v. 75, p. 3553-3557.

CALDER, J. A., HORVATH, G. J., SHULTZ, D. J., AND NEWMAN, J. W., 1974, Geochemistry of the stable carbon isotopes in some Indian Ocean sediments: *In* T. A. Davies, B. P. Luyendyk and others, Initial reports of the Deep Sea Drilling Project, v. 26: U.S. Government Printing Office, Washington, D.C., p. 613-617.

CALVERT, S. E., 1964, Factors affecting the distribution of laminated diatomaceous sediments in Gulf of California: *In* T. H. van Andel and G. G. Shor (eds), Marine geology of the Gulf of California: A symposium. Am. Assoc. Petroleum Geologists, Mem. 3, p. 311-330.

——, 1966, Accumulation of diatomaceous silica in the sediments of the Gulf of California: Geol. Soc. America Bull., v. 77, p. 569-596.

CIFELLI, R., 1969, Radiation of Cenozoic planktonic Foraminifera: Systematic Zoology, v. 18, p. 154-168.

COPLEN, T. B., AND SCHLANGER, S. O., 1973, Oxygen and carbon isotope studies of carbonate sediments from Site 168, Magellan Rise, Leg 17: *In* E. L. Winterer, J. I. Ewing and others, Initial reports of the Deep Sea Drilling Project, v. 17: U.S. Government Printing Office, Washington, D.C., p. 505-509.

CRAIG, HARMON, AND GORDON, L. I., 1965, Isotopic oceanography: Deuterium and oxygen variations in the ocean and marine atmosphere: Univ. Rhode Island, Occasional Pub. 3, p. 277-374.

DAMON, P. E., 1971, The relationship between late Cenozoic volcanism and tectonism and orogenic-epeirogenic periodicity: *In* Karl Turekian (ed.), Late Cenozoic glacial ages: Yale Univ. Press, New Haven, Connecticut, p. 15-36.

DAVIES, T. A., LUYENDYK, B. P., AND OTHERS, 1974, Initial reports of the Deep Sea Drilling Project, v. 26: U.S. Government Printing Office, Washington, D.C., 1129 p.

DEARNLEY, R., 1966, Orogenic fold-belts and a hypothesis of earth evolution: *In* L. H. Ahrens, Frank Press, S. K. Runcorn, and H. C. Urey (eds.), Physics and chemistry of the Earth, v. 7: Pergamon Press, New York, New York, p. 1-114.

DEVEREAUX, I., 1967, Oxygen isotope paleotemperature measurements on New Zealand Tertiary fossils: New Zealand Jour. Sci., v. 10, p. 988-1011.

DORF, ERLING, 1970, Paleobotanical evidence of Mesozoic and Cenozoic climatic changes: *In* E. I. Yochelson (ed.), Proceedings of the North American paleontological convention, v. 2: Allen Press, Lawrence, Kansas, p. 323-346.

DORMAN, F. H., 1968, Some Australian oxygen isotope temperatures and a theory for a 30-million-year world temperature cycle: Jour. Geology, v. 76, p. 297-313.

DOUGLAS, R. G., MOULLADE, M., AND NAIRN, A. E. M., 1973, Causes and consequence of drift in the South Atlantic: *In* D. H. Tarling and S. K. Runcorn (eds.), Implications of continental drift to the earth sciences, v. 1: Academic Press, New York, New York, p. 513-534.

—— AND SAVIN, S. M., 1971, Isotopic analysis of planktonic Foraminifera from the Cenozoic of the N.W. Pacific, Leg 6: *In* A. G. Fischer, B. C. Heezen and others, Initial reports of the Deep Sea Drilling Project, v. 6: U.S. Government Printing Office, Washington, D.C., p. 1123-1127.

—— AND ——, 1973, Oxygen and carbon isotope analyses of Cretaceous and Tertiary Foraminifera from the central North Pacific: *In* E. L. Winterer, J. I. Ewing and others, Initial reports of the Deep Sea Drilling Project, v. 17: U.S. Government Printing Office, Washington, D.C., p. 591-605.

—— AND ——, 1975, Oxygen and carbon isotope analyses of Tertiary and Cretaceous microfossils from Shatsky Rise and other sites in the North Pacific Ocean: *In* R. L. Larson, Ralph Moberly and others, Initial reports of the Deep Sea Drilling Project, v. 32: U.S. Government Printing Office, Washington, D.C., p. 509-521.

DUGDALE, R. C., 1972, Nitrogen cycle in the ocean: *In* R. W. Fairbridge (ed.), The Encyclopedia of geochemistry: Van Nostrand, Reinhold, New York, New York, p. 807-809.

DUPLESSY, J. C., LALOU, CLAUDE, AND VINOT, A. C., 1970, Differential isotopic fractionation in benthic Foraminifera and paleotemperatures reassessed: Science, v. 168, p. 250-251.

EDGAR, N. T., SAUNDERS, J. B., AND OTHERS, 1973, Initial reports of the Deep Sea Drilling Project, v. 15: U.S. Government Printing Office, Washington, D.C., 1137 p.

EDMOND, J. M., 1974, On the dissolution of carbonate and silicate in the deep ocean: Deep Sea Research, v. 21, p. 455-480.

EICHER, D. L., AND WORSTELL, PAULA, 1970, Cenomanian and Turonian Foraminifera from the Great Plains, U.S.: Micropaleontology, v. 16, p. 269-324.

EMILIANI, CESARE, 1954, Temperatures of Pacific bottom waters and polar superficial waters during the Tertiary: Science, v. 119, p. 853-855.

——, 1961, The temperature decrease of surface seawater in high latitudes and of abyssal-hadal water in open oceanic basins during past 75 m.y.: Deep Sea Research, v. 8, p. 144-147.

EPSTEIN, SAMUEL, BUCHSBAUM, RALPH, LOWENSTAM, H. A., AND UREY, H. C., 1951, Carbonate-water isotopic temperature scale: Geol. Soc. America Bull., v. 62, p. 417-426.

ERICSON, D. B., AND WOLLIN, GÖSTA, 1964, The Deep and the past: A. Knopf, New York, New York, 288 p.

EWING, MAURICE, WORZEL, J. L., AND OTHERS, 1969, Initial reports of the Deep Sea Drilling Project, v. 1: U.S. Government Printing Office, Washington, D.C., 672 p.

FISCHER, A. G., 1969, Geological time-distance rates: The Bubnoff Unit: Geol. Soc. America Bull., v. 80, p. 549-552.

——, HEEZEN, B. C., AND OTHERS, 1971, Initial reports of the Deep Sea Drilling Project, v. 6: U.S. Government Printing Office, Washington, D.C., 1329 p.

——, HONJO, SUSUMU, AND GARRISON, R. E., 1967, Electron micrographs of limestones and their nannofossils: Princeton Univ. Press, Princeton, New Jersey, 141 p.

FLESSA, K. W., AND IMBRIE, JOHN, 1973, Evolutionary pulsations: Evidence from Phanerozoic diversity patterns: *In* D. H. Tarling and S. K. Runcorn (eds.), Implications of continental drift to the earth sciences, v. 1: Academic Press, New York, New York, p. 245-284.

FRAKES, L. A., AND KEMP, E. M., 1973, Paleogene continental positions and evolution of climate: *In* D. H. Tarling and S. K. Runcorn (eds.), Implications of continental drift to the earth sciences, v. 1: Academic Press, New York, New York, p. 535-558.

FRERICHS, W. E., 1970, Paleobathymetry, paleotemperature and tectonism: Geol. Soc. America Bull., v. 81, p. 3445-3452.

FREY, R. W., 1971, Ichnology—The Study of fossil and Recent lebensspuren: *In* B. F. Perkins (ed.), Trace fossils; a field guide to selected localities in Pennsylvanian, Permian, Cretaceous and Tertiary rocks of Texas and related papers: School of Geosci., Louisiana State Univ., Baton Rouge, p. 91-125.

——, 1972, Paleoecology and depositional environment of Fort Hays Limestone member, Niobrara Chalk (Upper Cretaceous), west-central Kansas: Univ. Kansas Paleont. Contr., Art. 58, 72 p.

FRITZ, P., 1965, $^{18}O/^{16}O$ Isotopenanalysen und Paläotemperaturbestimmungen an Belemniten aus dem schwäbischen Jura: Geol. Rundschau, v. 54, p. 261-269.
FUNNELL, B. M., 1971, Post Cretaceous biogeography of oceans—with special reference to plankton: *In* F. A. Middlemiss, P. F. Rawson and G. Newall (eds.), Faunal provinces in space and time: Geol. Jour. Spec. Issue 4, p. 191-198.
GARDNER, J. V., AND BURKLE, L. H., 1975, Upper Pleistocene *Ethmodiscus rex* oozes from the eastern equatorial Atlantic: Micropaleontology, v. 21, p. 236-242.
GARRELS, R. M., AND PERRY, E. A., JR., 1975, Cycling of carbon, sulfur, and oxygen through geologic time: *In* E. D. Goldberg (ed.), The Sea, v. 5: John Wiley and Sons, New York, New York, p. 303-336.
GIGNOUX, MAURICE, 1955, Stratigraphic geology: W. H. Freeman and Co., San Francisco, California, 682 p.
GOERING, J. J., 1968, Denitrification in the oxygen minimum layer of the eastern tropical Pacific Ocean: Deep Sea Research, v. 15, p. 157-164.
——, RICHARDS, F. A., CODISPOTI, L. A., AND DUGDALE, R. C., 1973, Nitrogen fixation and denitrification in the ocean: Biochemical budgets: *In* Earle Ingerson (ed.), Proceedings of symposium on hydrogeochemistry and biochemistry, v. 2: Clarke Press, Washington, D.C., p. 12-27.
GRABAU, A. W., 1940, The rhythm of the ages: Henri Vetch Pub., Peking, China, 561 p.
HALLAM, A., 1975, Jurassic environments: Cambridge Univ. Press, London, England, 269 p.
HAQ, B. U., 1973, Transgressions, climatic changes and the diversity of calcareous nannoplankton: Marine Geology, v. 15, p. 25-30.
HARLAND, W. B., SMITH, A. G., AND WILCOCK, B. (EDS.), 1964, The Phanerozoic time scale: Geol. Soc. London, Quart. Jour., v. 120, p. 260-261.
HAY, W. W., 1970, Calcium carbonate compensation: *In* R. G. Bader, R. D. Gerard and others, Initial reports of the Deep Sea Drilling Project, v. 4: U.S. Government Printing Office, Washington, D.C., p. 672.
HAYS, J. D., AND PITMAN, W. C. III, 1973, Lithospheric plate motion, sea-level changes and climatic and ecological consequences: Nature, v. 246, p. 18-22.
HEATH, G. R., 1974, Dissolved silica and deep-sea sediments: *In* W. W. Hay (ed.), Studies in paleo-oceanography: Soc. Econ. Paleontologists and Mineralogists, Spec. Pub. 20, p. 77-93.
HEEZEN, B. C., MATTHEWS, J. L., AND OTHERS, 1973, Western Pacific guyots: *In* B. C. Heezen, I. D. MacGregor and others, Initial reports of the Deep Sea Drilling Project, v. 20: U.S. Government Printing Office, Washington, D.C., p. 653-729.
HERM, D., 1965, Mikropaläontologisch-stratigraphische Untersuchungen im Kreideflysch zwischen Deva and Zumaya (Provinz Guipuzcoa, Nordspanien): Deutsch. Geol. Gesell. Zeitschr., v. 15, p. 255-348.
HOLLISTER, C. D., EWING, J. I., AND OTHERS, 1972, Initial reports of the Deep Sea Drilling Project, v. 11: U.S. Government Printing Office, Washington, D.C., 1077 p.
——, JOHNSON, D. A., AND LONSDALE, P. F., 1974, Current-controlled abyssal sedimentation: Samoan Passage, equatorial West Pacific: Jour. Geology, v. 82, p. 275-300.
HURD, D. C., 1973, Interactions of biogenic opal, sediment and seawater in the central equatorial Pacific: Geochim. et Cosmochim. Acta, v. 37, p. 393-399.
JACKSON, E. D., AND SCHLANGER, S. O., 1976, Regional synthesis, Line Islands Chain, and Manihiki Plateau, central Pacific Ocean, DSDP Leg 33: *In* S. O. Schlanger, E. D. Jackson and others, Initial reports of the Deep Sea Drilling Project, v. 33: U.S. Government Printing Office, Washington, D.C., p. 915-927.
JOHNSON, D. A., 1972, Ocean-floor erosion in the equatorial Pacific: Geol. Soc. America Bull., v. 83, p. 3121-3144.
JUNGE, C. E., SCHIDLOWSKI, M., EICHMANN, R., AND PIETREK, H., 1975, Model calculations for the terrestrial carbon cycle: Carbon isotope geochemistry and evolution of photosynthetic oxygen: Jour. Geophys. Research, v. 80, p. 4542-4552.
KAUFFMAN, E. G., 1970, Population systematics, radiometrics and zonation—A new biostratigraphy: *In* E. I. Yochelson (ed.), Proceedings of the North American paleontological convention, v. 2: Allen Press, Lawrence, Kansas, p. 612-666.
KEITH, M. L., AND WEBER, J. N., 1964, Carbon and oxygen isotopic composition of selected limestones and fossils: Geochim. et Cosmochim. Acta, v. 28, p. 1787-1816.
—— AND ——, 1965, Systematic relationships between carbon and oxygen isotopes in carbonates deposited by modern corals and algae: Science, v. 150, p. 498-501.
KENNEDY, W. J., AND GARRISON, R. E., 1975, Morphology and genesis of nodular chalks and hardgrounds in the Upper Cretaceous of southern England: Sedimentology, v. 22, p. 311-386.
KENNETT, J. P., HOUTZ, R. E., AND OTHERS, 1974, Initial reports of the Deep Sea Drilling Project, v. 29: U.S. Government Printing Office, Washington, D.C., 1197 p.
KISTLER, R. W., EVERNDEN, J. F., AND SHAW, H. R., 1971, Sierra Nevada plutonic cycle: Part I, Origin of composite grantitic batholiths: Geol. Soc. America Bull., v. 82, p. 852-868.
KOLODNY, Y., AND KAPLAN, I. R., 1970, Uranium isotopes in sea-floor phosphorites: Geochim. et Cosmochim. Acta, v. 34, p. 3-24.
KRAMER, J. R., 1964, Sea Water: saturation with apatites and carbonates: Science, v. 146, p. 637-638.
KRUMBEIN, W. C., AND SLOSS, L. L., 1963, Stratigraphy and sedimentation: H. S. Freeman and Co., San Francisco, California, 660 p.
LAMBERT, R. ST. J., 1971, The pre-Pleistocene-Phanerozoic time-scale—Review: *In* W. B. Harland and E. H. Francis (eds.), The Phanerozoic time-scale—A supplement: Geol. Soc. London, Spec Pub. 5, p. 9-34.

LARSON, R. L., AND PITMAN, W. C. III, 1972, World-wide correlation of Mesozoic magnetic anomalies and its implications: Geol. Soc. America Bull., v. 83, p. 3645-3662.
——, MOBERLY, RALPH, AND OTHERS, 1975, Initial reports of the Deep Sea Drilling Project, v. 32: U.S. Government Printing Office, Washington, D.C., 980 p.
LEVINTON, J. S., 1970, The paleontological significance of opportunistic species: Lethaia, v. 3, p. 69-78.
LIPPS, J. H., 1970a, Plankton evolution: Evolution, v. 24, p. 1-22.
——, 1970b, Ecology and evolution of silicoflagellates: *In* E. I. Yochelson (ed.), Proceedings of the North American paleontological convention, v. 2: Allen Press, Lawrence, Kansas, p. 965-993.
LOWENSTAM, H. A., AND EPSTEIN, SAMUEL, 1954, Paleotemperatures of the post-Aptian Cretaceous as determined by the oxygen isotope method: Jour. Geology, v. 62, p. 207-248.
MACARTHUR, R. H., 1955, Fluctuations in animal populations as a measure of community stability: Ecology, v. 36, p. 533-536.
MACKENZIE, F. T., 1975, Sedimentary cycling and the evolution of sea water: *In* J. P. Riley and G. Skirrow (eds.), Chemical oceanography, v. 1: Academic Press, New York, New York, p. 309-364.
MAGUE, T. H., WEARE, N. M., AND HOLM-HASEN, O., 1974, Nitrogen fixation in the North Pacific Ocean: Marine Biology, v. 24, p. 109-119.
MANSFIELD, G. R., 1927, Geography, geology and mineral resources of part of southeastern Idaho: U.S. Geol. Survey, Prof. Paper 152, 409 p.
MARGALEF, R., 1968, Perspectives in ecological theory: Univ. Chicago Press, Chicago, Illinois, 111 p.
MAXWELL, A. E., VON HERZEN, R. P., AND OTHERS, 1970, Initial reports of the Deep Sea Drilling Project, v. 3: U.S. Government Printing Office, Washington, D.C., 806 p.
MCBRIDE, E. F., AND THOMSON, ALAN, 1970, The Caballos Novaculite, Marathon region, Texas: Geol. Soc. America, Spec. Paper 122, 129 p.
MCCRAE, S. G., 1972, Glauconite: Earth Sci. Rev., v. 8, p. 397-440.
MENZEL, D. W., AND RYTHER, J. H., 1968, Organic carbon and the oxygen minimum in the South Atlantic Ocean: Deep Sea Research, v. 15, p. 327-337.
MONECHI, SIMONETTA, AND PIRINI-RADRIZZANI, C., 1975, Nannoplankton from Scaglia Umbra Formation (Gubbio) at Cretaceous-Tertiary boundary: Riv. Italiana Paleontologia e Stratigrafia, v. 81, p. 45-87.
MOORE, R. C. (ed.), 1957, Treatise on invertebrate paleontology, Part K, Mollusca 4, Ammonoidea: Univ. Kansas Press, Lawrence, Kansas, 519 p.
MOORE, T. C., 1972, Deep Sea Drilling Project: Successes, failures, proposals: Geotimes, v. 17, p. 27-31.
MUNK, W. H., 1966, Abyssal recipes: Deep Sea Research, v. 13, p. 707-730.
OBRADOVICH, J. D., AND COBBAN, W. A., 1975, A time-scale for the Late Cretaceous of the Western Interior of North America: *In* W. G. E. Caldwell (ed.), The Cretaceous System in the Western Interior of North America: Geol. Assoc. Canada, Spec. Paper 13, p. 31-54.
PARSONS, T. R., AND LEBRASSEUR, R. J., 1970, The availability of food to different trophic levels in the marine food chain: *In* J. H. Steele (ed.), Marine food chains: Olliver and Boyd, Edinburgh, Scotland, p. 325-343.
PERCIVAL, S. F., AND FISCHER, A. G., 1977, Changes in calcareous nannoplankton in the Cretaceous-Tertiary biotic crisis at Zumaya, Spain: Evol. Theory, v. 2, p. 1-35.
PERRY, E. C., AND TAN, F. C., 1972, Significance of oxygen and carbon isotope variations in early Precambrian cherts and carbonate rocks of southern Africa: Geol. Soc. America Bull. v. 83, p. 647-664.
PIPER, D. Z., AND CODISPOTI, L. A., 1975, Marine phosphorite deposits and the nitrogen cycle: Science, v. 188, p. 15-18.
POSTUMA, J. A., 1971, Manual of planktonic Foraminifera: Elsevier, Amsterdam, The Netherlands, 420 p.
RAUP, D. M., 1972, Taxonomic diversity during the Phanerozoic: Science, v. 177, p. 1065-1071.
REID, J. L., JR., 1965, Intermediate water of the Pacific Ocean: Johns Hopkins Press, Baltimore, Maryland, 118 p.
REVELLE, R. R., 1944, Marine bottom samples collected in the Pacific Ocean by the Carnegie on its seventh cruise: Carnegie Inst. Washington, Pub. 556, pt. 1, 180 p.
RHOADS, D. C., AND MORSE, J. W., 1971, Evolutionary and ecologic significance of oxygen-deficient marine basins: Lethaia, v. 4, p. 413-428.
RICHARDS, F. A., 1965, Anoxic basins and fjords: *In* J. P. Riley and G. Skirrow (eds.), Chemical oceanography, v. 1: Academic Press, New York, New York, p. 611-645.
ROBINSON, P. L., 1973, Paleoclimatology and continental drift: *In* D. H. Tarling and S. K. Runcorn (eds.), Implications of continental drift to the earth sciences, v. 1: Academic Press, New York, New York, p. 449-474.
ROGERS, M. A., AND KOONS, C. B., 1969, Organic carbon $\delta^{13}C$ values from Quaternary marine sequences in the Gulf of Mexico: A reflection of paleotemperature changes: Gulf Coast Assoc. Geol. Socs. Trans., v. 19, p. 529-534.
——, VAN HINTE, J. E., AND SUGDEN, J. G., 1972, Organic carbon $\delta^{13}C$ values from Cretaceous, Tertiary, and Quaternary marine sequences in the North Atlantic: *In* A. S. Laughton, W. A. Berggren and others, Initial reports of the Deep Sea Drilling Project, v. 12: U.S. Government Printing Office, Washington, D.C., p. 1115-1126.
ROMER, A. S., 1966, Vertebrate paleontology: Univ. Chicago Press, Chicago, Illinois, 468 p.
RONA, P. A., 1973a, Worldwide unconformities in marine sediments related to eustatic changes of sea level: Nature, v. 244, p. 25-26.

——, 1973b, Relations between rates of sediment accumulation on continental shelves, sea-floor spreading, and eustacy inferred from the central North Atlantic: Geol. Soc. America Bull., v. 84, p. 2851-2872.

Ronov, A. B., 1958, Organic carbon in sedimentary rocks (in relation to the presence of petroleum): Geochimia USSR, v. 5, p. 510-536.

Ryther, J. H., 1969, Photosynthesis and fish production in the sea: Science, v. 166, p. 72-76.

Sackett, W. M., Eckelmann, W. R., Bender, M. C., and Bé, W. W. A., 1965, Temperature dependence of carbon isotope composition in marine plankton and sediments: Science, v. 148, p. 235.

Saito, Tsunemasa, and van Donk, Jan, 1974, Oxygen and carbon isotope measurements of Cretaceous and early Tertiary Foraminifera: Micropaleontology, v. 20 p. 152-177.

Sarjeant, W. A. S., 1967, The stratigraphical distribution of fossil dinoflagellates: Rev. Paleobotany and Palynology, v. 1, p. 323-343.

Savin, S. M., and Douglas, R. G., 1973, Stable isotope and magnesium geochemistry of Recent planktonic Foraminifera from the South Pacific: Geol. Soc. America Bull., v. 84, p. 2327-2342.

——, —— and Stehli, F. G., 1975, Tertiary marine paleotemperatures: Geol. Soc. America Bull., v. 86, p. 1499-1510.

Shackleton, N. J., and Kennett, J. P., 1974, Paleotemperature history of the Cenozoic and the initiation of Antarctic glaciation: Oxygen and carbon isotope analyses in DSDP Sites 277, 279, and 281: *In* J. P. Kennett, R. E. Houtz and others, Initial reports of the Deep Sea Drilling Project, v. 29: U.S. Government Printing Office, Washington, D.C., p. 743-755.

Shaw, H. R., Kistler, R. W., and Evernden, J. F., 1971, Sierra Nevada plutonic cycle: Part II, Tidal energy and a hypothesis for orogenic-epeirogenic periodicities: Geol. Soc. America Bull., v. 82, p. 869-896.

Smith, A. G., Briden, J. C., and Drewry, G. E., 1973, Phanerozoic world maps: *In* N. F. Hughes (ed.), Organisms and continents through time: Palaeontol. Assoc. Spec. Papers in Paleontology, v. 12, p. 1-42.

Spaeth, C., Hoefs, J., and Vetter, U., 1971, Some aspects of isotopic composition of belemnites and related paleotemperatures: Geol. Soc. America Bull., v. 82, p. 3139-3150.

Stewart, W. D. P., 1971, Nitrogen fixation in the sea: *In* J. D. Costlow (ed.), Fertility in the sea: Gordon, Breach, New York, New York, p. 537-564.

Stommel, Henry, 1958, The abyssal circulation: Deep Sea Research, v. 5, p. 80-98.

Strøm, K. M., 1939, Land-locked waters and the deposition of black muds: *In* P. D. Trask (ed.), Recent marine sediments: Am. Assoc. Petroleum Geologists, Tulsa, Oklahoma, p. 356-372.

Tappan, Helen, 1968, Primary production, isotopes, extinctions, and the atmosphere: Palaeogeography, Palaeoclimatology, Palaeoecology, v. 4, p. 187-210.

—— and Loeblich, A. R., 1971, Geobiologic implications of fossil phytoplankton evolution and time-space distribution: *In* R. Kosanke and A. T. Cross (eds.), Symposium on palynology of the Late Cretaceous and early Tertiary: Geol. Soc. America, Spec. Paper 127, p. 247-339.

Trask, P. D., 1932, Origin and environment of source sediments of petroleum: Gulf. Pub. Co., Houston, Texas, 323 p.

Umbgrove, J. H. F., 1947, The pulse of the earth, 2d ed.: Nijhoff, The Hague, The Netherlands, 358 p.

Vail, P. R., Mitchum, R. M., Jr., and Thompson, S. III, 1974, Eustatic cycles based on sequences with coastal onlap [abs.]: Geol. Soc. America Abs. with Programs, v. 6, p. 993.

Valentine, J. W., 1968, Climatic regulation of species diversification and extinction: Geol. Soc. America Bull., v. 79, p. 273-276.

——, 1971, Resource supply and species diversity patterns: Lethaia, v. 4, p. 51-61.

——, 1974, Evolutionary paleoecology of the marine biosphere: Prentice Hall, Englewood Cliffs, New Jersey, 498 p.

van Andel, T. H., 1964, Recent marine sediments of the Gulf of California: *In* T. H. van Andel and G. G. Shor, Jr., (eds.), Marine geology of the Gulf of California, a symposium: Am. Assoc. Petroleum Geologists, Mem. 3, p. 216-310.

——, 1975, Mesozoic/Cenozoic calcite compensation depth and the global distribution of calcareous sediments: Earth and Planetary Sci. Letters, v. 26, p. 187-195.

van Hinte, J. E., 1976, A Cretaceous time scale: Am. Assoc. Petroleum Geologists Bull., v. 60, p. 498-516.

Veeh, H. H., Burnett, W. C., and Soutar, A., 1973, Contemporary phosphorites on the continental margin of Peru: Science, v. 181, p. 844-845.

Veevers, J. J., Heirtzler, J. R., and others, 1974, Initial reports of the Deep Sea Drilling Project, v. 27: U.S. Government Printing Office, Washington, D.C., 1060 p.

Vinot-Bertouille, A. C., and Duplessy, J. C., 1973, Individual isotopic fractionation of carbon and oxygen in benthic Foraminifera: Earth and Planetary Sci. Letters, v. 18, p. 247-252.

von Stackelberg, U., 1972, Facies of sediments of the Indian Pakistan continental margin (Arabian Sea): Meteorol. Research Results, C, no. 9, 73 p.

Watkins, N. D., and Kennett, J. P., 1972, Regional sedimentary disconformities and upper Cenozoic changes in bottom water velocities between Australia and Antarctica: Am. Geophys. Union Antarctic Research Series, v. 19, p. 273-293.

—— and ——, 1973, Response of deep-sea sediments to changes in physical oceanography resulting from spearation of Australia and Antarctica: *In* D. H. Tarling and S. K. Runcorn (eds.), Implications of continental drift to the earth sciences, v. 2: Academic Press, New York, New York, p. 787-798.

Weber, J. N., 1967, Possible changes in the isotopic composition of the oceanic and atmospheric carbon reservoirs

over geologic time: Geochim. et Cosmochim. Acta, v. 31, p. 2343-2351.
—— AND RAUP, D. M., 1968, Comparison of $^{13}C/^{12}C$ and $^{18}O/^{16}O$ in the skeletal calcite of Recent and fossil echinoids: Jour. Paleontology, v. 42, p. 37-50.
WISE, D. U., 1974, Continental margins, freeboard and the volumes of continents and oceans through time: *In* C. A. Burk and C. L. Drake (eds.), Geology of the continental margins: Springer-Verlag, Heidelberg, Germany, p. 45-58.
WOLLAST, R., 1975, The silica problem: *In* E. D. Goldberg (ed.), The sea, v. 5: John Wiley and Sons, New York, New York, p. 359-392.
WORSLEY, T., 1974, The Cretaceous-Tertiary boundary event in the ocean: *In* W. W. Hay (ed.), Studies in paleo-oceanography: Soc. Econ. Mineralogists and Paleontologists, Spec. Pub. 20, p. 94-125.
WYRTKI, KLAUS, 1962, The oxygen minima in relation to oceanic circulation: Deep Sea Research, v. 9, p. 11-23.

SEPM SPECIAL PUBLICATION NO. 25, P. 51-81, NOVEMBER 1977

COMPARISON OF CONTINENTAL SLOPE AND SHELF ENVIRONMENTS IN THE UPPER CAMBRIAN AND LOWEST ORDOVICIAN OF NEVADA

HARRY E. COOK AND MICHAEL E. TAYLOR
U.S. Geological Survey, Menlo Park, California 94025, and
U.S. Geological Survey, Denver, Colorado 80225

ABSTRACT

Integrated sedimentological and paleontological analysis of the Whipple Cave Formation (Upper Cambrian and lowest Ordovician) and lower House Limestone (Lower Ordovician) of eastern Nevada and coeval parts of the Hales Limestone of central Nevada suggests that the depositional regime changes from shoal-water shelf in the east to deeper water slope to the west.

The Whipple Cave Formation and lower House Limestone consist of about 600 m of biostromal and biohermal lenses of algal stromatolites, fenestral limestone, flat-pebble conglomerate, lime grainstone, and other carbonate rocks normally associate with shoal-water environments. Sedimentation from the base of the section upwards records a progressive shallowing of water which reflects a westerly seaward progradation of the continental shelf. In contrast, 170 km to the west, the coeval lower part of the Hales Limestone is only a quarter as thick and consists of dark fine-grained limestone and interbedded coarse-grained sediment gravity-flow and slump deposits. The lower Hales Limestone is interpreted to have formed in a slope environment.

Slope sedimentation involved an interplay of depositional and erosional processes. The dark fine-grained limestone (lime mudstone and wackestone) represents hemipelagic sedimentation by suspension and other processes. Mass-flow and slump deposits reflect mobilization of semilithified and unconsolidated sediment from both slope and shoal-water shelf sites to the east and redeposition in deeper water environments to the west. Although submarine gravity movements were of a small scale (few cm to 10 m thick), they were a major process in the development of this environment. At least one-quarter of the section is composed of slump and sediment gravity-flow material.

On the basis of abundant small-scale slump and flow deposits in this and similar ancient, deeper-water settings, it is reasonable to speculate that resedimentation processes may be significant on modern continental slopes. The smooth undisturbed nature of some modern slopes as seen on continuous seismic profiles may be in part a function of the limited resolving power of much surface seismic equipment. By present surface seismic techniques, slump or flow features less than a quarter of a kilometer long and less than 10 m thick may not be recognized.

Trilobites in mass-flow deposits derived from the shelf show strong taxonomic differences from those in *in situ* slope deposits. Redeposited trilobites have affinities with faunas from shoal or higher slope habitats typical of areas of carbonate deposition in North America. *In situ* trilobite assemblages from slope sediments resemble those assemblages in deeper water deposits of east-central Asia, but not those in shallow-water deposits of either North America or Asia.

Interbedding of trilobites with Asian and North American affinities in the slope facies shows the Hales Limestone was not part of a lithospheric plate fragment left behind after a hypothetical collision between the Asian and North American plates and subsequent separation. Rather, biofacies differences developed in normal paleoecologic response to the paleogeographic relationship between the North American continental platform and adjacent ocean basin.

The paleogeographic model developed for the shelf-to-slope transition in Nevada is analogous to Holocene shelf-margin and slope sedimentation and biofacies patterns.

INTRODUCTION

Traditionally, North American Cambrian studies have concentrated on the use of trilobites for stratigraphic correlation. However, pioneer work by Wilson (1957), Lochman-Balk and Wilson (1958), and Palmer (1960) has recognized the division of North American Cambrian trilobite faunas into major coeval biofacies "realms" or "belts" that surround the North American craton. The earlier work provided understanding of the broader patterns of Cambrian history and, in large part, has provided the conceptual framework for most subsequent North American Cambrian research. Our approach here and elsewhere (Cook and Taylor, 1975; Taylor and Cook, 1976) is to expand earlier facies studies and pose questions that are amenable to an integrated sedimentological and paleontological analysis. Interpretations are compared with models of Holocene faunal distribution and sedimentary processes. Results provide implications for biological and physical processes that were operating during the Late Cambrian and earliest Ordovician. Thus, an integrated analysis may be synergistic.

GEOLOGIC FRAMEWORK

The broad classification of lower Paleozoic rocks of the Great Basin in Roberts and others

(1958) includes an eastern assemblage of miogeosynclinal carbonate rocks, a western assemblage of eugeosynclinal siliceous rocks, and a transitional assemblage of intermediate composition. More recent syntheses have compared these facies assemblages with Holocene tectonic and depositional models. The paleogeographic model that emerges is one in which lower Paleozoic rocks of the Great Basin reflect a broad continental shelf and either an adjacent marginal ocean-basin and volcanic-arc system (Burchfiel and Davis, 1972; Churkin, 1974) or an adjacent open oceanic basin (Dietz and Holden, 1966; Stewart, 1972; Stewart and Poole, 1974).

Our objectives are to understand better the paleogeographic significance of the eastern assemblage thick carbonates and the transitional assemblage thin carbonates and shales, and to relate them to proposed paleogeographic models. Two sections of Upper Cambrian and lowermost Lower Ordovician rocks were selected for detail study: (1) the Whipple Cave Formation (Kellogg, 1963) and the lower part of the House Limestone in the central Egan Range of eastern Nevada, which represent the eastern assemblage carbonates; and (2) the Hales Limestone (Ferguson, 1933) in the Hot Creek Range of central Nevada, which represents the transitional assemblage. Figure 1 shows the general location of the central Egan Range and Hot Creek Range sections in relation

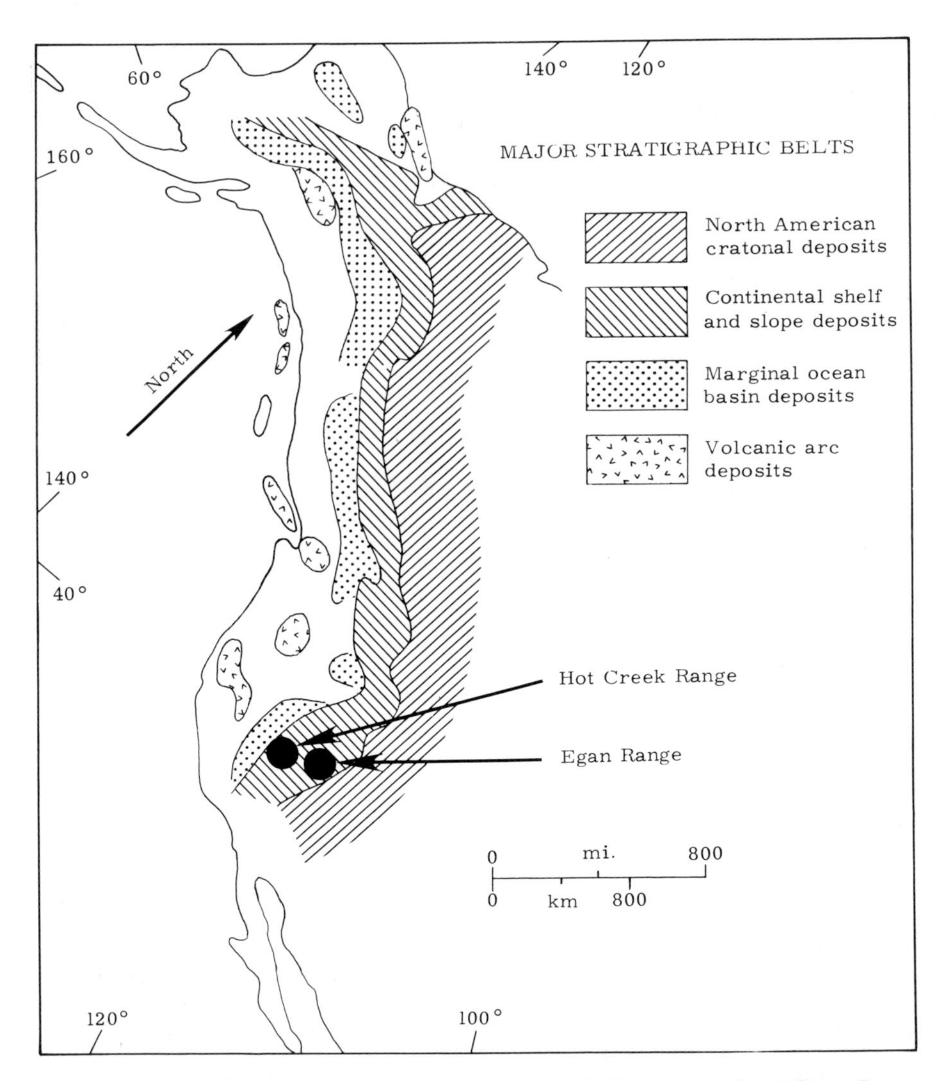

FIG. 1.—General location of sections in the Hot Creek Range and central Egan Range, Nevada, in relation to major regional stratigraphic belts. Map generalized from Churkin (1974).

to major stratigraphic belts in western North America. For convention the first will be referred to herein as the Egan section, the second as the Hot Creek section.

In an earlier summary paper (Cook and Taylor, 1975) and a later paleontologically oriented paper (Taylor and Cook, 1976), we concluded that the Whipple Cave Formation and lower House Limestone represent shoal-water deposition on a shallow carbonate shelf, whereas the Hales Limestone represents deeper water depositional environments on an ocean-facing slope. Here we focus on the lithologic characteristics that contrast the shelf and slope rocks, sedimentary processes that characterized the slope environment and some aspects of the trilobite faunas that bear on paleogeographic significance of the shelf-to-slope transition in the Great Basin.

SHOAL WATER ENVIRONMENTS—CENTRAL EGAN RANGE

The Whipple Cave Formation and the lower part of the House Limestone comprise about 600 m of marine carbonate rock (Figs. 2, 3) that is composed of argillaceous lime wackestone, algal buildups, algal grainstone, fenestral limestone and limestone conglomerate.

For convenience of discussion, the Egan Range section is here divided into three informal units (Fig. 2) on the basis of lithologic and faunal characteristics considered to have environmental significance. The lower unit consists of the lower 200 m of the Whipple Cave Formation. It is composed primarily of argillaceous limestone in beds 1 to 2 cm thick that form recessive slopes. This interval is characterized by wavy bedding surfaces, sedimentary boudinage (Fig. 4), patchy and unevenly distributed clay mudstone lenses

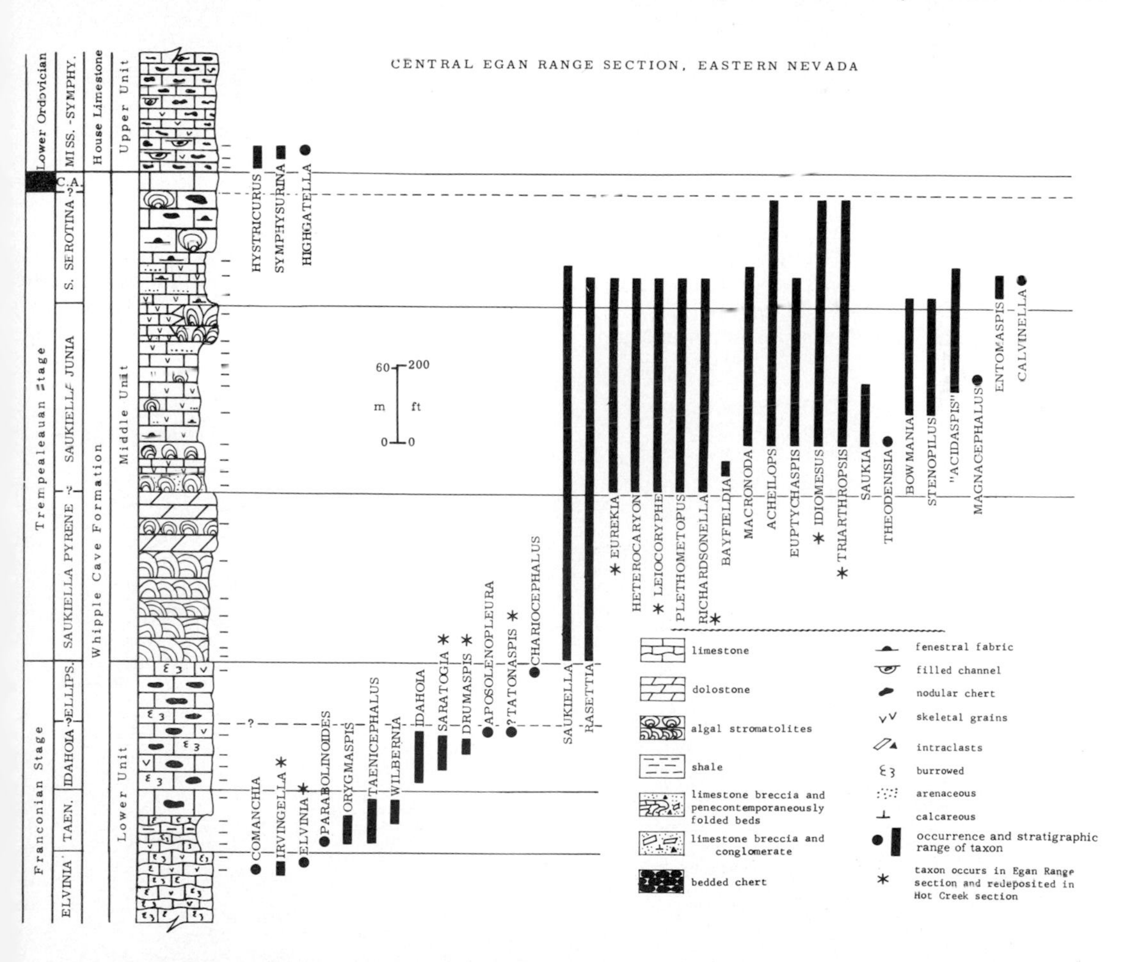

FIG. 2.—Range of polymeroid trilobite generally recognized from the Whipple Cave Formation and lower House Limestone in the central Egan Range, Nevada. The section was measured on the west face of the Egan Range near Lund, Nevada. *C.A.*, *Corbinia apopsis* Subzone of *Saukia* Zone; not recognized in measured section, but known to occur 170 km north in Cherry Creek Mountains. Dashes on right side of column show position of fossil samples.

FIG. 3.—Central Egan Range section of Whipple Cave Formation, near Lund, Nevada. Section was measured along the prominent spur near center of photograph. View toward east.

(Fig. 5), and abundant bioturbation (Fig. 6). Texturally the rocks are mudstone and wackestone. The main fabric constituents are trilobites, possible sponge spicules, lime mud, and peloids.

The lower unit probably represents deposition on a shallow subtidal, low-energy open shelf seaward of algal banks and tidal flats with little influence by wave action. The dominant faunal elements are trilobites typical of the North American Faunal Province. We concur with Wilson

FIG. 4.—Lower unit of Whipple Cave Formation, Egan Range section. Dark gray argillaceous lime mudstone and wackestone showing sedimentary boudinage bedding. Light-colored irregular lenses contain more abundant clayey material than do darker beds. Tape is 9 inches long.

FIG. 5.—Bedding-plane view of same beds shown in Fig. 4. Showing patchy and uneven distribution of claystone lenses (light).

(1969, p. 17) that the kind of sedimentary boudinage common in the lower unit suggests deposition in shallow subtidal shelf waters below active wave base but within well-oxygenated water. It is not a rock fabric typically seen either in deeper water carbonate settings or in shelf-lagoon or tidal-flat environments. This interpretation is further strengthened by the stratigraphic position of the unit directly below and gradational with rocks having distinctive shoal water features, and by its field characteristics that clearly contrast with those of contemporaneous sediments in the Hales Limestone that were deposited farther oceanward.

The upper 380 m of the Whipple Cave Formation constitutes the middle unit, which crops out as light-colored resistant cliffs along the west face of the central Egan Range (Fig. 7). Much of this interval contains algal buildups composed of laterally linked hemispheroids. Individual algal heads are as much as 30 cm in diameter (Fig. 8). Three types of buildups occur. The most common one is sheetlike and ranges from 1 to 2 m in thickness and 75 m or more in length (Fig. 9). Also common are pillow-shaped buildups

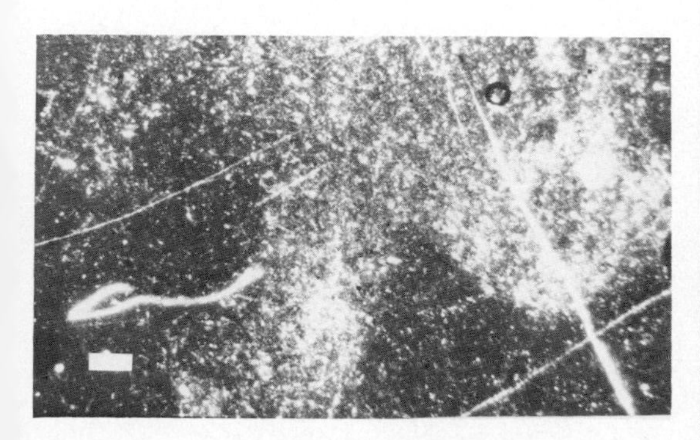

FIG. 6.—Photomicrograph of lime wackestone from lower unit, Egan section. Note abundant bioturbation and lighter colored burrow linings. Whipple Cave Formation. Bar scale equals 0.5 mm.

FIG. 7.—Middle unit of Whipple Cave Formation, Egan Range section. Skyline is the westerly sloping spur shown in Fig. 3. View toward southeast.

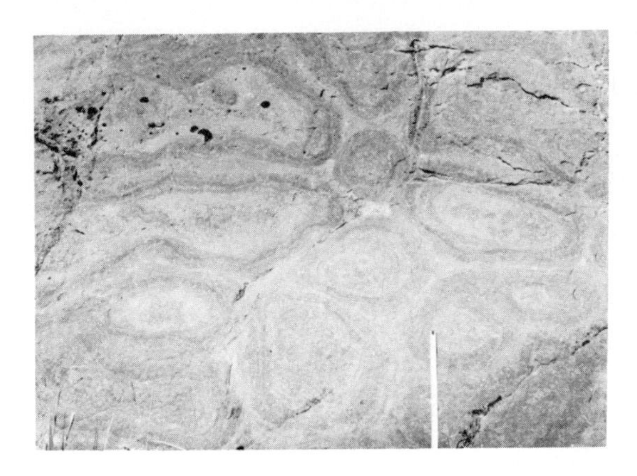

FIG. 8.—High-relief algal stromatolites from middle unit of Whipple Cave Formation, Egan Range section. View normal to bedding. Length of tape is 15 inches.

about 1 to 2 m thick and 1 to 5 m long (Fig. 10). Total areal extent of both types is unknown. A third, less common type consists of single isolated stromatolites as much as 30 cm high and 50 cm long (Fig. 11). Total relief on any of these algal buildups probably did not exceed more than a few tens of centimeters at any one time. In cross section, the buildups can be seen to interfinger with centimeter-thick beds of clastic limestone (Fig. 10). Also, on bedding-plane surfaces the tops of laterally linked algal heads have no more than a few centimeters of relief which is filled in with clastic limestone debris (Fig. 12).

Sediment types interfingering and interbedded with algal buildups consist of fenestral limestone and ribbon limestone (lime mudstone interbedded with lime packstone and grainstone in centimeter-thick beds). Grain-supported fabrics are composed of skeletal grains of echinoderms, gastropods, trilobites, the possible alga *Nuia*, and peloids and oncolites (Figs. 13, 14, 15, 16). Ribbon limestones have scour and fill features, cross-bedding, and occasional pebble-size intraclasts of lime mudstone.

Depositional environment of the middle unit is inferred to have been one of moderately high energy where algal mats flourished in peritidal habitats. Fenestral or birdseye fabrics suggest intertidal environments (Shinn, 1968), as does ribbon limestone (Ebanks, 1967). Ribbon limestone, scour and fill, and cross-bedding indicate a moderate amount of current activity. At times, small phototrophic algal hemispheroid mounds were subjected to adverse conditions and ceased to grow. These conditions probably occurred during storms when sudden influxes of sediment

FIG. 9.—Part of algal buildup in middle unit of Egan section. Typical form is sheetlike, 1 to 2 meters thick, composed of coalescing and stacked hemispherical algal stromatolites interbedded with thin-bedded calcarenites. Whipple Cave Formation. Pick is about 12 inches long.

FIG. 10.—Algal buildups in middle unit of Egan section. Light-colored massive lenses in upper part of photograph are composed of coalescing hemispherical algal stromatolites. Thinly bedded lime grainstone layers lap over and interfinger with algal buildups. Lime grainstone is composed of peloids and skeletal grains of trilobites, gastropods, echinoderms, and the possible alga, *Nuia*. Bar scale equals approximately 1 m. Whipple Cave Formation.

buried the algal mats. This is suggested by normally graded grainstone and wackestone in beds as much as 20 cm thick that occasionally are draped over the stromatolite heads without any evidence of upward algal growth through the bed (Fig. 17).

The upper unit consists of the lower 50 m of the House Limestone and contains abundant beds

FIG. 11.—Part of single isolated stromatolite about 30 cm high. Surrounding sediments are lime packstone and grainstone. White circle is 2 cm in diameter. Middle unit Egan section, Whipple Cave Formation.

FIG. 13.—Lime grainstone interbedded with algal stromatolites from middle unit of Egan Range section. Sand-size grains are composed of *Nuia* (alga ?), echinodermal plates, and peloids. Large dark-colored pebbles are lime mudstone. Whipple Cave Formation.

of flat-pebble breccia and conglomerate that alternate with dolomitic lime mudstone and fenestral limestone (Fig. 18). Scattered throughout the unit is light-brown and pink chert in lenses and complexly interconnected nodules. Breccia and conglomerate commonly fill shallow channels (Fig. 19) or form thin sheets having flat-parallel lower and upper contacts (Fig. 20). Clasts in the breccia are lime mudstone or dolomitized lime mudstone fragments (Fig. 21). Skeletal debris and burrows are less common in the upper unit.

The lower House Limestone of the central Egan Range is interpreted to represent deposition on tidal flats. This environment commonly contains fenestrae, flat pebbles, channels, laminated lime mudstone, and early-dolomitized lime mudstone (Roehl, 1967; Shinn, 1968; Shinn and others, 1969; Bathurst, 1971; Cook, 1972).

In summary, the Egan section was deposited in environments that ranged from shallow subtidal to supratidal. From the base of the section upward, the sedimentation history is one of progressively shallowing water that records a westerly seaward progradation of the carbonate platform.

FIG. 12.—Same locality as Fig. 10. Two laterally linked algal heads in bedding plane. Pick is lying on bedding plane. Tops of algal heads have a few centimeters of relief as exhibited by draping of clastic limestone debris. Middle unit Egan section, Whipple Cave Formation.

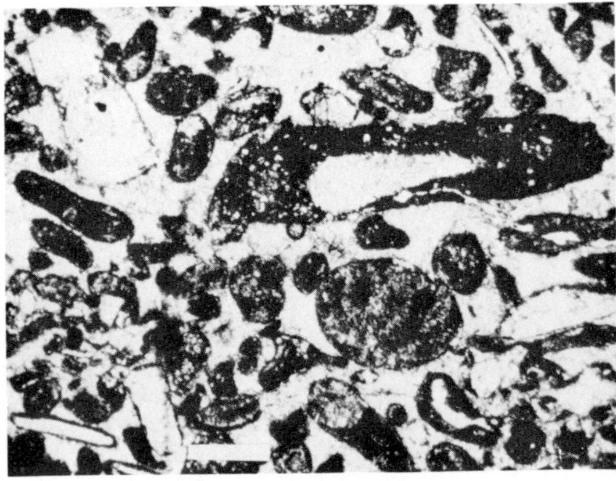

FIG. 14.—Photomicrograph of lime grainstone shown in Fig. 13. Grains are *Nuia* (alga ?) and coated echinodermal plates. Middle unit, Egan section, Whipple Cave Formation. Bar scale equals 0.5 mm.

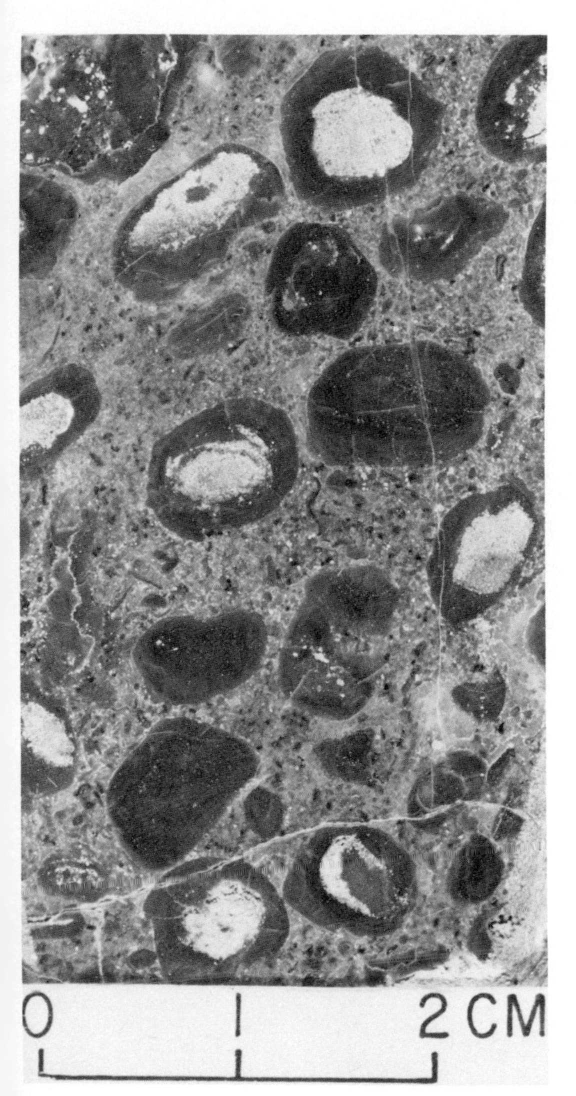

FIG. 15.—Polished sample of oncolites set within a lime packstone to grainstone matrix. Middle unit of Egan section, Whipple Cave Formation.

DEEPER WATER ENVIRONMENTS—HOT CREEK RANGE

The Hales Limestone in the Hot Creek Range contains vastly different rocks and faunas from those in the coeval Whipple Cave and lower House formations 170 km to the east (Figs. 2, 22, 23). Not only is the Hales Limestone different in these respects but net sediment accumulation during the Upper Cambrian part of the Hales was only about one-quarter of that of the Whipple Cave Limestone (Figs. 2, 23). The lower 200 m of the Hales Limestone consists of dark fine-grained limestone, much of which shows evidence of soft sediment deformation. This limestone is interbedded with abundant allochthonous limestone conglomerate and calcarenite.

Fine-Grained Limestone

About 75 percent of the rock in the Hot Creek section is dark grey to black limestone. This limestone is mostly lime mudstone and lime wackestone (Fig. 24) with lesser amounts of interbedded argillaceous lime mudstone (Fig. 25). It is further characterized by its dark color, thin bedding to millimeter-thick laminae, and scarcity of burrowing (Fig. 26). This sediment exhibits bedding contacts that range from planar and nearly parallel to more undulatory and discontinuous. Constituents in the sediment are composed largely of fine-grained carbonate (and spar) and minor amounts of pyrite and terrigenous siliciclastics of clay and silt size. The associated fauna consists mainly of abundant sponge spicules and trilobites. These trilobites are unknown from either the shallow-shelf environments in the Egan Range or elsewhere in the North American Faunal Province (Taylor, 1976).

Fine-grained limestone of the Hales is interpreted to have formed in a deep water setting. This environment is suggested by its dark color, planar and parallel bedding, millimeter-thick laminae, scarcity of burrowing, lack of wave produced sedimentary structures, a trilobite fauna unknown from any shallow-water environments in the North American Faunal Province and its regional setting seaward of coeval shoal-water limestone. Additionally, about one-quarter of the section is composed of interbedded coarse-grained allochthonous carbonate and soft sediment deformation structures.

Allochthonous Limestone

Five types of allochthonous deposits have been recognized interbedded with the dark thinly-bedded limestone: (1) deformed thinly-bedded limestone that appears to have moved only a short distance; (2) deformed thinly-bedded limestone that laterally along strike takes on a distinct clastic texture; (3) limestone conglomerate in sheet and channel form whose clasts consist wholly of the deep water dark, fine-grained limestone; (4) limestone conglomerate in sheet and channel-form composed of both deep water dark, fine-grained limestone and grain-supported clasts like that found in the *in situ* shoal-water carbonate of the Egan section, and (5) thin, well-sorted calcarenite whose main constituents are grains of the possible alga *Nuia* like those found in the *in situ* shoalwater carbonate of the Egan section.

The above first four types of allochthonous carbonate can be grouped into a sequence representing progressive loss of cohesion during movement (resedimentation) with variants for admixing

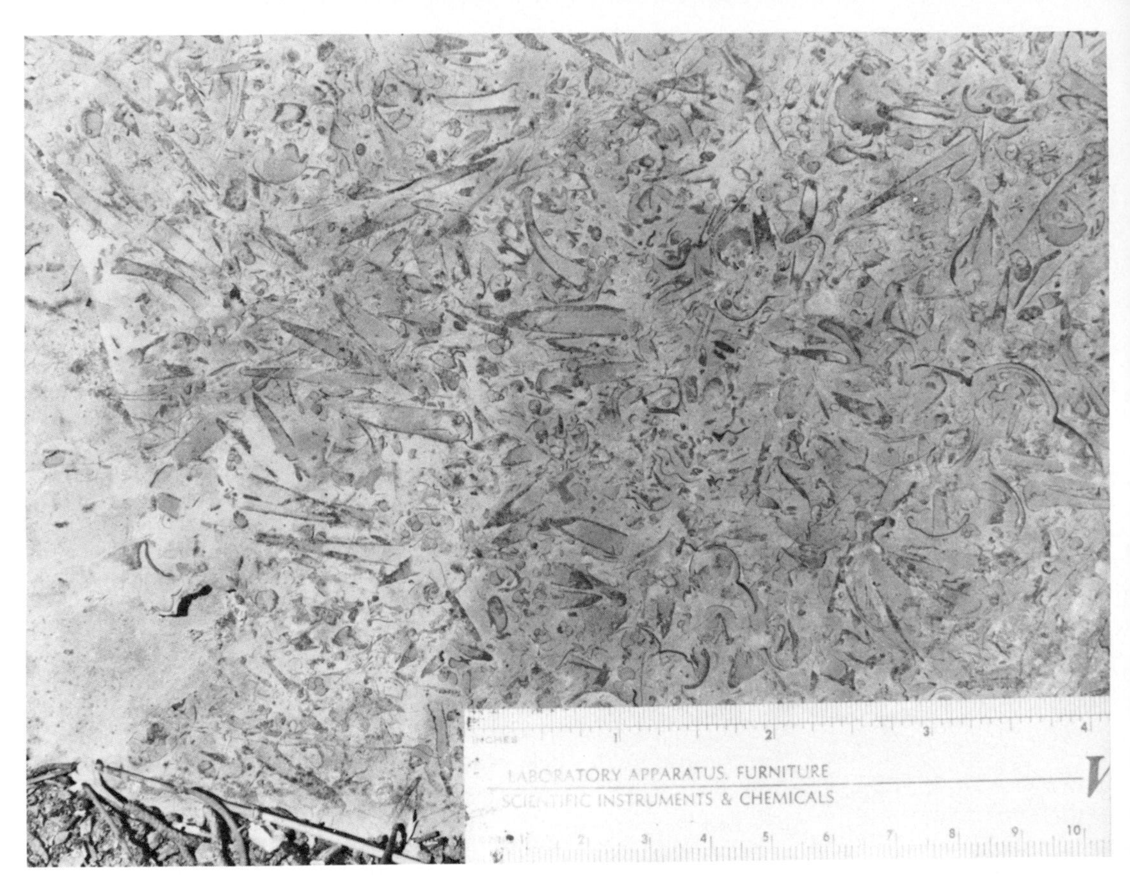

FIG. 16.—Bedding-plane view of lime packstone-grainstone that is interbeded with algal buildups in middle unit of Egan section. Whipple Cave Formation. Bed contains abundant hyolithoids, gastropods, and trilobite skeletal debris. Scale in inches and centimeters.

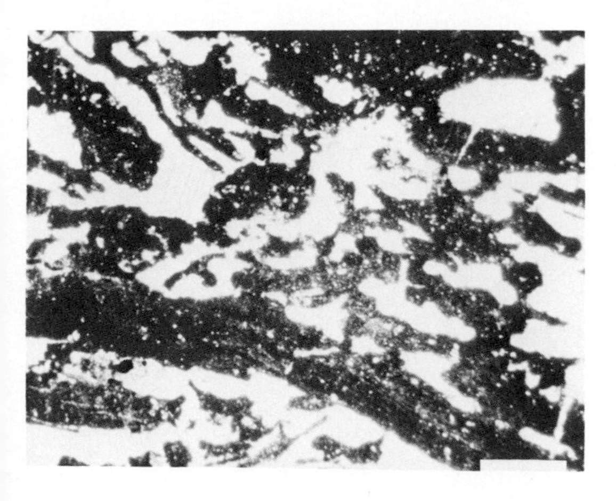

FIG. 18.—Photomicrograph of fenestral fabric in limestone. Upper unit of Egan section, House Limestone. Bar scale equals 0.5 mm.

of shallow-water and deeper water debris.

Criteria for the recognition of coarse-grained allochthonous carbonate deposits in deeper water environments have been summarized by Cook and others (1972). The mechanics of flow and deposition of submarine sediment gravity flow phenomena has recently been discussed by Cook and others (1972), Middleton and Hampton (1973), Carter (1975) and Hampton (1975). Massflow deposits in the Hales Limestone are varied in terms of their primary sedimentary structures, clast size and shape, sorting, and contact relation with *in situ* deep water deposits. They probably represent sediment moved under various conditions ranging from slumping to viscous debris flow to dilute turbidity flow. Detailed analysis of their possible transport and depositional processes is under study.

Slumps.—A large proportion of the dark, laminated *in situ* lime mudstone and wackestone show different degrees and scales of soft-sediment deformation inferred to represent slumping and sliding. In some cases only thin stratigraphic intervals are slumped (Fig. 27); other slumps, however, occupy stratigraphic intervals up to 8 m thick (Fig. 28). In all cases the beds are highly contorted but the shear strength of the mass was not exceeded. Apparently they did not become mobile enough to breakup into individual clasts, entrain water and mud, and actually flow.

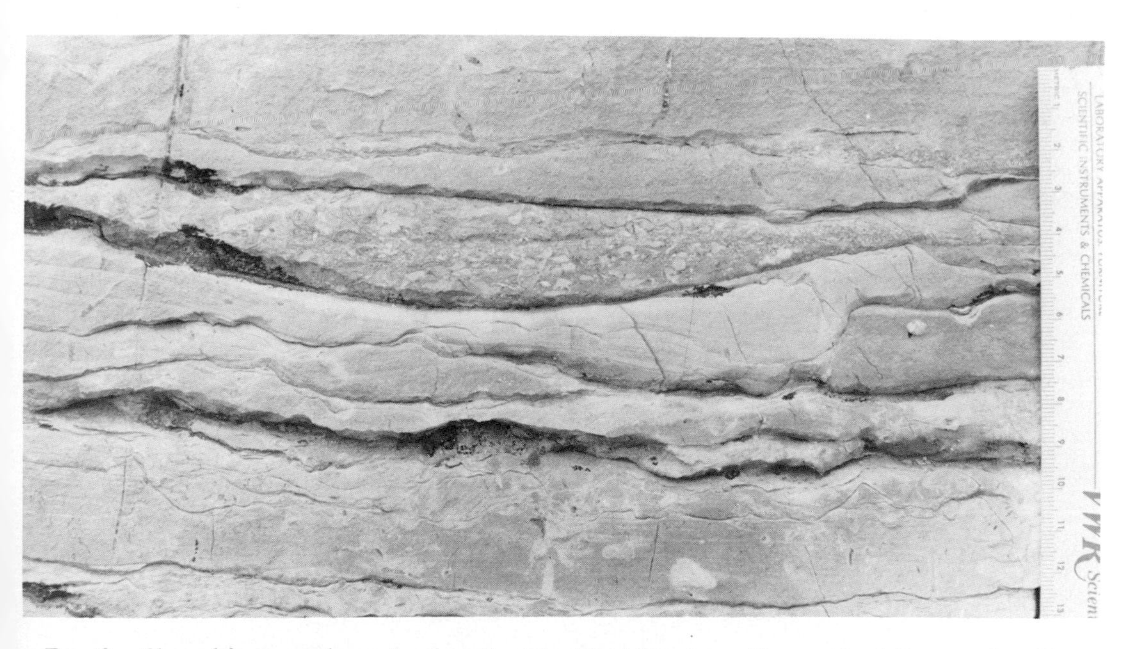

FIG. 19.—Channel-form conglomerate of partly dolomitized limestone. Upper unit of Egan section, House Limestone. Scale in centimeters.

FIG. 17.—Algal stromatolite (convex-up mound at right center of photograph) with about 20 cm of relief that is completely drapped by a normally graded clastic limestone bed. Middle unit of Egan section at same locality as Fig. 10, Whipple Cave Formation. Tape is 19 inches long.

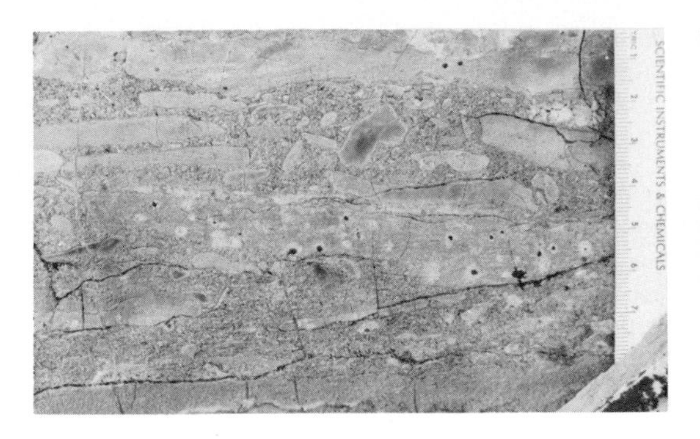

FIG. 20.—Partly dolomitized flat-pebble conglomerate with lime grainstone matrix. Upper unit of Egan section, House Limestone. Scale in centimeters.

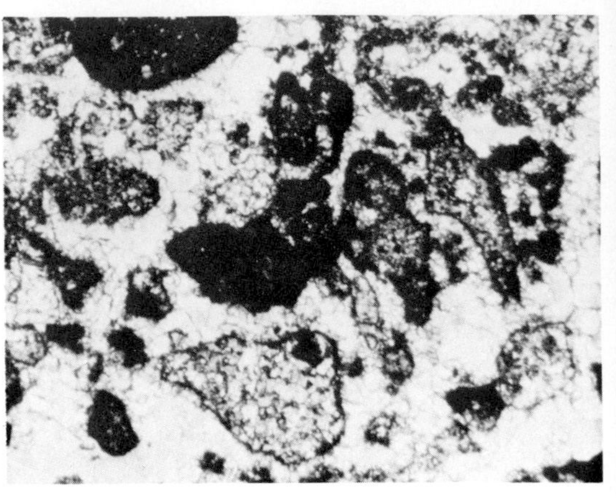

FIG. 21.—Photomicrograph of bed in Fig. 20. Rock is a mixture of dolomitic limestone clasts (light colored with dark rims) and limestone clasts (dark colored). Bar scale equals 0.5 mm; House Limestone.

Slumps transformed into flows.—In contrast to the preceeding examples the shear strength of some slumps was exceeded and the slump became mobile enough to transform into highly viscous sediment gravity flows. Figure 29 is a unit about 2 m thick that exhibits soft sediment slumping and overfolding. On closer inspection (Fig. 30) the core of one of the overfolded areas is seen to be in the beginning stages of breaking up into individual clasts. As this same 2 m thick slump is traced laterally along strike, it gradually takes on a clastic texture until it is composed of individual clasts (Fig. 31). Several flow mechanisms may have been involved from debris flow to highly viscous turbidity flow. Figure 31 shows part of this flow that exhibits tabular rounded clasts that have a suggestion of imbrication and normal size grading. The regional paleoslope was from right to left in Figure 31, and thus this inclination of the clasts appears to be a normal type of imbrication rather than the down-current inclination of clasts that Hubert and others (this volume) document convincingly. In figure 31 both the clasts and matrix between the clasts are composed of deeper water lime mudstone and wackestone. The trilobite fauna in the mud matrix of these deposits is unlike that found in the Egan Range or elsewhere in the North American Province (Taylor, 1976) but is like that which occurs in the nondeformed *in situ* deep water limestone.

Conglomerate of deeper water material.—Much

FIG. 22.—Hot Creek Range section of Hales Limestone on north side of Tybo Canyon, near Warm Springs, Nevada. Base and top of measured section (Fig. 23) bracketed by arrows. Resistant beds are allochthonous sediment gravity-flow and slump deposits. Recessive beds are *in situ* hemipelagic limestone.

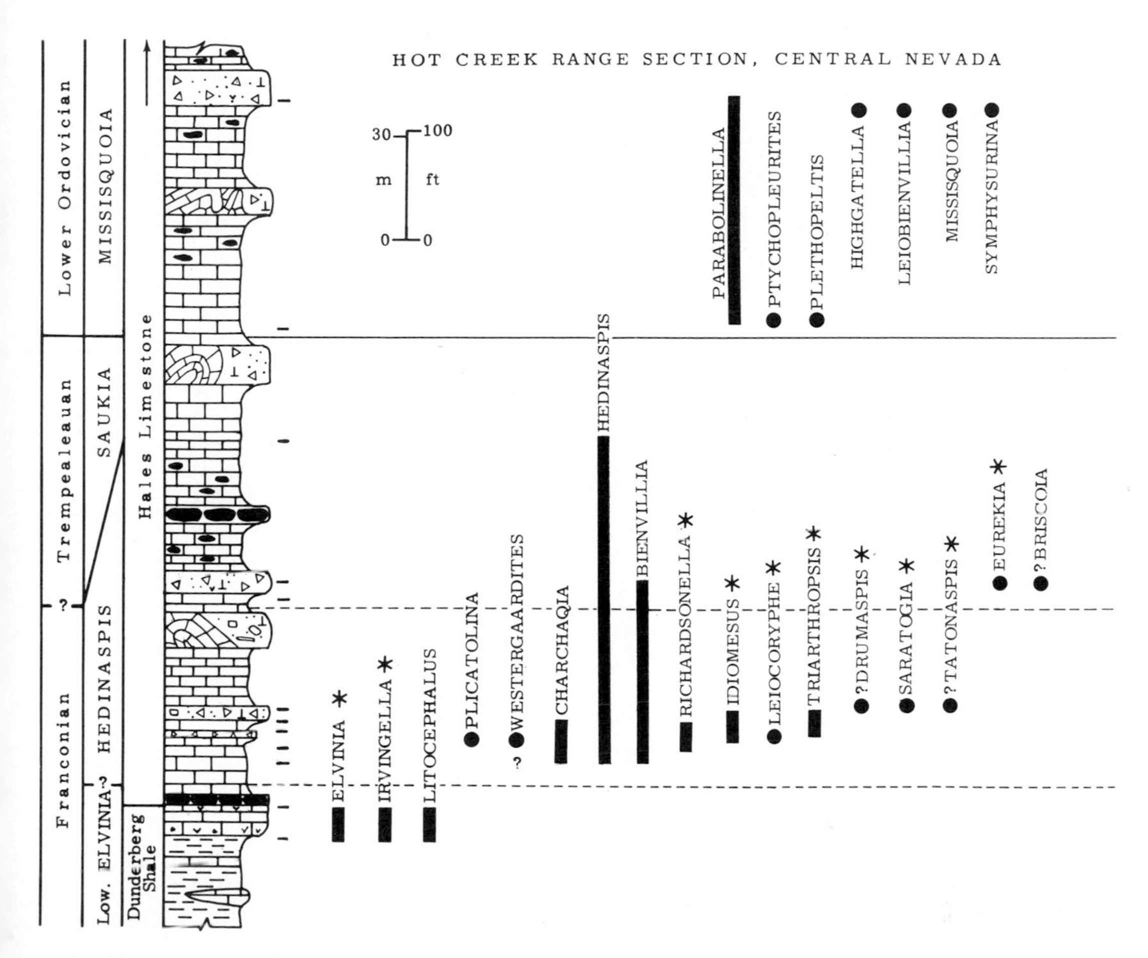

FIG. 23.—Range of polymeroid trilobite genera from the lower part of the Hales Limestone in the Hot Creek Range, Nevada. More detailed taxonomic data are given in Taylor (1976). Explanation of lithologic symbols given in Fig. 2. Dashes on right side of column show position of fossil samples.

FIG. 24.—*In situ* slope deposits of black lime mudstone and wackestone of the Hales Limestone, Hot Creek section. Bedding planes are parallel and fairly planar; individual beds can be traced for a few tens of meters.

FIG. 25.—*In situ* slope deposits of black argillaceous lime mudstone and wackestone of the Hales Limestone, Hot Creek section. This limestone has paper-thin laminations and is fissile when weathered.

FIG. 26.—Oblique view of polished slab from Hot Creek section, showing (left of bar scale) articulated specimen of *Hedinaspis regalis* (Troedsson, 1937) and fabric of entombing deeper water sediment. Note finely laminated parallel bedding and pervasive sponge spicules (small light-colored blebs) in black mud matrix. Larger white-appearing spherules are authigenic pyrite. Bar scale equals 2 cm, Hales Limestone.

FIG. 27.—Small slump overfold of hemipelagic lime mudstone and wackestone 15 cm thick, Hales Limestone. Bar shows upper and lower limits of slump.

FIG. 28.—Part of a large slump overfold of hemipelagic lime mudstone and wackestone about 8 m thick. Upper and lower limits of slump not in photograph. Hales Limestone.

of the conglomerate that contains lithic and faunal debris of only a deeper water type cannot be traced to any obvious slump or slide. These deposits commonly occur in sheets and channels a few tens of centimeters thick to 10 m thick. Figures 32, 33, 34 and 35 are from different parts of a 10 m thick massively bedded debris deposit composed wholly of deeper water clasts and matrix. This bed can be traced laterally for at least several hundred meters where it thins to 1 m (Fig. 33). The clasts are tabular shaped (0.5 to 2 cm by 1 to 20 cm) lime mudstone and wackestone. They are commonly rounded at their ends with some clasts showing smoothly bent terminations. Smoothly deformed ends probably reflect the semiconsolidated nature of the clasts just prior to disruption and redeposition (Fig. 34). Maximum clasts size observed in this 10 m thick debris bed was 1 by 2 m across (Fig. 35). This large clast is near the top of the bed in a finer grained muddy conglomerate.

Clasts in this debris are poorly sorted, exhibit random to subparallel orientation with respect to bedding and are usually devoid of any obvious normal or reverse grading (Figs. 32, 33). Locally the upper few centimeters may show crude normal grading.

Contacts with the enclosing host slope facies commonly are poorly exposed. Where these contacts are seen, the upper contact is slightly hummocky (Fig. 33) and the lower contact is channelized or parallel to the underlying *in situ* beds.

Even though these deposits cannot be traced to slumps or slump scars it is reasonable to infer that they originated as slumps and slides. Their textures, clast and matrix types and close resemblance to debris beds traceable to slumps make this a plausible interpretation. Deposits of this

FIG. 29.—Thin-bedded hemipelagic lime mudstone and wackestone showing soft-sediment slumping in Hales Limestone. Semiconsolidated beds were broken into individual intraclasts and later lithified. Scale is 15 inches long.

type that are massively bedded and are capable of supporting large blocks above the base of the flow suggest transport by some type of viscous debris flow mechanism. Flows of this type probably behave as non-Newtonian fluids and have significant yield strength (Hampton, 1975). Some deposits a meter or less in thickness are normally graded and have cross laminated tops. This type of debris may have been transported by turbulent flow.

Conglomerate of shoal water and deeper water material.—Sheet and channel-form carbonate debris containing allochthonous fragments ranging from small pebble size to 25 cm form distinct beds 0.3 to 1.5 m thick (Fig. 36). Some of the beds are entirely massive and show no textural changes from base to top. More commonly, however, the clasts are normally graded (Fig. 36). Tabular clasts are generally oriented subparallel to bedding, (Fig. 37), although some are imbricated in an up-slope direction near the top of the bed (Fig. 38). More rarely the beds are capped by laminated and cross-laminated sand-sized carbonate particles (Figs. 38, 39). The clasts are a mixture of both shoal and deeper water origin. Shoal-water derived clasts consist of peloid packstone and grainstone, algal-echinodermal grainstone and other lithologies similar to those in the Egan Range. Deeper water clasts are dark spicular lime mudstone and wackestone. Interclast matrix (Fig. 40) contains skeletal debris consisting of *Nuia* grains, echinodermal plates, and other skeletal material similar to that found in the coeval shoal-water rocks of the Egan Range. Additionally in the matrix these shoal-water grains are admixed with deeper water spicular lime mud and shoal and deeper water trilobite debris.

Many of these beds occur in broad, low-relief channels about 10 to 20 cm deep and 10 to 30 m wide. The bases of the channels exhibit local scouring of a few centimeters of the underlying *in situ* lime mudstone and wackestone. However, most of the debris beds of a mixed shoal- and deeper water origin occur in thin sheets whose

FIG. 30.—Close-up of central part of Fig. 29 showing tight overfold axis with semiconsolidated lime mudstone and wackestone beds starting to break up into individual clasts. Hales Limestone section. Bar scale equals 5 cm.

geometry has not yet been determined.

A rare type of deposit consists of 1 to 10 cm thick beds of almost entirely trilobite skeletal debris (Fig. 41). The trilobites have affinities with shoal-water faunas typical of the North American Faunal Province, and were redeposited from shelf sites. These beds are laminated, have scoured bases and some have convoluted laminae at their top.

Beds in the section that exhibit a graded base capped by laminated and/or cross-laminated intervals (Bouma sequences ABC, AB or, AC; Bouma, 1962) were probably transported by fairly viscous turbulent flow. Massively bedded debris may represent viscous turbulent or debris flow or a combination of flow mechanisms.

Calcarenite.—These form a type of redeposited shoal-water limestone that is easily overlooked in this section, but that may be volumetrically important. These calcarenites have grainstone fabrics, occur in layers less than 2 cm thick, are laminated, have small foreset bedding and ripple forms, and have sharp lower and upper contacts with the enclosing dark lime mudstone and wackestone (Figs. 42, 43). These grainstones consist of shoal-water derived grains of *Nuia*, echinoderm particles, and quartz from an unknown source. The grains are well-sorted and have virtually no interparticle matrix; inter-particle void space is filled with sparry calcite (Fig. 44).

This calcarenite does not appear to be the product of muddy turbidity currents, but rather

Fig. 31.—Same stratigraphic interval as Figs. 29 and 30. Here the shear strength of the slump mass was exceeded, and became mobile enough to be transformed into a highly viscous sediment gravity flow with a distinct clastic texture. Flow shows tabular clasts oriented subparallel to bedding. A slight imbrication and normal grading of clasts is seen in the right side of photograph. Flow westerly downslope from right to left. Scale is in inches.

the near perfect hydraulic sorting, and sharp lower and upper contacts suggest a different origin. The calcarenites may represent the product of winnowing of previously resedimented material by bottom-hugging oceanic currents. Thus they may be contourites, the product of so-called contour currents. The principle characteristics of contourites versus turbidites has recently been summarized by Bouma and Hollister (1973, p. 95). Alternatively, these bed forms somewhat resemble modern features off Oregon that are interpreted to represent levee deposits (Hans C. Nelson, personal commun.). These latter type of beds form when turbidity deposits fill their channels and spill over into interchannel areas forming levees. Sufficient paleocurrent data has not been collected on these deposits to help discriminate between contourites and turbidite levee deposits. Paleocurrent directions for contourites could be expected to be normal to the paleoslope and probably unidirectional whereas levee deposits should probably be unidirectional down slope or bimodal on a larger scale if strong currents form away from the channel. If these calcarenites are contourites or levee deposits, the processes that formed them could have been more active in this environment that we are presently aware. Their presence in outcrop is largely made evident by the fact they are coarser grained than the enclosing lime mudstone. Whether or not some or much of the finer textured dark limestone is the product of contour currents and/or turbidity currents is not known at this time.

DEPOSITIONAL MODEL

Major facies changes occur between the central Egan Range and the Hot Creek Range. During the Late Cambrian about 540 m of limestone accumulated in shoal water environments whereas only about 130 m of sediment formed in deeper water settings. Thus, sediment thickness decreases oceanward by a factor of at least four.

Fig. 32.—Massively bedded debris about 10 m thick in Hales Limestone. The bed is composed of slope-derived lime mudstone and wackestone clasts set within a pervasive dark lime mud matrix. Clasts are rounded and virtually all are tabular, 0.5 to 2 cm thick and 1 to 20 cm long. Clasts tend to be randomly oriented or subparallel to bedding with no obvious size grading from the base to top. Arrows point to top and base of beds.

FIG. 33.—Same bed as in Fig. 32. Here bed is about 1 m thick, massively bedded, clasts are tabular and poorly sorted. The upper contact appears to be slightly hummocky with up to 10 cm of local relief. Arrow points to upper contact. Basal contact is poorly exposed at hammer head. Hales Limestone.

At least 25 percent of the deeper water section consists of slump masses and sediment gravity-flow deposits. When it is considered that a large part of this allochthonous material was derived from shoal water environments the ratio between coeval shoal water carbonate and *in situ* deeper water carbonate is even greater. Dark, thin-bedded, laminated, spicule-rich, lime mudrock prevails oceanward, whereas light-colored grain-supported and stromatolitic fabrics dominate in the shoalward rocks.

We suggest a model that includes shoal-water carbonate environments in eastern Nevada representing the eastern assemblage of Roberts and others (1958), and slope environments in central Nevada representing the transitional assemblage (Fig. 45). Shallow-subtidal to peritidal environments in the central Egan Range seem well established. Sediment deposited under these conditions is well documented throughout the geologic column from many parts of the world. An excellent comparative analysis between modern and ancient shoal-water carbonate environments has been given by Roehl (1967).

Interpretation of a slope environment of deposition for the Hales Limestone is based on integration of the regional geologic framework (Roberts and others, 1958; Kellogg, 1963; Burchfiel and Davis, 1972; Stewart, 1972; Churkin, 1974; Stewart and Poole, 1974), comparison with other ancient deeper water environments (Garrison and Fischer, 1969; Thomson and Thomasson, 1969; Wilson, 1969; Cook, 1972; Cook and others, 1972; Mutti and Ricci-Lucchi, 1972; Walker and Mutti, 1973; Reinhardt, 1974), and comparison with modern continental margin transitions (Uchupi and Emery, 1963; Worzel, 1968; Stanley and Silverberg, 1969; Rona, 1970; Lewis, 1971; Stanley and Unrug, 1972).

Features of the Hales Limestone that suggest

Fig. 34.—Close-up of debris at Fig. 32 locality, Hales Limestone. Dark gray laminated lime mudstone clast about 1 by 6 cm across shows probable soft-sediment deformation when ripped apart (tip of pencil).

a slope rather than subtidal open shelf or submarine-fan environment include abundant slump masses involving stratigraphic thicknesses as great as 10 m, numerous viscous sediment gravity-flow deposits as much as 10 m thick containing clasts several meters across, occurring in relatively narrow channels, and an extremely attenuated section when compared with the coeval shoal-water section in the Egan Range. Thus, if the shoal-water derived deposits are discounted, sediment accumulation rate for *in situ* hemipelagic sediments would be less than one-fifth that of the Egan Range section.

Mutti and Ricci-Lucchi (1972) and Walker and Mutti (1973) summarized characteristics of facies changes between slope and submarine-fan environments in ancient rocks. Stanley and Unrug (1972) and Lewis (1971) present lithologic and geophysical data for the nature of modern continental slope and base-of-slope environments. The instability of the slope is reflected by the abundance of large amounts of slumped or displaced clastic beds. Seismic reflection profiles and side-scan sonar records show that slumps and

Fig. 35.—Lime mudstone to wackestone clast about 1 by 2 m in same debris bed shown in Fig. 32, Hales Limestone. Arrows point to upper and lower limits of clast. Clast is oriented subparallel to bedding and is near the top of this 10 m thick bed. Clast is embedded in a muddy conglomerate matrix.

FIG. 36.—Bed approximately 50 cm thick in Hales Limestone. The bed shows normal grading of clasts and contains a mixture of clasts derived from both shoal- and deeper water environments. Top of bed at zero footage on tape; base of bed at 21 inch mark on tape. Scale in inches.

slides are considerably more abundant on modern slopes and base of slopes than on submarine fans. Associated with slump deposits are relatively narrow channels filled with massively-bedded pebble to boulder deposits with single beds ranging from 1 to 10 m thick. Interbedded with mass movement deposits is abundant fine-grained, laminated pelagic and hemipelagic sediment. In contrast, on the inner and middle parts of submarine fans slumping is commonly a minor feature. Channels can meander more easily on a fan than on a slope, and thus fan channel deposits tend to be wider, up to several kilometers. The proportion of turbidites and other redeposited debris is greater in the fan environment than on slopes. Further, fans can attain considerably greater thickness of sediment than slopes because of the relatively larger proportion of redeposited to *in situ* beds. The characteristics of the lower Hales Limestone appear to be in general accord with a slope environment of deposition. This interpretation is also in accord with Stewart and Poole (1974), p. 42) who conclude that the Upper Cambrian Harmony Formation oceanward from the Hales Limestone may represent continental rise deposits.

Slope sedimentation in the Hot Creek Range involved an interplay between depositional and erosional processes. The dark laminated spicule-rich mudrocks largely represent hemipelagic sedimentation by suspension and other normal pelagic processes. Submarine gravity movements, although not of a large scale, were a major process in the development of the slope facies. At least one quarter of the section is composed of sediment gravity-flow and slumped sediment. Transportation by slumping probably ranged from relatively short distances for slump masses that retain their stratigraphic identity to greater distances where shear strengths were exceeded and slump masses were transformed into highly viscous debris and turbidity flows. These types of slumps and flows originated and were deposited in a deeper water environment. Other types of submarine sediment gravity flows, characterized by a mixture of shoal-water and deeper water constituents, had a more complex origin. This kind of flow apparently originated in a shoal-water setting and incorporated a variety of material during transport oceanward.

Although a significant change in depositional regime took place between the near-sea-level environments of the Egan section and the slope environments of the Hot Creek section, water depth of the slope environment is not known. However, depth can be estimated if certain assumptions are made. Palinspastic reconstructions by Steward and Poole (1974) suggest that the Egan and Hot Creek sections are in original spatial relationship to each other. Modern continental shelves are inclined seaward at an average angle of approximately 0° 07′ (Shepard, 1973, p. 277). A uniform inclination of this magnitude between the Egan and Hot Creek sections would yield a depth of approximately 350 m. This calculated depth assumes that the Hot Creek section was fortuitously located at the shelf-slope break and the Egan section was near the seaward margin of shoal-water deposition. Thus, 350 m is probably a minimum estimate, and actual depth may have been greater. Perhaps a more realistic estimate of depth can be calculated by assuming a shelf-slope break midway between the Egan and Hot Creek sections. The inclination of the continental slope may not necessarily have been great. Slumping on modern continental slopes inclined at 1° to 4° is probable (Lewis, 1971), and transport

FIG. 37.—Central part of debris bed 1 m thick in Hales Limestone. Clasts are a mixture of shoal- and deeper water types. Note strong preferred orientation of clasts parallel to bedding.

of thick boulder-sized debris can be initiated and sustained across low-angle lime mud bottoms (Cook and others, 1972). Therefore, if an angle of 0° 07′ is assumed for the shelf part, and an angle of 1° (a conservative estimate) for the slope part, then a calculated depth would be approximately 1600 m. In any case, water depth over the slope was very likely greater than that necessary for development of a permanent thermocline, a conclusion supported independently by faunal evidence (Cook and Taylor, 1975; Taylor, 1977).

BIOFACIES

Two kinds of trilobite assemblages can be recognized in the Hales Limestone (Taylor and Cook, 1976; Taylor, 1976). One, associated with some of the shoal-water derived sediment gravity-flow deposits, consists of disarticulated, broken and abraded fossils (Fig. 46a). These assemblages are most similar to autochthonous assemblages in coeval shallow-water rocks of the Whipple Cave Formation and from other sites of shelf and shelf-marginal deposition in North America (Taylor and Halley, 1974). A second type of trilobite occurrence consists of articulated and fragile fossils that are found parallel to bedding in dark lime mudstone (Figs. 26, 46b). These assemblages are interpreted to represent parauthochthonous associations indigenous to the deeper water slope environments. This fauna occurs in the Hales and in other deeper water settings in Asia (Taylor, 1976). Absence of the fauna from North American shallow shelf sites suggests that it was adapted for living in the deeper environments and lacked ecologic tolerance for shelf conditions. The fauna is assigned to the Chiangnan Faunal Province (Taylor, 1976). Locations of the North American and Asian faunal provinces are shown in Fig. 47. Data on Asian faunas and environments and their relation to the North American Faunal Province are documented and discussed more fully elsewhere (Taylor, 1976; Taylor and Cook, 1976).

Distribution patterns of living marine isopod crustaceans provide an analog for interpretation of trilobite distribution patterns (Taylor, 1977). The world pattern of isopod distribution shows that deep-sea faunas, and taxonomically most similar cold-shelf faunas, are geographically widely distributed in the polar regions and below the permanent thermocline in low latitudes. Low

FIG. 38.—Upper part of a 1.5 m thick, normally graded sediment gravity flow deposit from Hales Limestone. Shows rounded and tabular clasts of dark slope derived spicular mudstone and light-colored shoal-water grainstone. The clasts are oriented subparallel at the base and middle of the bed and are imbricated near the top. A capping bed is composed of planar- to cross-laminated *Nuia* (alga?) grainstone and spicular mudstone. Flow westerly downslope from right to left. Zero footage on tape at top of bed. Scale in inches.

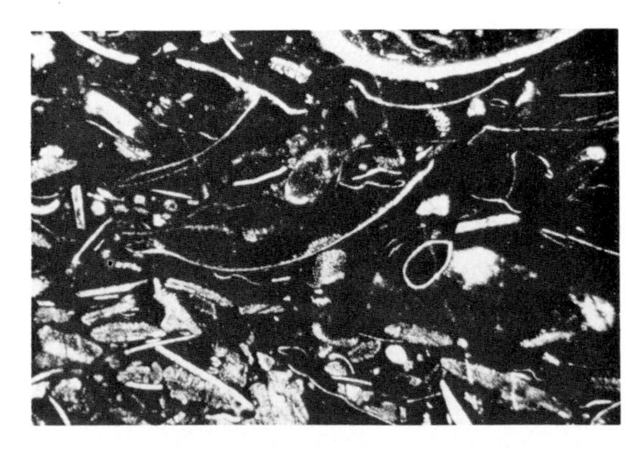

FIG. 40.—Photomicrograph showing matrix of bed in Fig. 36. The matrix contains abundant shoal-water derived skeletal grains of *Nuia* (alga?), echinoderms, trilobites, and dark lime mud. Bar scale equals 0.5 mm.

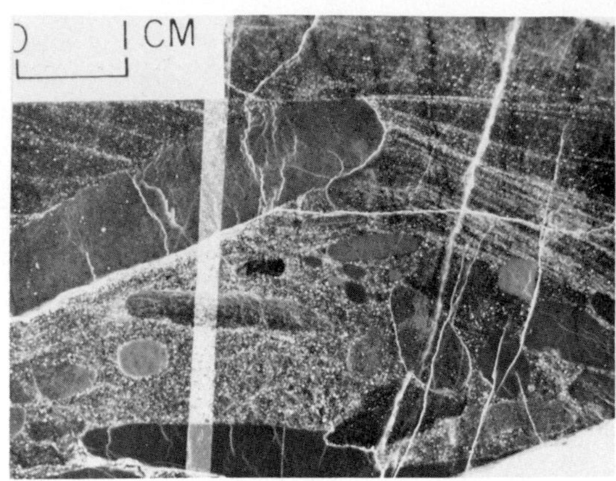

FIG. 39.—Polished slab from top few centimeters of bed in Fig. 38. Hales Limestone. Tabular light-colored clast inclined to the left is a shoal-water derived peloid, *Nuia* (alga?) packstone to grainstone. Dark-colored clasts are deeper water spicular lime mudstone and wackestone. Matrix is a mixture of shoal water *Nuia* (alga?) and echinodermal fragments and deeper water sponge spicules and lime mud. Top few centimeters are laminated and cross-laminated.

FIG. 41.—Polished slab of sediment gravity flow deposit from lower part of Hales Limestone, showing skeletal debris dominated by trilobites derived from shoal or higher slope habitats. Arrows point to upper and lower contacts with enclosing *in situ* slope rocks.

FIG. 42.—Thin, laminated, well-sorted lime grainstone (dark layer at base of rock) that shows foreset bedding and ripple forms. Grains are *Nuia* (alga?), echinoderm, and quartz. Lime mudstone weathers light color. Possible contourite, Hales Limestone.

FIG. 43.—Outcrop of same grainstone layer shown in Fig. 42. Arrows point to three ripple forms that have about 9 cm periods and 0.5 cm amplitudes. Both top and base have fairly sharp contacts with enclosing lime mudstone. Possible contourite, Hales Limestone. Scale at left in centimeters.

and intermediate latitude shelves tend to have more endemic faunas that show low degrees of faunal resemblance with adjacent slope faunas below the thermocline (Kussakin, 1973; Menzies, George, and Rowe, 1973). These distributional patterns of living isopod faunas are important to biofacies analyses of trilobites because they demonstrate that (1) continental margins are areas of rapid lateral (down depositional dip) biofacies changes, (2) geographically widespread faunas occur associated with the cold isothermal regions below the thermocline on low and intermediate latitude continental slopes, (3) benthic faunas on low and intermediate latitude areas above the thermocline generally tend to be more endemic, and (4) the widespread slope and endemic shelf benthic biofacies may lie adjacent to one another without intervening great distances or topographic highs.

Analogous relationships between trilobite and isopod biofacies patterns on continental margins are shown by a comparison of trilobite faunas along a shelf-to-basin profile in the western United States and isopod faunas across the Atlantic

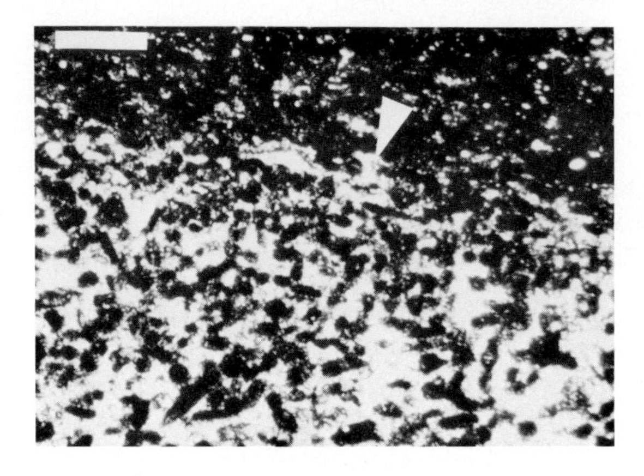

FIG. 44.—Photomicrograph of calcarenite in Fig. 42. Grainstone texture with shoal-water derived grains of *Nuia* (alga?), echinodermal plates and quartz. Interparticle void space filled with sparry calcite. Arrow points to fairly sharp contact with overlying *in-situ* hemipelagic lime mudstone and wackestone. Possible contourite, Hales Limestone. Bar scale equals 0.5 mm.

continental margin from North Carolina to Bermuda (Fig. 48). Orientation of the North Carolina-Bermuda transect has been reversed in Fig. 48 to facilitate visual comparison with the east-to-west trilobite profile in the upper graph. The analysis is based on a twofold classification of trilobite and isopod genera that occur along the respective transects. The solid line represents the geographic distribution of taxa; high values indicate greater proportions of geographically widespread taxa, and lower values indicate that endemic genera predominate. The dotted line represents the proportions of taxa that are unrestricted in habitat distribution. High values denote faunas that are relatively eurytopic, whereas low values denote faunas that are relatively stenotopic. In both trilobite and isopod analyses, faunas associated with slope and deeper environments are typically stentopic but geographically widespread, whereas shelf faunas are relatively more eurytopic but geographically more restricted in distribution. The point where the two curves

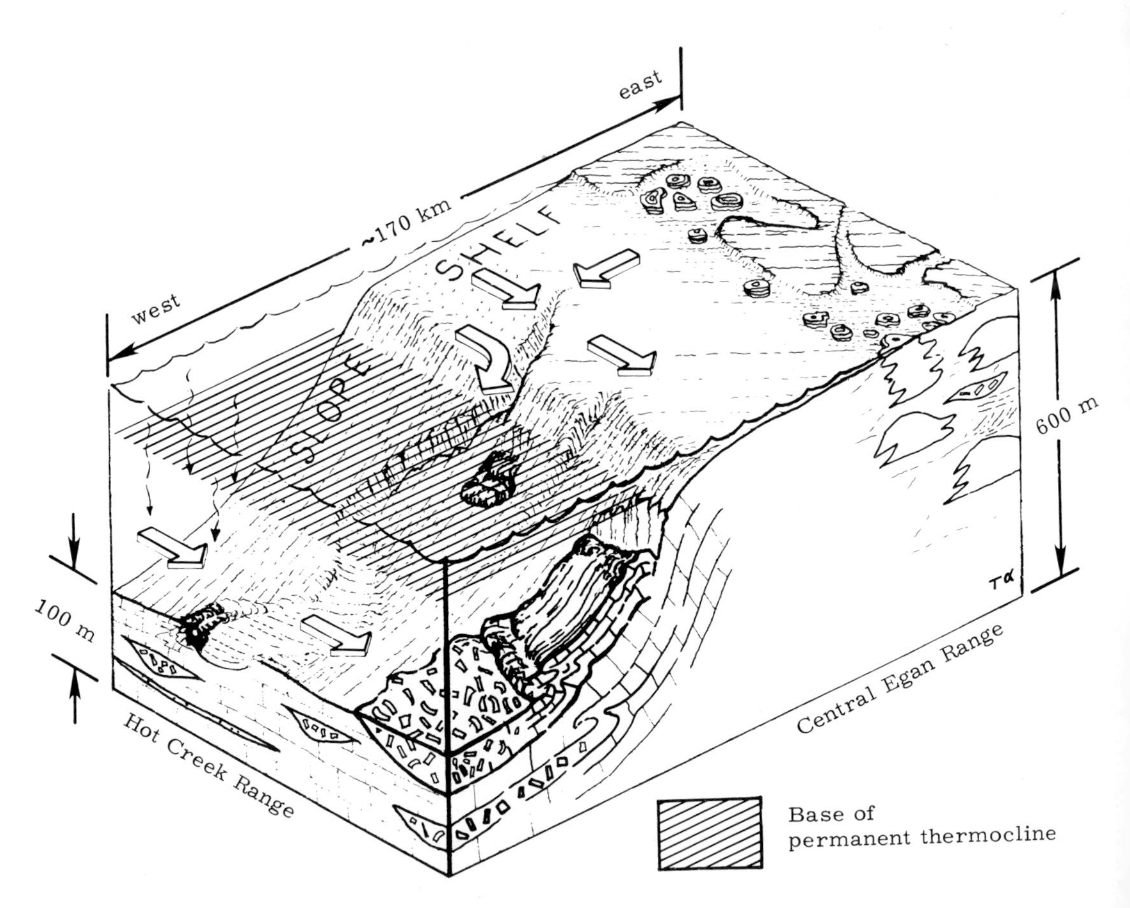

FIG. 45.—Graphic model of inferred shelf-slope transition in the Late Cambrian and earliest Ordovician of Nevada. See text for discussion. Not to scale. Modified from Cook and Taylor (1975, Fig. 2). Drawn by Tau R. Alpha.

FIG. 46.—Modes of occurrence of trilobites in the Hales Limestone, Hot Creek section. *A*—Redeposited trilobites: lime packstone containing disarticulated, broken, and abraded skeletal debris consisting of genera typical of the North American Faunal Province. USGS locality 6850-CO. Bar scale equals 2 cm. *B*—*In-situ* trilobites: dark lime mudstone containing thin articulated exoskeletons of species that show affinities with those of southeastern Asia. USGS locality 7522-CO. Both views are bedding surfaces. From Taylor and Cook (1976, text-fig. 26).

cross is environmentally significant. In the case of the isopod faunas off the northeastern coast of the United States, the cross-over is associated with the base of the permanent thermocline, which at the latitude of this transect is approximately 33°N. The permanent thermocline is associated with a transitional zone between the warm northerly-flowing Gulf Stream and cold southerly-flowing Western Boundary Undercurrent as they impinge on the Atlantic Continental Slope.

As a working model, we suggest that the restriction of trilobite biofacies with Asian affinities to deeper water settings adjacent to shallow, warm carbonate shelves is analogous to the abrupt separation of Holocene marine isopod faunas on either side of the permanent thermocline in low

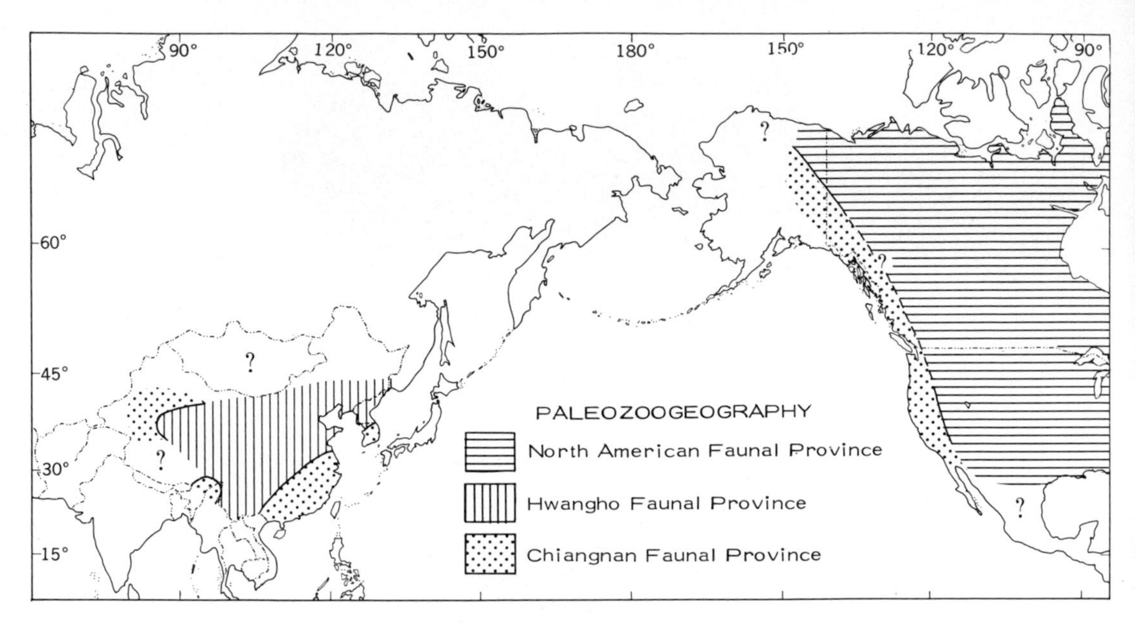

FIG. 47.—Late Cambrian (late Franconian and Trempealeauan) paleozoogeography of parts of North American and southern Asia. From Cook and Taylor (1975, Fig. 3).

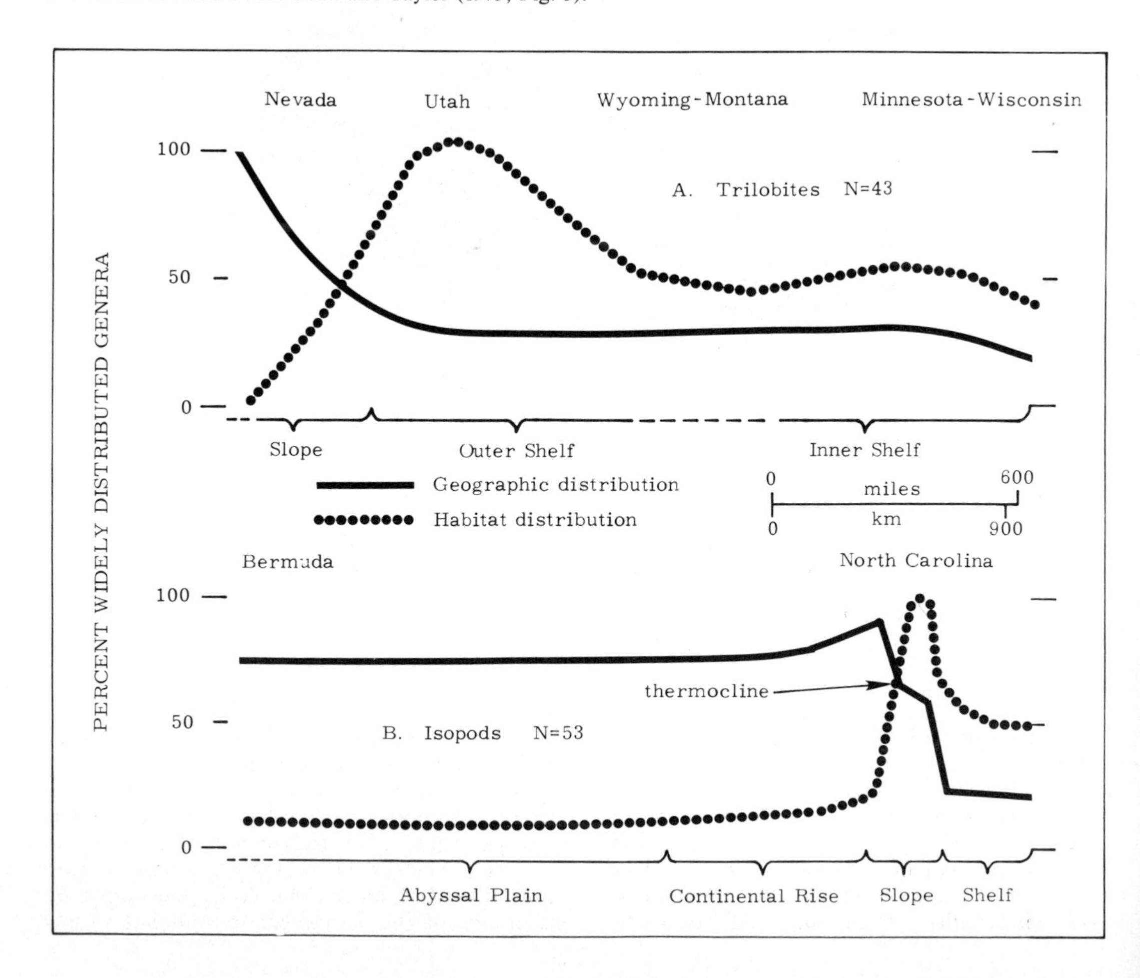

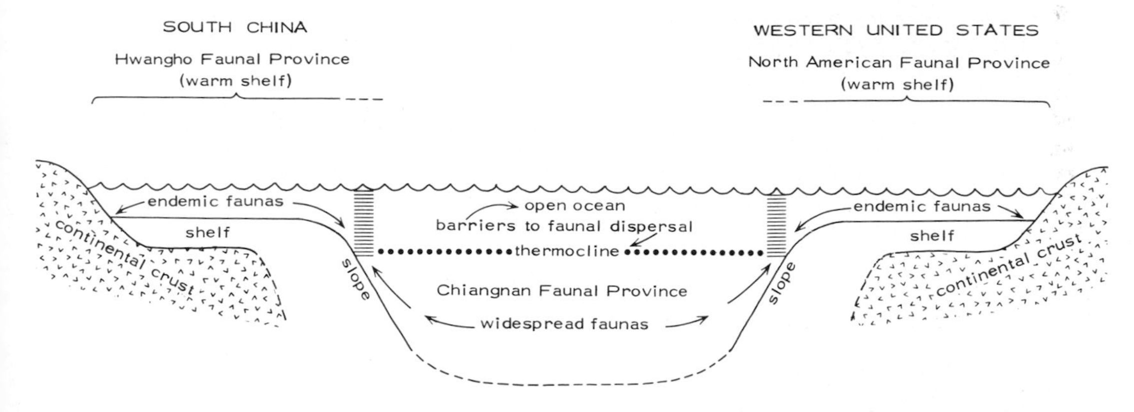

FIG. 49.—Inferred environmental relation of the Hwangho, Chiangnan, and North American trilobite faunal provinces. Not drawn to scale. From Cook and Taylor (1975, Fig. 4).

latitudes (Fig. 49). If this conclusion is correct, then biofacies and lithofacies analyses provide independent evidence for a continental slope origin of the lower Hales Limestone.

CONCLUSIONS AND DISCUSSIONS

During the Late Cambrian and earliest Ordovician of Nevada a shoal-water carbonate shelf existed to the east and adjacent deeper water slope environments to the west.

Slope processes and implications to modern slope processes.—Slope sedimentation involved an interplay between depositional and erosional processes. The dark laminated mudstone represents hemipelagic sedimentation by suspension, bottom currents and other normal pelagic processes. Submarine gravity movements, including slumps, debris and turbidity flows, although not of a large scale, were a significant process in the development of this environment. At least one quarter of this section is composed of this type of sediment. On the basis of abundant small-scale slump and flow deposits in this and other ancient deeper water settings it is reasonable to suggest that where continuous seismic profiles reveal a smooth undisturbed slope along some modern continental margins, the smoothness may not be entirely real. The smoothness may be, at least in part, a function of the resolving power of present surface seismic equipment.

Late Cambrian continental margin.—The similarity of trilobite and isopod distribution curves suggests that a thermocline probably existed between the Egan and Hot Creek Range sections during the Late Cambrian (Fig. 45). Further, the presence of geographically widespread trilobites in the slope habitat, but not in shallow-water habitats, and the apparent absence of coeval shallow-water rocks west of the Hot Creek Range suggest that the Hales Limestone was deposited on an ocean-facing continental margin during early Paleozoic time. This interpretation is consistent with conclusions of Stewart and Poole (1974, p. 42) that at least some Upper Cambrian rocks (Harmony Formation) oceanward from the Hales outcrops may represent continental rise deposits.

Implications to regional tectonics.—Occurrence of trilobites with Asian affinities interbedded with trilobites with North American affinities in central Nevada shows that the deeper water facies were not simply part of an Asian plate fragment left behind after a hypothetical Asian and North American collision with a subsequent separation. This interpretation might be inferred if one were to only consider the presence of faunas with Asian affinity. Rather, biofacies relations developed in normal response to environmental differences between a continental shelf and deeper water slope setting. If a North American plate and an Asian plate collided, separation occurred before the Late Cambrian. The general question of using trilobite faunal resemblance data for plate

FIG. 48.—Distributional characteristics of Late Cambrian (late Franconian and Trempealeauan) trilobites (A) compared with those of Holocene marine isopod crustaceans (B) In both analyses, greatest biofacies changes occur within the slope habitat. From Taylor and Cook (1976, text-fig. 31). See text and Taylor (1977) for further discussion. From Taylor and Cook (1976), text-fig. 31).

tectonic reconstructions has been discussed in more detail elsewhere (Taylor and Cook, 1976, p. 211).

ACKNOWLEDGMENTS

Our manuscript has benefited considerably from critical readings by Paul Enos, State University of New York at Binghamton, and James V. Gardner, U.S. Geological Survey, Menlo Park, California. We also thank C. Hans Nelson and William R. Normark for helpful discussions concerning some of the ideas presented in this paper. Taylor's initial field studies in the Hot Creek Range were supported by Geological Society of America Grants 1082-66 and 1221-69. Field work during 1973 was aided significantly by the capable assistance of L. James Vigil.

REFERENCES

Bathurst, R. G. C., 1971, Carbonate sediments and their diagenesis: Elsevier, Amsterdam, The Netherlands, 620 p.

Bouma, A. H., 1962, Sedimentology of some flysch deposits: Elsevier, Amsterdam, The Netherlands, 168 p.

—— and Hollister, C. D., 1973, Deep ocean basins in sedimentation: *In* G. V. Middleton and A. H. Bouma (eds.), Turbidites and deep-water sedimentation: Soc. Econ. Paleontologists and Mineralogists Pacific Sec., Los Angeles, California, p. 79-118.

Burchfiel, B. C., and Davis, G. A., 1972, Structural framework and evolution of the southern part of the Cordilleran orogen, western United States: Am. Jour. Sci., v. 272, p. 97-118.

Carter, R. M., 1975, A discussion and classification of subaqueous mass-transport with particular attention to grain-flow, slurry-flow, and fluxoturbidites: Earth-Sci. Rev., v. 11, p. 145-177.

Churkin, Michael, Jr., 1974, Paleozoic marginal ocean basin-volcanic arc systems in the Cordilleran foldbelt: *In* R. H. Dott, Jr, and R. H. Shaver (eds.), Ancient and modern geosynclinal sedimentation: Soc. Econ. Paleontologists and Mineralogists, Spec. Pub. 19, p. 174-192.

Cook, H. E., 1972, Miette platform evolution and relation to overlying bank ("reef") localization, Upper Devonian, Alberta: Bull. Canadian Petroleum Geology, v. 20, p. 375-411.

——, McDaniel, P. N., Mountjoy, E. W., and Pray, L. C., 1972, Allochthonous carbonate debris flows at Devonian bank ('reef') margins Alberta, Canada: Bull. Canadian Petroleum Geology, v. 20, p. 439-497.

—— and Taylor, M. E., 1975, Early Paleozoic continental margin sedimentation, trilobite biofacies, and the thermocline, western United States: Geology, v. 3, p. 559-562.

Dietz, R. S., and Holden, J. C., 1966, Miogeoclines (miogeosynclines) in space and time: Jour. Geology, v. 74, p. 566-583.

Ebanks, J. W., Jr., 1967, Recent carbonate sedimentation and diagenesis Ambergris Cay, British Honduras: Ph. D. Dissert., Rice University, Houston, Texas, 189 p.

Ferguson, H. G., 1933, Geology of the Tybo district, Nevada: Univ. Nevada Bull., v. 27, 61 p.

Garrison, R. E., and Fischer, A. G., 1969, Deep-water limestones and radiolarites of the Alpine Jurassic: *In* G. M. Friedman (ed.), Depositional environments in carbonate rocks: Soc. Econ. Paleontologists and Mineralogists, Spec. Pub. 14, p. 20-56.

Hampton, M. A., 1975, Competence of fine-grained debris flows: Jour. Sed. Petrology, v. 45, p. 834-844.

Kellogg, H. E., 1963, Paleozoic stratigraphy of the southern Egan Range, Nevada: Geol. Soc. America Bull., v. 74, p. 685-708.

Kussakin, O. G., 1973, Peculiarities of the geographical and vertical distribution of marine isopods and the problem of deep-sea fauna origin: Marine Biology, v. 23, p. 19-34.

Lewis, K. B., 1971, Slumping on a continental slope inclined at 1°-4°: Sedimentology, v. 16, p. 97-110.

Lochman-Balk, C., and Wilson, J. L., 1958, Cambrian biostratigraphy in North America: Jour. Paleontology, v. 32, p. 312-350.

Menzies, R. J., George, R. Y., and Rowe, G. T., 1973, Abyssal environment and ecology of the world oceans: Wiley-Interscience, John Wiley and Sons, New York, New York, 488 p.

Middleton, G. V., and Hampton, M. A., 1973, Sediment gravity flows mechanics of flow and deposition: *In* G. V. Middleton and A. H. Bouma (eds.), Turbidites and deep-water sedimentation: Soc. Econ. Paleontologists and Mineralogists Pacific Sec., Los Angeles, California, p. 1-38.

Mutti, Emiliano, and Ricci-Lucci, Franco, 1972, Le torbiditi dell'Appennino settentrionale: Introduzione all'analisi di facies: Soc. Geol. Italiana, Mem. 11, p. 161-199.

Palmer, A. R., 1960, Some aspects of the early Upper Cambrian stratigraphy of White Pine County, Nevada, and vicinity: Intermountain Assoc. Petroleum Geologists, Guide to 11th Ann. Field Conf., p. 53-58.

Reinhardt, Juergen, 1974, Stratigraphy, sedimentology and Cambro-Ordovician paleogeography of the Frederick Valley, Maryland: Maryland Geol. Survey, Rept. Inv. 23, 73 p.

Roberts, R. J., Hotz, P. E., Gilluly, James, and Ferguson, H. G., 1958, Paleozoic rocks of north-central Nevada: Am. Assoc. Petroleum Geologists Bull., v. 42, p. 2813-2857.

Roehl, P. O., 1967, Stony Mountain (Ordovician) and Interlake (Silurian) facies analogs of recent low-energy marine and subaerial carbonates, Bahamas: Am. Assoc. Petroleum Geologists Bull., v. 51, p. 1979-2031.

Rona, P. A., 1970, Submarine canyon origin on upper continental slope off Cape Hatteras: Jour. Geology, v. 78, p. 141-152.

Shepard, F. P., 1973, Submarine geology, 3rd. Ed.: Harper and Row, New York, New York, 517 p.

Shinn, E. A., 1968, Practical significance of birdseye structures in carbonate rocks: Jour. Sed. Petrology, v. 38, p. 215-223.

——, Lloyd, R. M., and Ginsburg, R. N., 1969, Anatomy of a modern carbonate tidal-flat, Andros Island, Bahamas: Jour. Sed. Petrology, v. 39, p. 1202-1228.

Stanley, D. J., and Silverberg, N., 1969, Recent slumping on the continental slope off Sable Island bank, southeast Canada: Earth and Planetary Sci. Letters, v. 6, p. 123-133.

—— and Unrug, R., 1972, Submarine channel deposits, fluxoturbidites and other indicators of slope and base-of-slope environments in modern and ancient marine basins: *In* J. K. Rigby and W. K. Hamblin (eds.), Recognition of ancient sedimentary environments: Soc. Econ. Paleontologists and Mineralogists, Spec. Pub. 16, p. 287-340.

Stewart, J. H., 1972, Initial deposits in the Cordilleran Geosyncline: Evidence of a Late Precambrian (850 m.y.) continental separation: Geol. Soc. America Bull., v. 83, p. 1345-1360.

—— and Poole, F. G., 1974, Lower Paleozoic and uppermost Precambrian Cordilleran Miogeocline, Great Basin, western United States: *In* W. R. Dickinson (ed.), Tectonics and sedimentation: Soc. Econ. Mineralogists and Paleontologists, Spec. Pub. 22, p. 28-57.

Taylor, M. E., 1976, Indigenous and redeposited trilobites from Late Cambrian basinal environments of central Nevada: Jour. Paleontology, v. 50, p. 668-700.

——, 1977, Late Cambrian of western North America: Trilobite biofacies, environmental significance, and biostratigraphic implications: *In* E. G. Kauffman and J. E. Hazel (eds.), Concepts and Methods of Biostratigraphy: Dowden, Hutchison and Ross, Inc. (eds.), Stroudsburg, Penn., p. 397-425.

—— and Cook, H. E., 1976, Continental shelf and slope facies in the Upper Cambrian and lowest Ordovician of Nevada: *In* R. A. Robison and A. J. Rowell (eds.), Cambrian paleontology and environments of western North America, a symposium: Brigham Young Univ. Geol. Stud., v. 23, pt. 2, p. 181-214.

—— and Halley, R. B., 1974, Systematics, environment, and biogeography of some Late Cambrian and Early Ordovician trilobites from eastern New York State: U.S. Geol. Survey, Prof. Paper 834, 38 p.

Thomson, A. F., and Thomasson, M. R., 1969, Shallow to deep water facies development in the Dimple Limestone (Lower Pennsylvanian), Marathon region, Texas: *In* G. M. Friedman (ed.), Depositional environments in carbonate rocks: Soc. Econ. Paleontologists and Mineralogists, Spec. Pub. 14, p. 57-78.

Troedsson, G. T., 1937, On the Cambro-Ordovician faunas of western Quruq Tagh, eastern T'ien-Shan: Paleont. Sinica, New Series B, No. 2, Whole Series no. 106, 74 p.

Uchupi, E., and Emery, K. O., 1963, The continental slope between San Francisco, California and Cedros Island, Mexico: Deep Sea Research, v. 10, p. 397-447.

Walker, R. G., and Mutti, Emiliano, 1973, Turbidite facies and facies associations: *In* G. V. Middleton and A. H. Bouma (eds.), Turbidites and deep-water sedimentation: Soc. Econ. Paleontologists and Mineralogists Pacific Sec., Los Angeles, California, p. 119-157.

Wilson, J. L., 1957, Geography of olenid trilobite distribution and its influence on Cambro-Ordovician correlation: Am. Jour. Sci., v. 255, p. 321-340.

——, 1969, Microfacies and sedimentary structures in "deeper water" lime mudstones: *In* G. M. Friedman (ed.), Depositional environments in carbonate rocks: Soc. Econ. Paleontologists and Mineralogists, Spec. Pub. 14, p. 4-19.

Worzel, J. L., 1968, Survey of continental margins: *In* D. T. Donovan (ed.), Geology of shelf seas: 14th Inter-University Geol. Congr. Proc., Oliver and Boyd, Edinburgh and London, p. 117-154.

SEPM Special Publication No. 25, p. 83-112, November 1977

CAMBRIAN OFF-SHELF SEDIMENTATION, CENTRAL APPALACHIANS

JUERGEN REINHARDT
U.S. Geological Survey, Reston, Virginia 22092

ABSTRACT

Cambrian carbonate rocks in the central Appalachian region were deposited both on an extensive carbonate shelf and in a deeper water slope and basin which was generally east of the shelf. Modern sedimentary environments provide analogues for specific parts of the shelf carbonate rocks and to a lesser extent for the basinal carbonate rocks.

The deeper water carbonate rocks are composed of thin-bedded micritic limestone and thickly to massively bedded peloidal limestone and breccia. These rocks result primarily from clastic deposition of shelf-derived material. Evidence for subaqueous slumps and debris flows is present in the composition and stratification of the coarse-grained limestone. Centimeter-thick argillaceous limestone beds are graded and in rare cases contain complete Bouma cycles; these suggest deposition by turbidity currents. Relative amounts of biogenic debris, bioturbation, and vertical variability of sedimentary structures are qualitative bathymetric indicators. Stratigraphic proximity to shoal-water carbonate rocks and the thickness of slope deposits suggest that water depths were as great as 500 meters in the basin.

Erosive structures ranging from tidal channels to submarine canyons are local features at or near the carbonate shelf margin. The composition and organization of breccia deposits and surrounding rocks in the Shady Dolomite (Lower and Middle Cambrian) near Austinville, Virginia, in the Kinzers Formation (Lower and Middle Cambrian) near Thomasville, Pennsylvania, and in the Frederick Limestone (Upper Cambrian) near Frederick, Maryland, represent three distinct deeper water environments adjacent to the Cambrian carbonate shelf.

The shelf and basin in the Frederick and possibly the Austinville areas were connected by a carbonate depositional slope which extended seaward at a very low angle from mean sea level to depths of 500 m or more. Near York the shelf margin was highly dissected in at least one area and higher angle slopes persisted from Early to Middle Cambrian time.

INTRODUCTION

Carbonate rocks of late Early Cambrian to Middle Ordovician age that crop out in the Great Valley section of Pennsylvania, Maryland, Virginia, and West Virginia were deposited in a shallow epeiric sea (Matter, 1967; Sando, 1957; Root, 1964, Donaldson, 1969, Reinhardt and Wall, 1975). Reviews of the Cambrian lithostratigraphic framework and the level of biostratigraphic control in the entire Appalachian miogeocline, extending from Newfoundland to Alabama, have been presented by Rodgers (1956) and Palmer (1971).

These Cambrian and Ordovician carbonate rocks consist of several belts of differing rock type, which surround the North American craton. The gross lithofacies changes from the craton across the carbonate platform to the basin have been described by Wilson (1952), Rodgers (1956, 1968), Palmer (1971), and Harris (1973). Predominantly coarse siliciclastic rocks are closest to or on the craton; dolomite, limestone, and calcareous shale belts follow, in turn, from west to east in the Appalachians. These lithofacies are the products of varying water depths and isolation from clastic sources overprinted by diagenetic environments. Many of the sedimentary rocks are the product of tidal-flat complexes migrating across a Cambrian platform 800 to 1600 km wide.

The transition from shelf to basin occurs, however, in a much narrower belt and has been documented in only a few areas. Elsewhere, structural complications have obscured or eliminated these transitional rocks. Rodgers' paper (1968) provided a stimulus for looking at these basinal limestones. At least three separate studies—Keith (1974) in New York; Reinhardt (1974) in Maryland; and Gohn (1976) in Pennsylvania—were stimulated by Rodgers' "eastern edge of the North American continent" concept. The modern analog chosen by Rodgers was Tongue of the Ocean an oceanic reentrant in the Bahama platform. The variations in geometry along modern and ancient shelf-basin margins allow for a number of possible models (Wilson, 1974; Ginsburg and James, 1974). The major purpose of this paper is to describe and interpret three different lithologic assemblages at or near the Cambrian platform margin in the central Appalachians.

GEOLOGIC FRAMEWORK

The specific localities discussed in this paper (Fig. 1) include: (1) the Frederick Valley in central Maryland; (2) the Thomasville area, a part of the Conestoga Valley in south-central Pennsylvania; and (3) the Austinville area, along the eastern margin of the Great Valley in southwestern Virginia. The stratigraphic terminology for these

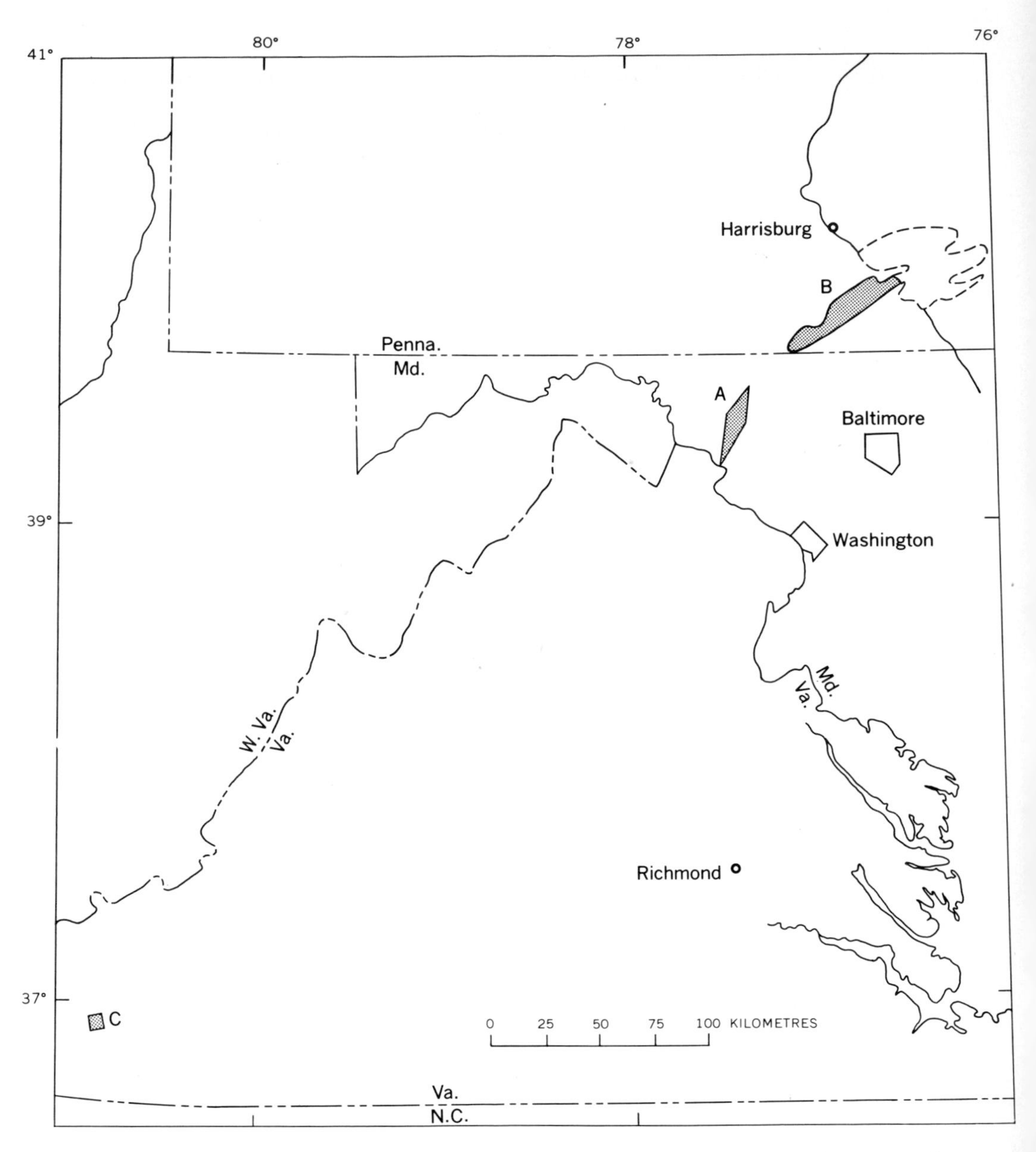

FIG. 1.—Index map of areas discussed: *A*, Frederick Valley; *B*, Hanover-York portion of the Conestoga Valley (eastern portion of the Valley is delineated by a dashed line); *C*, Austinville area.

areas is taken from Reinhardt (1974), Gohn (1976), and Butts (1940) for the Maryland, Pennsylvania, and Virginia sections, respectively (Table 1). In each area the Cambrian and Ordovician carbonate sequence is underlain by Lower Cambrian siliciclastic rocks which fine from coarse sands to silts eastward across strike (Schwab, 1970; Gohn, 1976).

The Frederick Valley is underlain by Cambrian and Lower Ordovician rocks which crop out in a narrow belt 10 km or less in width between the Blue Ridge province and the western Piedmont (Figs. 2, 3). The valley is an asymmetrical synclinorium, as eastward extension of the South Mountain structure described by Fellows (1943) and Cloos (1947). The structure is relatively simple. Two distinct periods of folding have been documented; major normal and (or) reverse faults have

TABLE 1.—STRATIGRAPHIC FRAMEWORK

	Frederick Valley, Maryland (Reinhardt, 1974)		Conestoga Valley, Pennsylvania (Gohn, 1976)		Great Valley, Virginia (Butts, 1940)	
Lower Ordovician	Grove Formation[1] (450 m)		Conestoga		Beekmantown Formation (700 m)	
Upper Cambrian	Frederick L.S. (800 m)	Lime Kiln Member (180 m) Adamstown Member (325 m) Rocky Springs Station Member (300 m)	Limestone (300–700 m)		Conococheague Formation (700 m) Elbrook Formation	
Middle Cambrian	Araby Formation (>100 m)		Ledger Dolomite (30–350 m) Kinzers Formation	Longs Park Mbr. (40–55 m) Thomasville Mbr.	(700 m) Rome Formation (150–300 m)	Ivanhoe Mbr. (100–500 m)
Lower Cambrian			(200–400 m) Vintage Formation (5–300 m) Antietam Formation (60–70 m)	(100–310 m) Emigsville Mbr. (30–70 m)	Shady Dolomite[2] (300–1200 m) Erwin Formation (50 m)	Austinville Mbr. (300–400 m) Patterson Mbr. (300 m)
Additional clastic rock units, variously referred to L€ or Late P€ underlie each area.						

[1] Contains Upper Cambrian trilobite faunas in lower 30 m (M. E. Taylor, written comm., 1976).

[2] Members and their thicknesses are defined only for the Austinville area. Modified to Lower and Middle Cambrian to include faunas described by Willoughby (1976).

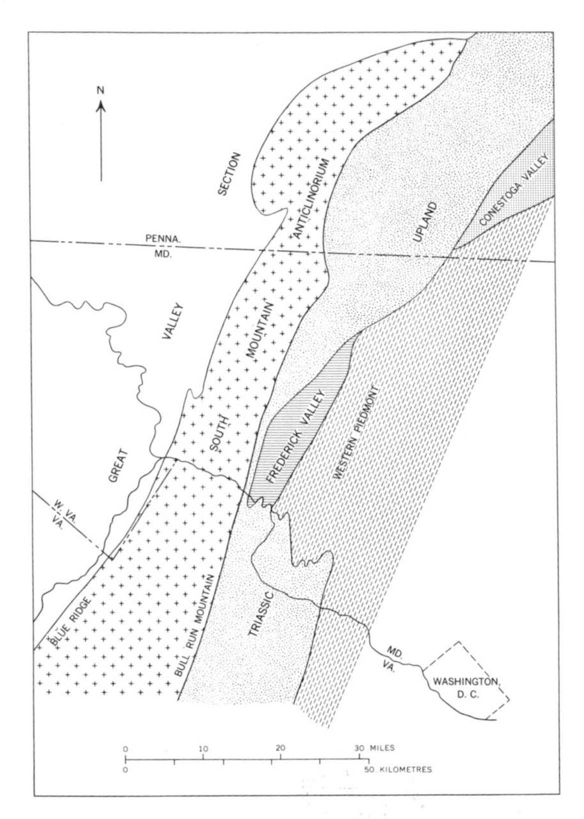

FIG. 2.—Major physiographic provinces in part of the central Appalachians. Note the position of both the Frederick and Conestoga Valleys east of the Blue Ridge province (South Mountain Anticlinorium).

not been recognized in the Frederick Valley (Reinhardt, 1974, pl. 1). Two major Triassic basins pinch out along the western margin of the Frederick Valley but overlap the Cambrian and Ordovician carbonate rocks to the north and south (Figs. 2, 3).

The stratigraphic terminology of Jonas and Stose (1936) as revised by Reinhardt (1974) is shown in Table 1. The revision consisted of: (1) recognition of a considerably thicker section of Upper Cambrian and Lower Ordovician carbonate rocks (as much as 1250 m rather than 400 m); and (2) elimination of an unconformity between the Araby Formation (Antietam Sandstone of Jonas and Stose, 1936) and the Frederick Limestone.

The Araby Formation[1] (Lower to Middle Cambrian) is composed of phyllite and metasiltstone with few primary bedding features. Silt to fine sand grains float in a chlorite-sericite matrix. The rocks are typically mottled and cleaved. The unit is at least 100 m thick and contains *Olenellus* sp. about 50 m from the top (Reinhardt, 1974).

Stratigraphic revision and refinement of the Frederick Limestone (Upper Cambrian) resulted from detailed mapping, better biostratigraphic control, and measured sections. The stratigraphic units from bottom to top are Rocky Springs Station Member, Adamstown Member, and Lime Kiln Member.[2] The Frederick Limestone is composed primarily of thin-bedded argillaceous limestone interbedded with thick-bedded breccias, peloidal limestones, oolitic limestones, and shales-slates. Variations in sedimentary structures, bedding characteristics, petrology of thick-bedded limestone, and biogenic components are the primary indicators of changing depositional environments within the basin (Fig. 4).

The Grove Limestone (Lower Ordovician) is dominated by thick-bedded, arenaceous, and stromatolitic limestone, and dolomite. These rocks differ sharply in composition, sedimentary structures, organization by cycles and depositional environment from the underlying Frederick Limestone.

The Conestoga Valley is a complex synclinorium approximately 100 km long along strike with the Frederick Valley (Figs. 2, 5). Local structural highs along both margins of the Valley include the Pigeon Hills, Hellam Hills, Mine Ridge, and Welsh Mountain. Complex structure includes imbricate thrusting in the eastern portions of the Conestoga Valley (Cloos and Heitanen, 1941) and two major periods of folding (Wise, 1970). Local high angle normal faults are present in many quarry exposures and are well documented at the Thomasville Quarry (Cloos, 1968).

The stratigraphic framework for the Conestoga Valley is based largely on nearly two decades of mapping by Stose and Jonas (see Jonas and Stose, 1930; Stose and Jonas 1939; Stose and Stose, 1944). Recent biostratigraphic work by Campbell (1971) indicates a Middle Cambrian age for the Ledger Dolomite and the upper portions of the Kinzers Formation.

Gohn (1976) made a regional study of the Conestoga Valley, has added biostratigraphic control for the Conestoga Limestone, and subdivided both the Kinzers Formation and the Conestoga Limestone. The Emigsville, Thomasville, and Longs Park Members of Gohn (1976) from base to top of the Kinzers correspond to the "lower, middle, and upper" informal members

[1]The Araby Formation of Reinhardt (1974) is herein adopted for U.S. Geological Survey usage.

[2]The Rocky Springs Station Member, Adamstown Member and Lime Kiln Member of Reinhardt (1974) are herein adopted for U.S. Geological Survey usage.

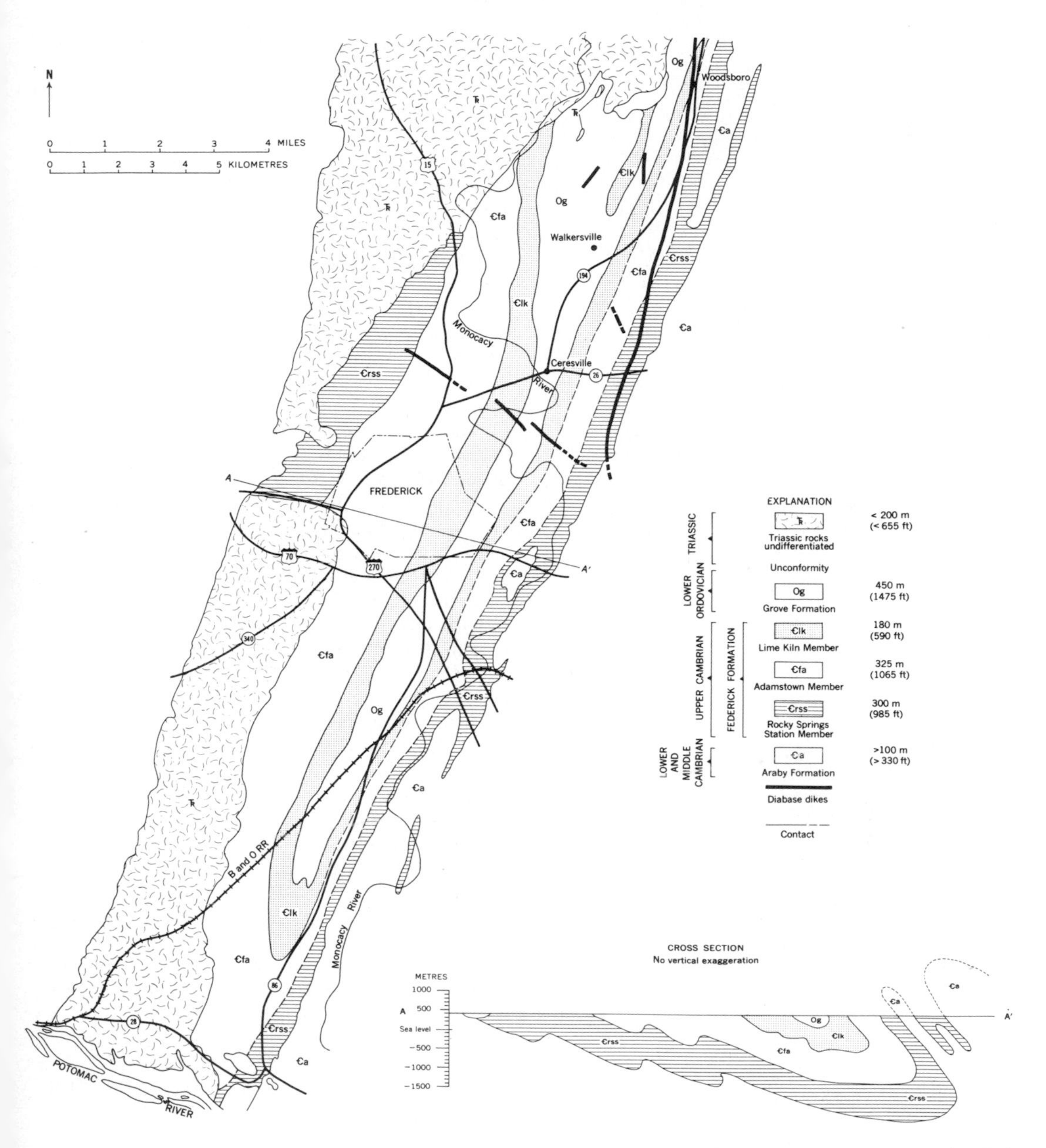

FIG. 3.—Generalized geologic map and cross-section of the Frederick Valley. Quaternary units are unlabeled. Modified from Reinhardt (1974).

of Stose and Stose (1944). The Kinzers Formation and especially the Thomasville Member of Gohn (1976) will be discussed in subsequent sections.

At Thomasville Stone and Lime Company Quarry the stratigraphy is well known from over 300 core holes and extensive surface exposures (Fig. 6). Of the 300 m of section nearly 200 m is composed of megabreccia deposits (Cloos, 1968). The two breccia sequences are separated by the "bottom black" (6–9 m thick) and "top black" (6–12 m thick) limestones of Cloos. The top of the Thomasville Member is not exposed in the quarry but the top 100 m is composed of thick-bedded to massive limestones and dolomites. Away from the Thomasville area this unit probably thins to less than 100 m (Gohn, 1976).

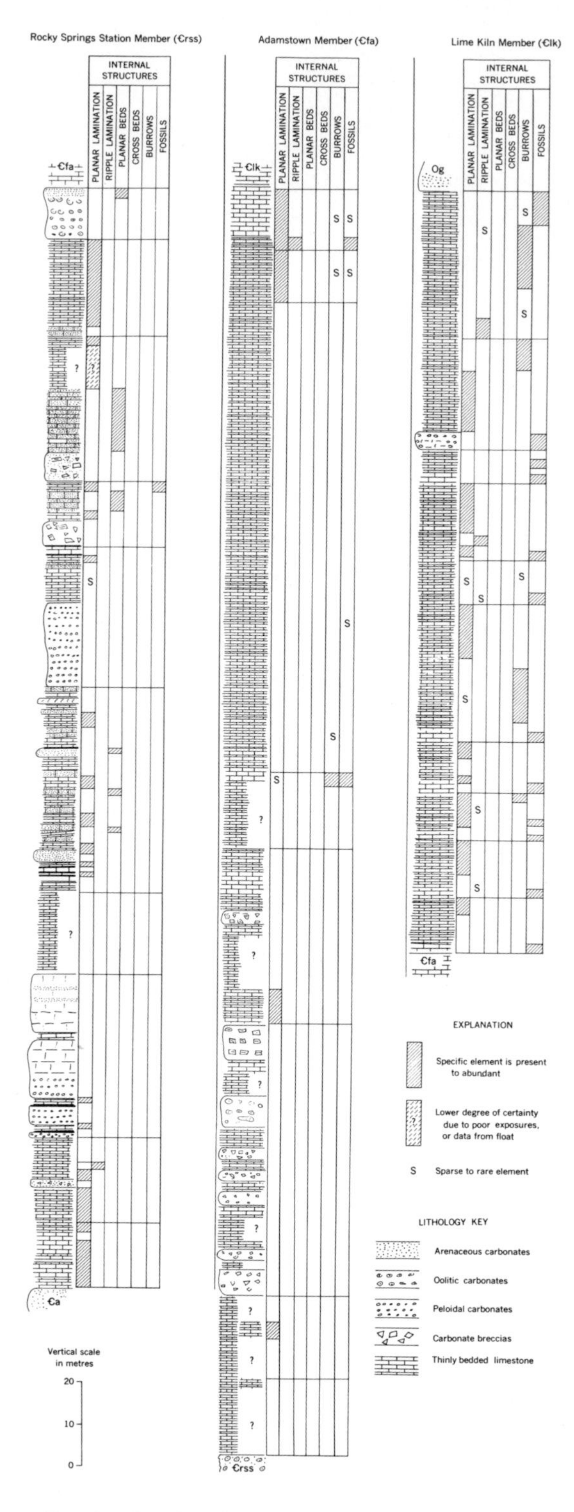

FIG. 4.—Stratigraphic columns of the three Frederick Limestone units showing the distribution of lithologies and sedimentary structures.

The Austinville area in southwestern Virginia has been considered a "problem area" by numerous workers (see Palmer, 1971 for a brief review). The Shady Dolomite (Lower and Middle Cambrian) crops out along the eastern edge of the Great Valley section adjacent to the Blue Ridge (Fig. 7). Complex imbricate thrusting has repeated the section and compressed the basin of deposition.

The Shady Dolomite is about 700 m thick and possibly up to 1200 m thick in the southeastern facies near Austinville (written comm., J. F. Read, 1976), which is somewhat anomalous compared to 300–500 m for the Shady in Northern Virginia (Butts, 1940; Gathright and Nystrom, 1974) and only 150 m for the equivalent Tomstown Dolomite in Maryland (Reinhardt and Wall, 1975). None of the other Cambrian and Ordovician units are atypically thick in the Austinville area (Table 1).

Despite the complex structure a detailed local stratigraphy has been worked out as a result of surface and subsurface mapping in the Austinville-Ivanhoe lead-zinc district (Brown and Weinberg, 1968). A refined stratigraphic column (Fig. 8) with inferred facies relationships results from recent stratigraphic work by Russell Pfeil (Pfeil and Read, 1976; Pfeil, 1977), a graduate student at Virginia Polytechnic Institute and State University. Another V.P.I. student (Willoughby, 1976), has described an early Middle Cambrian fauna in the "Upper Shady Dolomite" or Post Taylor Marker of Brown and Weinberg (1968), which is in part temporally equivalent to the Rome Formation.

The abrupt lithologic transitions from the "north-limb" to "south-limb" facies of Brown and Weinberg (1968) may in part be the result of telescoping along thrust faults, but the lithologic changes as shown by Pfeil (1977) result from differing primary depositional environments near the carbonate shelf margin. Breccia units are prominant both in the surface Post Taylor Marker (Upper Shady of Pfeil) and in the subsurface Patterson Member (Brown and Weinberg, 1968).

LITHOLOGIC FACIES

The two major rock types comprising the off-shelf carbonate rocks are: (1) thin-bedded carbonate rocks, typically "dark and dirty" limestones 1 to 3 cm thick, but, including rocks which range in composition from calcareous or dolomitic shales to arenaceous or pure limestones, and (2) thick-bedded carbonate rocks, including detrital (allodapic) limestone, breccia, and megabreccia deposits. Detrital limestone beds are thin (3-5 cm) to thick (1-10 m); breccia and megabreccia deposits are typically thicker than 3 m.

The thin-bedded limestones are characterized by such features as: dominance of lime mud,

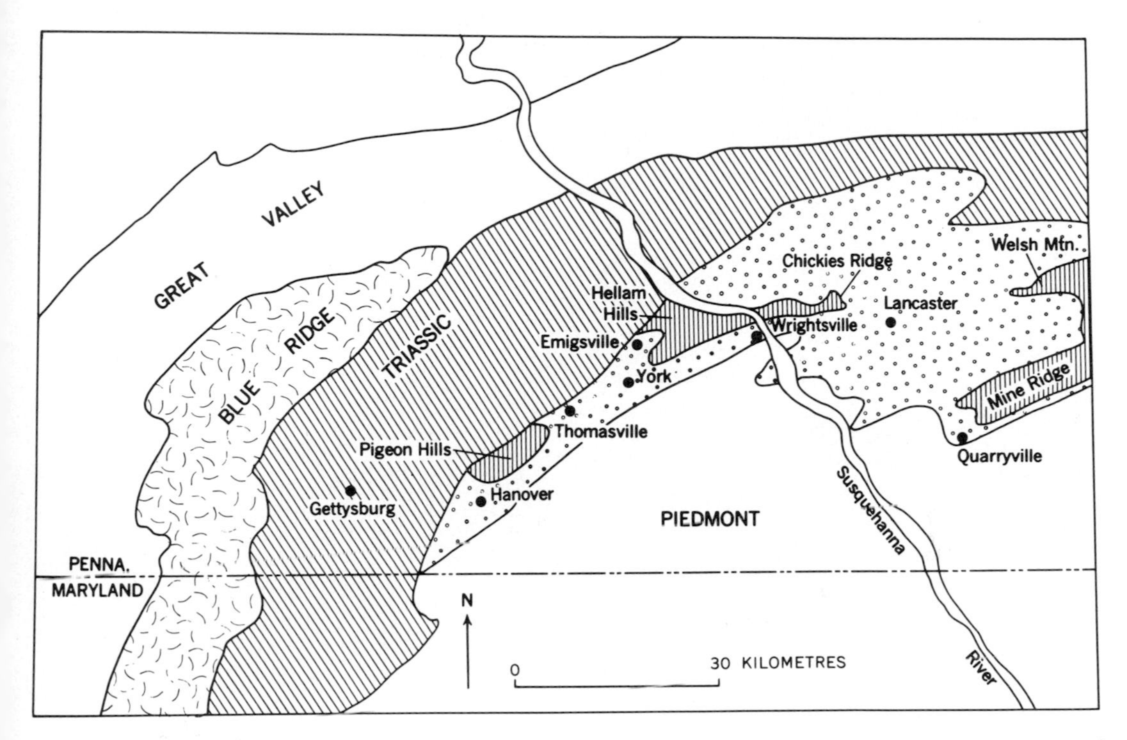

FIG. 5.—Physiographic provinces adjacent to or relating to the Conestoga Valley. Note the position of Thomasville, the area under discussion within the Conestoga Valley (dotted pattern) Vertical lined areas are structural highs (see text).

planar millimeter lamination, lack of bioturbation; such features have been considered distinctive of "deeper water" limestones (Wilson 1969, Thomson and Thomasson, 1969; and many others). These rocks result from deposition by turbidity currents and hemipelagic fall-out, and possibly reworking by bottom currents. The thick-bedded rocks are allochtonous deposits, involving such transport processes as submarine rock fall, slump or slide, mass or debris flow, and turbidity flow (Dott, 1963; Middleton and Hampton, 1973). One or more transport mechanism involving variable sediment, water, and gravitational components have been involved in the genesis of these deposits. Both thin- and thick-bedded rocks were deposited in environments basinward of a shallow-water carbonate shelf. These depositional sites include shelf-margin to basin-margin environments (slope or submarine canyon environments) and basinal environments (including starved basin) adjacent to or within a submarine-fan complex.

Both major rock groups will be discussed for the Frederick Valley, while only breccias and megabreccias will be discussed in detail for the Austinville and Thomasville areas.

Frederick Valley

Patterns in sedimentation.—Most of the rocks in the lower part of the Frederick Limestone are either thin-bedded, argillaceous, laminated limestone (in sequences tens of meters thick) or thick-bedded detrital limestone or breccia (as much as 20 m thick). The gross differences in sediment particle size and sedimentary structures indicate different sedimentation conditions and energy regimes.

At several localities, good exposures provide an opportunity to observe the relationship between thin- and thick-bedded limestone. A measured section near the top of the Rocky Spring Station Member at the Staley quarry (Fig. 9) contains two fining-upward sequences separated by an erosive contact.

Lithologic packaging is somewhat more systematic and repetitious in the Lime Kiln Member. Ten cycles, which resemble the one shown schematically in Figure 10, have been delineated at the type section for the Lime Kiln (Alpha Portland Quarry at Lime Kiln, Maryland) where the section is completely exposed. These sequences are characteristically finer grained, more dolomitic, and thinner bedded to laminated upwards.

FIG. 6.—Stratigraphy of the Thomasville Member of the Kinzers Formation at Thomasville Stone and Lime Company Quarry as seen in the quarry walls. The massive light-colored units are the two Thomasville breccias and the interbedded thin darker-colored units are the "top and bottom blacks" of Cloos (1968).

Thin-bedded limestone.—Limestone beds 1 to 3 cm thick constitute more than 80 percent of the Frederick Limestone. Most beds can be traced over the entire length of outcrops (as much as tens of meters) without measurable change in thickness. Each bed is characterized by a sharp base, micritic or microspar texture, and a dolomitic-argillaceous top. Most beds appear to be normally graded; in thin section, however, recrystallization of fine-grained carbonate makes identification of original size grading difficult.

The sedimentary structures within the thin-bedded limestone change from structureless or planar-laminated beds in the lower part of the Frederick Limestone to interbedded planar-laminated, cross-laminated, and bioturbated beds toward the top of the formation. Although grading is not easily demonstrable for most thin beds composed entirely of fine-grained carbonate, those beds containing quartz or carbonate sand-size particles normally are graded. A few beds (Figs. 11 and 12) in the Rocky Springs Station Member are not only graded, but also contain complete Bouma cycles (Bouma, 1962).

Thin beds higher in the stratigraphic column (Adamstown Member of the Frederick Limestone) contain abundant, low-angle cross-laminae, isolated asymmetrical ripples, and scour-and-fill structures on several scales (Fig. 13) in addition to the planar-laminated and graded beds characteristic of the Rocky Springs Station Member.

Biogenic debris, primarily of trilobites and echinoderms, is a key petrologic element (Fig. 14); both the amount and the diversity of fossil debris increase from the base to the top of the Frederick Limestone. Only very sparse trilobites are present in the Rocky Springs Station Member (Rasetti, 1961). In addition to the increased abundance of trilobites, brachipods, cephalopods, and pelmatozoans are found in the Adamstown and Lime Kiln Members. In addition to the more abundant body fossils, burrows (open spirals) are found within the Adamstown Member (Fig. 15). This is the first conclusive evidence of a bottom community within the basin, as the body fossils may all be transported.

In the Lime Kiln Member (the uppermost 180 m of the Frederick Limestone), bedding-plane burrows and mottled textures are common throughout the thinly bedded limestone (Fig. 16).

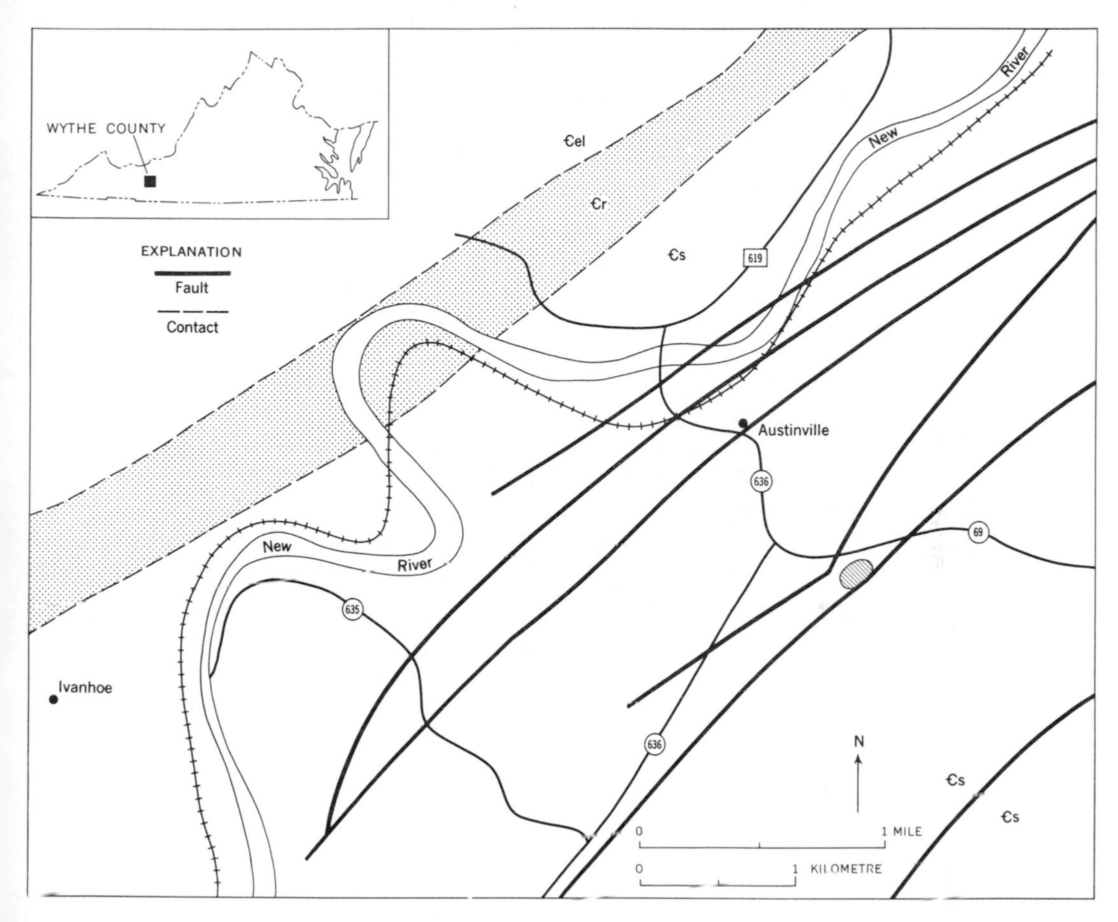

FIG. 7.—Index map of the Austinville area showing distribution of the stratigraphic units: Єel, Elbrook Limestone; Єr, Rome Formation; Єs, Shady Dolomite; Єe, Erwin Formation; high-angle reverse faults (southeast side up in each case), and the area of breccia outcrops (lined area adjacent to Virginia Rt. 69). Modified from Brown and Weinberg, 1968.

Concentration and alignment of fossil debris, planar laminae, and scour-and-fill structures are the dominant expression of current activity within the Lime Kiln Member (Figs. 16, 17). Planar laminae are uncommon near the Frederick Limestone-Grove Limestone boundary and are replaced by wavy argillaceous laminae in the upper half in beds 3 to 5 cm thick. Associated with these wavy laminated limestones are isolated stromatolitic mounds with poorly developed internal laminae (Fig. 18).

Thick-bedded limestone.—The two major types of thick-bedded to massive limestone within the Frederick Limestone are detrital limestone and breccia. In the lower half of the Frederick Limestone both of these lithic types are quite distinct petrologically from the interbedded thin-bedded limestones. Differences in bedding thickness and petrology between thin- and thick-bedded limestones are less pronounced in the upper half of the Frederick Limestone (especially in the Lime Kiln Member).

The detrital limestone beds are 10 cm to 20 m thick and are composed primarily of coarse well-rounded sand-sized calcareous and dolomitic lithoclasts (peloids) and quartz grains. Oolites, spherulites, dolomitic clusters, coated grains, and bioclasts are less abundant (Fig. 19). Thick beds commonly contain three or more grain types, although several beds as much as 10 m thick contain single types of allochems (Fig. 20). The cement in these thick-bedded limestones is microspar, probably in part recrystallized from a fine-grained matrix (Fig. 20). Detrital limestones are either massive or are composed of amalgamated planar tabular sets, which are defined by a

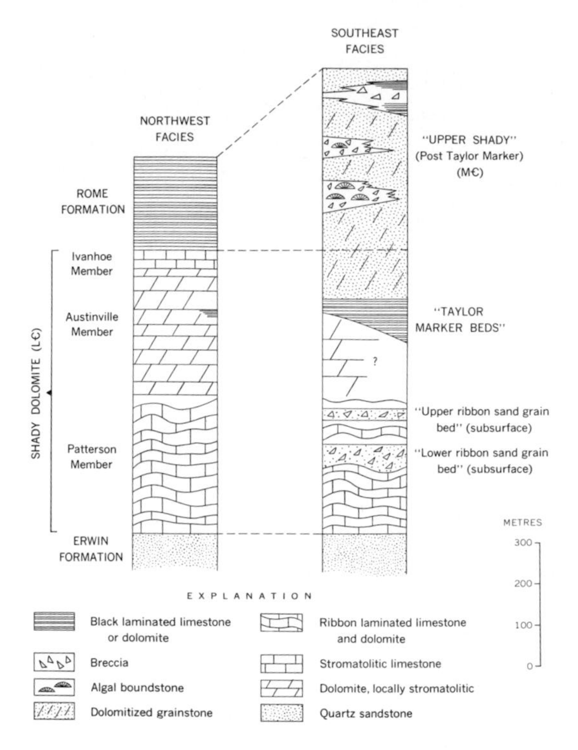

FIG. 8.—Stratigraphic column of the Shady Dolomite at Austinville showing the facies relationships from northwest to southeast. Modified from Pfeil (1977); subsurface data to Pfeil courtesy of E. L. Weinberg.

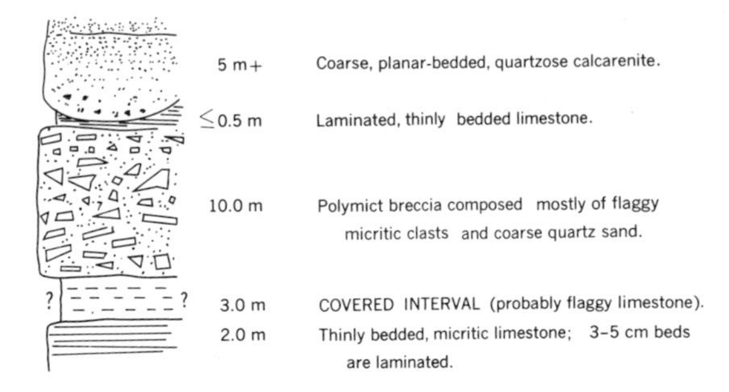

FIG. 9.—Measured section near the top of the Rocky Springs Member of the Frederick Limestone at Staley Quarry. Two fining-upward sequences are partially exposed.

difference in grain size or composition (Fig. 21).

Breccia in the Frederick Limestone is present at several stratigraphic levels. Coarse polymict breccia as much as 5 m thick occurs at several horizons in the Rocky Springs Station Member.

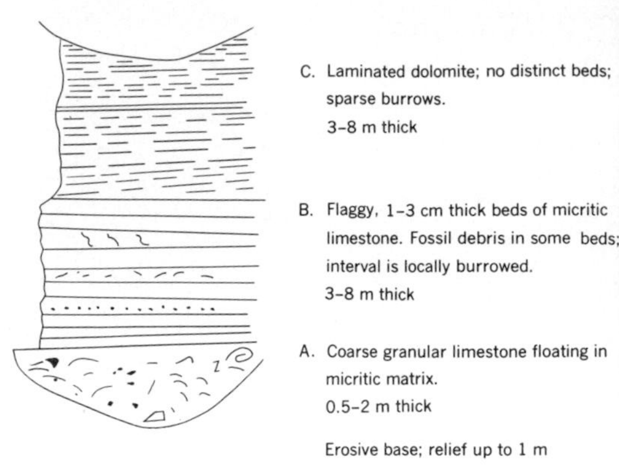

FIG. 10.—Lime Kiln cycle. Average thickness is 18 meters.

Breccia composed of relatively homogeneous, angular to rounded, centimeter-size clasts occurs in the lower half of the Adamstown Member. No breccia, except for lenses less than 50 cm thick, has been found in the Lime Kiln Member.

The polymict breccia in the Rocky Springs Station Member is dominated by clasts composed of laminated micritic limestone; however, clasts of several varieties of laminated limestone, lithoclastic limestone, oolitic and arenaceous limestone, and dolomite are present (Fig. 22). The matrix of this breccia is primarily micrite, but contains floating quartz grains, oolites, and peloids. The flatter breccia blocks occasionally show a weak imbrication at a low angle (about 20°) to bedding. The matrix content increases toward the top of a breccia interval; clast size and diversity correspondingly decrease.

Breccia intervals in the Adamstown Member are 3 to 5 cm thick and have poorly defined top and bottom contacts with thin-bedded limestones. These breccias are characterized by (1) total dominance of microspar clasts, (2) angular clast boundaries, often matching adjacent clasts, (3) fracture-filling cements (Fig. 23), (4) interbedded poorly sorted detrital limestones (Fig. 24) and (5) crude size-grading near the top of breccia intervals.

Upper boundary lithologies.—The Frederick Valley section is topped by thick-bedded arenaceous limestone and dolomite of the Grove Limestone. The Grove Limestone is composed of three major lithofacies that are cyclically repeated (Fig. 25). Each part of the cycle persists throughout the Grove Limestone without major changes in composition or sedimentary structures.

FIG. 12.—An etched slab of a thin limestone bed from the Rocky Springs Station Member bounded sharply at the base by a shale-slate (E). Normal Bouma cycle transition from (A) structureless interval, containing carbonate peloids, through (D) planar laminae.

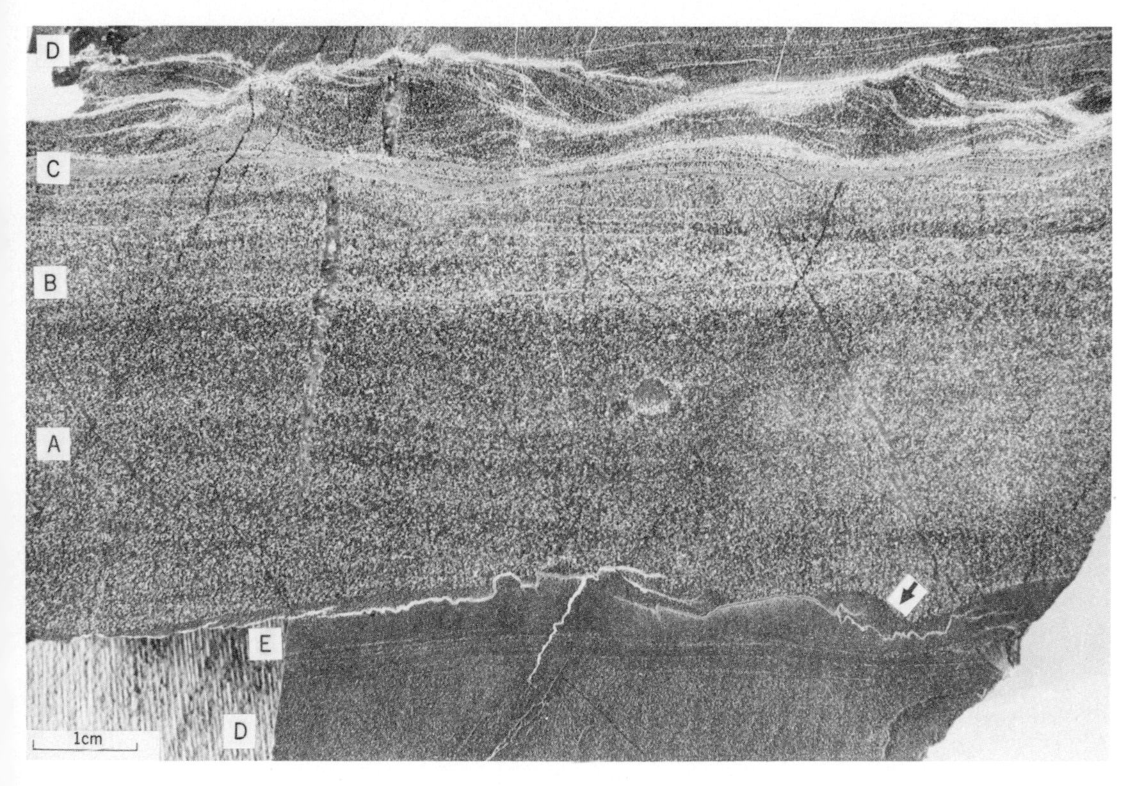

FIG. 11.—Slab of an atypically thick bed from a section of thin-bedded limestone in the Rocky Springs Station Member. Letters correspond to Bouma cycle intervals. The erosive base (arrow) cuts a structureless argillaceous limestone (E); note that a stylolite roughly follows the base.

FIG. 13.—Flaggy limestone near the top of the Adamstown Member of the Frederick Limestone. Beds have planar to undulatory erosive bases are are typically graded. A transition from low-angle, ripple cross-laminae to planar laminae (lower arrow) is coincident with grain size reduction. Microscours filled by peloids cut planar-laminated limestone (upper arrow).

The composition and organization of the Grove Limestone cycle is similar to that of other Appalachian Cambrian and Ordovician limestones, especially within the Conococheague Group (Root, 1964; Reinhardt and Hardie, 1976). The basal part of the Grove cycle is composed of arenaceous, oolitic, and peloidal limestones, which are characteristically ''herringbone'' cross-bedded (Fig. 25a). The multidirectional cross-bed sets are as much as 30 cm thick; foresets are typically inclined less than 20°. Near the base of the Grove Limestone uninterrupted intervals of cross-bedded to planar-bedded limestone as much as 10 m thick are present locally.

The middle part of a Grove cycle is dominated by large, poorly laminated, mound-shaped stromatolites; thrombolites are rare cryptalgal components in this part of the cycle. Contacts between the cross-bedded limestone and the stromatolitic limestone are typically sharp, generally defined by a ragged stylolite (Fig. 25b). These cryptalgal structures are outlined by coarse intermound fills, composed principally of carbonate interclasts and quartz grains (Fig. 26) and by dolomitic clots and laminae within and along the margins of the stromatolites.

The Grove cycles are capped by a thin-bedded, to massive, fine-grained, planar-laminated dolomite (Fig. 25c) or coarse-grained, irregularly laminated arenaceous dolomite (Fig. 27). Coarse irregular laminates are more abundant lower in the section; fine-grained, planar to crenulated laminates are more common near the top of the Grove Limestone.

Lower Cambrian Depostis

The off-shelf limestone deposits of the Kinzers Formation (Lower and Middle Cambrian) near Thomasville, Pennsylvania, and of the Shady Dolomite (Lower and Middle Cambrian) near

FIG. 15.—Oblique view of open-spiral burrows (arrows) in thinly bedded limestone. Within this part of the Adamstown Member, burrows are typically at or near the top of beds.

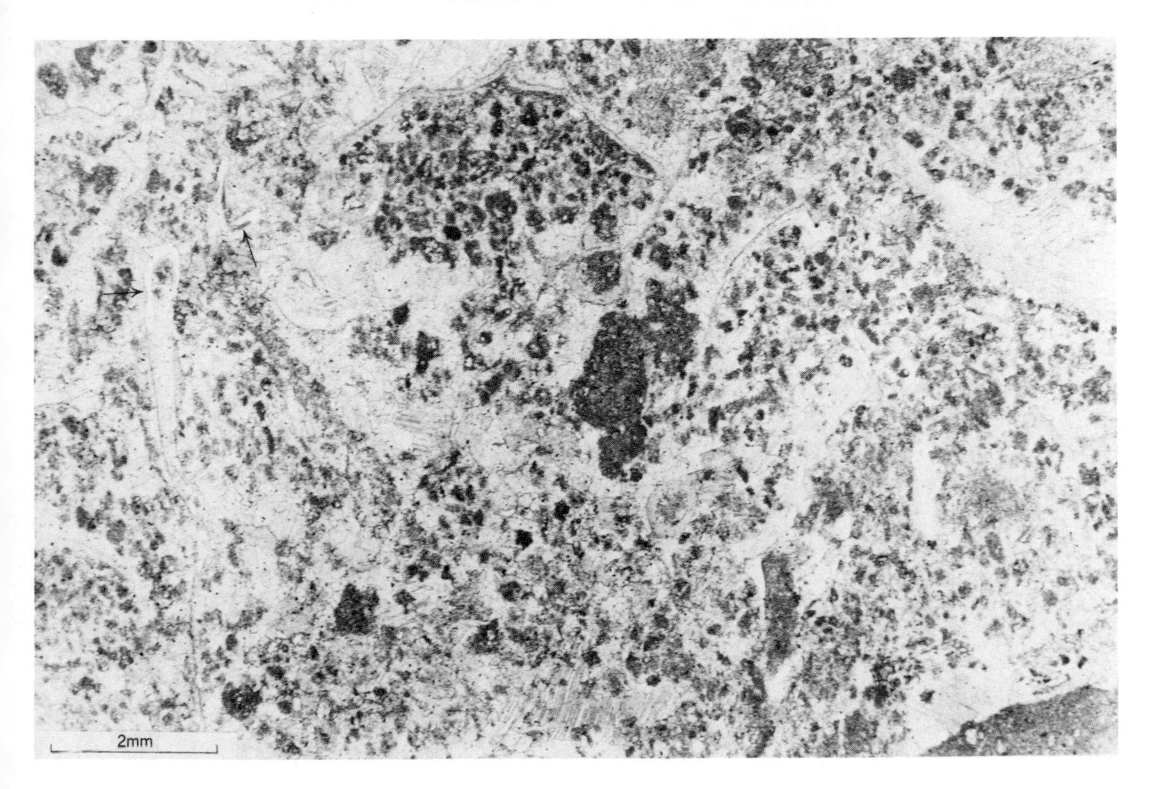

Fig. 14.—Photomicrograph of a mottled thin-bedded limestone (Adamstown Member). Peloids, pelmatozoan debris, and partially silicified trilobites (arrows) are major allochems.

FIG. 16.—Mottled limestone in the Lime Kiln Member containing a planar-laminated bed. Laminated bed is scoured at the top; subsequent fill is draped on left side and unstratified on right.

FIG. 18.—High-relief, crudely laminated stromatolite encased in wavy limestone. Dolomitic laminae, within fine-grained limestone emphasize wavy bedding and cryptalgal structure. Stromatolitic head indicates deposition within the photic zone; sediment size suggests a low-energy environment at the top of the Lime Kiln Member of the Frederick Limestone.

Austinville, Virginia, contain some lithologies similar to the Frederick Limestone. This discussion, however, is directed primarily to the contrasting features. Two megabreccia zones 20 to 100 m thick in the Kinzers Formation are exposed in quarries at Thomasville, Pennsylvania, and several breccia horizons in the Shady Dolomite are exposed both on the surface and in the subsurface in the Austinville area. Although the breccia units in the Shady are similar in thickness to those in the Frederick Limestone, the clast petrology and the interbedded sediments contrast sharply with the Rocky Springs Station and Adamstown breccia units.

Thomasville area.—The intent of this section is not to evaluate critically the rocks and paleogeographic setting of the entire Cambrian and Ordovician carbonate section in the Conestoga Valley as done by Gohn (1976) but to concentrate on the Thomasville Member of Gohn (1976) in the Kinzers Formation at Thomasville Stone and Lime Company. The stratigraphy is dominated by two very thick intervals of breccia, containing blocks as large as 15 m across. The breccia intervals (20 to 100 m thick) are separated by sections 10 m thick of thin-bedded, argillaceous limestone ("bottom black" and "top black" of Cloos, 1968). The predominance of breccia over thin-bedded limestone constitutes an off-shelf assemblage quite different from the Frederick Limestone.

The external geometry of the breccia deposits is unclear; the bases appear to be erosive and broadly undulatory. The variation in thickness of the breccia units from 15 to 90 m indicates a lensoid shape, yet these do not appear to be channel fills. Channel forms up to 30 m across in the Thomasville Quarry are filled by skeletal, laminated, micritic limestones (Gohn, 1976).

The breccia blocks are angular to poorly rounded, poorly sorted, and loosely packed in either a dolomitic sand or a micrite matrix (Fig. 28). The largest blocks are composed of white, aphanitic limestone and light-gray oolitic limestone; smaller clasts and chips are second-cycle limestone breccias and dark-gray argillaceous limestone. The mega-breccia deposits are poorly stratified to chaotic; normal grading of breccia blocks is rare and confined to the tops of breccia beds. Inverse grading is seen in 1 m thick beds (Gohn, 1976). Apparent imbrication of clasts in some areas of the quarry results from tectonic elongation of the breccia blocks.

Austinville area.—Breccia units are restricted to the "south limb facies" of Brown and Weinberg (1968). Although breccia intervals are seen at many locations within the active mine, surface exposures just south of Austinville provide an excellent site for study of the shape and petrology of the breccia horizons. The breccia units can be traced laterally for 10 to 30 m and are thought to be pod-shaped bodies. The rocks of the Post Taylor Marker are locally composed of amalgamated breccia deposits separated by thin intervals of laminated, argillaceous to banded, detrital limestone. Stromatolitic and oolitic clasts constitute 50 to 70 percent of the clast population (Fig.

←

FIG. 17.—Mottled beds in the Lime Kiln Member containing pods of fine-grained limestone and wispy dolomitic laminae, interbedded with layers containing trilobite debris. Note lag deposit at base.

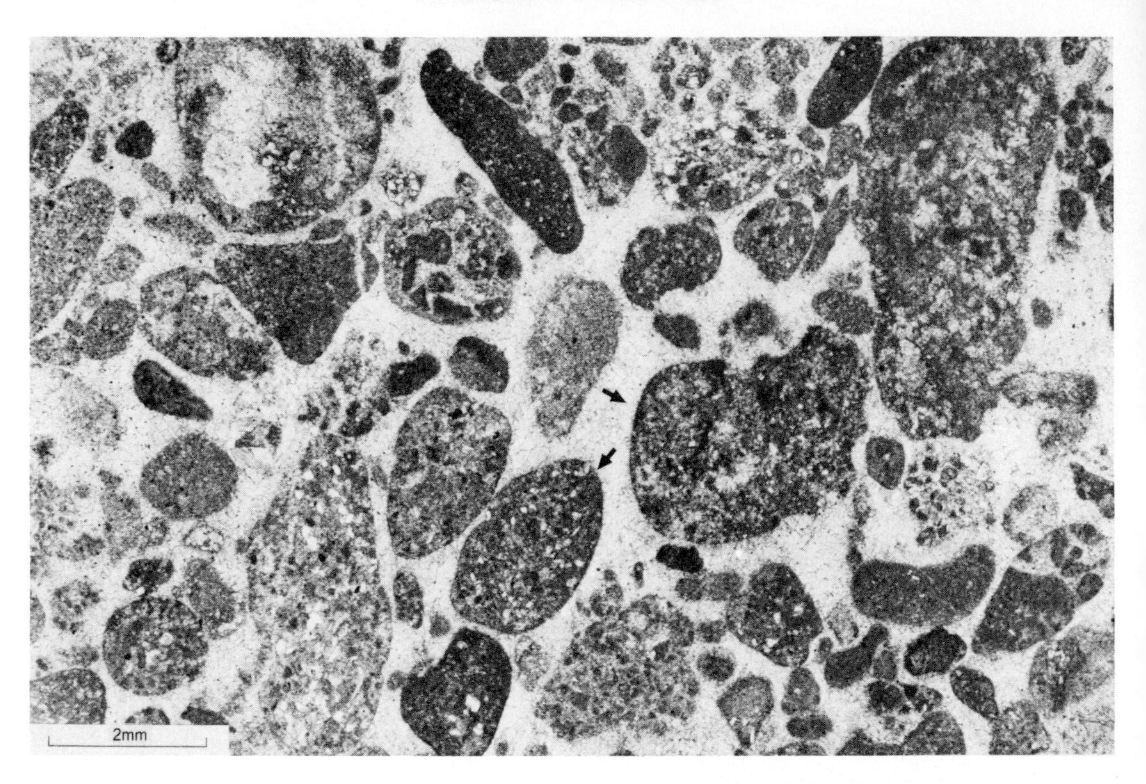

FIG. 19.—Photomicrograph of a polymict detrital limestone. Microspar cement fills the interstices of this grain-supported lithology; fine-grained cement "crusts" (arrows) on most grains.

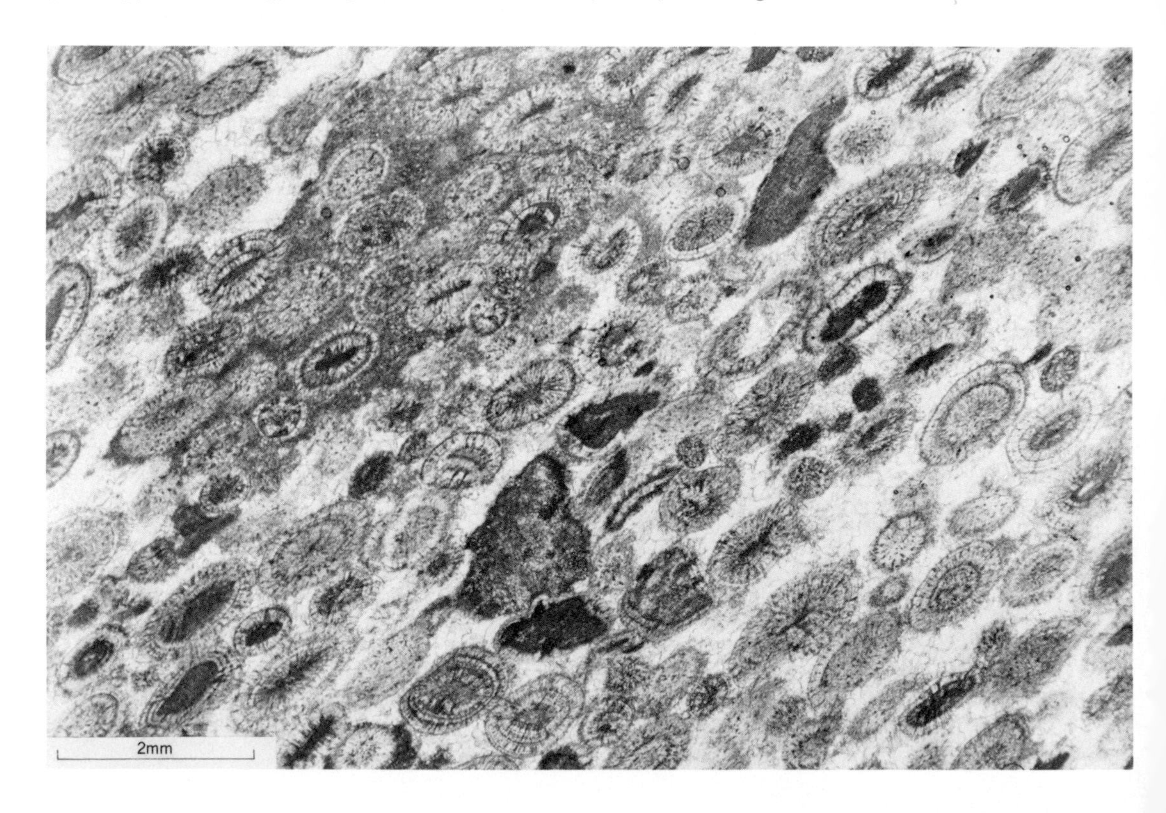

FIG. 21.—Thick-bedded, arenaceous limestone stratified in planar-tabular beds. This lithology, composed of coarse quartz sand and peloids, tops the Rocky Springs Station sequence illustrated in Figure 8.

29). Pressure solution has modified the margins of most larger clasts, but soft-sediment deformation of many flat clasts is still evident. The matrix of the breccia deposits is coarse and sand sized, but is poorly sorted both in size and composition; the breccia deposits are typically framework supported.

Observations by Pfeil (oral commun. 1976, Pfeil and Read, 1976) in new surface quarries southeast of Austinville, also in the "Upper Shady," indicate interfingering and flanking relationships between beeccias and calcareous algal bioherms. The breccias are composed of calcareous algal boundstone, lime grainstone, and black shale fragments typically 3 to 30 cm long and rare blocks up to 10 m long (Pfeil and Read, 1976).

ENVIRONMENTAL INTERPRETATION

Criteria for off-shelf depositional environments are still in the process of development. Individual lithologies can hardly be assigned to a unique depositional environment; however, suites of lithologies and sedimentary structures, coupled with basin analysis enable some level of confidence in environmental interpretation even in areas of considerable structural complexity.

The Frederick Valley contains a thick sequence of carbonate rocks which in the lower two-thirds contain no quantitative bathymetric indicators. The top of the sequence, specifically the Grove Limestone, however, is composed of shallow-water rocks, which can be compared with modern shallow-water and tidal-flat carbonate sediments. Each element in the Grove Limestone cycle has a modern analogue, to be found on the Florida-Bahama platform, the Persian Gulf, or Shark Bay, Western Australia.

The cross-bedded, oolitic and arenaceous limestone, which constitutes the basal lithologic unit in the Grove cycle (Fig. 25), is analogous to the

FIG. 20.—Photomicrograph of a moderately deformed, massively bedded oolite from the top of the Rocky Springs Station Member. Size and compositional sorting is much better than that in the sample shown in Figure 18. Micritic matrix is partially neomorphosed to spar cement.

FIG. 22.—Poorly bedded, polymict breccia in the Rocky Springs Station Member. Granular matrix and many stylolitic clast margins are apparent in center of photo. Poorly defined imbrication of clasts is inclined to the right at a low angle to bedding.

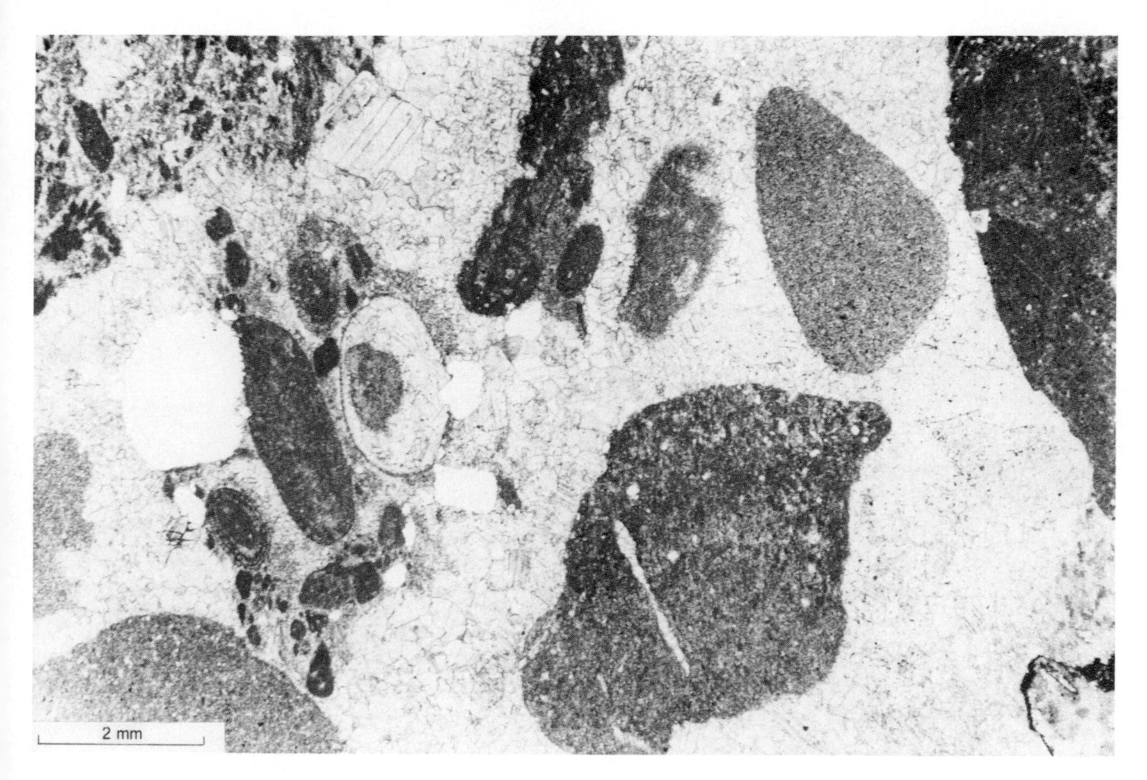

FIG. 24.—Photomicrograph of a detrital limestone interbedded with Adamstown Member breccias. Diverse framework components are poorly defined because of recrystallization of matrix to spar.

marginal sand bodies described by Ball (1967). On the modern Florida and Bahama platforms, sand bodies, such as oolitic shoals, typically occur at the break in slope either along the edge of a carbonate shelf or at the ends of deep embayments. The thick local accumulation of cross-bedded and planar-tabular strata at or close to the base of the Grove Limestone is a strong indication of a shelf margin environment. The arenaceous sand bodies throughout the Grove Limestone probably represent deposition on shoals or sand bars; those containing abundant oolite probably accumulated in less than 5 m of water.

The large, poorly laminated stromatolites, which characterize the middle lithologic unit of a Grove cycle, have modern analogues in Shark Bay, Western Australia. Descriptions of the stromatolites in and along Hamelin Pool (Logan 1961; Logan and others, 1974) include a wide range of structures, from well laminated to poorly organized, which can be matched with many examples of stromatolites in Paleozoic and Proterozoic rocks. Colloform-mat structures, now forming in the subtidal zone (about 2 m below mean tide level), and smooth-mat structures and fabric within the intertidal zone are the two mat types which most closely resemble the stromatolitic forms and fabrics in the middle part of the Grove cycles. The coarse intra-stromatolite fill (Fig. 25b) in the Grove Limestone may be analogous to the intraclast pavements described from the lower supratidal to upper intertidal subzones of Hutchinson embayment (Hagan and Logan, 1974, Fig. 19).

The massive, finely laminated or irregularly laminated dolomite, which caps the Grove Limestone cycles, is probably the product of a variety of supratidal environments. The range of algal laminates from the Persian Gulf, (Kendall and Skipwith, 1968), the Bahamas (Hardie and Ginsburg, 1977; Monty, 1967; Black, 1933), and Shark Bay (Davies, 1970; Hagan and Logan, 1974) include nearly flat, crenulated, and scalloped laminae. These reflect an interplay of algal species, sedimentation, and wetting-desiccation. A com-

FIG. 23.—Limestone breccia in the Adamstown Member composed of clasts having similar compositions in a uniform dolomitic matrix.

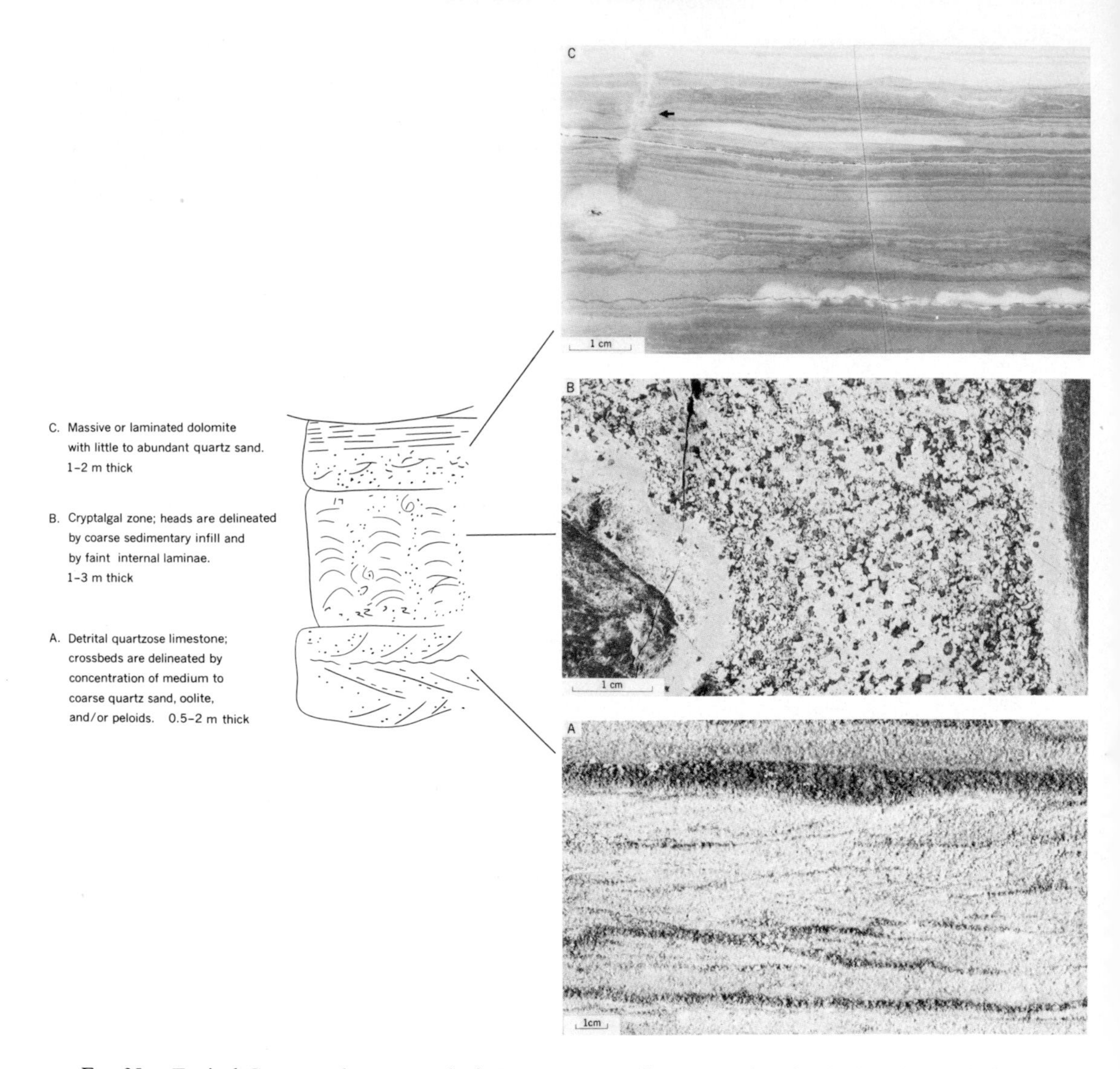

FIG. 25.—Typical Grove cycle composed of: (A) arenaceous limestone with "herringbone" crossbedding, (B) stromatolites (plan view) delineated by intraclasts (filling intrastrome areas) and poorly laminated, dolomitic margins, and (C) fine-grained dolomitic laminite containing sparse mudcracks (arrow).

parison between scallop fabrics in the Grove Limestone (Fig. 27) and similar fabrics produced by tufted algal mats (Hagan and Logan, 1974, Fig. 19c) growing on beach ridges is striking. Another possible modern analogue for the irregular dolomitic laminates is a crenulated to tufted smooth mat from the levee backslope subzone; the example in Figure 30 is from Sugarloaf Key, Florida. The sandy pods are composed of fossil debris, which accumulates as storm lag; the fine-grained continuous laminae result from sediment trapping by blue-green algae.

In summary, the rocks of the Grove Limestone were deposited in shallow subtidal to supratidal environments close to the shelf margin during the Early Ordovician. Each Grove cycle records sedimentation in progressively shallower environments indicating that the sedimentation rate was always greater than the rate of subsidence. The cycle was completed or terminated either by erosion partially obliterating the previous cycle or by subsidence reducing the area of positive relief to below mean sea level, and another cycle began.

FIG. 26.—SH-type stromatolites above cross-bedded arenaceous limestone. The contact is defined by a ragged stylolite. Note the discontinuity in cross-beds, resulting from large-scale pressure solution.

The transition from the "deeper water" carbonate rocks in the Frederick Limestone to the shoal water carbonates in the Grove Limestone is a single large-scale shallowing-upward sequence that resulted from a major seaward progradation of the shelf. The inferred stratigraphic relationships are shown in Figure 31. The depositional environments as expressed by sedimentary structures within the thin-bedded lithologies of the Frederick Limestone are quite gradational. The rocks in the Rocky Springs Station Member indicate an abiotic basin lacking bottom currents. Higher in the stratigraphic sequence sedimentation barely exceeds bioturbation in the depositional basin. Although precise bathymetry is lacking, the increases in biological activity and the evidence for bottom currents support the model of a shallowing basin during the Late Cambrian. Comparative sedimentology allows analysis of the upper part of the basin fill, but the depositional environments lower in the carbonate section lack well-documented modern analogues.

The rocks in the Lime Kiln Member stratigraphically below and paleogeographically seaward of the high-energy shelf are characterized by fine-grained sediments, mottled textures, and an absence of peritidal features in cycles 10–20 m thick (Fig. 10). The cyclic pattern is thought to result from the long term migration of channels across a subtidal below wave-base aggrading surface. The sediments are largely accumulations by hemipelagic fallout of fine particles probably winnowed from the high-energy adjacent shelf. The poorly sorted cycle bases are probably a combination of a coarse traction lag and infiltration of fine sediments either during intervals of slack current or after channel abandonment.

The maximum depth of accumulation is not known, but proximity to peritidal carbonate rocks of the Grove Limestone and the appearance of stromatolites at the top of the Lime Kiln suggests an outer shelf to upper slope environment.

The absence of cyclic sedimentation, sparse burrowing, and presence of breccias in the Adamstown suggest a slightly different depositional environment. The dominant lithology is an even-laminated or cross-laminated thin-bedded limestone. Erosive bases and graded bedding indicate waning, mostly likely turbudity currents as a dominant sedimentary process. The cross-laminated beds composed of well-sorted, sand size sediment suggest sedimentation or at least reworking by traction rather than fallout.

The breccias in the Adamstown Member are less obviously sedimentary than those of the Rocky Springs Station Member; indeed, they were previously considered tectonic in origin (Reinhardt, 1974). Similarity to intraformational slump breccias in a very similar suite of rocks in the Great Basin (Cook and Taylor, this volume) has led to reconsideration of these deposits. Faunal comparisons with the Great Basin biofacies assemblages suggest a shelf-edge to basin-margin environments for the transported forms found in the Adamstown (M. E. Taylor, written commun., 1973).

These lines of evidence suggest that the Adamstown Member is, at least in part, a slope deposit. The breccias are thought to be intraformational slumps and the major thin-bedded accumulations are turbidites. The degree of sedimentation or resedimentation by bottom currents within this unit has not been evaluated, but sediment transport by bottom currents is thought to have been a probable slope environment process.

The Rocky Springs Station Member is comprised of sediments deposited in lower slope to basinal environments. The laminated micritic sediments which constitute the dominant lithofacies probably result from a number of sedimentary

FIG. 27.—Irregularly laminated dolomite composed of arenaceous pods and continuous argillaceous laminae. Note the scallop textures between peaked highs.

mechanisms including turbidity currents which would account for complete "Bouma sequences" within individual beds (Figs. 11, 12) and the prevalence of graded beds. *In situ* sedimentation from hemipelagic fallout accounts for even-laminated argillaceous limestones to calcareous shales.

Extrabasinal components are rare in the thick, normally graded breccia deposits in the Frederick Limestone. Oolite clasts are typically small and constitute only a small portion of the deposits. The coarse, well-rounded quartz and oolite sand matrix is probably extrabasinal, as are allochems of the thick-bedded detrital limestone (Fig. 9c). These sediments and their vertical arrangement are consistent with existing models for submarine-fan and channel-fill systems (Stanley and Unrug, 1972; Walker, 1975). On the basis of good stratification and normal grading, the breccias were probably deposited in the inner fan close to the toe of slope. These channel systems were apparently highly erosive based on the composition of the channel fills. The thick-bedded detrital limestone units are characteristic of channeled suprafan deposits; the laminated micritic sediments represent interchannel overbank deposits and/or sedimentation from pelagic fallout. The turbidites which dominate the lower portion of the Rocky Springs Station Member are the middle- and outer-fan deposits. The Araby Formation which underlies the entire carbonate sequence represents the starved clastic basin.

The dimensions of the basin can be estimated from the thickness and composition of the sediments. An estimate of water depth based on the thickness of slope deposits has application in a basin fill resulting from progradation (Asquith, 1970). The composition of the sediments and the shoaling-upward pattern within the Frederick Limestone indicate that the basin was more than 500 m deep during the early Late Cambrian (Dresbachian Stage). This estimate is based on the entire thickness of the Adamstown Member (325 m) and parts of both the Rocky Springs Station Member (the remainder being basinal) and the Lime Kiln Member (the balance being subtidal low-energy shelf).

The rocks and consequently the interpreted depositional environments of the two Early Cambrian off-shelf areas discussed are somewhat different. The contrasts, especially in the scale

FIG. 28.—*Left*—Upper breccia unit in the Thomasville Member of Gohn composed of polymict angular blocks floating in a dolomitic sand matrix. Pen for scale in lower center of photograph. *Right*—The upper breccia unit in the Thomasville Member overlain by thin-bedded limestone ("top black"). The oolitic block is at least 5 m long.

and composition of the breccia deposits, lead us to consider the environmental parameters which might lead to such differences. This assumes that the breccia of the Frederick and Shady of Thomasville and megabreccia deposits were transported by a similar sediment-gravity mechanism. The major differences in the deposits probably relate to such factors as: (1) nature and composition of the shelf and shelf margin (emergent-submerged; cemented-unconsolidated; channeled-undissected), (2) composition and coherence of the slope, (3) slope angle, (4) proximity to source(s), and (5) tectonics of the basin margin. The composition and scale of the breccia units at Thomasville quarry and the regional relationships of the Thomasville Member as described by Gohn (1976) suggest greater proximity to source than in the Frederick Limestone breccias. Poor stratification, enormous clast size, clast variability, lack of soft-clast deformation, and lack of deformation within beds suggest massive debris flows. Gohn (1976) has suggested that the thick accumulation of coarse basinal sediment in the Thomasville area may result from the physical geometry of the shelf margin. A major embayment with associated submarine canyons (sediment funnels) could allow for thick, coarse sediment accumulations. The modern analog of Tongue of the Ocean suggested by Rodgers (1968) might be applied to this area based on such criteria as clast size and multi-cycle transport down a relatively high-angle slope (Andrews and others, 1970).

The nature of the shelf margin as seen from the breccia deposits in the Austinville area contrasts sharply with both the Frederick and Thomasville rocks. The average clast size and matrix composition of the Shady breccias is similar to the breccias near the top of the Rocky Springs Station. The abundance of algal boundstone, stromatolitic, and oolitic clasts, however, indicates the importance of the shelf-margin contribution. Pfeil and Read (1976) have documented

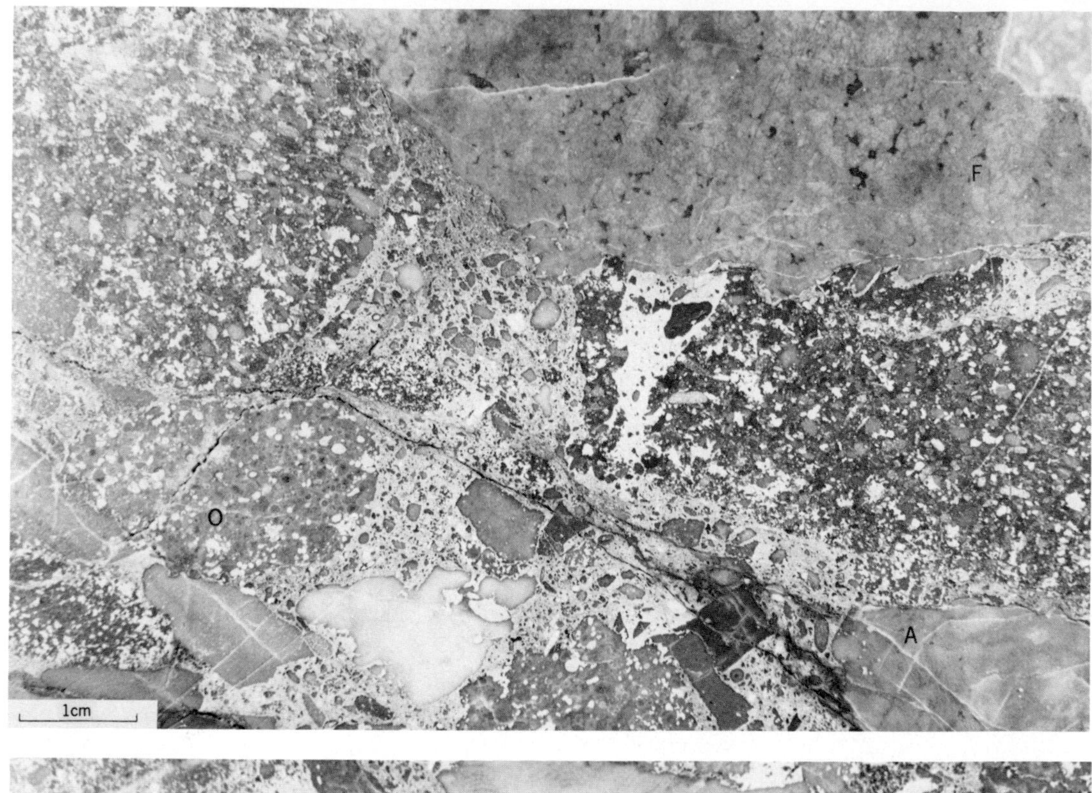

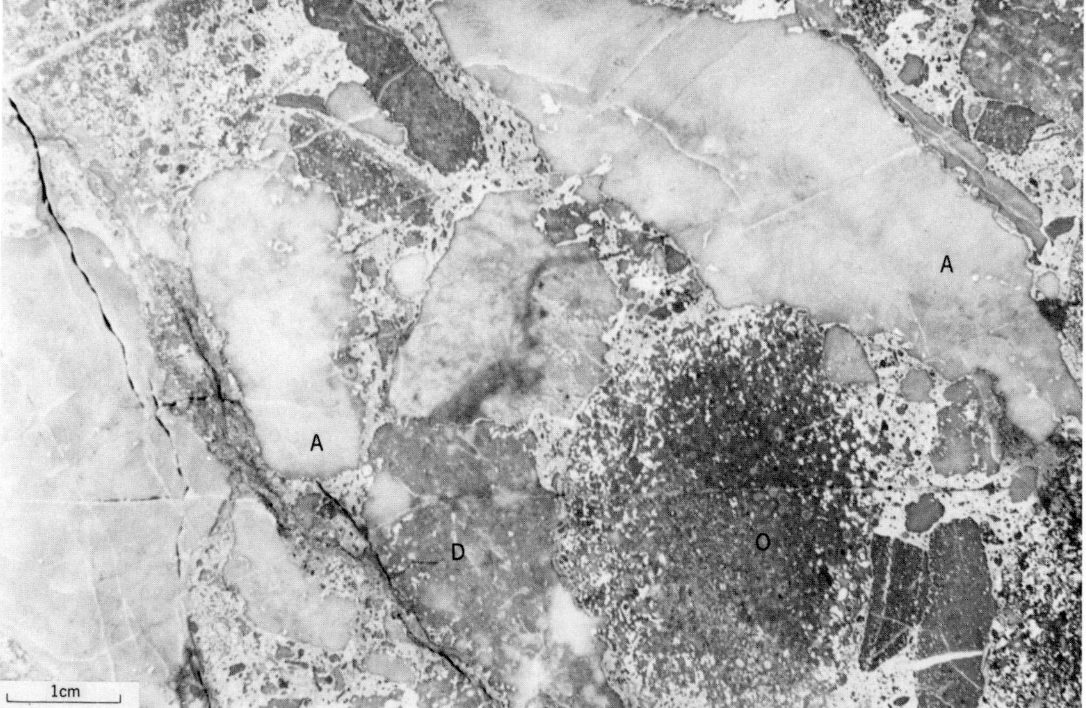

FIG. 29.—*Bottom*—Etched slab of polymict breccia from the Shady Dolomite. Clasts are composed of stromatolitic limestone (A), oolitic limestone (O), detrital limestone (D), and calcareous shale. Clast margins are highly stylotized. *Top*—Etched slab of Shady Dolomite composed of oolitic limestone (O), stromatolitic limestone (A), fenestral limestone (F), and micritic limestone (M). Note the partially dolomited coarse matrix.

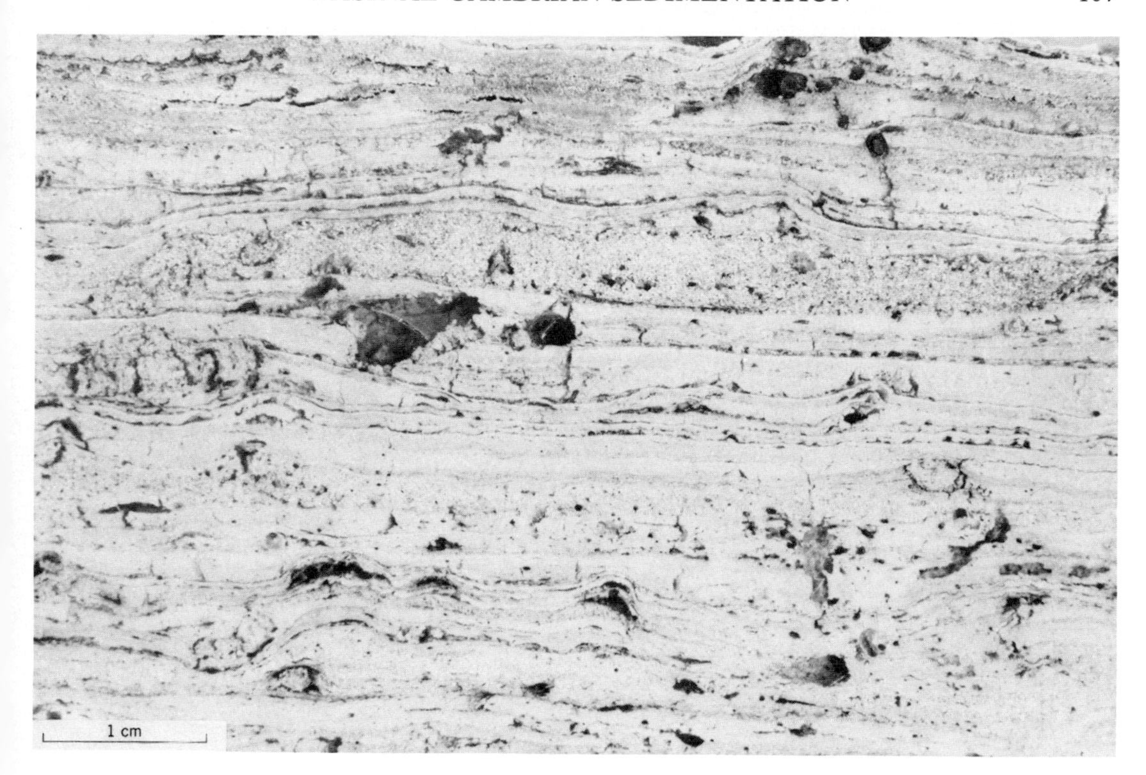

FIG. 30.—Modern algal laminate from the levee backslope zone, Sugar Loaf Key, Florida. Lenses of coarse skeletal debris (storm lags) are interbedded with continuous fine-grained drapes (products of algal trapping). Large vugs result from mangrove roots or algal blisters.

interfingering relationships between algal bioherms and breccias. Some of the largest clasts are algal boundstone. Recent reinterpretation of the Devonian algal reefs in eastern Australia by Conaghan and others (1976) and Canada by Cook and others (1972) as allochthonous blocks raises similar questions about possible basinward transport of the Austinville reefal carbonate rocks. Juxtaposition with peritidal carbonate rocks in the "north limb" facies suggests proximity to shelf margin, yet the abundance of interbedded argillaceous black limestones suggests a quiet basinal setting. A tentative conclusion is that the Austinville breccia deposits were deposited just seaward of a partially cemented shelf margin on a quiet low-angle slope.

PLATFORM MARGIN MIGRATION

The Cambrian and Ordovician carbonate platform was maintained at near sea-level conditions for about 130 million years in the Appalachians. The position of the platform margin during most of that time is unknown or unclear because it was not stationary and the diagnostic rocks have been stripped or covered by structural events. The specific mechanisms for progradation and regression are not known, but differential rates of sedimentation and subsidence on the platform and in the basin, and fluctuations in sea level were probably the controlling factors.

The west-to-east migration of the shelf margin in the Frederick area has been documented and is schematically shown in Figure 32. Although carbonate sediment production is almost entirely on the carbonate platform, sedimentation rates on the slope or in the basin may be higher. Sedimentation on the slope and in the basin results from bypassing of the platform, spillover of loose allochems (quartz sand, oolites), and sedimentation of platform lithologies. Much of the micritic sediment in the basin was probably winnowed by tidal currents from close to the platform margin. Grains as large as coarse sand were probably swept off the platform either as a result of storms (see Hayes, 1967) or normal circulation patterns (see Stanley and others, 1972). These processes probably contributed more to platform progradation than slumping of platform lithologies along the shelf margin.

The evolution of the Conestoga Valley basin from late Early Cambrian to Middle Cambrian time has been discussed by Gohn (1976). Rock

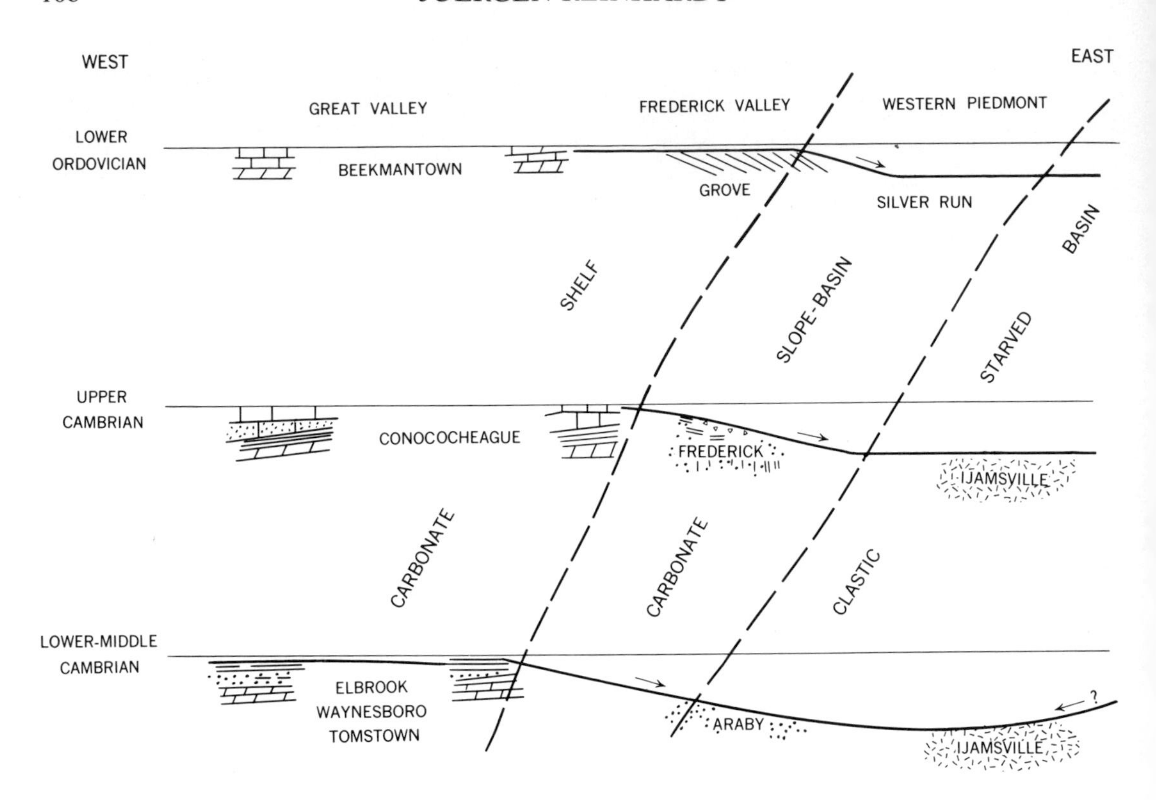

FIG. 31.—Schematic relationships between the Great Valley and the Frederick Valley sections. The lower lithofacies line defines the clastic-carbonate rock transition. The upper contact marks the transition from primarily fine-grained carbonate rocks (Elbrook-Frederick) to high-energy shelf carbonate rocks (Conococheague-Grove). The distribution of lithofacies indicates an eastward progradation of environments.

facies representing basin, slope, and shelf-margin environments define a shoaling-upward sequence analogous to the Frederick Valley basin fill.

This progradational sequence has been documented on the north side of the Conestoga Valley as Antietam Formation, Vintage Dolomite, and Kinzers phyllite (basinal); Thomasville and Longs Park Members of the Kinzers and lower Ledger Formation (basin margin and slope); and upper Ledger (shelf to shelf margin) (Gohn 1976). Lithologies along the south side of the valley are all basinal. Thinning of the carbonate rocks from above 1000 m along the north side of the valley to about 80 m along the south side may result in part from tectonic thinning, but may in large part be controlled by sedimentation rates (Fig. 33).

In the Austinville area, the physical stratigraphy implies an eastward progradation of the carbonate platform. Not only is the Shady Dolomite unusually thick at or near the platform margin, but the northeastern facies is stratigraphically overlain by shallow-water to subaerial carbonate deposits in the Rome Formation and Elbrook Limestone (Butts, 1940; Pfeil and Read, 1976).

The magnitude of shelf margin progradation at any one of these localities is somewhat speculative. Palinspastic reconstruction of the Frederick Valley synclinorium indicates that the minimum seaward progradation of the carbonate shelf must have been about 10 kilometres. On the basis of regional geology (position of the shelf margin on the western margin of South Mountain in the late Early Cambrian), clast size, and petrology of the Rocky Springs Station breccias, a more realistic estimate would be about 15 to 25 kilometres. Shelf

FIG. 33.—Restored stratigraphic section with inferred basinal (1), slope (2), and shelf margin (3) lithologies in the Conestoga Valley shoaling-upward sequence. Time lines are inferred and represented by bold dashes. The thinned section on the south side is thought to result from lower sediment supply, but may also result from erosion prior to deposition of the overlying rocks. Modified from Gohn, 1976.

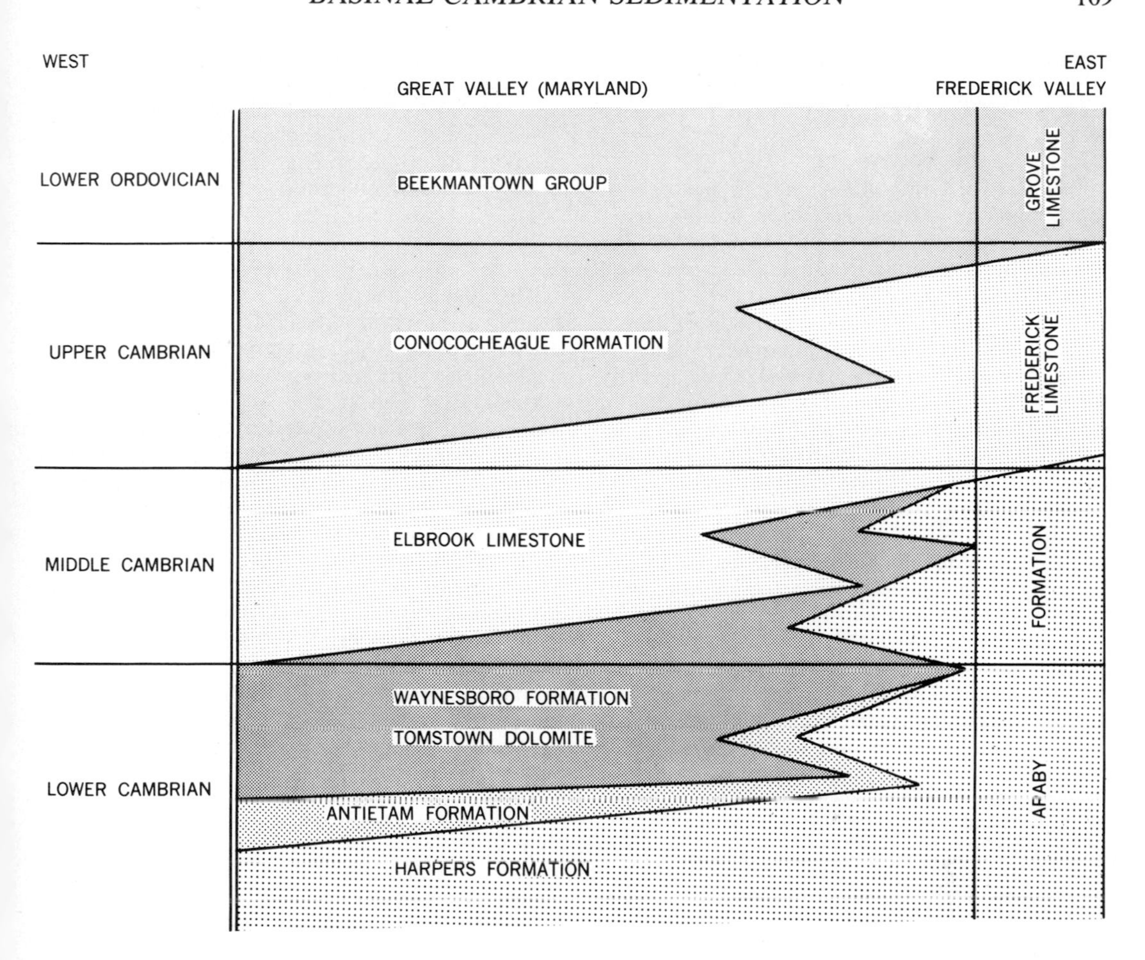

FIG. 32.—Schematic progradation and eastward migration of the carbonate shelf of the Frederick Valley from Middle Cambrian to Early Ordovician. Similar progradation and migration may have taken place during the Early Cambrian.

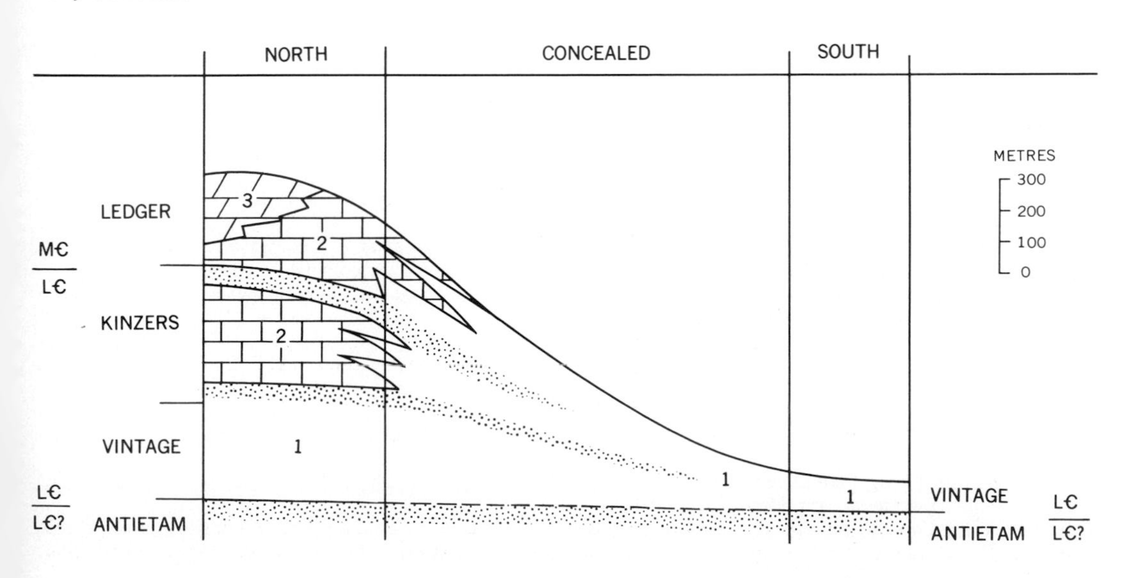

migration with similar orders of magnitude have been postulated for the Early and Middle Cambrian in the Conestoga Valley by Gohn (1976).

CONCLUDING REMARKS

Cambrian off-shelf limestones in the central Applachians have been described and discussed for three distinct paleogeographic areas. Only the rocks of the Frederick area had been discussed in terms of synchronous shelf and basin sedimentation (Wilson, 1952) prior to Rodgers' (1968) "Eastern edge of the North American continent." Although Rodgers' model has been challenged by and modified by Reinhardt (1974), Gohn (1976) and Gohn and Reinhardt (1976) the model acted as the stimulus to critically evaluate these basinal rocks, set in tectonically complex settings and characterized by extremely poor outcrop conditions.

The models developed for shelf-margin migration in the Appalachians require comparison with the mirror image rocks in the Cordilleran basin. Such comparisons require better biostratigraphic and lithostratigraphic control than is currently available in the Appalachians. A careful evaluation of the similarity in sedimentation patterns at or near the eastern and western margins of the North American plate might result in an understanding of the major controls on sedimentation during the Cambrian and Ordovician. Was the North American plate rigid and passively acted upon by eustatic changes in sea level or did plate margin tectonics play an important role in continental freeboard sedimentation?

ACKNOWLEDGEMENTS

The discussion of the Frederick area draws heavily upon parts of a Ph.D. dissertation completed at John Hopkins Univerity. Instrumental in completion of that work were the late Ernst Cloos, G. W. Fisher, L. A. Hardie, and F. J. Pettijohn. Financial support by the German Society of Maryland and the Maryland Geological Survey are acknowledged. Ernst Cloos and especially Gregory S. Gohn aided my understanding of the stratigraphy of the Thomasville area. I thank E. L. Yochelson, Ralph Willoughby, and E. L. Weinberg for an introduction to the rocks and problems of the Austinville area. Russell Pfeil, J. F. Read, and G. S. Gohn gave access to unpublished information and have provided stimulating discussion. P. A. Scholle, J. B. Epstein, Paul Enos, and H. E. Cook read preliminary drafts of this paper; their comments and criticism were very much appreciated.

REFERENCES

Andrews, J. E., Shepard, F. P., and Hurley, R. J., 1970, Great Bahama Canyon: Geol. Soc. America Bull., v. 81, p. 1061-1078.

Asquith, D. O., 1970, Depositional topography and major marine environments, Late Cretaceous, Wyoming: Am. Assoc. Petroleum Geologists Bull, v. 54, p. 1184-1224.

Ball, M. M., 1967, Carbonate sand bodies of Florida and the Bahamas: Jour. Sed. Petrology, v. 37, p. 556-591.

Black, Maurice, 1933, Carbonate sand bodies of Florida and the Bahamas: Royal Soc. London Pholos. Trans., Ser. B, v. 222, p. 165-192.

Bouma, A. H., 1962, Sedimentology of some flysch deposits: Elsevier, Amsterdam, The Netherlands, 168 p.

Brown, W. H., and Weinberg, E. L., 1968, Geology of the Austinville-Ivanhoe District, Virginia: *In* I. D. Ridge (ed.), Ore deposits of the United States 1933-1967: Am Inst. Mining, Metallurgical and Petroleum Engineers, New York, New York, p. 169-186.

Butts, Charles, 1940, Geology of the Appalachian Valley in Virginia: Virginia Geol. Survey, Bull. 52, pt. 1, 568 p.

Campbell, L. D., 1971, Occurrences of "*Ogygopsis* shale" fauna in southeastern Pennsylvania: Jour. Paleontology, v. 45, p. 436-440.

Cloos, Ernst, 1947, Oolite deformation in the South Mountain fold, Maryland: Geol. Soc. America Bull., v. 58, p. 843-917.

——, 1968, Thomasville Stone and Lime Company, Thomasville, Pennsylvania: *In* Geology and mineral deposits in south-central Pennsylvania: Guidebook 33rd Field Conf. Pennsylvania Geologists, p. 16-21.

—— and Heitanen, A. M., 1941, Geology of the "Martic overthrust" and the Glenarm Series in Pennsylvania and Maryland: Geol. Soc. America, Spec. Paper 35, 207 p.

Conaghan, P. J., Mountjoy, E. W., Edgecomb, D. R., Talent, J. A., and Owens, D. E., 1976, Nubrigyn algal reefs (Devonian), eastern Australia: Allochthonous blocks and megabreccias: Geol. Soc. America Bull., v. 87, p. 515-530.

Cook, H. E., McDaniel, P. N., Mountjoy, E. N., and Pray, L. C., 1972, Allochthonous carbonate debris flows at Devonian bank ('reef') margins, Alberta, Canada: Bull. Canadian Petroleum Geology, v. 20, p. 439-497.

Davies, G. R., 1970, Algal-laminated sediments, Gladstone Embayment, Shark Bay, Western Australia: Am. Assoc. Petroleum Geologists, Mem. 13, p. 169-205.

Donaldson, A. C., 1969, Stratigraphic framework of stromatolites in Lower Ordovician of the central Appalachians: *In* A. C. Donaldson (ed.), Some Appalachian coals and carbonates: Models of ancient shallow-water deposition:

West Virginia Geol. and Econ. Survey, Morgantown, West Virginia, p. 357-384.

Dott, R. H., Jr., 1963, Dynamics of subaqueous gravity depositional processes: Am. Assoc. Petroleum Geologists Bull., v. 47, p. 104-128.

Fellows, R. E., 1943, Recrystallization and flowage in Appalachian quartzite: Geol. Soc. America Bull., v. 54, p. 1399-1432.

Gathright, T. M., and Nystrom, P. G., 1974, Geology of the Ashby Gap Quadrangle, Virginia: Virginia Div. Mineral Resources, Rept. Inv. 36, 55 p.

Ginsburg, R. N., and James, N. P., 1974, Holocene carbonate sediments of continental shelves: *In* C. O. Burk and C. L. Drake (eds.), The geology of continental margins: Springer-Verlag, New York, New York, p. 137-156.

Gohn, G. S., 1976, Sedimentology, stratigraphy, and paleogeography of lower Paleozoic carbonate rocks, Conestoga Valley, southeastern Pennsylvania: Ph.D. dissert., Univ. Delaware, Newark, Delaware, 315 p.

—— and Reinhardt, Juergen, 1976, Platform and basin lithofacies in Lower Cambrian carbonate rocks—Tomstown and Vintage Dolomites, central Appalachians [abs.]: Geol. Soc. America Abs. with Programs, v. 8, p. 182.

Hagan, G. M., and Logan, B. W., 1974, History of Hutchinson Embayment tidal flat, Shark Bay, Western Australia: Am. Assoc. Petroleum Geologists, Mem. 22, p. 283-315.

Hardie, L. A., and Ginsburg, R. N., 1977, Layering: The origin and environmental significance of lamination and thin bedding: *In* L. A. Hardie (ed.) Sedimentation on the modern carbonate tidal flats of northwest Andros Island, Bahamas: Johns Hopkins Studies in Geology No. 22, Johns Hopkins Press, Baltimore, Md., p. 50-123.

Harris, L. D., 1973, Dolomitization model for the Upper Cambrian and Lower Ordovician rocks in the eastern United States: U.S. Geol. Survey Jour. Research, v. 1, p. 63-78.

Hayes, M. O., 1967, Hurricanes as geological agents: Case studies of hurricanes Carla, 1961, and Cindy, 1963: Univ. Texas Bur. Econ. Geology, Rept. Inv. 61, 56 p.

Jonas, A. I., and Stose, G. W., 1930, Geology and mineral resources of the Lancaster Quadrangle: Pennsylvania Geol. Survey, Atlas 168, 106 p.

—— and ——, 1936, The reclassification of the Frederick Valley (Maryland) limestones: Geol. Soc. America Bull., v. 47, p. 1657-1674.

Keith, B. D., 1974, Recognition of a Cambrian paleoslope and base-of-slope environment, Taconic sequence, New York and Vermont: Ph.D. Dissert., Renssalaer Polytechnic Institute, Troy, New York, 215 p.

Kendall, C. J., and Skipwith, P. A. D'E., 1968, Recent algal mats of a Persian Gulf lagoon: Jour. Sed. Petrology, v. 38, p. 1040-1058.

Logan, B. W., 1961, *Cryptozooan* and associated stromatolites from the Recent, Shark Bay, Western Australia: Jour. Geology, v. 69, p. 517-533.

——, Hoffman, Paul, and Gebelein, C. D., 1974, Algal mats, cryptalgal fabrics and structures, Hamelin Pool, Western Australia: Am. Assoc. Petroleum Geologists, Mem. 22, p. 140-194.

Matter, Albert, 1967, Tidal flat depostis in the Ordovician of western Maryland: Jour. Sed. Petrology, v. 37, p. 601-609.

Middleton, G. V., and Hampton, M. A., 1973, Sediment gravity flows: Mechanics of flow and deposition: *In* G. V. Middleton and A. H. Bouma (eds.), Turbidites and deep water sedimentation: Soc. Econ. Paleontologists and Mineralogists Pacific Sec. Los Angeles, California, p. 1-38.

Monty, Claude, 1967, Distribution and structure of recent stromatolitic algal mats, eastern Andros Island, Bahamas: Soc. Geól. Belgique Annales, v. 90, p. 55-100.

Palmer, A. R., 1971, The Cambrian of the Appalachian and eastern New England region: *In* C. H. Holland (ed.), Cambrian of the New World: Wiley-Interscience, London, England, p. 169-217.

Pfeil, R. W., 1977, Stratigraphy and sedimentology, Cambrian Shady Dolomite: M.Sc. Thesis, Virginia Polytechnic Institute and State Univ., Blacksburg, Virginia, 137 p.

—— and Read, J. F., 1976, Reefal carbonates and associated lithologies, Cambrian Shady Dolomite, Austinville region, Virginia [abs.]: Geol. Soc. America Abs. with Programs, v. 8, p. 244.

Rasetti, Franco, 1961, Dresbachian and Franconian trilobites of the Conococheague and Frederick Limestones of the central Appalachians: Jour. Paleontology, v. 35, p. 104-124.

Reinhardt, Juergen, 1974, Stratigraphy, sedimentology and Cambro-Ordovician paleogeography of the Frederick Valley, Maryland: Maryland Geol. Survey, Rept. Inv. 23, 74 p.

—— and Hardie, L. A., 1976, Selected examples of carbonate sedimentation, central Appalachians: Maryland Geol. Survey, Field Trip Guidebook 5, 53 p.

—— and Wall, Edward, 1975, The Tomstown Dolomite (Lower Cambrian), central Appalachians and the habitat of *Salterella conulata*: Geol. Soc. America Bull., v. 86, p. 1555-1559.

Rodgers, John, 1956, The known Cambrian deposits of the southern and central Appalachian Mountains: *In* John Rodgers (ed.), El Sistema Cámbrico su paleogeogafía y el problema de su base—Symposium, pt. 2: 20th Internat. Geol. Congr., Mexico City, Mexico, p. 353-384.

——, 1968, The eastern edge of the North American continent during the Cambrian and Early Ordovician: *In* E-An Zen and others (eds.), Studies of Appalachian geology, northern and maritime: Wiley-Interscience, New York, New York, p. 141-149.

Root, S. I., 1964, Cyclicity of the Conococheague Formation: Pennsylvania Acad. Sci. Proc., v. 38, p. 157-160.

Sando, W. J., 1957, Beekmantown Group (Lower Ordovician) Maryland: Geol. Soc. America, Mem. 68, 161 p.

Schwab, F. L., 1970, Origin of the Antietam Formation (late Precambrian-Lower Cambrian) central Virginia: Jour. Sed. Petrology, v. 40, p. 345-366.

Stanley, D. J., Swift, D. I., Silverberg, N. P., James, N. P., and Sutton, R. G., 1972, Late Quaternary progradation and sand spillover on the outer continental margin off Nova Scotia, southeast Canada: Smithsonian Contr. Earth Sci., no. 8, 88 p.

—— and Unrug, Rafael, 1972, Submarine channel deposits, fluxoturbidites and other indicators of slope and base-of-slope and base-of-slope environments in modern and ancient marine basins: *In* J. K. Rigby and W. K. Hamblin (eds.), Recognition of ancient sedimentary environments: Soc. Econ. Paleontologists and Mineralogists, Spec. Pub. 16, p. 287-340.

Stose, G. W., and Jonas, A. I., 1939, Geology and mineral resources of York County, Pennsylvania: Pennsylvania Geol. Survey, Bull. C67, 199 p.

—— and Stose, A. J., 1944, Geology of the York-Hanover district, Pennsylvania: U.S. Geol. Survey, Prof. Paper 204, 84 p.

Thomson, A. F., and Thomasson, M. R., 1969, Shallow to deep water facies development in the Dimple Limestone (Lower Pennsylvanian), Marathon region, Texas: *In* G. M. Friedman (ed.), Depositional environments in carbonate rocks: Soc. Econ. Paleontologists and Mineralogists, Spec. Pub. 14, p. 57-77.

Walker, R. G., 1975, Generalized facies models for resedimented conglomerates of turbidite association: Geol. Soc. America Bull., v. 86, p. 737-748.

Willoughby, Ralph, 1976, Lower and Middle Cambrian fossils from the Shady Formation, Austinville, Virginia [abs]: Geol. Soc. America Abs. with Programs, v. 8, p. 301-302.

Wilson, J. L., 1952, Upper Cambrian stratigraphy in the central Appalachians: Geol. Soc. America Bull, v. 63, p. 275-322.

——, 1969, Microfacies in sedimentary structures in "deeper water" lime mudstones: In G. M. Friedman (ed.), Depositional environments in carbonate rocks: Soc. Economic Paleontologists and Mineralogists, Spec. Pub. 14, p. 4-19.

——, 1974, Characteristics of carbonate platform margins: Am. Assoc. Petroleum Geologists Bull., v. 58, p. 810-824.

Wise, D. U., 1970, Multiple deformation, geosynclinal transitions and the Martic problem in Pennsylvania: *In* G. W. Fisher and others (eds.), Studies in Appalachian geology—central and southern: Wiley-Interscience, New York, New York, p. 317-333.

SEPM Special Publication No. 25, p. 113-124, November 1977

ACCUMULATION OF A MIDDLE CAMBRIAN, DEEP-WATER LIMESTONE DEBRIS APRON ADJACENT TO A VERTICAL, SUBMARINE CARBONATE ESCARPMENT, SOUTHERN ROCKY MOUNTAINS, CANADA

IAN A. McILREATH
Shell Canada Resources Limited, Calgary, Alberta T2P 2K3

ABSTRACT

The boundary limestone (Middle Cambrian), exposed in the Main Ranges of the Southern Canadian Rocky Mountains, accumulated in a deep-water (≈200 m) embayment adjacent to a near vertical, shelf-margin reef, the Cathedral escarpment. Prior to accumulation of the boundary limestone, reef growth was terminated and the adjacent basin, which had been a site of exclusive carbonate deposition, received argillaceous sediments (Stephen Formation). The boundary limestone accumulated during the only significant period of restored carbonate sedimentation to interrupt the argillaceous basin filling.

The boundary limestone accumulated along a broad front from the deposition of lime mud and granular carbonate sediment swept basinward off the adjacent shelf. Reef growth was not re-established on the adjacent escarpment; rather it acted as a sediment by-pass slope, contributing rare redeposited material either as small submarine talus blocks or clasts in thin debris flows. The resulting deposit is a proximal bench of allochthonous debris, approximately 100 m thick, abutting the escarpment. This proximal bench is flanked on its basinward side by a mud wedge tapering to less than 12 m thick. In outcrop, the proximal bench exceeds 180 m in width and its basinward flanking wedge exceeds 3 km in width.

There is a distinct difference in the lithofacies of the proximal bench compared to those of the distal wedge. Bench lithofacies consist of medium to dark-grey, thin-bedded to massively bedded intervals of heterogeneously mixed, bioclastic peloidal wackestone to grainstone interbedded with uniformly thin-bedded, dark-grey, basinal lime mudstone intervals. Many of the peloids were derived by micritization of such shallow-water grains as oolites, oncolites and coated bioclastic fragments. In contrast, the interfingering distal wedge consists of dark-grey, thin-bedded, "normal" basinal lime mudstone that is interrupted only by widely spaced thin beds of fine allochthonous calcarenite.

Submarine sliding was important in extending the flanking wedge of mudstone further basinward. Periodic slope failure in mud accumulating along the upslope portion of the distal wedge resulted in sliding along mappable slip surfaces. Repetition of these events produced a basinward-thinning stack of lime mudstone slide masses of which the youngest extends furthest into the basin.

At carbonate shelf margins dominated by growing reefs, the redeposited margin material interrupting adjacent basin mudstone will include talus blocks and debris-flow deposits. However, if reef growth is terminated, the shelf margin becomes a sediment by-pass slope and the resulting basinal deposit is a debris apron of allochthonous, shelf-derived mud and granular sediment lacking reef debris; such is the case of the boundary limestone. The overall wedge-shaped geometry of the debris apron distinguishes this type of accumulation from the cones formed in submarine channel-fan systems that breach a shelf margin.

INTRODUCTION

The purpose of this paper is to: (1) show how a particular Middle Cambrian limestone accumulated by dispersal of carbonate shelf sediments into a "deep-water" basinal environment, below a vertical, submarine carbonate escarpment, (2) document the nature of the resultant "deep-water" lithofacies, (3) discuss dispersal processes responsible for these lithofacies, and (4) document excellent exposures of submarine slide masses in "deep-water" lime mudstones.

The study is based on surface work conducted in 1973 in eastern British Columbia, just west of the Continental Divide marking the boundary with Alberta and straddling the northerly trending boundary between Eastern and Western Main Ranges of the Southern Canadian Rocky Mountains (Fig. 1). Within this study area, the exposed stratigraphic succession is predominantly Middle Cambrian and involves a regional facies change from an eastern carbonate sequence to an argillaceous western sequence (Cook, 1970).

This area is known for its fossil localities of Middle Cambrian trilobites (Rasetti, 1951) and soft-bodied fauna, especially from the Burgess Shale (Whittington, 1971). A reinvestigation of the stratigraphy, paleontology and paleoecology of the Burgess Shale by the Geological Survey of Canada (Aitken, Fritz and Whittington, 1967; Aitken and Fritz, 1968) led to the interpretation that this shale accumulated in "deep-water" near the edge of a steep submarine carbonate escarpment bounding a regional carbonate shelf (Fritz, 1971). This "deep-water" interpretation is based on the occurrence of the upper limit of the *Glossopleura* Zone at the top of the carbonate

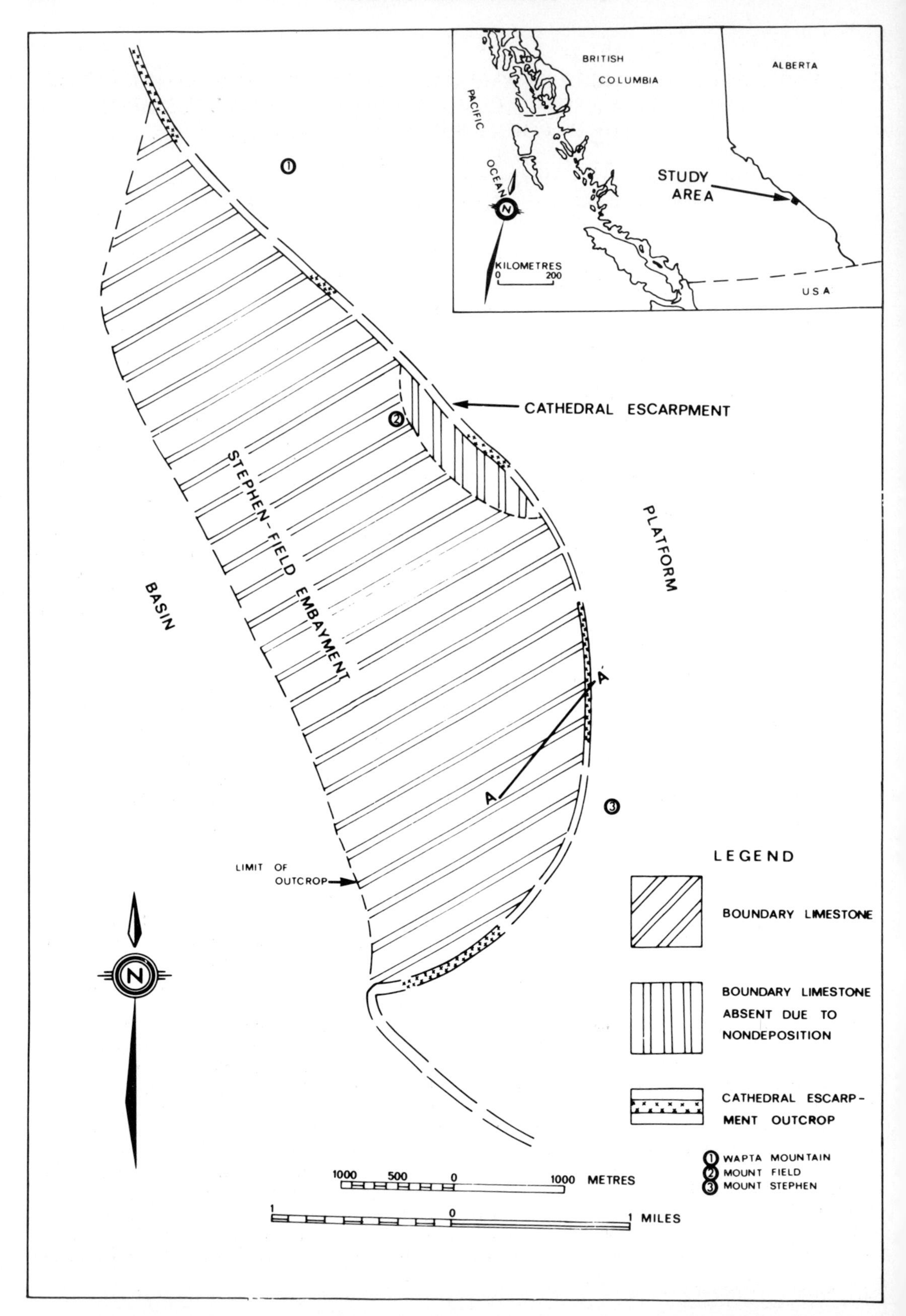
BRITISH
COLUMBIA
ALBERTA
PACIFIC
OCEAN
STUDY
AREA
KILOMETRES
0 200
USA
CATHEDRAL ESCARPMENT
STEPHEN-FIELD EMBAYMENT
PLATFORM
BASIN
A
A'
LIMIT OF
OUTCROP
LEGEND
BOUNDARY LIMESTONE
BOUNDARY LIMESTONE
ABSENT DUE TO
NONDEPOSITION
CATHEDRAL ESCARP-
MENT OUTCROP
1 WAPTA MOUNTAIN
2 MOUNT FIELD
3 MOUNT STEPHEN
1000 500 0 1000 METRES
1 0 1 MILES

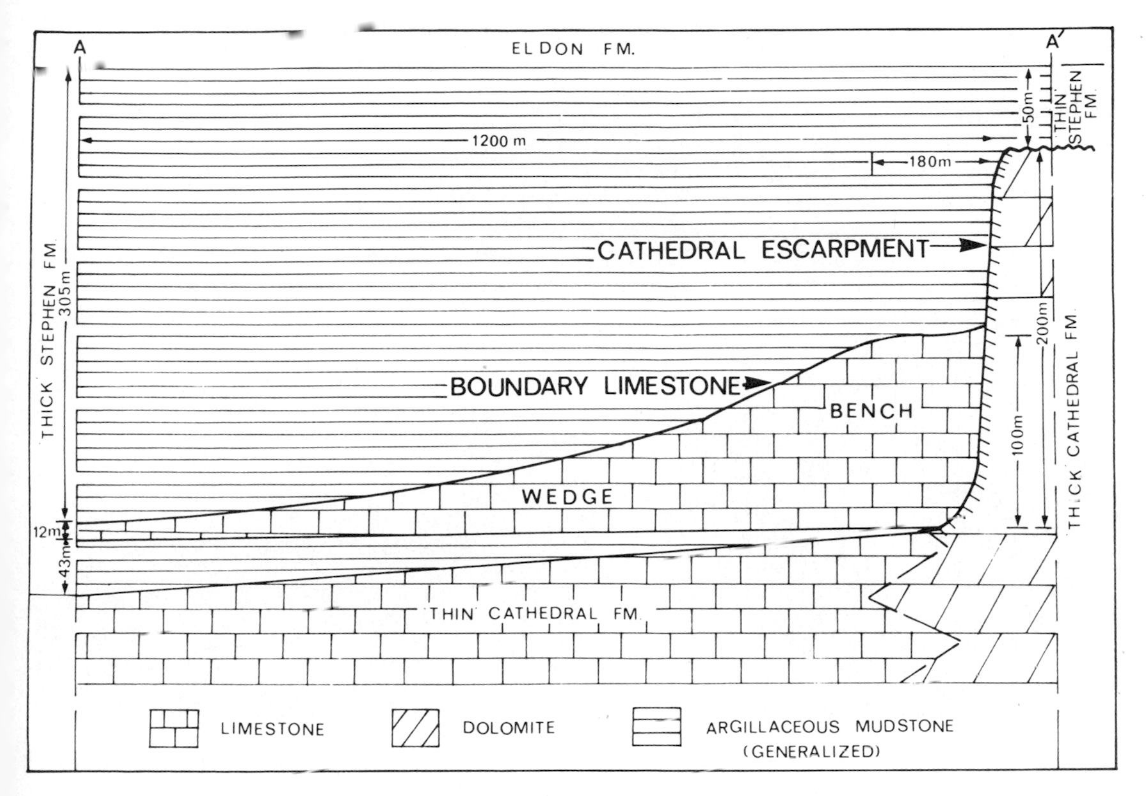

FIG. 2.—A schematic cross section across the Stephen-Field embayment (location in Fig. 1), based on the excellent exposure on the north face of Mount Stephen. The composite bench to tapering flank wedge form of the boundary limestone is shown. The "thick" Stephen Formation includes 43 m of argillaceous mudstone beneath the boundary limestone. The Cathedral escarpment is nearly vertical in outcrop. Not to scale, representative measurements are indicated.

shelf and lower in the adjacent basin. The actual topographic relief separating shelf basinal occurrences is approximately 200 m as evidenced from measured sections (Fritz, 1971).

In the basin, the boundary between the *Glossopleura* Zone and the succeeding *Bathyuriscus-Elrathina* Zone occurs in a thick limestone unit below the Burgess Shale that Fritz (1971) informally named the "boundary limestone." The present study examines the nature of the stratigraphy and sedimentology of the boundary limestone in terms of its depositional environment. Initial results are reported elsewhere (McIlreath, 1975).

STRATIGRAPHY

The Middle Cambrian boundary limestone accumulated within the basinal Stephen-Field embayment of the Cathedral escarpment (Fig. 1). A schematic cross section in this embayment (Fig. 2), based on the excellent exposure along the north face of Mount Stephen (Fig. 3), illustrates the stratigraphic relationships of the boundary limestone relative to the Cathedral escarpment. Prior to accumulation of the boundary limestone, the underlying Cathedral Formation formed a nearly vertical submarine escarpment of variable height (120-300 m), separating an embayed carbonate shelf ("thick" Cathedral Formation) from the basin ("thin" Cathedral Formation).

The Cathedral Formation in the escarpment is entirely white, coarsely crystalline dolomite. However, in the former Cathedral basin area, margin-derived debris is limestone with preserved relic texture. This debris occurs either as individual talus blocks (up to 30 m diameter) or as individual blocks up to 20 m diameter incorporated into mass breccia flows. Throughout a total of

FIG. 1.—Outcrop limit of the Middle Cambrian boundary limestone and its stratigraphic setting during accumulation. A-A′ locates the cross section in Figure 2.

Fig. 3.—Looking south at the north face of Mount Stephen (3200 m) illustrating the occurrence of the boundary limestone. Field work was conducted from a helicopter-established fly camp (C).

350 m of Cathedral basinal lime mudstone, the margin-derived debris is predominantly algal boundstone consisting almost exclusively of *Epiphyton* which bound lime mud to form a porous framework (identical to the algal boundstone illustrated in Fig. 6C). Original porosity in the algally bound lime mud is now filled with a thin, isopachous lining of calcite cement having a relic fibrous habit (submarine cement?) and interlocking, blocky, clear, spar calcite filling the central portion of each former void.

The Cathedral shelf-basin transition across the nearly vertical escarpment is fully exposed, particularly on the south face of Mount Field and the opposite north face of Mount Stephen (Fig. 1). The escarpment has been traced from west of Wapta Mountain (Fig. 1) southwards beyond the limit of the study area for a distance of 16 km. Where it is exposed along this trend (Fig. 1), the escarpment profile is consistently vertical and there is no outcrop evidence that the wall was ever breached by submarine canyons or channels.

Near the end of *Glossopleura* time, carbonate sedimentation in the basin was replaced by deposition of argillaceous mud ("thick" Stephen Formation—Fig. 2) without redeposited carbonate debris. Basin filling by argillaceous mud was interrupted by one significant period of carbonate deposition during which the boundary limestone accumulated (Fig. 2).

The boundary limestone displays a composite form consisting of a bench approximately 180 m wide and 100 m thick intertongued basinward with a wedge that tapers to approximately 12 m thick within 1.2 km of the escarpment (Fig. 2). Further basinwards, the flanking wedge is 10–12 m thick to its outcrop limit (Fig. 1).

In the basinal edge of the bench, trilobites of *Glossopleura* Zone are overlain by those from *Kootenia* sp. 1 and *Ogygopsis klotzi* faunules of the *Bathyuriscus-Elrathina* Zone whereas approximately 1.2 km further basinward the wedge contains only the younger *Ogygopsis klotzi* faunule (Fig. 4). In this more distal portion of the wedge, *Glossopleura* has been collected 10 m below the base of the wedge, from the underlying argillaceous mudstone (Fig. 4). Here, the intervening similar argillaceous mudstone is nonfossiliferous so that it is not possible to demonstrate equivalence of these strata to the basal portion of the carbonate bench. However, the distribution of trilobite faunules (Fig. 4) suggests basinal carbonate sedimentation began directly below the es-

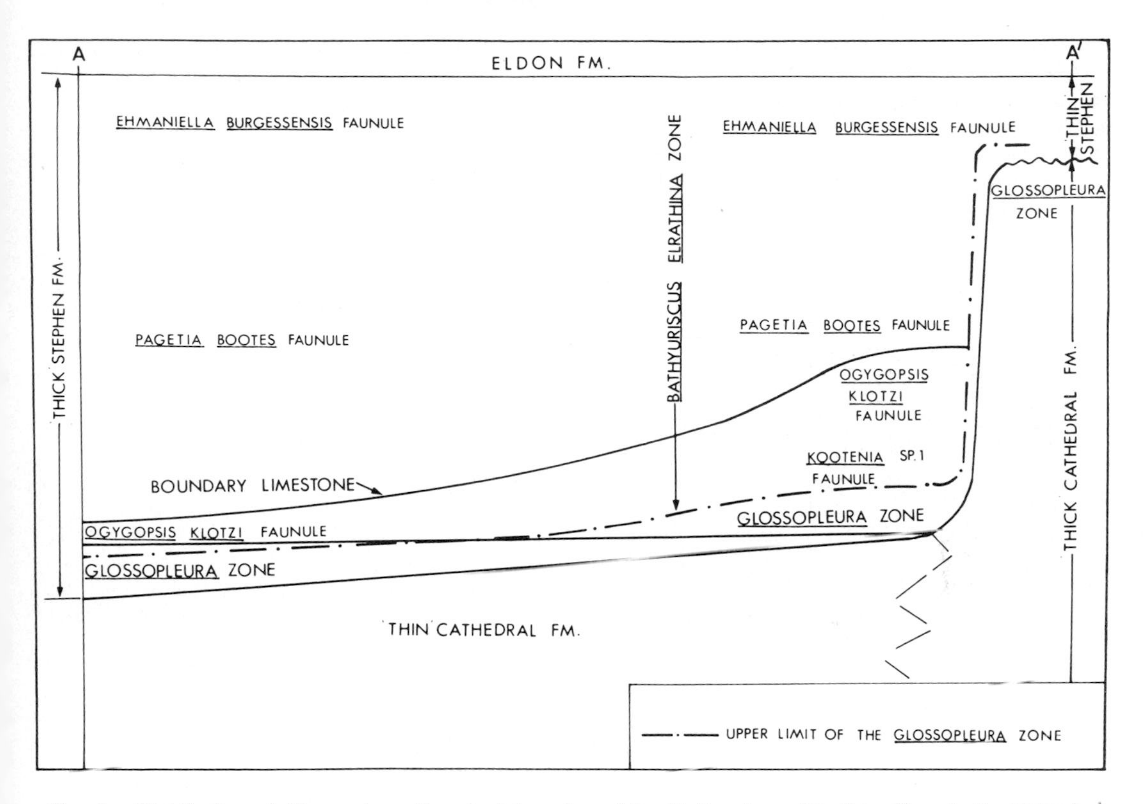

FIG. 4.—Distribution of *Glossopleura* Zone and faunules of the *Bathyuriscus-Elrathina* Zone in the Cathedral and Stephen Formations. Trilobite specimens were identified by W. H. Fritz, Geol. Surv. Canada.

carpment and later extended basinwards.

If the basin shelf profile is drawn for the end of *Glossopleura* time (Fig. 4), it would indicate as suggested by Fritz (1971) that the basal portion of the boundary limestone accumulated in at least 200 m of water. However, absolute water depth may have been much more, since there is an absence of Stephen sediments containing fauna of the lower three faunules of the *Bathyuriscus-Elrathina* Zone (Fig. 4) over the former Cathedral shelf area, implying an unconformity at the "thin" Stephen-"thick" Cathedral contact. Therefore, *Glossopleura* occurring at the top of the "thick" Cathedral Formation may be older than those collected from the upper limit of their occurrence in the basin where continuous deposition took place. Thus erosion may have intervened to reduce the height of the Cathedral escarpment before deposition of the "thin" Stephen Formation. Alternatively, the "thin" Stephen-"thick" Cathedral contact could represent a surface of nondeposition with currents continuously winnowing sediments representing this stratigraphic interval into the basin. This would mean that the escarpment never was much higher than it is today.

LITHOFACIES

Lithofacies in the bench consists of redeposited shoal-water debris (approximately 40 m thick) and interbedded *in situ* lime mudstone (approximately 60 m thick) whereas the more distal wedge consists predominantly of thin-bedded, basinal lime mudstone.

Bench lithofacies.—In the bench, allochthonous carbonate debris occurs in irregular, thin- to medium-bedded intervals (2.5 cm–10 m thick) interbedded with intervals (1–10 m thick) of *in situ* planar and nearly parallel thin-bedded mudstone. There is no evidence of any cone-shaped sediment accumulation in the bench which would suggest the presence of an original submarine fan or even coalescing fans. In addition, there is no evidence of the adjacent escarpment being breached by a submarine channel. Therefore, deposition appears to have occurred along a broad front adjacent to at least a locally continuous escarpment rather than by coalescing of submarine fans to form a submarine bajada.

Allochthonous bench limestone is wackestone to grainstone consisting of shallow-water carbonate grains heterogeneously mixed with variable percentages of micrite matrix and calcite cement. A typical example of allochthonous bench lithofacies (Fig. 5A) contains shallow-water derived oolite grains, coated bioclastic fragments, and intraclasts of previously cemented oolites. Another allochthonous lithofacies is oncolite packstone with a bioclastic peloid wackestone to packstone matrix (Fig. 5B). The most common allochthonous bench lithofacies is bioclastic, peloid wackestone to grainstone with predominantly random fabric (Fig. 5E) and occasionally planar aligned fabric (Fig. 5C,D). Many peloids show traces of microstructure suggesting derivation by micritization of bioclastic grains, coated grains and oolites (Fig. 5E). The process of micritization is not exclusively shallow-water (Friedman and others, 1971); however, enough microstructure is preserved to identify oolites etc. of shallow-water

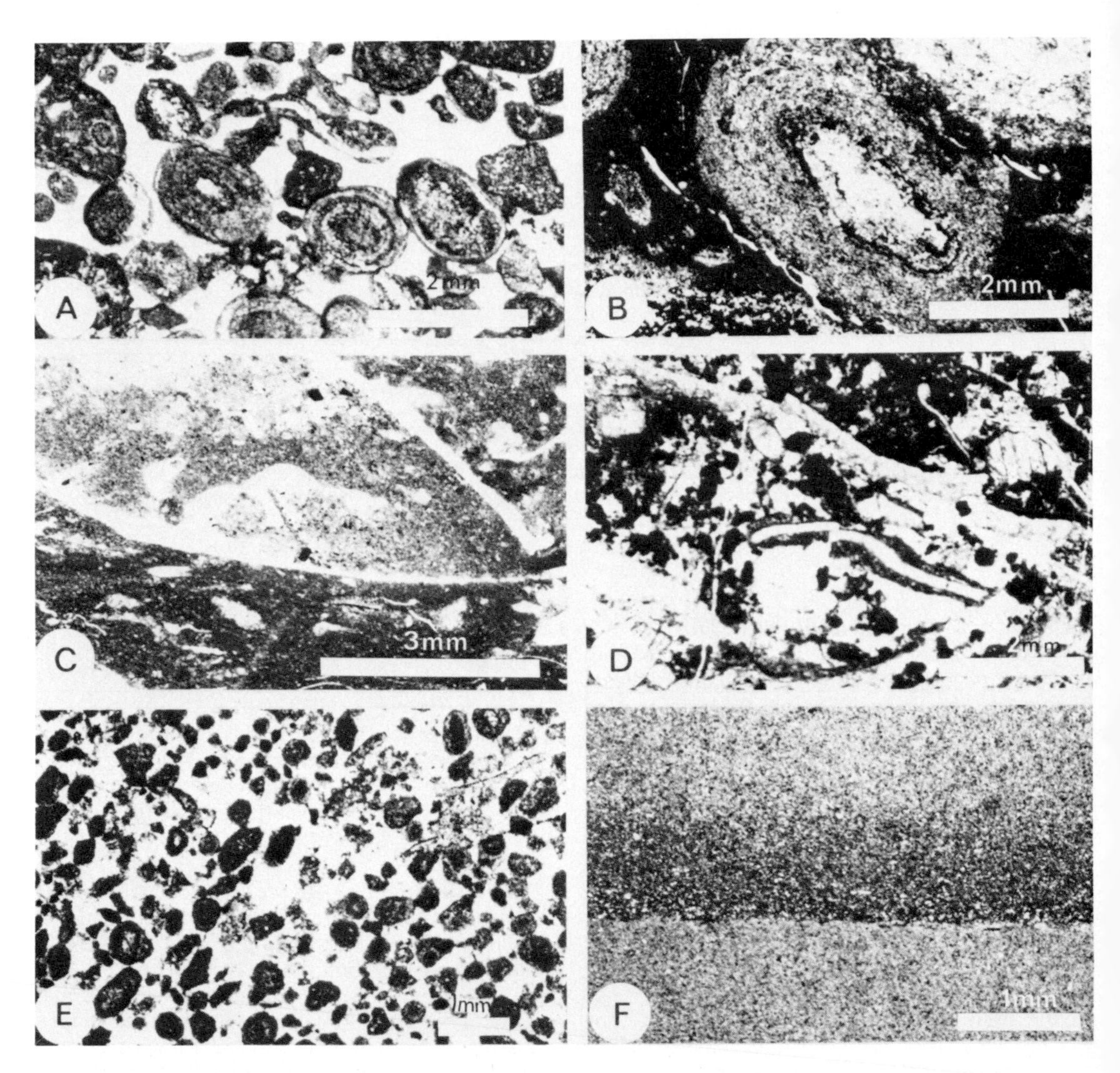

FIG. 5.—Photomicrographs of thin-sections, viewed in plane-polarized light, from the bench lithofacies (A-E) and the flanking wedge lithofacies (F) of the boundary limestone. *A*—Oolites, coated bioclastic grains, and rounded intraclast of several previously cemented oolites in a grainstone. *B*—Compressed oncolites with a bioclastic, peloid wackestone to packstone matrix. *C*—Trilobite, bioclastic wackestone. *D*—Bioclastic-peloid packstone; note pelmatozoans and trilobites. *E*—Bioclastic-peloid grainstone; most abundant bench calcarenite; many peloids show relic internal structure derived from micritization of oolites and skeletal (or "other") grains. *F*—Internally planar-laminated lime mudstone of the flanking wedge.

origin and therefore establish their allochthonous nature.

Many relatively complete trilobites of the *Bathyuriscus-Elrathina* faunules exist parallel to bedding planes in the *in situ* lime mudstone of the bench. These indigenous fossils (Fritz, person. comm., 1975) are in contrast to the disarticulated, rounded bioclastic fragments (predominantly trilobites and minor amounts of pelmatozoans, brachiopods, and hyolithids, Fig. 5D) in the allochthonous bench lithofacies. Also, trilobites of the *Glossopleura* Zone are found in the lime mudstones only as disarticulated specimens and those may be allochthonous. The lack of bioturbation structures in the *in situ* beds suggests the absence of an effective infauna in the bench.

In addition to the above allochthonous debris, there are also rare occurrences of individual submarine talus clasts and debris flow deposits (in the terminology of Cook and others, 1972) limited to the bench facies. Individual debris flow deposits are less than 1 m thick and consists of sand-sized to boulder-sized, angular limestone clasts that occur randomly oriented and grain supported in lime mudstone matrix (Fig. 6A). Submarine talus clasts vary from pebble to small boulder size (Fig. 6B). These allochthonous talus clasts and those in the debris flows are predominantly *Epiphyton* algal boundstone, possibly submarine cemented, from the adjacent Cathedral escarpment (Fig. 6C). The paucity of allochthonous margin-derived debris in the boundary limestone of the Stephen Formation compared to its occurrence in the underlying "thin" Cathedral Formation, suggests that growth of the type that had previously characterized the Cathedral shelf margin was not re-established. Rather, the escarpment appears to have been essentially a resistant by-pass slope during accumulation of the boundary limestone.

Flanking wedge (basinal) lithofacies.—In contrast to the calcarenitic lithofacies in the bench, the basinward flanking wedge consists of dark-grey lime mudstone in planar parallel thin beds with internal planar lamination (Fig. 5F). Bioclastic remains are limited to rare occurrences of

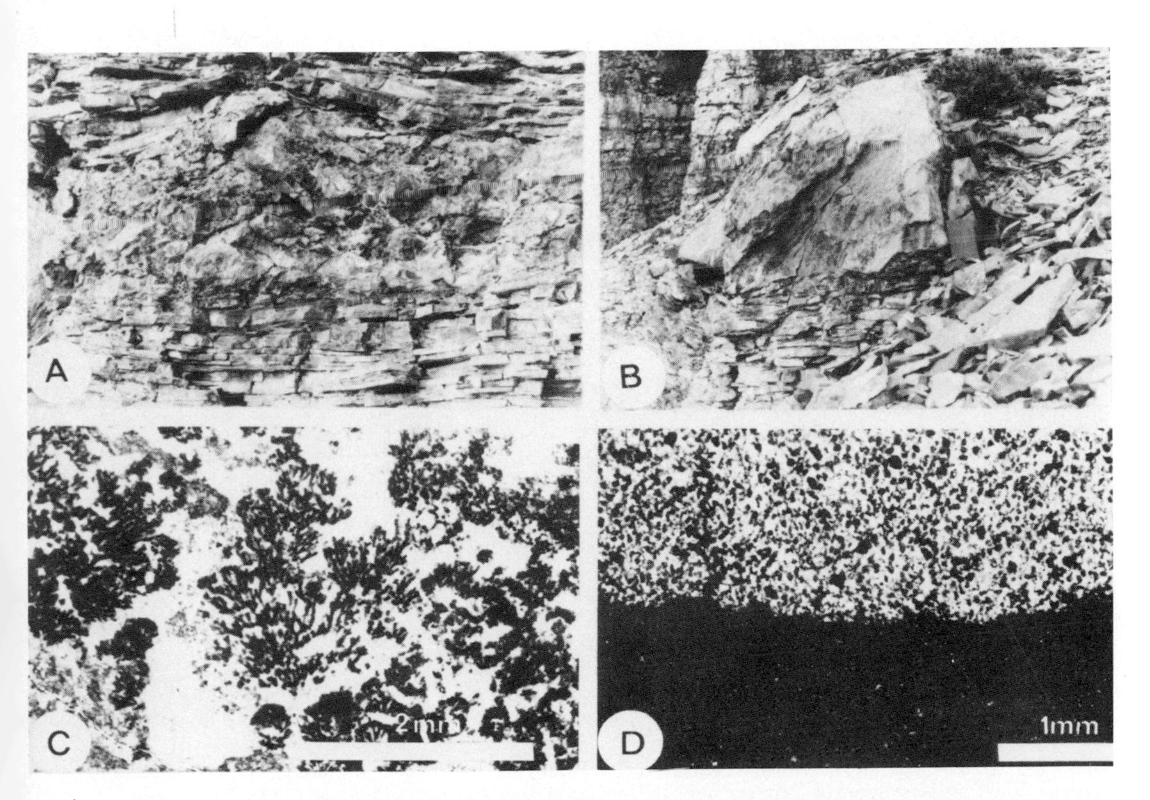

Fig. 6.—Sediment gravity flow deposits. *A*—Limestone debris flow (45-61 cm thick) in thin-bedded bench sediments. *B*—Individual submarine talus block in bench sediments. *C*—Photomicrograph of thin-section, viewed in plane-polarized light, of *Epiphyton* algal boundstone clast derived from the Cathedral escarpment and occurring in bench. *D*—Photomicrograph, viewed in plane-polarized light, of contact between well-sorted, allochthonous calcarenite overlying basinal lime mudstone.

complete trilobite exoskeletons parallel to bedding. The excellent preservation of stratification and lack of biogenic structures in this mudstone suggests the absence of an active infauna.

The mudstone in the wedge is quite similar to typical "deep-water" lime mudstone described by Wilson (1969). The mechanism of origin proposed by Wilson is highly plausible in the present instance; the lime mud of the boundary limestone was probably derived by basinward dispersal of lime mud produced on the adjacent shelf. Such an origin would account for the rapid, basinward decrease in total flank wedge thickness (Fig. 2) as a reflection of increasing distance from the shelf source area. This *in situ* lime mudstone is interbedded with rare thin beds of fine-grained allochthonous calcarenites consisting of shelf-derived, peloid packstones and grainstones (Fig. 6D).

Lithofacies in the flanking wedge are cut by

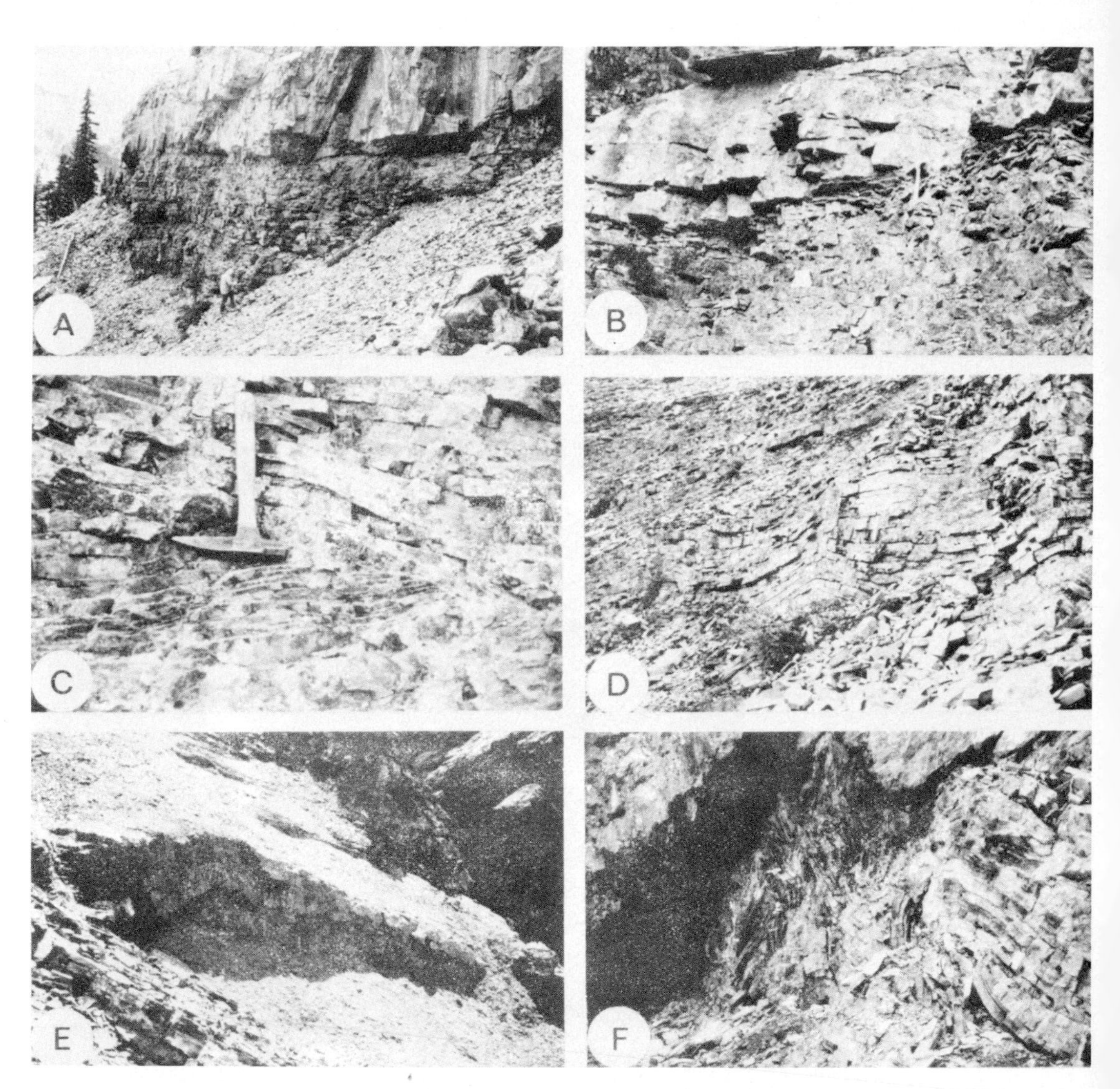

FIG. 7.—*A, B*—Separate outcrops in gullies incised into the north face of Mount Stephen illustrating the concave-down cross section of intraformational truncation surfaces in the boundary limestone wedge. *C-F*—Separate outcrops illustrating various degrees of deformation of beds underlying truncation surfaces from the proximal to distal portion of the wedge on Mount Stephen. *C*—Sharp truncation of underlying beds, typical of most contacts. *D*—Disharmonic folding of beds below a truncation surface. *E*—Distal portion of the boundary limestone wedge cut by truncation surface. *F*—Close-up of E showing local intense deformation in beds underlying the truncation surface.

intraformational truncation surfaces which are very similar in shape to the penecontemporaneous, cut-and-fill structures in typical "deep-water" lime mudstone (Wilson, 1969). These truncation surfaces are particularly well-exposed on the north face of Mount Stephen. There, the mountain face parallels the long dimension of the surfaces whereas gullies truncating the north face expose sections perpendicular to the long dimension. Individual surfaces can be traced up to 300 m basinward and define lenticular bodies with pod-shaped cross sections exceeding 30 m in width (Fig. 7A,B).

The intraformational truncation surfaces predominantly truncate underlying beds with only minor interaction (Fig. 7C), especially at their proximal end. Further basinward, slightly more interaction with underlying beds has been observed as seen in minor disharmonic folding (Fig. 7D). In the distal portion of the wedge, localized, spectacular deformation occurs in a few places in the underlying beds (Fig. 7E,F). Variation in the degree of interaction between overlying and underlying beds does not occur along every truncating surface traced basinwards. Most show typical, sharp truncation along their entire length.

GENESIS OF THE BOUNDARY LIMESTONE

Boundary limestone lithofacies originated by current-settling processes, sediment gravity flows and by submarine sliding. The first two mechanisms operated in the proximal bench while all three were responsible for forming the distal wedge.

Current Settling

To explain the origin and distribution of most of the bench and wedge lithofacies, a simple model is proposed whereby individual carbonate grains and mud originated on the adjacent shelf and were dispersed by currents basinwards across the margin area (Fig. 8). The decrease in grain size and percentage of carbonate granular components across the bench and the tapering wedge geometry of the basinal lime-mudstone flank suggests that deposition of both granular debris and mud decreased substantially seaward of the toe-of-slope; rather sediment accumulated predominantly along

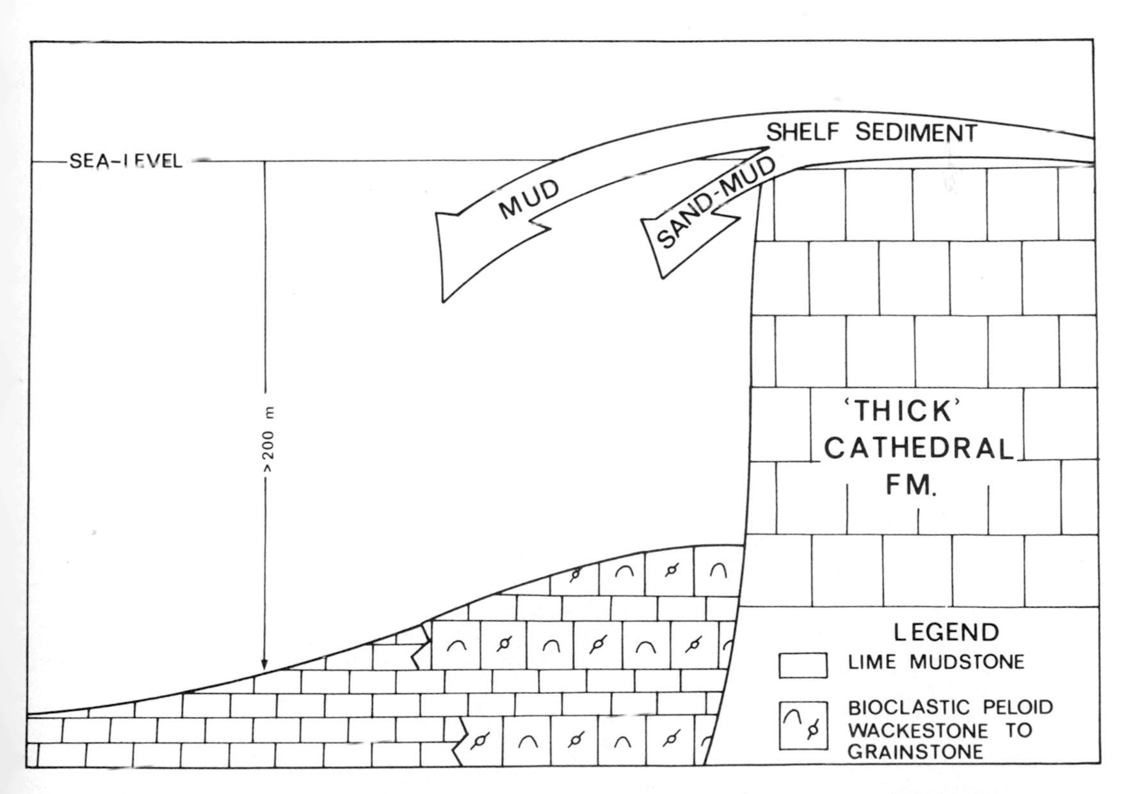

FIG. 8.—Schematic diagram illustrating the dispersal of carbonate shelf sediments into the adjacent basin bypassing the organically dead margin. Crude sorting results in granular debris being concentrated in the bench and mud in the flanking wedge.

a narrow front, parallel to a near vertical submarine by-pass slope.

Sediment Gravity Flows

Individual talus clasts.—These angular, small boulders are rare and limited in occurrence to the bench. Since they consist of *Epiphyton* boundstone, the clasts are interpreted to have been derived by submarine erosion of the adjacent escarpment face with subsequent gravitational falling, rolling, or sliding into the basin.

Debris-flow deposits.—These deposits are also thin, rare, and limited to the bench. Individual clasts in these deposits consist predominantly of *Epiphyton* boundstone derived by brecciation of the adjacent escarpment. The cause of such brecciation and subsequent mixing of the clasts with lime mud is not understood; however, once initiated, their method of gravitational movement was probably similar to that described for debris flows by Cook and others (1972).

Allochthonous calcarenites.—It is probable that some of the allochthonous calcarenites, especially the thin beds of massive grainstone in the flanking wedge, originated by some other mechanism than current settling. Turbidity currents are a possible transport mechanism although Bouma divisions, sole markings, and even graded bedding are lacking. The origin of these deposits remains uncertain.

Submarine Sliding

Intraformational truncation structures.—These structures which are limited to the wedge are interpreted to be submarine slide surfaces resulting from the transport of a coherent mass of mudstone further basinward along a discreet shear plane, without discernible internal deformation. The resulting deposit is properly termed a submarine slide (Dott, 1963, Cook and others, 1972) and not a slump feature.

Slides in the boundary limestone probably originated by mud accumulating on the upper portion of the flank slope until instability resulted in slope failure and the basinward transport of a slide along a slip plane. In laboratory experiments, Einsele and others (1974) have shown that at low sedimentation rates and critical slope angles, water-rich muds can break up into sharp-bounded blocks parallel to bedding planes and shift downslope over underlying undeformed material.

The process of submarine sliding was significant because it modified the configuration of the wedge. By tracing out individual slides occurring in the wedge on Mount Stephen, it can be demonstrated that the wedge consists of a series of

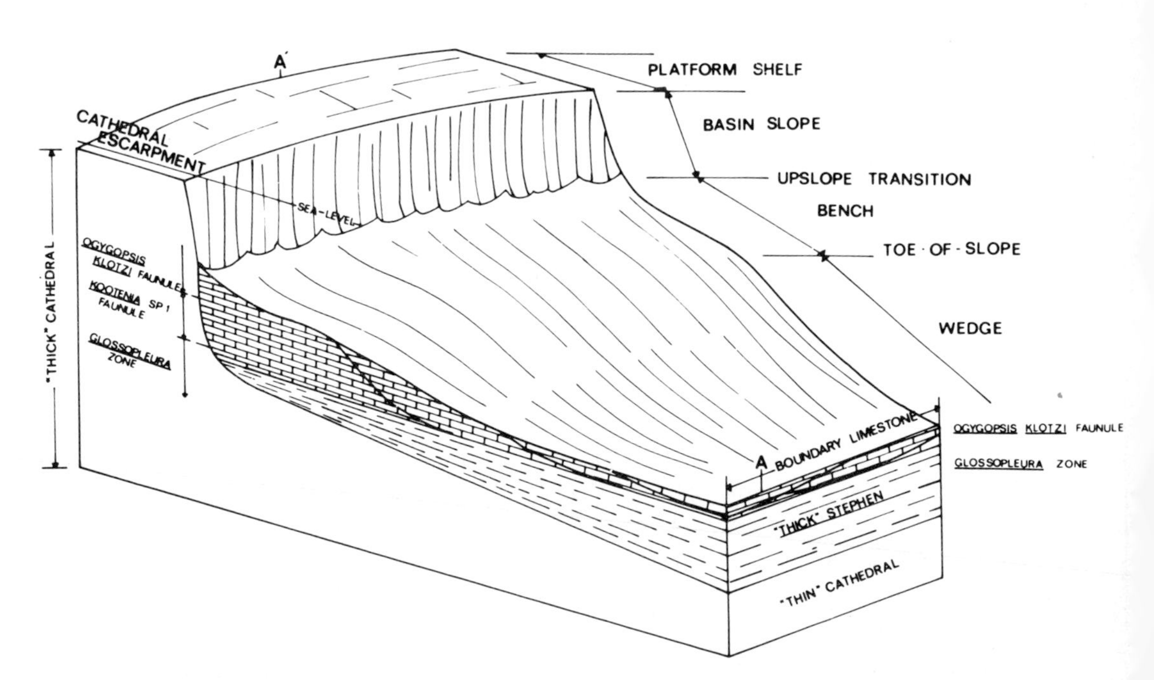

FIG. 9.—Block diagram showing the nature of the boundary limestone in the Stephen-Field embayment at the end of *Ogygopsis klotzi* time.

stacked or overlapping slide masses, prograding basinwards. Repetition of sliding has produced much of the tapered wedge configuration of the basinal flank by simple elongation.

DEPOSITION MODEL

A model is presented for the accumulation of the Middle Cambrian boundary limestone within a deep-water (approximately 200 m) embayment adjacent to the nearly vertical, Cathedral escarpment (Fig. 9). This limestone accumulated primarily by carbonate shelf sediment settling through the water column after it had been swept basinwards off the shelf by current activity. As this sediment descended through the water column, crude sorting developed so that sand-size sediment settled seaward of the escarpment to form a bench of debris. In contrast, mud originating on the shelf accumulated farther out, on the wedge-shaped, basinal flank of the bench.

The basin slope was a continuous, nearly vertical wall and was a sediment by-pass. Submarine erosion of the escarpment provided clasts for the rare allochthonous debris flows and talus blocks which accumulated on the bench. Submarine slides extended the flanking wedge by transporting large coherent masses of mudstone further basinward in a prograding fashion.

CONCLUSIONS

The boundary limestone is an example of a basinal limestone accumulating adjacent to a steep, shelf margin that acted as a by-pass slope. The resulting deposit, a debris apron, consisted primarily of (1) interbedded carbonate mud and granular material derived from the adjacent shelf in a proximal bench, and (2) carbonate mud in a distal wedge. Since the boundary limestone accumulated along a broad front below an unbreached submarine wall, the geometry of the resulting debris apron is distinct from the point-source, radial geometry of an accumulation from a canyon into a submarine fan system.

The type of deposit characterized by the boundary limestone is unlike basinal limestone associated with a growing reef-dominated shelf margin, such as occurs in the underlying Cathedral Formation. In the latter situation, substantial amounts of margin-derived, redeposited material in the form of talus blocks and debris flow deposits in addition to allochthonous calcarenite are interbedded with basinal lime mudstone.

ACKNOWLEDGEMENTS

The above research was carried out as part of a Ph.D dissertation at the Geology Department, University of Calgary, while the author was a recipient of a National Research Council Scholarship. I am most grateful to Dr. J. D. Aitken, Prof. N. C. Wardlaw and Prof. A. E. Oldershaw for their critical reviews of this manuscript. Sincere thanks are also due to Dr. W. H. Fritz for identification and establishment of biochronological relationships of trilobite specimens collected by the writer. Special appreciation is expressed to my wife Susan for drafting of the figures. The author gratefully acknowledges the financial and logistical support from the Geological Survey of Canada. Finally, I thank Messrs. R. Rokosh and J. Mitchell for field assistance.

REFERENCES

AITKEN, J. D., AND FRITZ, W. H., 1968, Burgess Shale project, British Columbia: Canada Geol. Survey, Paper 68-1, pt. A, p. 190–192.

——, ——, AND WHITTINGTON, H. B., 1967, Stratigraphy, palaeontology and palaeoecology of the Burgess Shale: Canada Geol. Survey, Paper 67-1, pt. A, p. 52.

COOK, D. G., 1970, A Cambrian facies change and its effect on structure, Mount Stephen-Mount Dennis area, Alberta-British Columbia: Geol. Assoc. Canada, Spec. Paper 6, p. 27–39.

COOK, H. E., MCDANIEL, P. N., MOUNTJOY, E. W., AND PRAY, L. C., 1972, Allochthonous carbonate debris flows at Devonian bank ('reef') margins, Alberta, Canada: Bull. Canadian Petroleum Geology, v. 20, p. 439–497.

DOTT, R. H., JR., 1963, Dynamics of subaqueous gravity depositional processes: Am. Assoc. Petroleum Geologists Bull., v. 47, p. 104–128.

EINSELE, G., OVERBECK, R., SCHWARZ, H. U., AND UNSÖLD, G., 1974, Mass physical properties, sliding and erodibility of experimentally deposited and differently consolidated clayey muds: Sedimentology, v. 21, p. 339–372.

FRIEDMAN, G. M., GEBELEIN, C. D., AND SANDERS, J. E., 1971, Micritic envolopes of carbonate grains are not exclusively of photosynthetic algal origin: Sedimentology, v. 16, p. 89–96.

FRITZ, W. H., 1971, Geological setting of the Burgess Shale: *In* E. I. Yochelson (ed), Proceedings of the North American Paleontological Convention, v. 2: Allen Press, Lawrence, Kansas, p. 1155–1170.

MCILREATH, I. A., 1974, Stratigraphic relationships at the western edge of the Middle Cambrian carbonate facies belt, Field, British Columbia: Canada Geol. Survey, Paper 74-1, pt. A, p. 333–334.

——, 1975, Stratigraphic relationships at the western edge of the Middle Cambrian carbonate facies belt, Field, British Columbia: Canada Geol. Survey, Paper 75-1, pt. A, p. 556–557.

RASETTI, FRANCO, 1951, Middle Cambrian stratigraphy and faunas of the Canadian Rocky Mountains: Smithsonian Misc. Colln., v. 116, no. 5, 277 p.

WHITTINGTON, H. B., 1971, The Burgess Shale: History of research and preservation of fossils: *In* E. I. Yochelson (ed.), Proceedings of the North American Paleontological Convention, v. 2: Allen Press, Lawrence, Kansas, p. 1170-1201.

WILSON, J. L., 1969, Microfacies and sedimentary structures in "deeper water" lime mudstones: *In* G. M. Friedman (ed.), Depositional environments in carbonate rocks: Soc. Econ. Paleontologists and Mineralogists, Spec. Pub. 14, p. 4-19.

SEPM SPECIAL PUBLICATION NO. 25, P. 125–154, NOVEMBER 1977

THE COW HEAD BRECCIA: SEDIMENTOLOGY OF THE CAMBRO-ORDOVICIAN CONTINENTAL MARGIN, NEWFOUNDLAND

J. F. HUBERT, R. K. SUCHECKI AND R. K. M. CALLAHAN
University of Massachusetts, Amherst, 01002; University of Texas at Austin 78712; Sun Oil Company, Houston, Texas 77001

ABSTRACT

The Cow Head Breccia in western Newfoundland is a slope sequence within a Taconic klippe transported from the southeast. The Cow Head Breccia accumulated on the western side of the proto-Atlantic ocean from the Middle Cambrian through early Middle Ordovician. The 310 m sequence consists of limestone breccia and thin beds of lime mudstone, calcarenite, green silty shale, marl, and radiolarian-sponge spicule chert. Paleotopographic maps for the Cambrian and ordovician parts of the Cow Head Breccia show that a stable paleoslope configuration existed for 70 million years. As the 10 by 75 km klippe is oriented, the regional paleocontours of the paleotopography trend northwest-southeast with a paleoslope that dips northeast. The spectacular megabreccias contain algal-rich, oolitic, fossiliferous limestone boulders that reach 60 by 150 m in size. These breccias were deposited by gravity-controlled viscous mass flows that travelled downslope from narrow carbonate platforms that trended northwest-southeast near Cow Head and Martin Point. Bottom currents consistently flowed southeast, parallel to the paleocontours. The southeast-flowing contour currents deposited mostly nongraded beds of calcarenite, but also some graded beds in the Upper Cambrian. U-tube trace fossils of *Arenicolites* are oriented parallel to the contour currents.

As the proto-Atlantic narrowed during the Middle Ordovician, the continental margin rapidly subsided leading to deposition of 200 m of red shale, followed by more than 400 m of volcanogenic sandstone and gray shale. Within the basin, the northeast-dipping paleoslope shifted to dip northwest. During the Taconic Orogeny the klippe was transported to the northwest by gravity sliding within the basin.

STRATIGRAPHY

Introduction.—This report presents the paleogeography and sedimentary history of the Cambrian and Ordovician rocks in the Cow Head klippe in western Newfoundland (Fig. 1). The allochthonous sequence consists of the Cow Head Breccia (310 m thick) of Middle Cambrian through early Middle Ordovician age and the overlying Middle Ordovician "red shale" (200 m) and "green sandstone" (400 m) (Kindle and Whittington, 1958). The rocks accumulated on the margin of an early Paleozoic ocean, the "proto-Atlantic" or "Iapetus," that lay between the North American and Baltic shields. The depositional site was near the paleoequator in the world carbonate belt (Williams, 1973).

The Bay of Islands klippe was emplaced during the Taconic Orogeny in the late Llandeilian or early Caradocian (Stevens, 1970; Fåhraeus, 1973). The Cow Head klippe was probably emplaced at about the same time.

Autochthonous sequence.—A 1200 m thick autochthonous shelf sequence beneath the Cow Head klippe unconformably rests on "Grenville" basement. This sequence begins with shallow-water feldspathic sandstone, limestone and shale of the Lower Cambrian Labrador Group. The overlying Lower Ordovician St. George Dolomite reflects peritidal and shallow marine environments (Swett and Smit, 1972). The Table Head Limestone of Middle Ordovician age accumulated in deeper water on the outer shelf and top of the continental slope; the upper part of the formation commonly contains beds of limestone breccia and slump sheets (Whittington and Kindle, 1963).

The continental shelf and slope subsided in the Middle Ordovician in response to progressive narrowing of the proto-Atlantic. The Table Head Limestone grades up through 10 m of black shale into at least 100 m of volcanogenic sandstone and gray shale deposited in a rapidly developing basin. The Cow Head klippe was subsequently emplaced as a single structural slice into the basin by gravity sliding as the proto-Atlantic narrowed in the Taconic Orogeny (Rodgers and Neale, 1963; Bird and Dewey, 1970). The base of the klippe is covered so that it is not known whether sedimentary wildflysch or tectonic mélange breccia are present in the volcanogenic sandstone and gray shale beneath the klippe.

Allochthonous sequence.—The Cow Head Breccia is characterized by limestone breccia interstratified with green silty shale, calcarenite, and lime mudstone (Figs. 2, 3A). There are lesser amounts of quartzose sandstone, yellow siltstone, gray marl and radiolarian-sponge spicule chert. The sequence has long been recognized as a slope facies because of the distinctive association of (1) limestone megabreccia, (2) thin-bedded lime mudstone and shale with graptolites, (3) thin beds

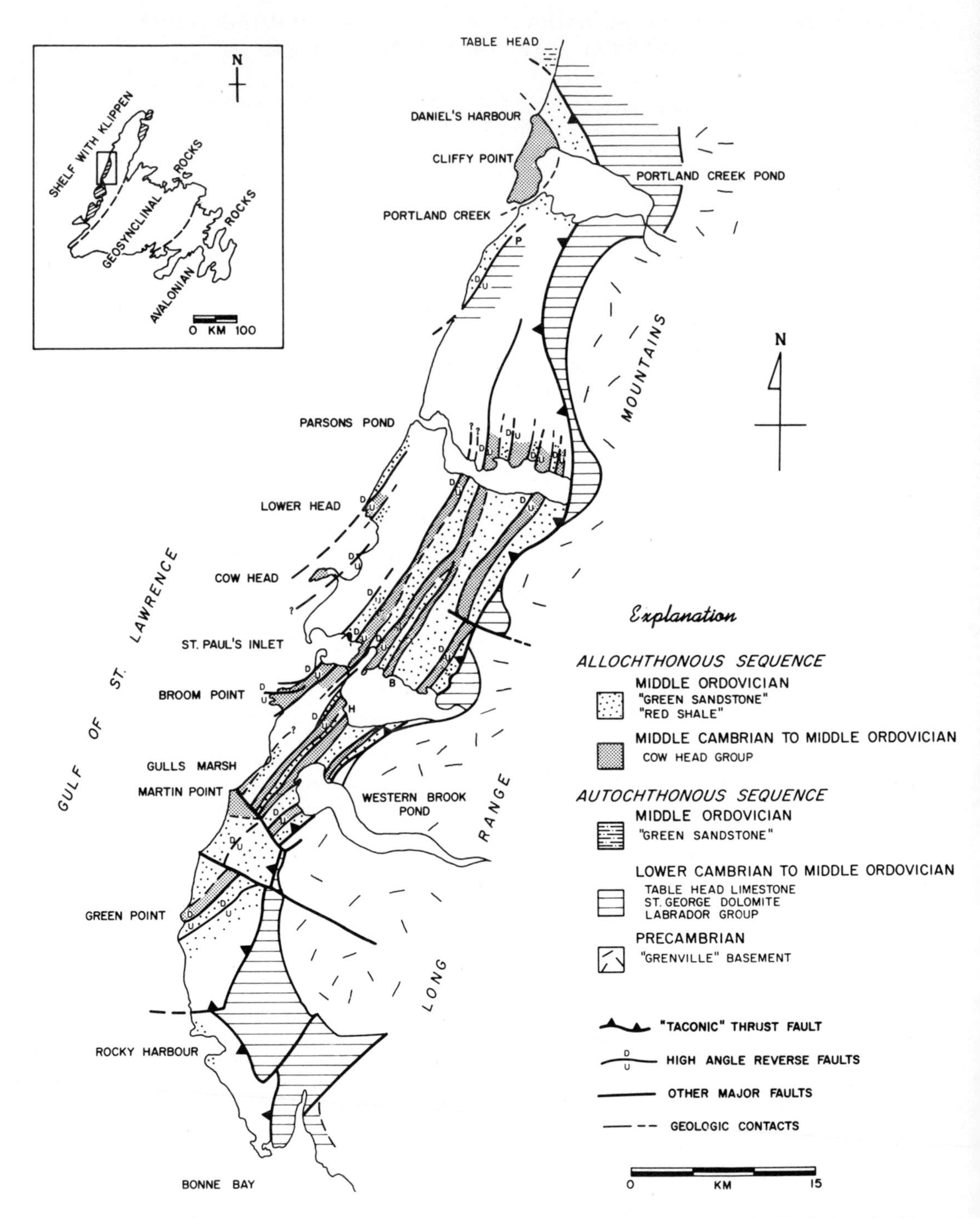

FIG. 1.—Geologic map of the Cow Head klippe. Letters indicate Berry Head (H), Black Brook (B), and Portland Hill (P). Geology modified from Oxley (1953). The klippe was transported from the southeast.

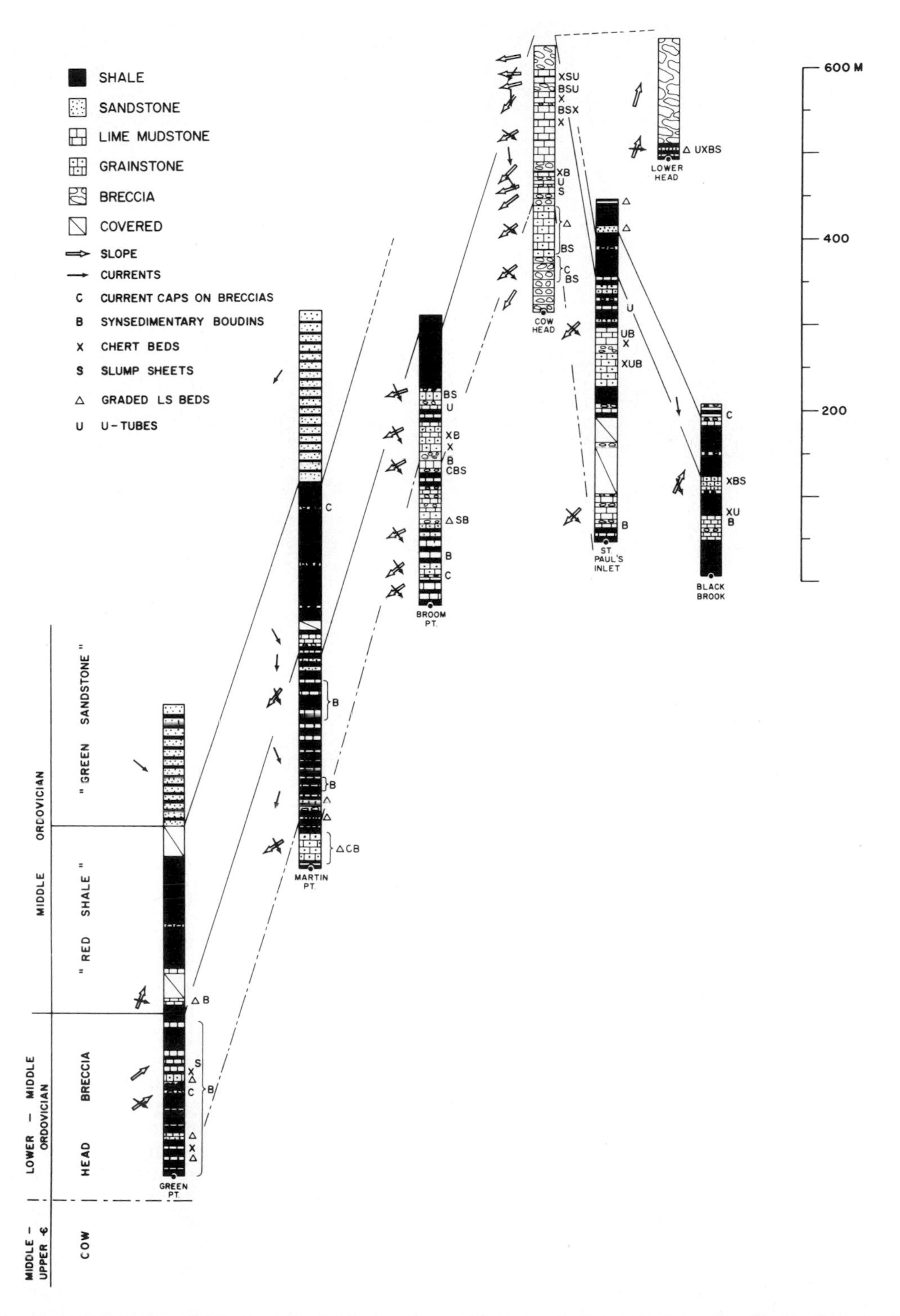

FIG. 2.—Stratigraphy of the Cow Head klippe. The sections are plotted at their geographic locations in the klippe (Fig. 1). Time lines follow biozones established for trilobites and graptolites (Kindle and Whittington, 1958) and conodonts (Nowlan, 1973). Each paleoslope and paleocurrent arrow is the mean azimuth with north vertical.

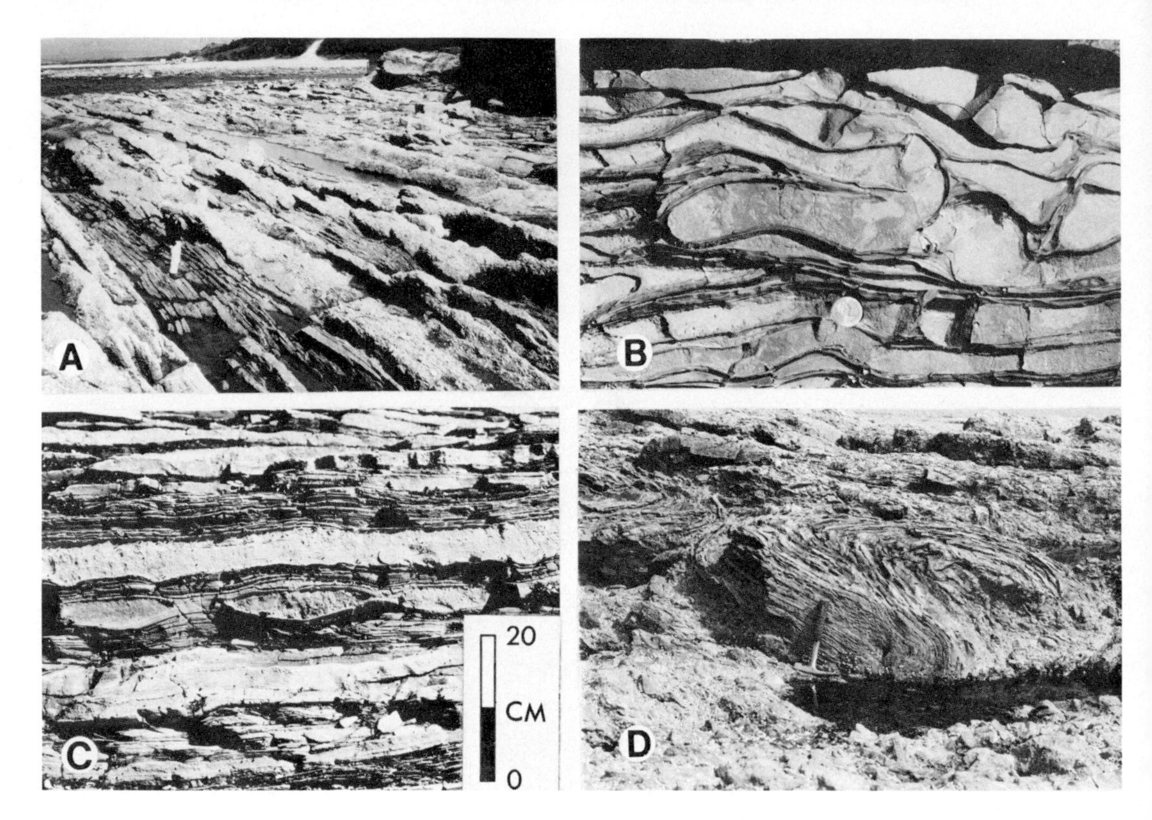

FIG. 3.—*A*—Sequence of breccia and thin-bedded limestone and shale in the Cow Head Breccia of Middle Cambrian age at Broom Point. *B*—Slump sheet of thin-bedded limestone and shale; the sheet moved to the right. Cow Head Breccia, Middle Ordovician at Green Point. *C*—Synsedimentary boudins of limestone interbedded with green shale. Cow Head Breccia, Middle Ordovician at Green Point. *D*—Soft-sediment fold in boulder of limestone and shale in Cow Head breccia, Middle Cambrian at Cow Head.

of radiolarian-sponge spicule chert, (4) numerous slump horizons, and (5) rarity of *in situ* shelly fossils (Kindle and Whittington, 1958; Baird, 1960; Stevens, 1970). The strata accumulated on the slope seaward of the autochthonous shelf carbonates (Williams and others, 1973).

With Middle Ordovician subsidence of the continental margin, red mud rapidly spread over the Cow Head Breccia, followed by volcanogenic sandstone and gray shale. The "red shale" of the allochthonous sequence is absent in the autochthonous sequence.

METHODS FOR DETERMINING PALEOSLOPE DIRECTIONS

Knowledge of the paleoslopes is necessary to interpret the origin of the limestone breccia and to map the paleogeography. Paleoslope directions were measured using slump sheets, synsedimentary boudins, and soft-sediment folds in boulders of thin-bedded limestone and shale in the breccia.

Slump sheets.—Synsedimentary slump sheets are common in the thin-bedded limestone and shale interstratified with the Cow Head breccias in both the Cambrian and Ordovician sequences. The strata slid at the sediment-water interface because the folds are beveled by erosion beneath the overlying undeformed strata (Fig. 3B). The slump sheets are mostly 0.3 to 0.6 m thick.

Hansen (1966) discussed the general solution for determining the slip line of a slump sheet. He emphasized that an outcrop located anywhere in a slump sheet can be used to determine the direction of movement of the slide. The paleoslope direction is determined by plotting the orientation and sense of rotation of the fold axes on a Schmidt net after the tectonic tilt, including plunge where present, is removed by rotating the beds to a horizontal position. By convention, the clockwise, or counterclockwise, rotation of each fold axis is plotted looking down the plunge. Although there is substantial scatter in the orientation of the axes, the paleoslope direction is easily determined by the rotation sense of the folds. The fold axes form two groups characterized by clockwise and counterclockwise rotation. The paleoslope azimuth bifurcates the groups in the downslope direction shown by the rotation of the folds. The two groups of axes lie along the paleocontour direction. Axes that lie nearly parallel to the slip line are especially useful because they more closely define the slip line.

This method is illustrated by a slump sheet in the Lower Ordovician strata at Cow Head. The fold axes form two clusters that define the NW-SE paleocontour direction (Fig. 4). The rotation sense of the axes shows that the paleoslope azimuth is southwest at 251°.

The paleoslope at each of the sections of the Cow Head Breccia was remarkably consistent from the Middle Cambrian through the early Middle Ordovician (Fig. 2). Slump horizons at six levels in the 310-m section of the Cow Head Breccia at Cow Head yielded readings for 68 fold axes (Figs. 2, 5). The average paleoslope was to the southwest. Significant differences in paleoslope direction among the outcrops of Cow Head Breccia define the regional paleotopography.

Synsedimentary boudins.—The thin beds of limestone interbedded with green shale are deformed into synsedimentary boudins at many horizons throughout the Cow Head Breccia (Figs. 2, 3C). The cylindrical boudins vary in thickness from a few centimeters to about 15 cm and have length-width ratios between 2 and 4 to 1. After the tectonic tilt is removed, the axes of the boudins plot on a Schmidt net in two clusters that define a horizontal line. The line is always parallel to the paleocontour direction independently determined by the axes of slump folds. For example, Figure 4 shows the near coincidence of the axes of 20 boudins and 12 slump folds for a part of the Lower Ordovician sequence at Cow Head. The entire section at Cow Head yielded 67 boudin axes whose orientation is similar to the 68 axes of slump folds (Fig. 5). The rotation sense of the slump folds shows that the paleoslope was to the southwest.

The boudins are not of tectonic origin because the orientation of the boudin axes is independent of the geometry of the post-Taconic (Acadian?) anticlines and synclines and their associated minor folds (Williams and others, 1973; Smith, 1973; Suchecki, 1975). The major anticlines and synclines are drag features on reverse faults that strike subparallel to the reverse and thrust faults along the western margin of the "Grenville" core of the Long Range Mountains. The faults may be the most western expression of the Acadian Orogeny whose effects are seen in central New-

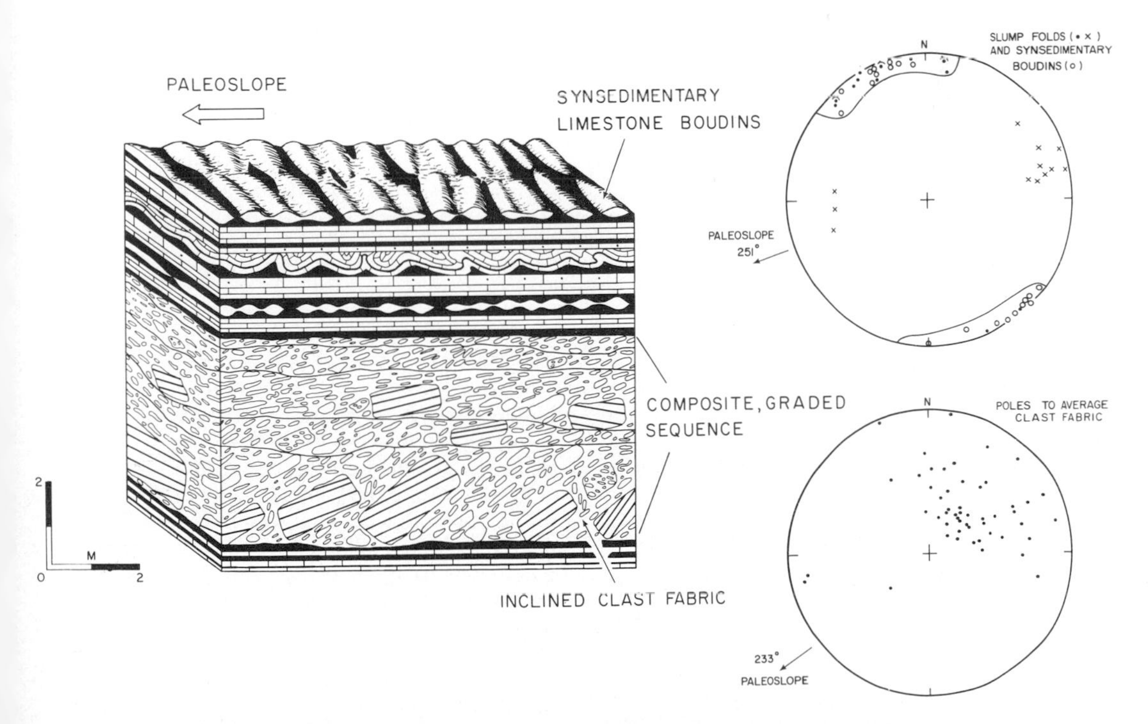

FIG. 4.—"Composite graded sequence" of breccias of Early Ordovician age, Cow Head Breccia at Cow Head. The thin-bedded limestone and shale contain slump sheets and synsedimentary boudins. Throughout this report, the data on Schmidt nets are always plotted with the tectonic tilt of the beds removed.

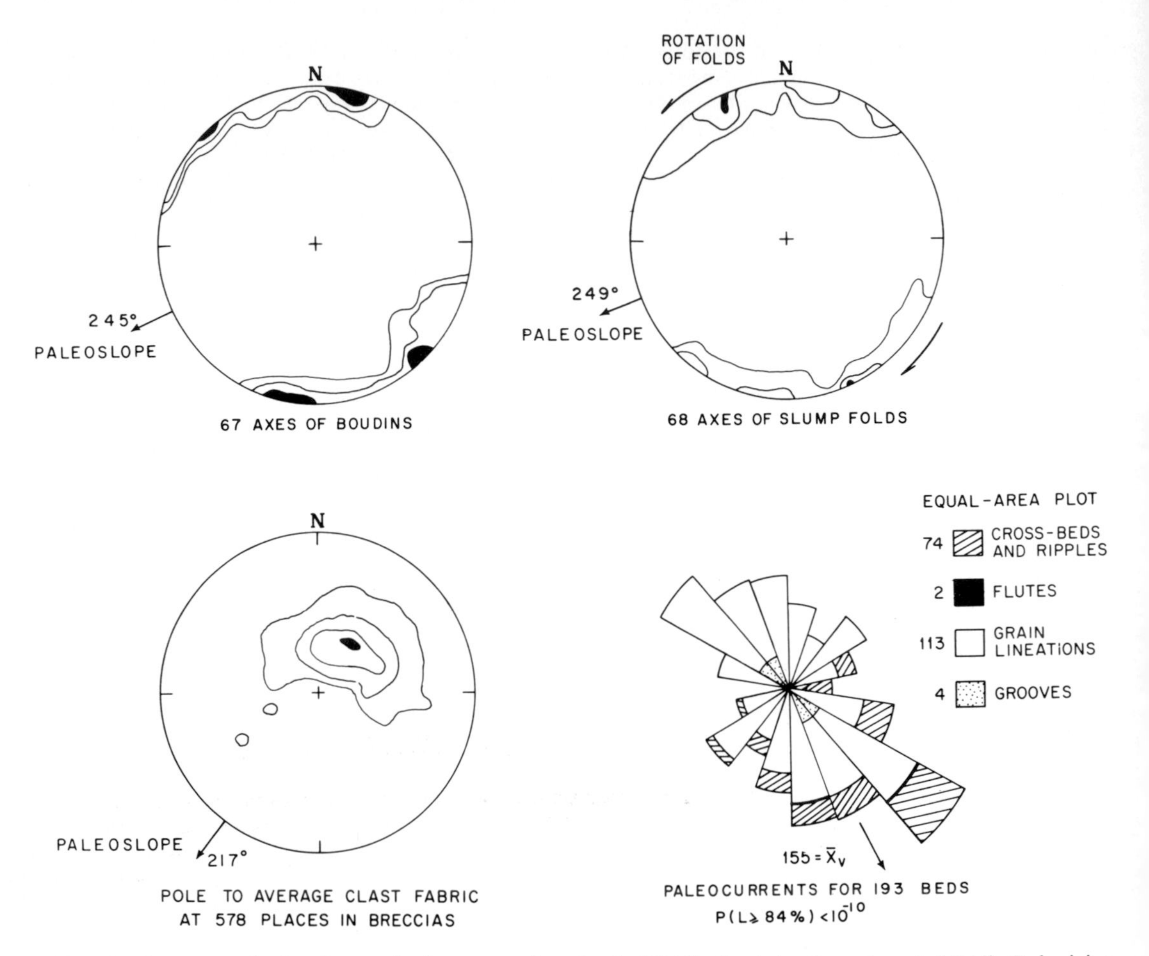

FIG. 5.—Summary of paleoslope and paleocurrent data for the Middle Cambrian through early Middle Ordovician type section of the Cow Head Breccia at Cow Head. The paleoslope was southwest whereas the paleocurrents flowed southeast. The contours on the Schmidt net diagrams are: synsedimentary boudins—2, 6 and 8 percent; axes of slump folds—2, 6 and 12 percent; inclined clast fabric—2, 4, 6 and 8 percent. Throughout this report, each reading for clast fabric is the average for a small area of a breccia, not the orientation of an individual clast.

foundland in folding, intrusion, and metamorphism. The direction of Acadian (?) stress was consistently northwest throughout the Cow Head klippe.

Furthermore, the limestone in the Cow Head Breccia shows no petrographic evidence of regional metamorphism and the shale lacks slaty cleavage. None of the folds can be attributed to Ordovician emplacement of the klippe (Williams and others, 1973).

Soft-sediment folds in boulders in the breccia.—Many breccias contain boulders of thin-bedded limestone and shale with soft-sediment folds (Fig. 3D). The lithology of the boulders is similar to the beds interstratified with the breccias. These boulders were semi-lithified slope deposits when incorporated into the mass of mud and clasts and became deformed during downslope transport.

To determine the paleoslope azimuth, the folds in the boulders were analyzed in the same manner as the folds in slump sheets. For example, a paleoslope azimuth of 20° was measured for an unusually thick and coarse breccia at Lower Head by plotting on a Schmidt net the orientation and sense of rotation of the axes of 21 folds in boulders (Fig. 6). The azimuth of 20° closely agrees with the paleoslope azimuth of 13° determined using the axes of 20 slump folds and 26 synsedimentary load and boudin structures in the underlying thin-bedded limestone and shale.

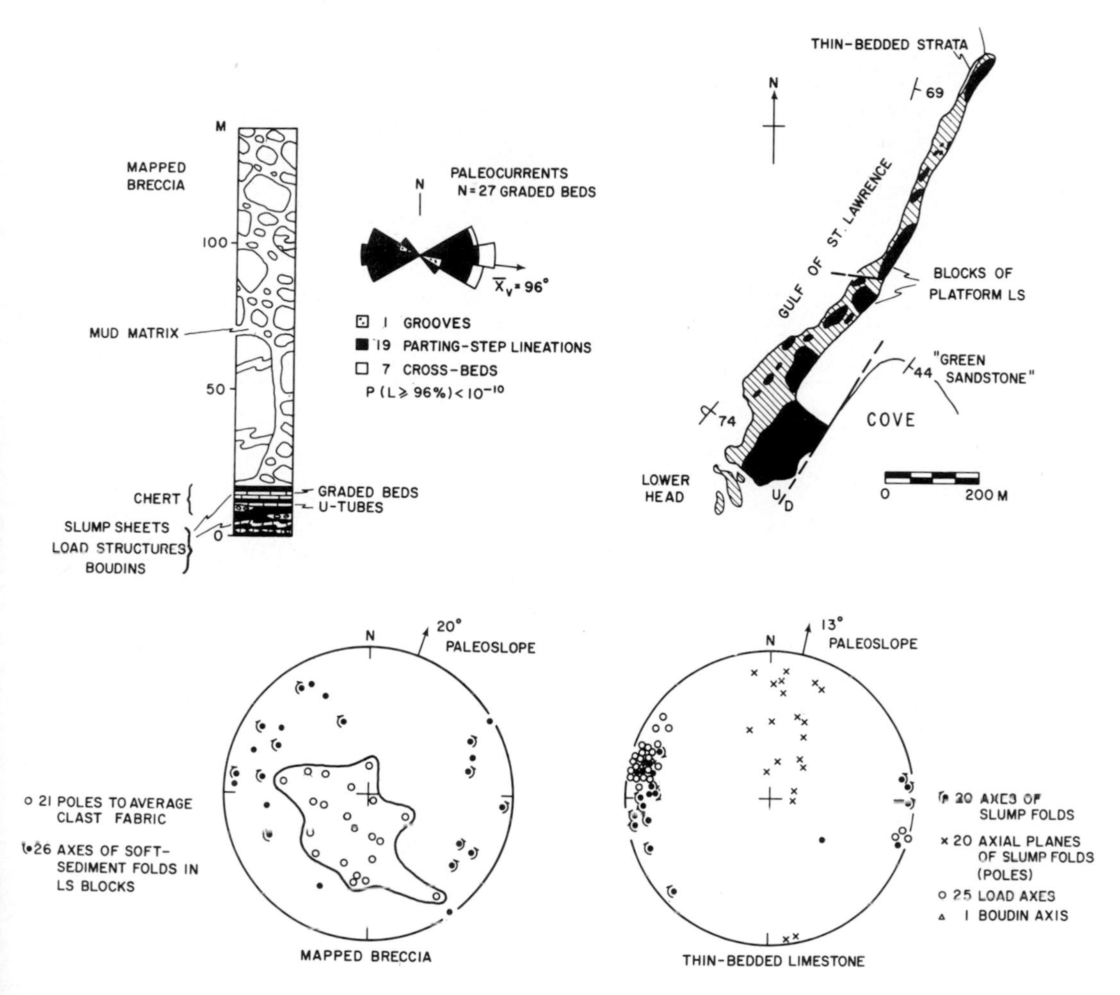

FIG. 6.—Middle Ordovician sequence of the Cow Head Breccia at Lower Head. The mapped breccia (upper right diagram) has boulders of platform limestone that exceed 60 by 150 m in size implying that Lower Head was located close to the carbonate platform. The hatching symbol in the mapped breccia indicates mud matrix and clasts. The larger clasts are shown in black. The paleoslope was northeast and the paleocurrents flowed southeast.

SEDIMENTOLOGY OF THE COW HEAD BRECCIAS

Physical Characteristics of the Breccias

Composition of the clasts.—The clasts are entirely sedimentary with no igneous or metamorphic pebbles. About 90 percent are the same types of limestone interbedded with the breccias. Most abundant are lime mudstone, calcarenite, and boulders of interbedded limestone and shale. The limestone clasts are mostly plate-shaped with smooth edges. Some clasts are bent, implying partial, very early lithification of the calcareous sediment.

Peritidal and shallow-water limestone comprise about 10 percent of the clasts (Fig. 7A). The subspherical clasts were lithified to rock prior to incorporation in the breccia. Calcite fills both grain interstices and moldic porosity developed by solution of aragonitic fossils and ooids. The boulders contain blue-green and green algae, stromatolites, trilobites, pelmatozoans, brachiopods, ooids, quartzose sand, and peritidal "birds-eye" structure. Many of the limestone boulders of both Cambrian and Ordovician age contain layers of "dusty," inclusion-rich rhombs of dolomite, averaging 0.05 mm in size, that replaced pelleted carbonate mud.

Many of the Cambrian and Ordovician shallow-water limestone boulders are solution-collapse breccia formed by karstification. The fluctuations

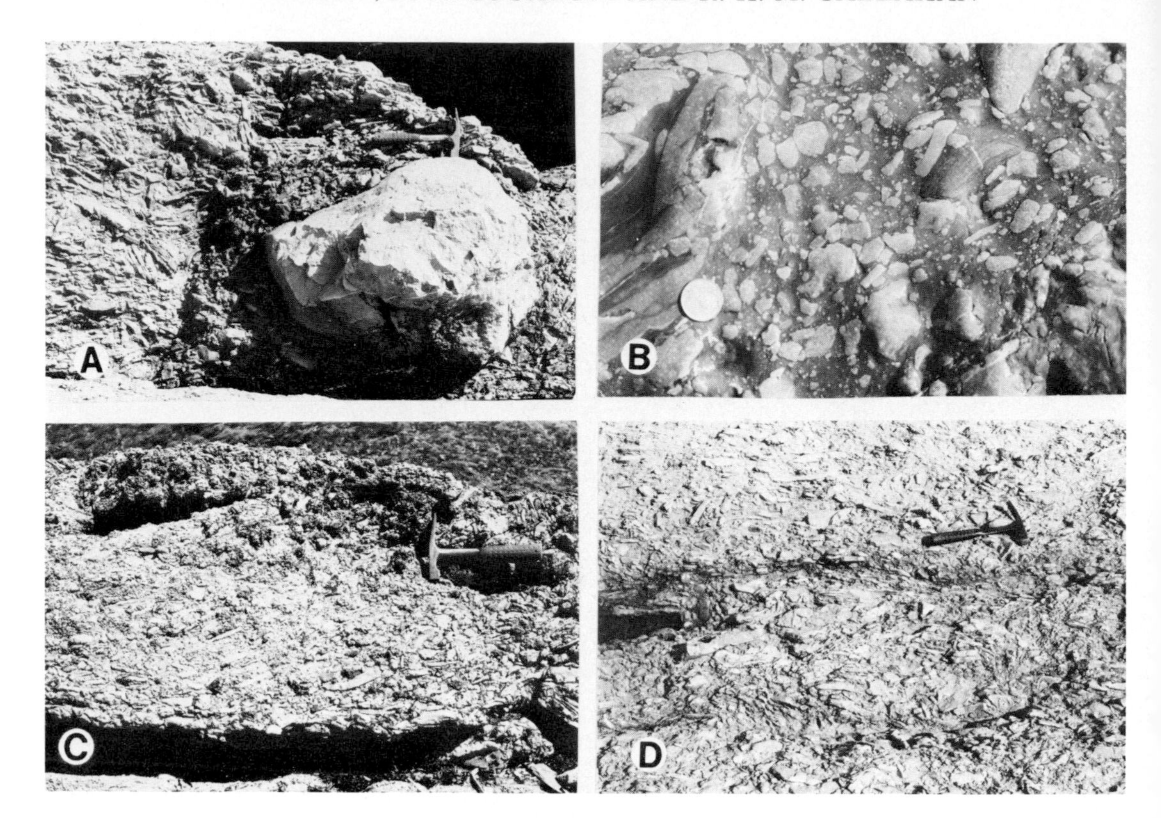

FIG. 7.—Physical characteristics of the Cow Head breccias. *A*—Breccia composed mostly of plate-shaped clasts of slope limestone. The large white boulder is platform oolitic limestone. Middle Cambrian at Broom Point. *B*—Terrigenous mud matrix in Middle Cambrian breccia at Cow Head. The wave-polished surface is parallel to the bedding. *C*—Inclined clast fabric in Middle Ordovician breccia at St. Paul's Inlet. The clasts dip downslope to the left. *D*—Shear surfaces in mud-rich breccia (also shown in Fig. 13). The flow moved to the right. Upper Cambrian, Broom Point.

in sea level cannot be directly related to glacial episodes because the glacial deposits in North Africa near the Ordovician south pole are restricted to the Upper Ordovician and hence younger than any rocks in the klippe (Allen, 1972).

The source of these shallow-water limestone boulders existed continuously from Middle Cambrian through early Middle Ordovician because the trilobites and conodonts in the boulders become progressively younger concommittantly with the graptolite and conodont faunas in the thin-bedded limestone and shale interbedded with the breccia (Kindle and Whittington, 1958; Nowlan, 1973).

A few clasts of cross-bedded, arkosic sandstone from the Labrador Group, and brown, coarsely crystalline St. George Dolomite, occur in Ordovician breccia in the northern part of the klippe between Portland Creek and Daniel's Harbour.

Matrix.—Every breccia has a mud matrix (Fig. 7B). The mud consists of terrigeneous clay and calcareous mud (now calcite microspar-pseudospar) mixed in all proportions, commonly with fossiliferous, oolitic carbonate sand. The mud matrix comprises 5 to 35 percent of the volume of the breccia. Most of the breccia is framework supported. Isolated patches in some breccias are matrix supported.

Composite graded sequences of breccia.—A few breccias form "composite graded sequences," as in a 4 m interval of five breccias at Cow Head (Fig. 4). The thickness of each breccia, and the size of the largest boulders, decreases upwards in the sequence implying a genetic relation among the flows. The time between flows was brief because of the absence of both current scour and layers of thin-bedded *in situ* limestone and shale.

Thickness.—The Cow Head Breccia is a "starved sequence" where only 310 m of strata accumulated in the approximately 70 million years from the Middle Cambrian through early Middle Ordovician. Most of the time is evidently represented by the intervals of thin-bedded shale and limestone, with rapid deposition of each breccia.

A continuous log of the type section of the Cow Head peninsula yielded 223 individual breccia flows, which averages one breccia each 300,000 years (Fig. 8). Individual breccias are commonly superposed into a compound bed; they probably were deposited in rapid succession. The modal thickness of the individual breccia flows is 20 to 30 cm with the distribution skewed to the thicker flows.

Soles.—The lower surfaces of the beds of breccia are nonerosional except for linear scours, commonly a few tens of centimeters deep and wide. Rarely the linear scours are quite large, as at Broom Point where one is about 50 m wide, measured perpendicular to the paleoslope, and about 1.5 m deep (Fig. 9). The linear scours trend in the downslope direction and contain some of the coarsest boulders in the breccia. The breccias do not occur in channels visible on the scale of the outcrops which extend hundreds of meters along the shoreline. The soles of a few breccias have linear load structures oriented perpendicular to the paleoslope. Flute marks are not present on the breccia soles.

Clastic fabric.—About 90 percent of the clasts are plate-shaped pieces of limestone which determined the resulting fabric characterized by parallel, close packing. Most commonly the clasts are aligned subparallel to the bedding or in low amplitude waves (Fig. 9). Also common is a fabric where the plate-shaped clasts are inclined to the bedding at a modal value of 15 to 35 degrees (Figs. 7C, 10). The dip of the modal clast fabric is down the paleoslope, as determined by independent indicators. This is the reverse of the familiar upstream dip of imbricated particles.

Many breccias show transitions among horizontal, wave form, and inclined clast fabrics both laterally and vertically (Fig. 9). All three clast fabrics are due to the downslope, surging motion of the mass of mud and limestone plates.

The breccia at the top of the sequence of the Cow Head Peninsula has an exceptional, rare clast fabric which is very steep, even locally overturned (Fig. 11). The 25 soft-sediment folds in the boulders of thin-bedded limestone and shale show that the paleoslope was southwest, perpendicular to the average strike of the plate-shaped clasts. The poles to the average clast fabric at 100 different places in the breccia form a broad NE-SW zone on the Schmidt net parallel to the paleoslope.

This breccia is 28 m thick and some boulders exceed 20 by 60 m in size. The minimum volume is 17.5 million cubic meters measured in exposures around a large cove. The exposed volume is probably much less than the total volume because the breccia does not thin along the outcrop.

Many breccias have wave forms outlined by plate-shaped clasts of lime mudstone, calcarenite, chert, and breccia. The best exposed breccia crops out continuously for 718 m along the coast at Gulls Marsh (Fig. 12). The upper surface of the breccia is a planar erosional surface that truncates the underlying wave forms. The average wave length of 56 wave forms is 2.4 m and the mean amplitude is 0.6 m. The thickness of the breccia varies from 1.7 to 2.4 m due to local subsidence

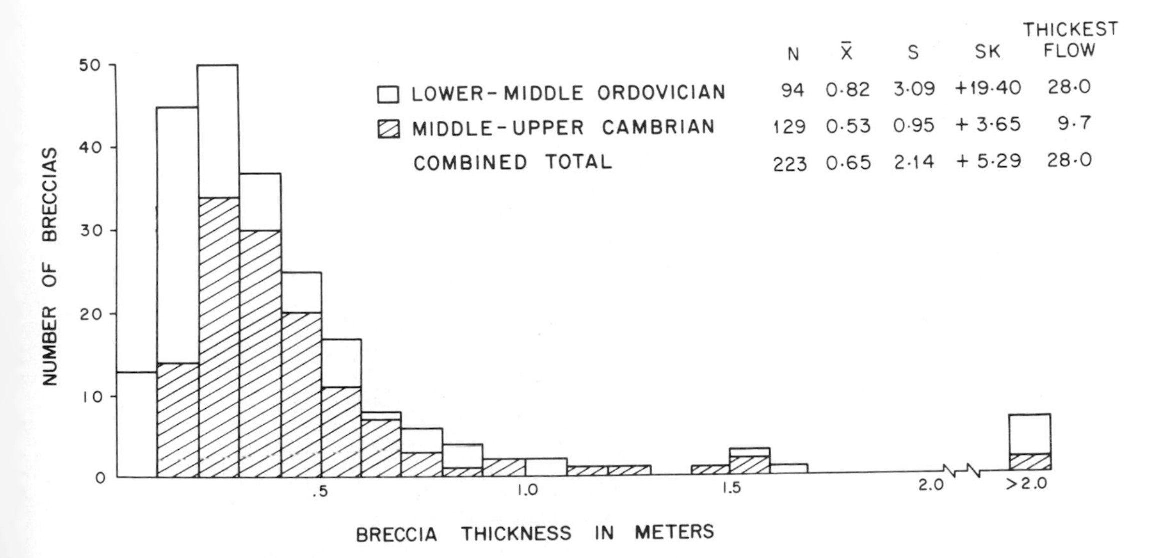

Fig. 8.—Thickness of individual breccias in continuous log of 310 m section of the Cow Head Breccia at Cow Head.

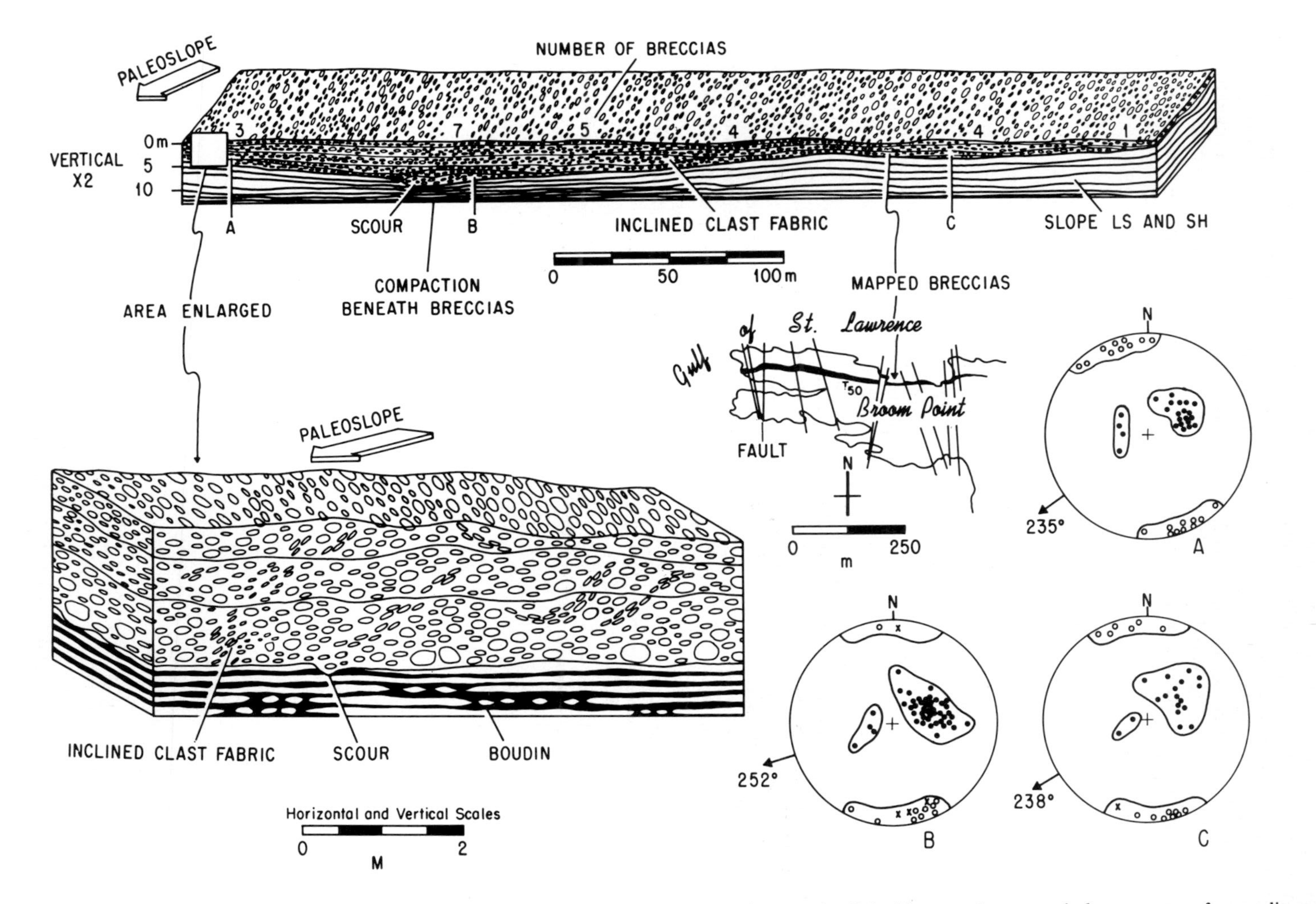

FIG. 9.—Cow Head breccias of Early Ordovician age mapped along the coast at Broom Point. On the Schmidt nets, the open circles are axes of synsedimentary boudins in thin-bedded limestone. Each x is the axis of a load structure on the base of a breccia. The arrows show the southwest paleoslope. The solid dots are the poles to the clast fabric of the breccias which dips mostly downslope. Schmidt nets A, B and C refer to the three areas shown on the diagram.

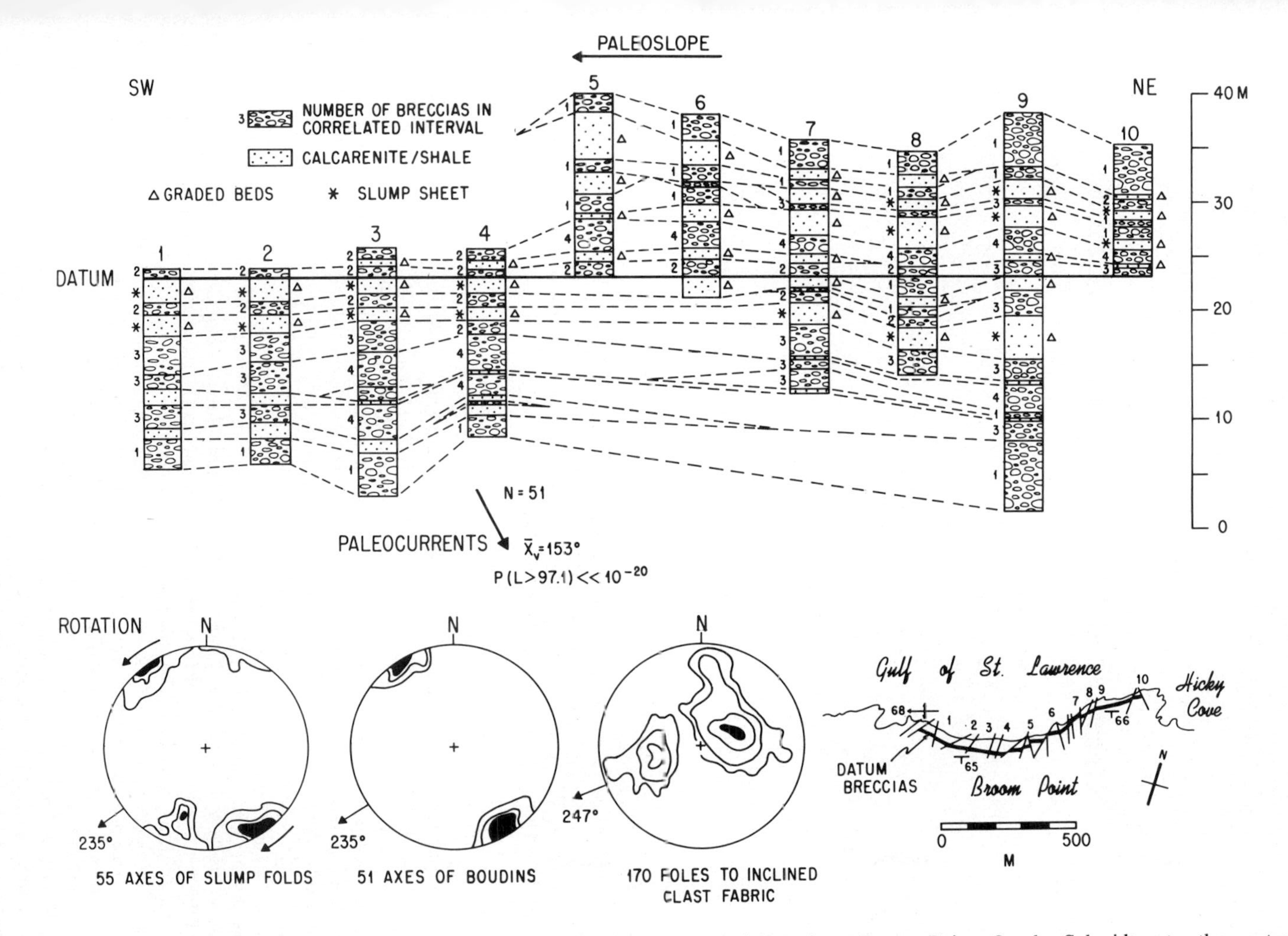

FIG. 10.—Breccias and graded beds of calcarenite in the Upper Cambrian Cow Head Breccia at Broom Point. On the Schmidt nets, the contours for 51 synsedimentary boudin axes of thin-bedded limestone are 1, 3 and 15 percent. For the 55 axes of slump folds the contours are 1,6 and 15 percent. The poles to the clast fabric of the breccias at 170 different places are contoured at 1, 3, 9 and 18 percent. The paleocurrent readings for 51 graded beds are sole marks, parting-step lineations and ripple-marks. The paleoslope was southwest whereas the paleocurrents flowed southeast.

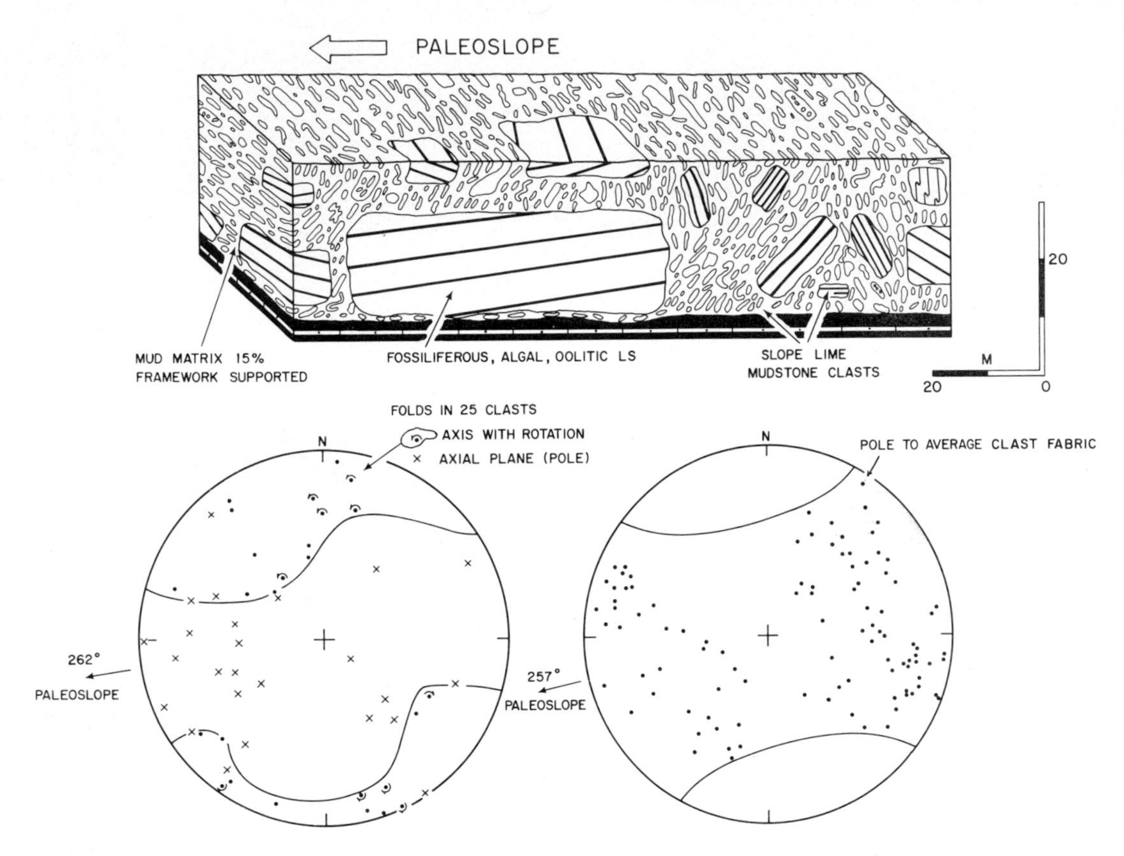

FIG. 11.—Breccia with steep to overturned clast fabric. The plate-shaped clasts of limestone strike perpendicular to the southwest flow direction shown by the orientation and sense of rotation of soft-sediment folds in 25 boulders. The poles to the clast fabric at 100 different places in the breccia form a broad zone parallel to the paleoslope. Cow Head breccia of Middle Ordovician age, Cow Head.

of the underlying strata caused by the weight of the breccia. The breccia overlies a thin bed of green shale for most of the outcrop showing that basal erosional scour is minor.

The anticlines in the wave forms of this breccia are overturned to the northeast in the paleoslope direction (Fig. 12). The limbs of the wave forms strike northwest-southeast when the bed is rotated to a horizontal position, so that the poles to the limbs form a NE-SW belt on the Schmidt net. The clast fabric in the lower part of this breccia, below the wave forms, dips down the paleoslope.

The largest boulders in most breccia beds tend to be near the base producing crude graded beds.

Shear surfaces.—Where a breccia has a large volume of mud matrix, exceeding about 25 percent, it tends to have lensing, tabular subdivisions separated by shear surfaces that developed during downslope flow of the mass of mud and clasts. A fine example is exposed for 450 m along the coast at Broom Point (Figs. 7D, 13). The soft-sediment folds in the boulders of limestone and shale show that the paleoslope was southwest. Most of the plate-shaped clasts are subparallel to the shear surfaces. In places an inclined clast fabric dips downslope at about 30°. There is also a subordinate, shallower clast fabric that dips upslope at about 15° (Fig. 13).

Deposition of the Breccias

Mass flows.—The Cow Head breccias have in common: (1) a terrigenous-mud carbonate matrix; (2) poor sorting from clay to huge boulders; (3) a downslope direction of travel; (4) crude grading; (5) an absence of traction-current antidunes, plane beds, cross-beds, and ripples; (6) nearly planar lower surfaces; and (7) volumes measured in millions of cubic meters. These features suggest the breccia formed as mass flows. Mass-flow breccia is a characteristic component of the slope facies of continental margins and carbonate platforms (Cook and others, 1972; Enos, 1973;

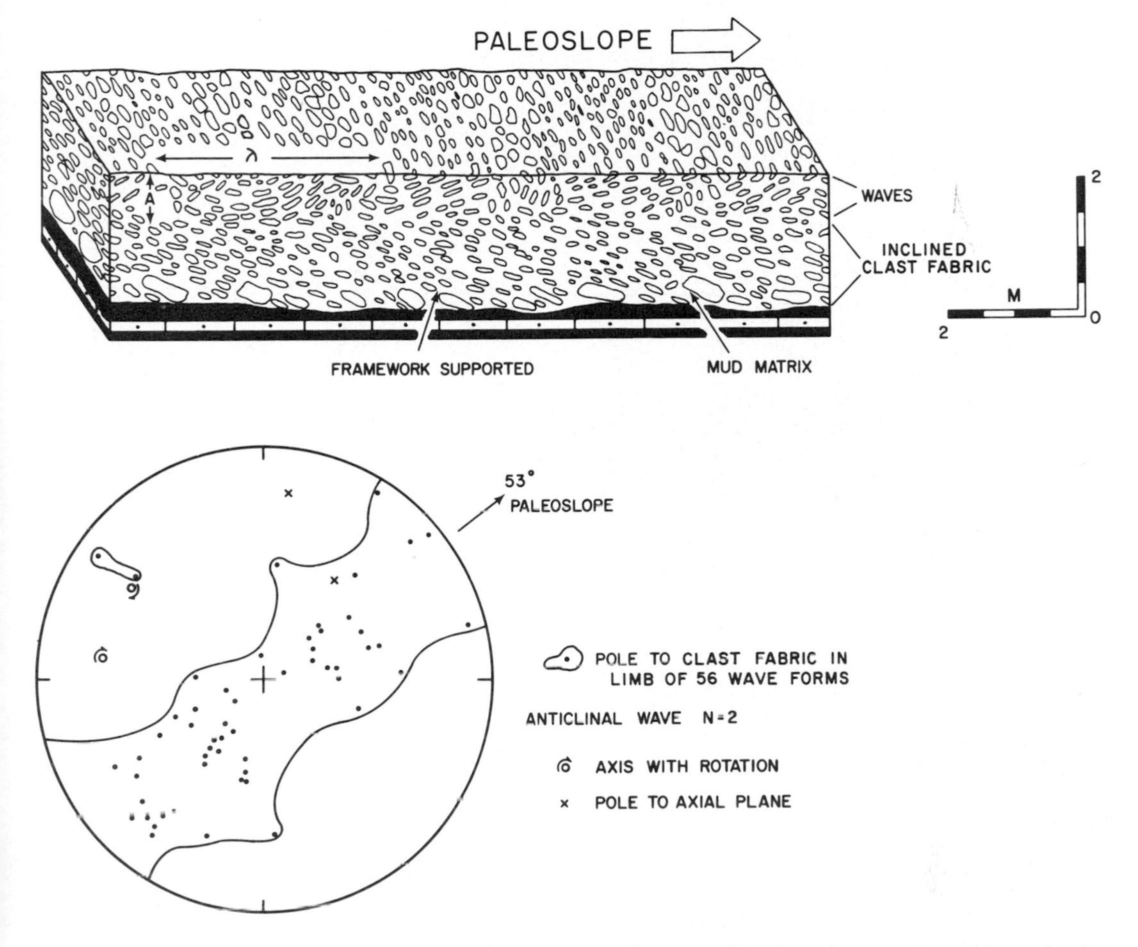

FIG. 12.—Wave forms at the top of a breccia. The mass flow travelled downslope to the northeast in the direction of overturning of the wave forms. The poles to the clast fabric in the limbs of the anticlinal and synclinal wave forms comprise a broad zone parallel to the paleoslope. The clast fabric in the lower part of the breccia dips downslope to the northeast. Cow Head Breccia of Cambrian age at Gulls Marsh.

Hendry, 1972, 1973; Hopkins, 1972, this volume; Mountjoy and others, 1972; Srivastava and others, 1972; Cook and Taylor, 1975, this volume).

Careful study of the breccia failed to locate shear surfaces restricted to the bottom and top of the breccias; the flows evidently did not transport passive, solid "plugs." Thinning of a breccia to zero thickness is always smoothly progressive, without the steep terminations expected when a subaqueous mass flow moves as a rigid "plug" with a low water content (Hampton, 1972).

Some possible mechanisms for initiating mass flows are: (1) earthquakes (Baird, 1960, p. 23); (2) rapid deposition on a slope producing underconsolidation of mud with consequent large excess pore pressures and low internal angles of friction (Morgenstern, 1967); (3) collapse of the platform edge due to patchy distribution of early cements greatly increasing the mass of the sediment (Cook and others, 1972; Hopkins, this volume); and (4) tsunamis (Cook and others, 1972). The platform near Cow Head was maintained for 70 million years, perhaps by active faults, so that earthquakes are an attractive hypothesis for initiating the mass flows.

Mechanism of flow.—Massive slope failure along the flank of a carbonate platform generated gravity-driven mass flows consisting of tens of millions of cubic meters of material. Lithified clasts of platform limestone and slope-facies semilithified lime mudstone, calcarenite, chert, and breccia were thoroughly shuffled with terrigenous-carbonate mud. The elongate flows were not confined to channels and were very thin

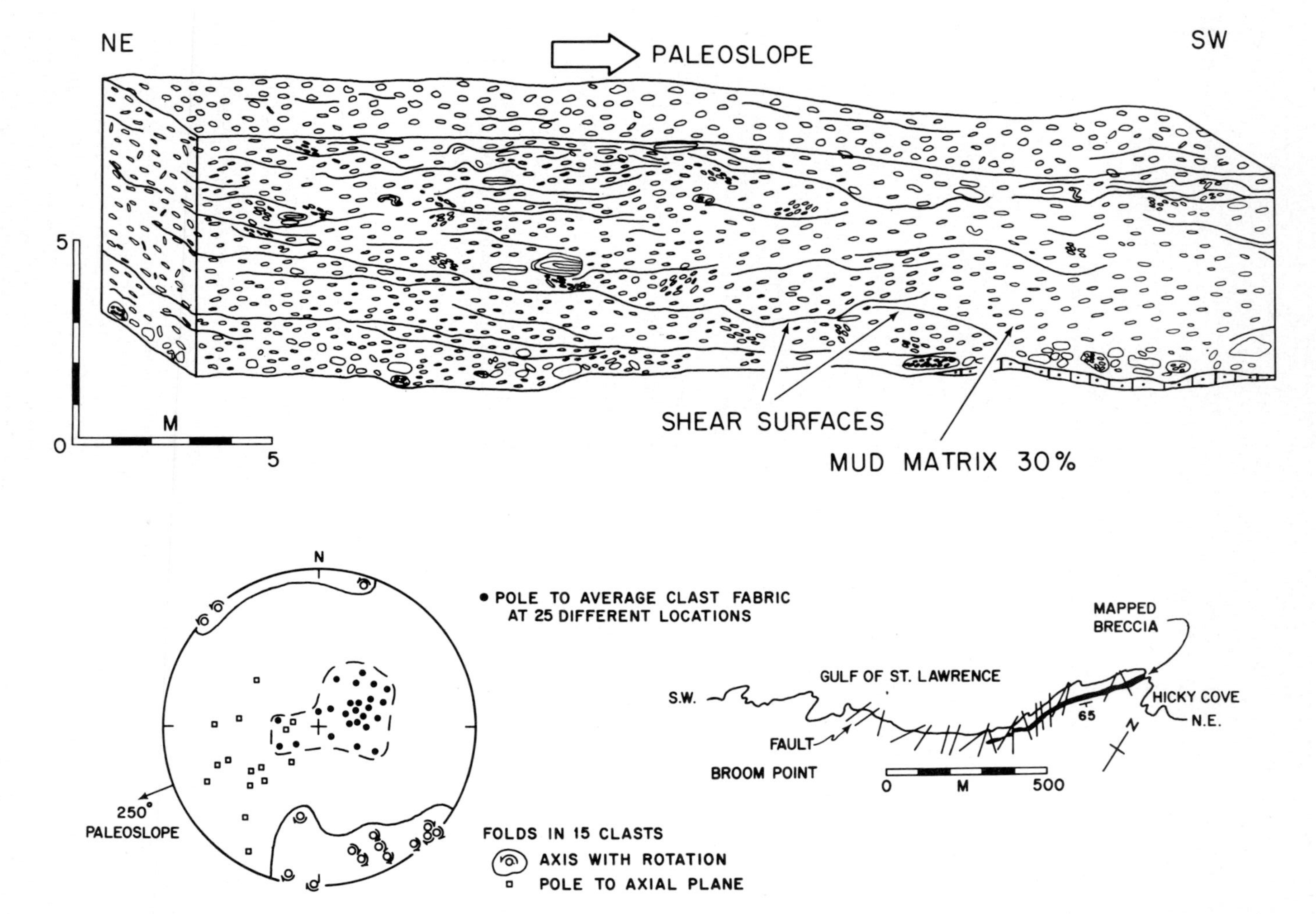

FIG. 13.—Breccia with internal shear surfaces. The mass flow travelled downslope to the southwest as shown by orientation and sense of rotation of soft-sediment folds in 15 boulders of limestone and shale. The dominant clast fabric dips downslope. Cow Head breccia of Cambrian age at Broom Point.

compared to their length and width.

The flows moved downslope as a wet, highly viscous, surging mass of mud and clasts. The matrix strength generally supported the large clasts (Middleton and Hampton, 1973). Some of the largest boulders sank through the flows to form crudely graded breccia. Carried along in the flows were blocks of platform limestone and slope-facies limestone and shale that exceed 50 m in length and commonly project through the upper surfaces of the flows. The plate-shaped clasts of limestone slid smoothly past one another in the wet mud to produce the dominant horizontal and waveform clast fabrics. Some flows with a large amount of mud matrix developed shear surfaces subparallel to the bedding.

The inclined clast fabric present in some breccias resulted from shear within the flow produced by friction that increased toward the base. The plate-shaped clasts of limestone were rotated through various angles in the direction of downslope movement. This suggests highly viscous flow; otherwise upslope imbrication of clasts during cessation of movement might be expected.

An upslope orientation of clasts occurs where clasts lie on the upslope side of low amplitude waves and also where clasts were forcefully thrust up through the flow. An inclined clast fabric that dips upslope at low angles is associated with local concentrations of mud matrix; the inclined fabric evidently offered minimum resistance to flow in a manner analogous to terrigenous gravels deposited by streams.

Composite graded sequences of breccia.—These sequences occur sparingly throughout the Cow Head Breccia and are best developed at Cow Head (Fig. 4). A possible origin is progressive failure of the source material, perhaps unconsolidated layers of boulders, calcareous sand, and calcareous mud in a fan at the base of the carbonate platform. After the initial slump, the area of the slump scar perhaps had excess pore pressure due to removal of the sediment mass. Several successive, closely spaced mass flows might then relieve the pressure (Hendry, 1973). Another possibility is each compound bed reflects successive surges of a single mass flow that progressively overrode one another downslope.

CAMBRIAN HISTORY OF THE COW HEAD BRECCIA

The Cambrian strata are a slope sequence of mass-flow breccia interstratified with thin beds of lime mudstone, calcarenite, green silty shale, and gray marl (Fig. 2).

Paleocurrents and Paleoslopes

Two narrow carbonate platforms that trended northwest-southeast were present near Cow Head and Martin Point as the klippe is now oriented (Fig. 14). No part of the platforms is now preserved. The eleven Cambrian sections are concentrated near the two platforms and define them well. The northeast-dipping regional paleoslope is less well shown and is in part inferred from the thirty Ordovician sections.

The paleocurrents consistently flowed to the southeast parallel to the paleocontours, indicating a stable system of contour currents (Figs. 14, 15).

In considering paleoslope and paleocurrent patterns, the question arises whether Newfoundland should be rotated 30 degrees clockwise to remove a possible rotation imparted during the Acadian Orogeny (Black, 1964). The answer seems negative because no structures are known to support the proposed rotation, and the paleomagnetic evidence is ambiguous (Deutsch, 1969).

The amount of lateral movement along NW-SW trending faults of possible Carboniferous age is unknown. It would seem to be minor because of the consistent paleocurrent patterns observed throughout the Cow Head klippe for the Cambrian and Ordovician rocks in the Cow Head Breccia after the beds are restored to a horizontal attitude.

At Cow Head, which was close to the carbonate platform, the Middle Cambrian sequence is mostly breccia with about 10 percent thin beds of limestone and shale. The average paleoslope dips southwest, with the direction of travel of the mass flows at 17 stations varying from southeast to northwest (Fig. 16). Either the flows fanned out away from the platform, or the sea floor was uneven, perhaps due to the accumulation of successive breccias. The thickest breccia in unit 3 averages 9.7 m in thickness with an exposed volume of 2.3 million cubic meters (Fig. 16). The large boulders of limestone and shale exceed 10 by 50 m and project slightly above the hummocky surface of the breccia.

Calcarenite Layers on the Breccias

At most localities, calcarenite layers lie directly on the upper surfaces of a few of the breccias, especially in the Cambrian (Figs. 15, 17A). The layers average 30 to 40 cm in thickness, ranging up to about a meter, and commonly lens rapidly. Most of the layers contain plane beds and festoon cross beds. Only a few are graded. These are similar to calcarenite layers on the upper surfaces of breccia described by Cook and others (1972) and Cook and Taylor (this volume, Fig. 38).

In the example from Broom Point (Fig. 18), the paleoslope is southwest as shown by slump sheets and soft-sediment folds in boulders in the breccia. The southeast-flowing currents scoured the surface of the breccia and deposited the layer of quartzose calcarenite. The across-slope current pattern shows that the sand did not settle from

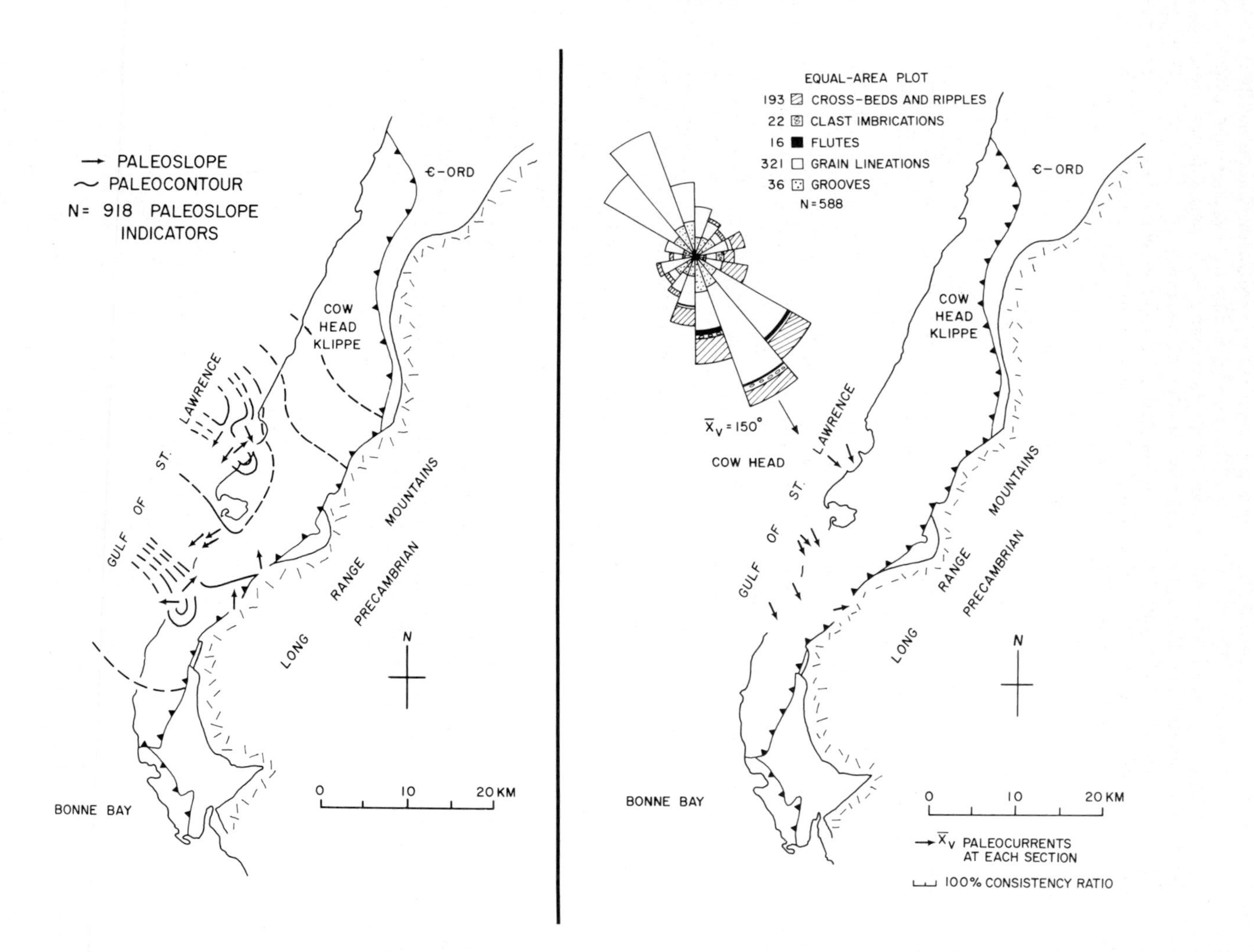

FIG. 14.—Paleoslopes (left) and paleocurrents (right) for the Cambrian strata in the Cow Head Breccia. Each paleoslope and paleocurrent arrow is the mean azimuth for that section. The paleocontour lines are drawn perpendicular to the paleoslope directions at each section. The paleocontour lines are stylized and only intended to suggest relative steepness of paleoslope. Each paleocurrent vector mean is statistically significant at the 95 percent level when tested for preferred orientation by the Rayleigh statistic L.

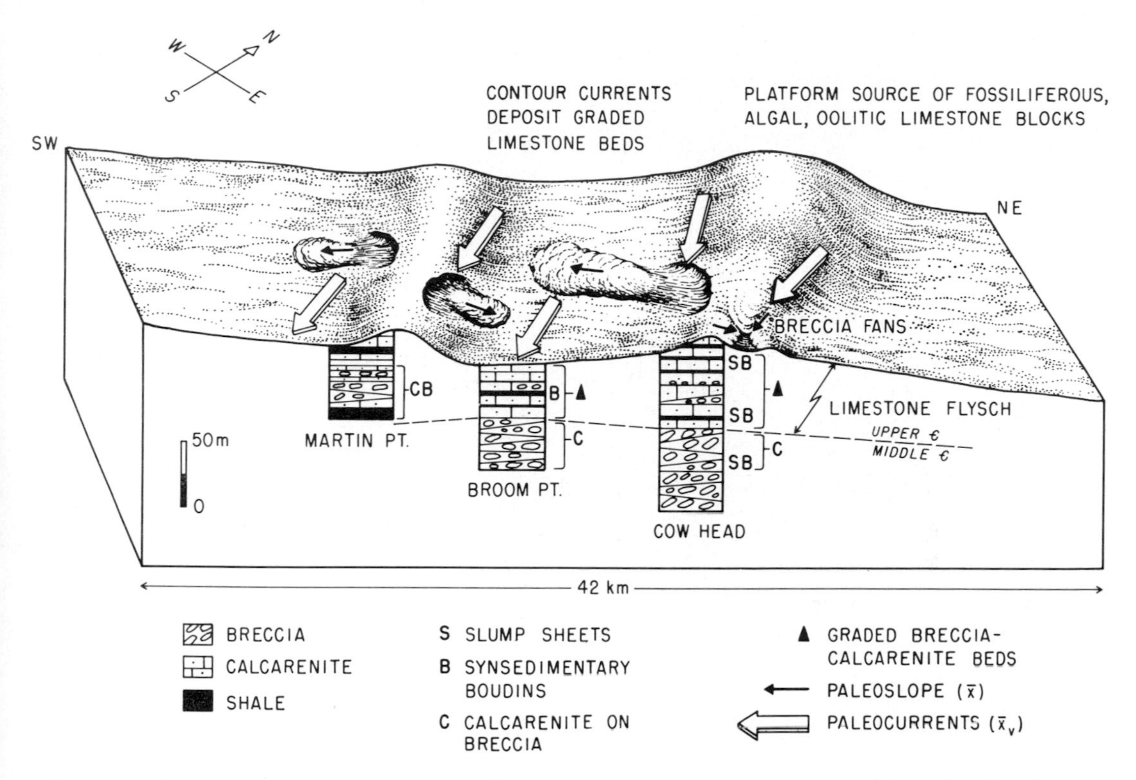

FIG. 15.—Paleotopography and stratigraphy of the Cambrian strata in the Cow Head Breccia. The diagram shows the central part of the klippe omitting the southwest and northeast ends. The paleotopography follows the schematic paleocontours of Figure 14.

suspension in an entrained layer dragged along by the mass flow as it travelled downslope.

Thoughout the Cow Head Breccia, the calcarenite layers were deposited by southeast-flowing currents, the same contour currents that deposited the nongraded beds of limestone throughout the Cambro-Ordovician sequence (Figs. 14, 20).

Graded Beds of Calcarenite in the Upper Cambrian

Description of the graded beds.—The Upper Cambrian consists of mass-flow breccia, graded beds of quartzose calcarenite, lime mudstone and green shale that give the sequence a somewhat flysch-like appearance (Figs. 17C, D). The graded beds of calcarenite correspond to the Bouma sequence if modified by a division of festoon cross-beds between the B-division of horizontal lamination and C-division of ripples (Fig. 19). Cross-beds in graded beds of limestone have been observed in the Ordovician of Scotland (Hubert, 1966) and Pennsylvanian of West Texas (Thomson and Thomasson, 1969) but in general are rare in the rock record.

At Cow Head the graded beds of quartzose calcarenite occur through a 60 m sequence. Most of the graded beds begin with the A or B divisions of coarse calcarenite without a mud matrix. Compound beds, flutes, and grooves are common. There is one erosional channel, 2 by 15 m in cross section. At Broom Point, the sequence is 20 m in thickness, the graded beds are finer grained, and both festoon cross-beds and the A-division are less common (Fig. 15). At Martin Point there are quartzose calcarenite layers on many of the breccias, but only a few graded beds.

Graded beds and the hypotheses of deposition from contour currents versus turbidity currents.—The mechanism of deposition of the graded calcarenite poses a dilemma. The grading could imply deposition from downslope, gravity-controlled turbidity currents, but the paleocurrents consistently flow southeast perpendicular to the paleoslopes. Especially significant is that the paleocurrents flowed southeast across the paleoslopes that dipped both southwest and northeast away from the flanks of the northwest-southeast trending carbonate platforms near Cow Head and Martin Point. We interpret the paleo-

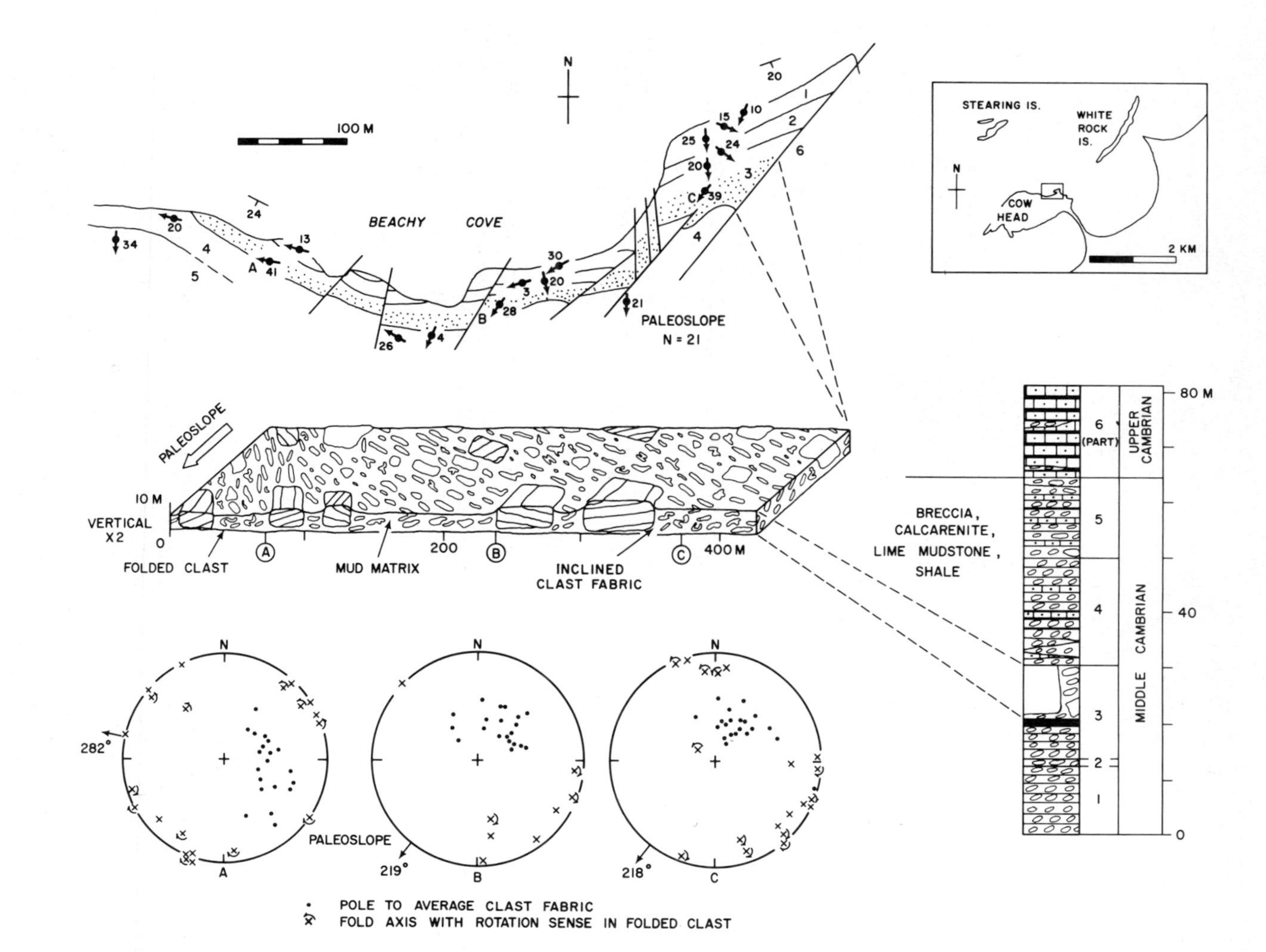

FIG. 16.—Middle and Upper Cambrian strata in the Cow Head Breccia at Cow Head. Stratigraphic units 1-6 are those of Kindle and Whittington (1958). The arrows show the downslope direction of movement of mass-flow breccia at seventeen locations. The three areas A, B and C in the mapped breccia are summarized on the Schmidt nets.

FIG. 17.—Calcarenite beds in the Cow Head Breccia. *A*—Calcarenite layer on mass-flow breccia. The ripples are eroded beneath the overlying mass-flow breccia. Middle Cambrian, Martin Point. *B*—Cross-bedded calcarenite interbedded with lime mudstone and shale. Middle Ordovician, Cow Head. *C*—Graded bed of breccia-calcarenite. Dish structure in division of horizontal lamination probably formed during dewatering. Upper Cambrian, Broom Point. *D*—Graded calcarenite with horizontal lamination followed by ripples. Upper Cambrian, Broom Point.

currents as contour currents because they travelled parallel to the topographic paleocontours known independently from slump sheets, soft-sediment folds in boulders of limestone in the breccias, and synsedimentary boudins in thin beds of limestone (Figs. 14, 15, 19).

An important consideration is that the contour currents deposited only nongraded beds of lime mudstone and calcarenite in the Middle Cambrian and Lower-Middle Ordovician, as well as some nongraded limestone in the Upper Cambrian. Thus the system of contour currents was stable for 70 million years. Also stable was the northeast dipping regional paleoslope with the northwest-southeast trending carbonate platforms. It seems logical to infer that the system of contour currents also deposited the graded beds of calcarenite in the Upper Cambrian.

Under our hypothesis, the graded beds of calcarenite were deposited by waning contour currents and not by turbidity currents. The grading requires a decreasing current velocity during deposition, but must it necessarily imply transport by a turbidity current? Perhaps contour currents are "threads" of water that produce a waning effect by lateral and vertical shifts. At any specific location, current velocity would decrease, allowing the grains to settle from suspension in a graded bed.

Modern contour currents along the flanks of carbonate platforms are strong and deep, especially where the flow is accelerated due to passing through a constriction formed by a strait. At a depth of 320 m along the west flank of the Bahama Platform, contour currents flow at velocities that exceed 50 cm/sec (Neumann and Ball, 1970). Our

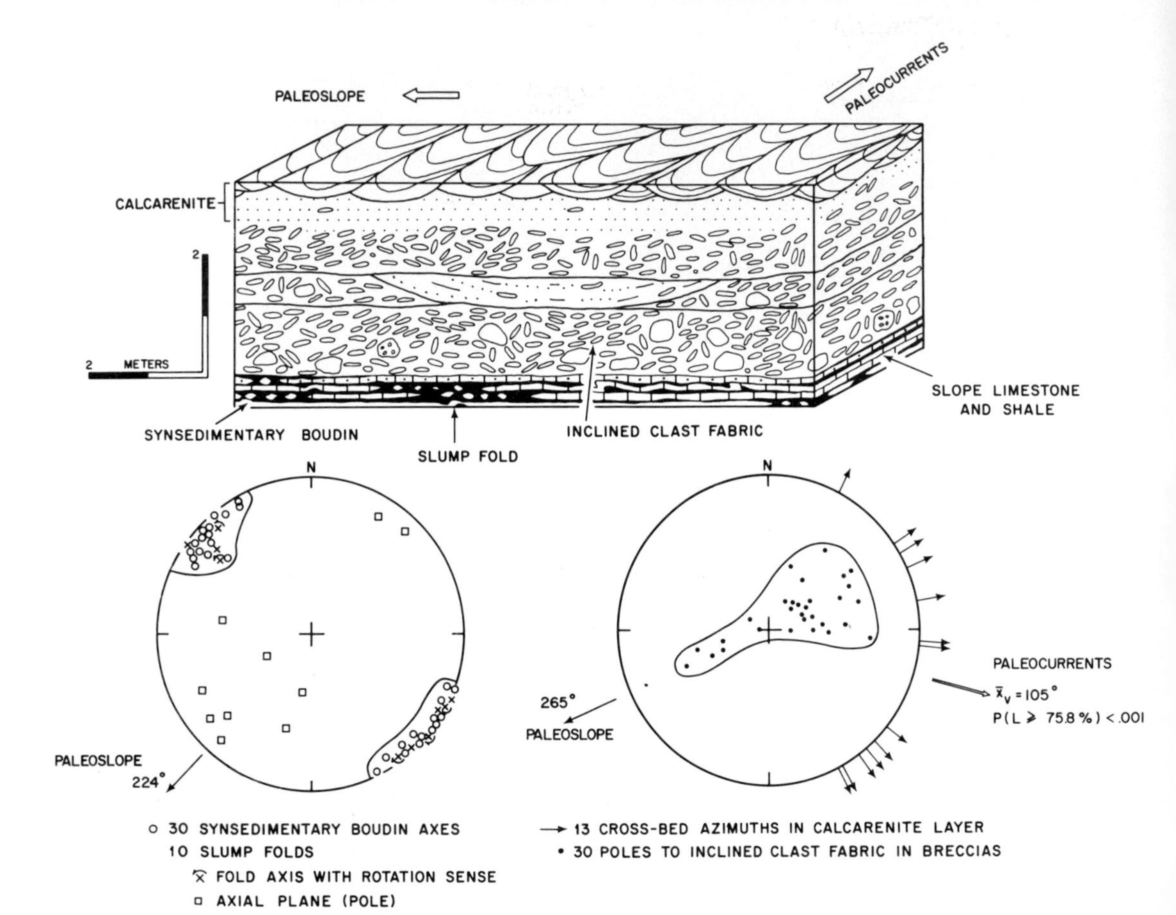

FIG. 18.—Calcarenite layer lying on a sequence of three mass-flow breccias, Middle Cambrian at Broom Point. The 0.7 m bed is graded from plane-bedded pebbly calcarenite to festoon cross-bedded calcarenite. The slump folds and synsedimentary boudins in the thin beds of limestone show that the paleoslope dipped southwest. The inclined clast fabric at 30 places in the breccias dips downslope. The calcarenite was deposited by southeast-flowing contour currents.

knowledge of these currents is based on only a few submersible dives so that current velocities at times are probably much greater than the measured values.

The graded beds occur at Cow Head, Broom Point, and Gulls Marsh, sections situated between the carbonate platforms near Cow Head and Martin Point (Fig. 15). A reasonable inference is that the bottom currents were accelerated due to the constriction between the platforms. The higher velocities could erode and transport the coarse calcareous sand present in some of the graded beds.

The southeast-flowing currents eroded the substrate to form flutes and grooves on the soles of the graded beds and also deposited the plane beds, cross-beds, and ripples. The sole marks thus could not have been formed by downslope turbidity currents with the sand later reworked by contour currents.

Contour currents along continental slopes and rises commonly exceed 25 cm/sec at depths down to several thousand meters (Zimmerman, 1972; Rabinowitz and Eittreim, 1974). Contour currents are inferred to be important in depositing some deep-sea sands (Hubert, 1964; Heezen and Hollister, 1971). Other sands may have been deposited by turbidity currents, especially sands on the abyssal plains away from the slope and rise.

ORDOVICIAN HISTORY OF THE COW HEAD BRECCIA

The Ordovician strata, like the Cambrian, are a slope sequence of breccia and thin beds of lime mudstone, calcarenite, green silty shale and gray marl. In contrast to the Cambrian, the Ordovician also contains beds of radiolarian-sponge spicule

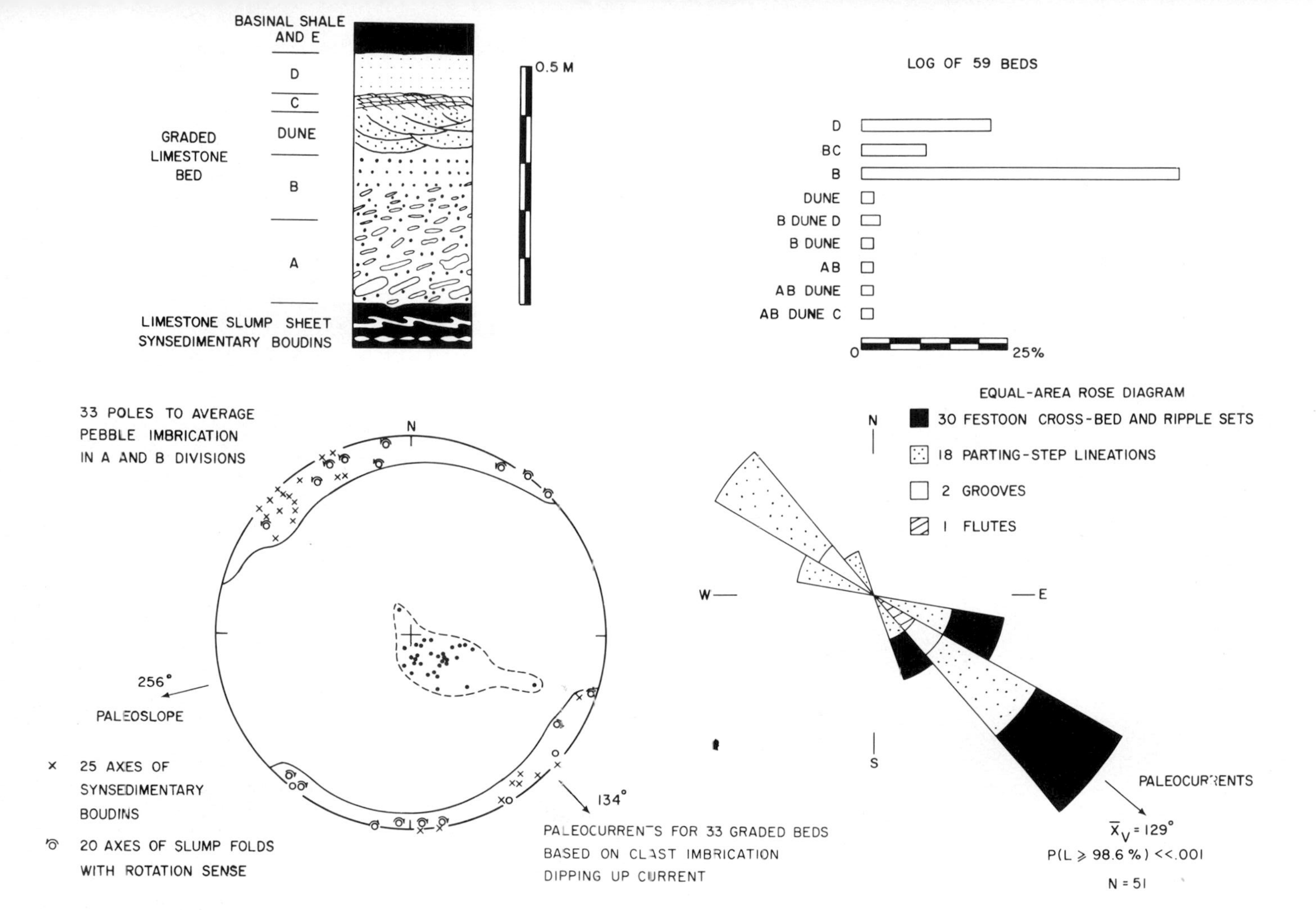

FIG. 19.—Upper Cambrian sequence in the Cow Head Breccia at Cow Head. Shown are the internal divisions in the graded beds of calcarenite, southeast-flowing paleocurrents, and southwest paleoslope.

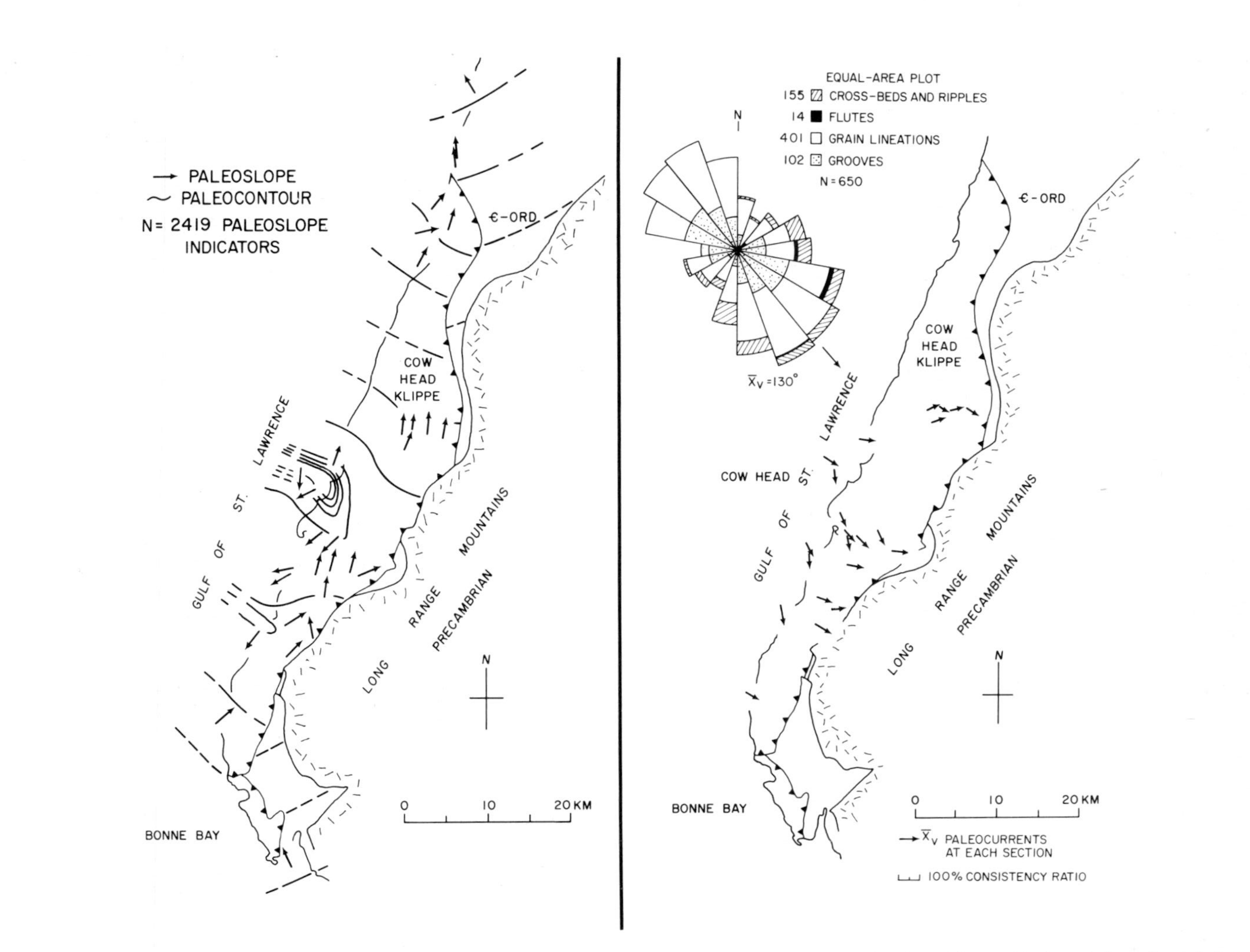

FIG. 20.—Paleoslopes (left) and paleocurrents (right) for the Ordovician strata in the Cow Head Breccia. See Figure 14 for explanation of how maps were constructed. Each paleocurrent vector mean is statistically significant at the 95 percent level when tested for preferred orientation using the Rayleigh statistic L.

chert, lacks graded beds of quartzose calcarenite, has a larger proportion of thin beds of limestone and shale, and has less breccia.

The Ordovician paleotopography was similar to that of the Cambrian (Figs. 20, 21). The regional paleoslope was northeast, as the klippe is now oriented. The carbonate platforms near Cow Head and Martin Point persisted from the Cambrian into the Ordovician.

The paleocurrents consistently flowed southeast as in the Cambrian (Fig. 20). These contour currents swept across the inclined sea floor depositing nongraded beds of lime mudstone and calcarenite. A few calcarenites at all sections contain planar and festoon cross-beds (Fig. 17B).

Boulders of platform limestone in breccia at Green Point, White Point on St. Paul's Inlet, and the eastern part of Parson's Pond (Fig. 1) could not have come from the carbonate platforms near Cow Head and Martin Point because the platforms did not lie upslope. There must have been additional sources, for example southwest of Green Point.

The carbonate platform near Cow Head strongly influenced the surrounding sediments. The largest boulders of platform limestone exceed 60 by 150 m and occur at Lower Head and Cow Head on opposite flanks of the carbonate platform (Figs. 6, 20, 21). These localities received more and thicker mass flows than the downslope area at Broom Point. The platform was a source of carbonate mud and sand so that the sections at Lower Head and Cow Head also have the highest ratios of thin-bedded limestone to shale. Sections away from the platform contain more shale and less limestone.

Some mass flows travelled at least 10 km, the distance downslope from Cow Head to Broom Point. Also the boulders of platform limestone become smaller and less numerous from Cow Head to Broom Point.

The beds of breccia have an elongate shape. The mass flows, however, may have been lobate with elongate downslope projections.

PALEOSLOPE IN THE AUTOCHTHONOUS SEQUENCE

The Table Head Limestone of early Middle Ordovician age in the shelf sequence below the Cow Head klippe has a regional paleoslope to the north-northwest (Fig. 20). The northerly paleoslope extended at least 216 km from Port-aux-Port (Fig. 22) to Table Point just north of the klippe. Each paleoslope azimuth was measured using slump sheets of limestone and shale in the

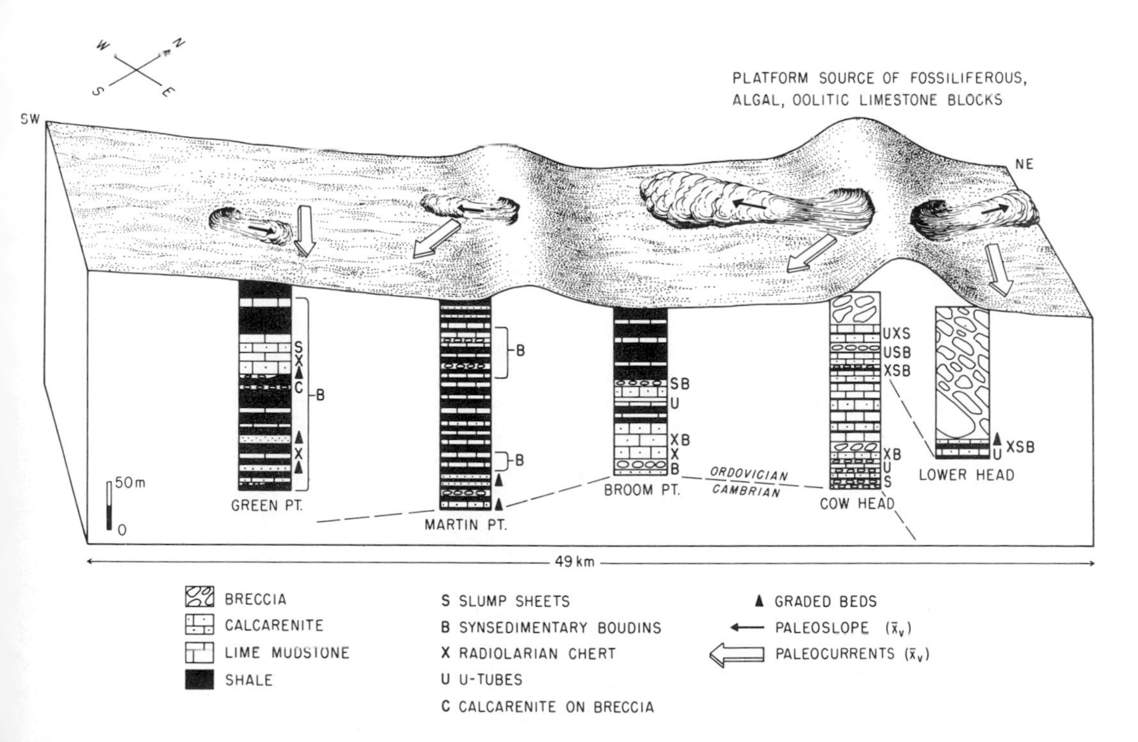

FIG. 21.—Paleotopography and stratigraphy of the Ordovician strata in the Cow Head Breccia. The diagram shows the center of klippe omitting the southwest and northeast ends. The paleotopography follows the schematic paleocontours of Figure 20.

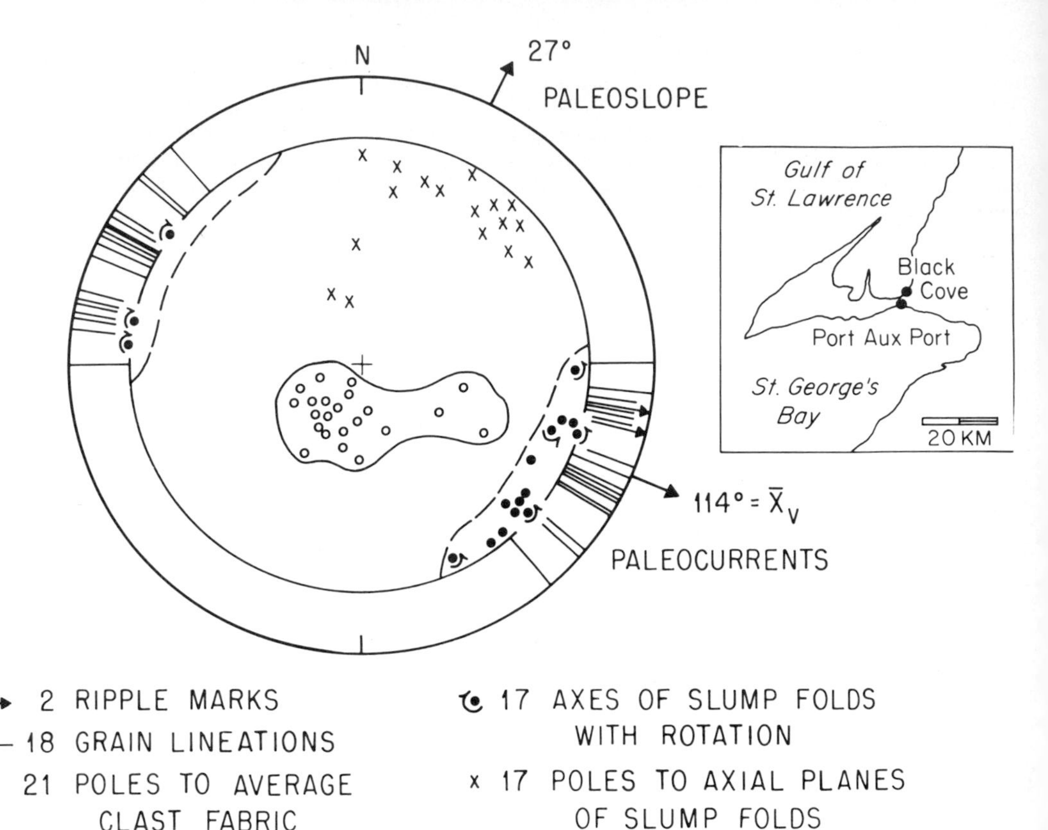

FIG. 22.—Northeast paleoslope and southeast paleocurrents in the autochthonous Table Head Limestone at Black Cove, North of Port-aux-Port, 140 km southwest of Bonne bay (Fig. 1).

upper part of the formation which was deposited at the edge of the continental shelf (Whittington and Kindle, 1963).

During the Middle Ordovician, the depositional strike of the Table Head Limestone and the Cow Head Breccia was northwest. If the strike is projected across the fault system on the east side of the Long Range, it abuts against volcanogenic sandstone, gray shale and pillow lava also of Middle Ordovician age in central Newfoundland. This relation implies substantial net dextral strike-slip movement on the fault system because of the regional northeast dip of the continental slope. Net dextral movement has also been interpreted from structural evidence (Belt, 1969; Webb, 1969).

PALEONTOLOGICAL EVIDENCE OF WATER DEPTHS

Ten genera of eyeless trilobites and two genera of trilobites with eyes occur in the Cambro-Ordovician thin-bedded limestone of the Cow Head Breccia (Kindle, 1969). The thin beds of lime mudstone and green shale contain graptolites (Kindle and Whittington, 1958), conodonts (Nowlan, 1973) and aglaspids, gastropods, trilobites, and ostracods occur *in situ* in lime mudstone only at Cliffy Point.

Some of the Ordovician thin beds of lime mudstone at Lower Head, Cow Head, Broom Point, and St. Paul's Inlet were intensively burrowed by an animal that constructed U-tubes (Figs. 2, 21). The mean diameter of 600 tubes is 0.6 cm, with a standard deviation of 0.08 cm. The mean distance between the centers of the two holes is 2.0 cm, with a standard deviation of 0.4 cm. The U-tubes extend about 8 cm below the surface of each bed.

Jansa (1974) identified the U-tubes as *Arenicolites*, a suspension-feeding polychaete worm. Due to the absence of *in situ* calcareous shelly fossils, he concluded that the worms lived in water depths somewhat greater than 200 m on the flank of the carbonate platform near Cow Head. *Arenicolites* previously has been thought to be restricted to very shallow marine and intertidal environments.

Each U-tube was a permanent domicile in which

a hypothetical line connecting the two holes is aligned subparallel to the bottom currents whose direction is independently known by grain lineation in the limestone (Fig. 23). One hole was upcurrent to take in nutrient-rich water, whereas the downcurrent opening disposed of waste products.

The restricted types of fossils, plus the evidence for accumulation on a slope, suggest water depths greater than the edge of the continental shelf, but not abyssal.

COLLOPHANE NODULES

Collophane nodules occur in calcarenite and breccia in the Cambrian and Ordovician sequences at all outcrops. They are most abundant, forming up to 10 percent of the grains, at Cow Head and Lower Head implying derivation from the flanks

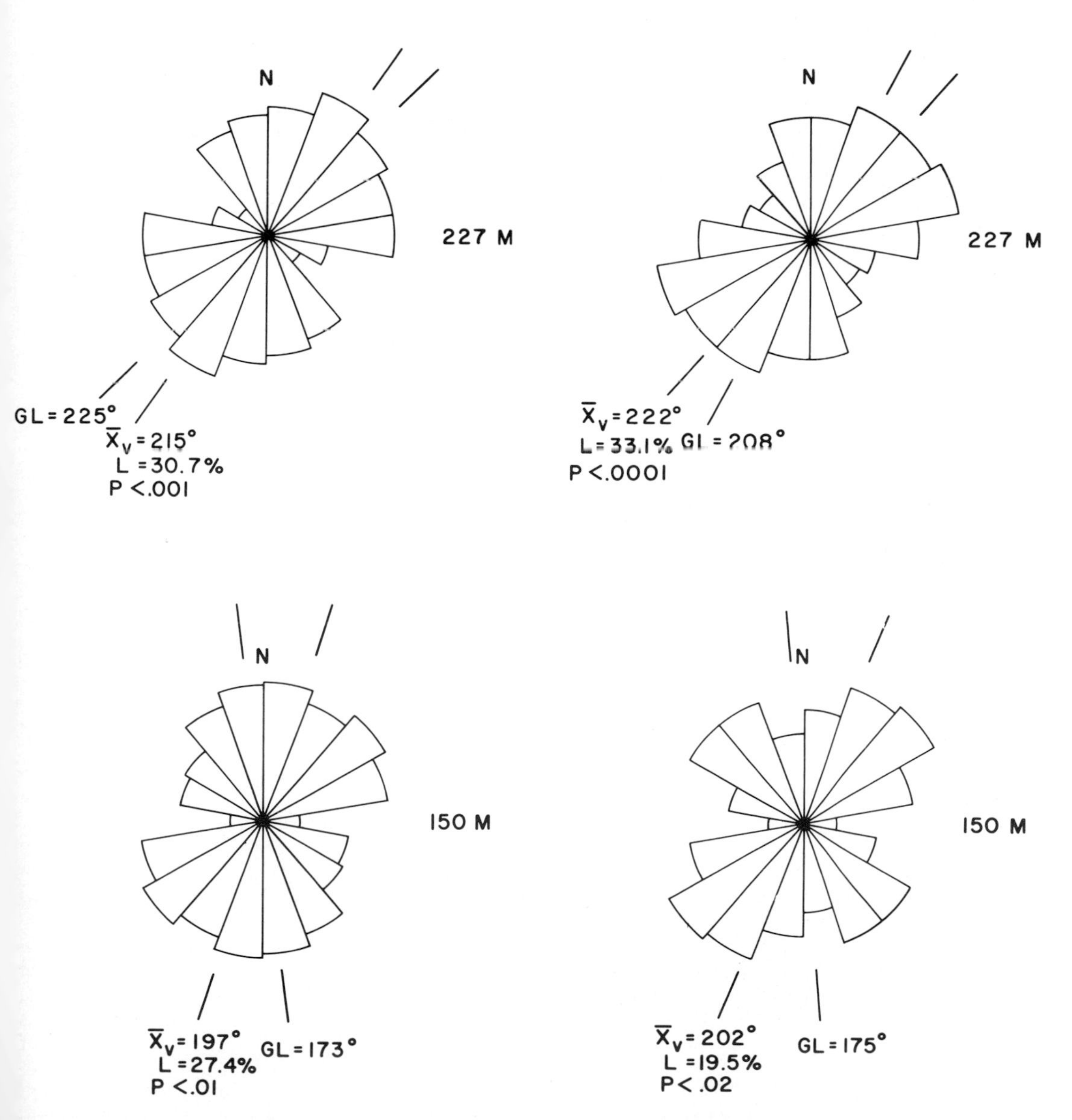

FIG. 23.—Equal-area geometric plots of orientation of 100 U-tubes in each of four limestone beds. Symbols are vector mean ($\bar{x}_v$), grain lineation (GL), consistency ratio (L), probability (P) by the Rayleigh test of there being no preferred orientation, and meters (M) above base of the Cow Head Breccia at Cow Head.

of the nearby carbonate platform. The nodules are mostly sand and pebbles with botryoidal surfaces. The dark surface stains and the interiors of the nodules are rich in iron oxides, but lack even traces of manganese as verified by X-ray fluorescence.

The Middle Tertiary to Holocene occurrences of collophane nodules on deep-water slopes have three common denominators: (1) the collophane formed in shallow water, up to about 300 m deep, on the continental slope and submarine ridges; (2) upwelling of cold, phosphate-rich water provided much of the phosphorous; and (3) sedimentation was slow (Emery, 1960; Gulbrandsen, 1969; Burnett, 1974).

These conditions seem to apply to the Cow Head area which was in the carbonate belt that paralleled the paleoequator. Marine upwelling of cold, phosphate-rich bottom water may have been important due to warm surface water being blown away from the carbonate platforms by the low latitude winds (Fig. 24). Collophane replaced particles of calcium carbonate, perhaps aided by decay of phosphate-rich phytoplankton. Framboidal pyrite is common in the lime mudstone and green shale implying abundant organic matter and perhaps high productivity of phytoplankton.

MIDDLE ORDOVICIAN "RED SHALE"

As the proto-Atlantic narrowed in the Middle Ordovician, the slope and carbonate platforms foundered so that the Cow Head Breccia grades upward into the "red shale" (Fig. 2). Minor lithologies interbedded with red shale are mass-flow carbonate breccia, green and purple shale, green and purple radiolarian-sponge spicule chert, lime mudstone, gray marl, calcarenite, and near the top of the unit, volcanogenic sandstone and gray shale. At Green Point and Martin Point, the "red shale" measures 217 and 204 m, thinning to less than 100 m at the Inner Tickle and Black Brook on St. Paul's Inlet, and on the north shore of Parson's Pond (Fig. 2).

The "red shale" at Martin Point contains Arenigian and Llanvirnian graptolites (Kindle and Whittington, 1958; Erdtmann, 1971). In contrast, the Arenigian and Llanvirnian rocks at Cow Head are limestone, green shale, breccia and chert, but not red shale (Kindle and Whittington, 1958). The area near Martin Point foundered before the carbonate platform at Cow Head.

The three available outcrops suggest that the southeast-flowing contour currents that existed during deposition of the Cow Head Breccia were maintained as the "red shale" accumulated (Fig. 2). At Green Point, for example, southeast-flowing paleocurrents laid down beds of calcarenite interbedded with red shale whereas the paleoslope dipped northeast as shown by slump sheets.

Carbonate particles, *e.g.*, ooids, pellets, intraclasts and fragments of calcareous fossils, are conspicuously absent in the beds of red shale, in contrast to the green silty shale of the Cow Head Breccia. The fauna in the red mud comprises

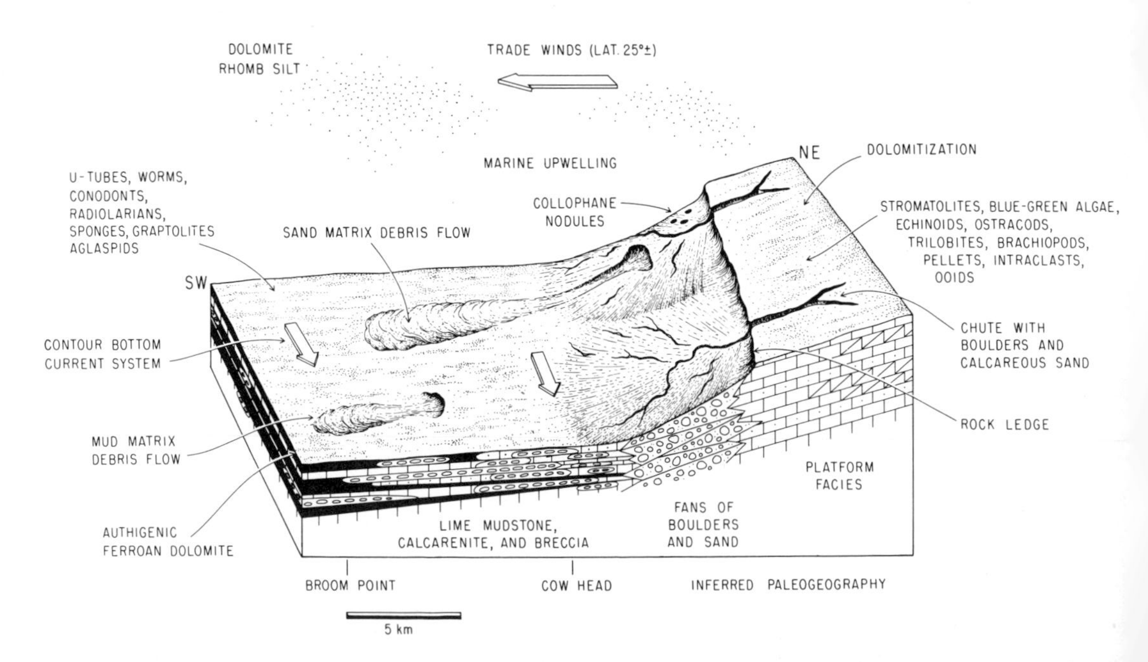

FIG. 24.—Depositional model of the Cow Head Breccia near Cow Head and Broom Point.

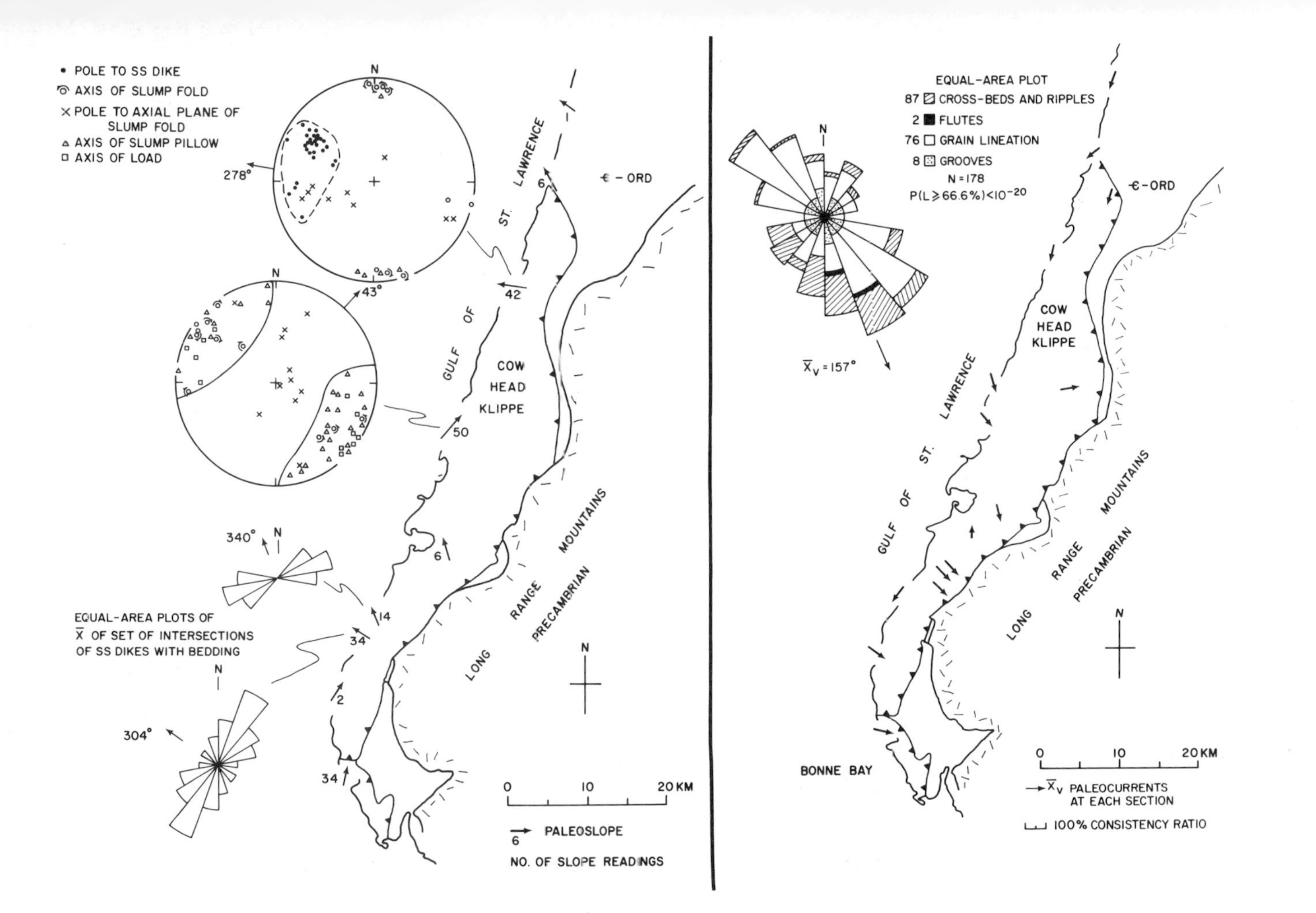

FIG. 25.—Paleoslopes (left) and paleocurrents (right) for the "green sandstone." Each paleoslope and paleocurrent arrow is the mean azimuth for that section. Each paleocurrent vector mean is statistically significant at the 95 percent level when tested for preferred orientation using the Rayleigh statistic L. The sandstone dikes are inferred to strike parallel to the paleocontours and to dip upslope.

graptolites, radiolaria, siliceous spicules, desma of sponges, and aglaspids. The "red shale" evidently accumulated in deeper water than the Cow Head Breccia.

MIDDLE ORDOVICIAN "GREEN SANDSTONE"

As the continental margin continued to subside, volcanogenic sandstone and gray shale transitionally succeeded both the "red shale" in the klippe and the Table Head Limestone in the shelf sequence. In north-central Newfoundland, the Lower-Middle Ordovician rocks are several thousand meters of volcanogenic sandstone and gray shale interbedded with pillow lava (Horne and Helwig, 1969). Progressive narrowing of the proto-Atlantic collapsed the continental margin producing a marginal basin between the craton and volcanic island arc (Williams and others, 1973).

The source areas for the "green sandstone" were of varied lithology. Modal thin-section analyses of 30 sandstones from the autochthonous and allochthonous sequences show that the sandstones have a feldspathic litharenite composition. About half the grains are from the plutonic-metamorphic "Grenville" basement, a third from volcanic-ophiolite complexes, and a tenth reworked grains from the shelf sequence of sandstone and carbonate. There are also clasts from the "red shale" and Cow Head Breccia.

Partial sections of the "green sandstone" give minimum thicknesses of 400 m in the klippe at Martin Point and 100 m in the autochthonous sequence at Table Point. The graptolites found in the beds of gray shale are Llanvirnian in age, both in the klippe and autochthonous sequences (Whittington and Kindle, 1969).

Nongraded beds of sandstone are common with festoon cross-beds randomly interstratified with plane beds and ripples. There are also some graded beds of sandstone. The beds average 20 to 40 cm in thickness, many exceeding a meter. Pebbly sandstone, scour surfaces and compound beds are common. The ratio of sandstone to gray shale is high, exceeding 6:1.

The "green sandstone" is poorly exposed, but paleoslope data are available for six locations in the klippe, two in the autochthonous sequence, and one in the Bay of Islands klippe at Bonne Bay (Fig. 25). The stratigraphically higher sequences show paleoslopes directed to the northwest whereas the lower sequences show northeast and north paleoslopes. With progressive narrowing of the proto-Atlantic, the continental shelf and slope foundered and the former northeast paleoslope present during accumulation of the Cow Head sediments shifted to the north and northwest. The paleocurrents at the eleven locations in the klippe flowed east, southeast and southwest with one exception (Fig. 25). At Berry Head on St. Paul's Inlet, a local topographic high may be recorded by 10 m of cross-bedded conglomerate and pebbly sandstone deposited by north-flowing paleocurrents. The pebbles are calcarenite, breccia, lime mudstone, and chert eroded from the Cow Head Breccia and "red shale."

Paleocurrents flowed southwest in the autochthonous sequence at two locations near Table Head just north of the klippe. In the Bay of Islands klippe at Bonne Bay, the paleocurrents flowed southeast.

With further closing of the ocean in the Taconic Orogeny, the Cow Head klippe became detached as a single structural slice, sliding northwest down the paleoslope of the basin to lie on the shelf sequence. The direction of klippe transport was approximately perpendicular to the northeast strike of the geosynclinal axis in central Newfoundland.

SUMMARY

The Cow Head Breccia is a slope sequence of carbonate breccia and thin beds of lime mudstone, green silty shale, gray marl, calcarenite, and radiolarian-sponge spicule chert that accumulated on the western margin of the proto-Atlantic. The paleotopography and paleocurrent system were stable from the Middle Cambrian through the early Middle Ordovician. As the klippe is now oriented, the regional paleoslope dipped northeast, interrupted by northwest-southeast trending carbonate platforms near Cow Head and Martin Point. Mass flows travelled away from the carbonate platforms to deposit the carbonate breccia. Contour currents flowed southeast, mostly depositing thin beds of nongraded limestone, but also graded beds of quartzose calcarenite during the Upper Cambrian. U-tube burrows of *Arenicolites* were aligned subparallel to the contour currents to aid suspension feeding.

As the proto-Atlantic narrowed in the Middle Ordovician, the slope and carbonate platforms foundered and were buried by red mud. Volcanogenic sandstone and gray shale then accumulated in a marginal basin that developed on the subsiding continental shelf and slope. During the Taconic Orogeny, the klippe slid to the northwest down the paleoslope of the basin. Subsequent net dextral strike-slip movement along the Long Range fault system was substantial.

ACKNOWLEDGEMENTS

We appreciate the many helpful suggestions of Harry Cook, Paul Enos, M. A. Hampton, G. V. Middleton, Lloyd Pray, R. G. Walker and Harold Williams who critically read the manuscript. David Morse, Alan Reed, Walter Rewinski and Mark

Smith, students at the University of Massachusetts, were cheerful and competent field assistants. The illustrations were drafted by Marie Litterer and Joseph K. Lucey. This research was supported by the Division of Earth Sciences, National Science Foundation, NSF Grant GA 31870. Student grants from the American Association of Petroleum Geologists and the Geological Society of America supported the thesis work of Robert Suchecki.

REFERENCES

ALLEN, PERCIVAL, 1972, Glacial sandstones—Ordovician of central Sahara: *In* F. D. Crawford (ed.), Arenaceous deposits, sedimentation, and diagenesis: Alberta Soc. Petroleum Geologists, Calgary, Alberta, p. 145-164.

BAIRD, D. M., 1960, Observations on the nature and origin of the Cow Head breccias of Newfoundland: Canada Geol. Survey, Paper 60-3, 26 p.

BELT, E. S., 1969, Newfoundland Carboniferous stratigraphy and its relation to the Maritimes and Ireland: *In* Marshall Kay (ed.), North Atlantic geology and continental drift: Am. Assoc. Petroleum Geologists, Mem. 12, p. 743-753.

BIRD, J. M., AND DEWEY, J. F., 1970, Lithosphere plate-continental margin tectonics and the evolution of the Appalachian orogen: Geol. Soc. America Bull., v. 81, p. 1031-1060.

BLACK, R. F., 1964, Paleomagnetic support of the theory of rotation of the western part of Newfoundland: Nature, v. 202, p. 945-948.

BURNETT, W. C., 1974, Origin of phosphorite from the continental margins of Peru and Chile [abs.]: Geol. Soc. America Abs. with Programs, v. 6, p. 280.

COOK, H. E., MCDANIEL, P. N., MOUNTJOY, E. W., AND PRAY, L. C., 1972, Allochthonous carbonate debris flows at Devonian bank ('reef') margins, Alberta, Canada: Bull. Canadian Petroleum Geology, v. 20, p. 439-497.

——, AND TAYLOR, M. E., 1975, Early Paleozoic continental margin sedimentation, trilobite biofacies, and the thermocline, western United States: Geology, v. 3, p. 559-562.

DEUTSCH, E. R., 1969, Paleomagnetism and North Atlantic paleogeography: *In* Marshall Kay (ed.), North Atlantic—Geology and continental drift: Am. Assoc. Petroleum Geologists, Mem. 12, p. 931-954.

EMERY, K. O., 1960, The sea of southern California: John Wiley and Sons, New York, New York, 366 p.

ENOS, PAUL, 1973, Channelized submarine carbonate debris flows, Cretaceous, Mexico [abs.]: Am. Assoc. Petroleum Geologists Bull., v. 57, p. 777.

ERDTMANN, B. D., 1971, Ordovician graptolite zones of western Newfoundland in relation to paleogeography of the North Atlantic: Geol. Soc. America Bull., v. 82, p. 1509-1528.

FÅHRAEUS, L. E., 1973, Depositional environments and conodont-based correlation of the Long Point Formation (Middle Ordovician), western Newfoundland: Canadian Jour. Earth Sci., v. 10, p. 1822-1833.

GULBRANDSEN, R. A., 1969, Physical and chemical factors in the formation of marine apatite: Geology, v. 69, p. 365-382.

HAMPTON, M. A., 1972, The role of subaqueous debris flow in generating turbidity currents: Jour. Sed. Petrology, v. 42, p. 775-793.

HANSEN, EDWARD, 1966, Methods of deducing slip-line orientations from the geometry of folds: Carnegie Inst. Washington, Ann. Rept. Geophys. Lab., p. 386-410.

HEEZEN, B. C., AND HOLLISTER, C. D., 1971, The face of the deep: Oxford Univ. Press, New York, New York, 659 p.

HENDRY, H. E., 1972, Breccias deposited by mass flow in the Breccia Nappe of the French pre-Alps: Sedimentology, v. 18, p. 277-292.

——, 1973, Sedimentation of deep water conglomerates in Lower Ordovician rocks of Quebec—Composite bedding produced by progressive liquefaction of sediment: Jour. Sed. Petrology, v. 43, p. 125-136.

HOPKINS, J. C., 1972, Petrography, distribution and diagenesis of foreslope, nearslope and basin sediments, Miette and Ancient Wall carbonate complexes (Devonian), Alberta: Ph.D. Dissert., McGill Univ., Montreal, Quebec, 234 p.

HORNE, G. S., AND HELWIG, JAMES, 1969, Ordovician stratigraphy of Notre Dame Bay Newfoundland: *In* Marshall Kay (ed.), North Atlantic—Geology and continental drift: Am. Assoc. Petroleum Geologists, Mem. 12, p. 388-407.

HUBERT, J. F., 1964, Textural evidence for deposition of many western North Atlantic deep-sea sands by ocean bottom currents rather than turbidity currents: Jour. Geology, v. 72, p. 757-785.

——, 1966, Sedimentary history of Upper Ordovician geosynclinal rocks, Girvan, Scotland: Jour. Sed. Petrology, v. 36, p. 677-699.

JANSA, L. F., 1974, Trace fossils from the Cambro-Ordovician Cow Head Group, Newfoundland, and their paleobathymetric implications: Palaeogeography, Palaeoclimatology, Palaeoecology, v. 15, p. 233-244.

KINDLE, C. H., 1969, Eyeless trilobites and flysch sediments [abs.]: Geol. Soc. Canada, Abs. Montreal Meeting, p. 25.

—— AND WHITTINGTON, H. B., 1958, Stratigraphy of the Cow Head region, western newfoundland: Geol. Soc. America Bull., v. 69, p. 315-342.

MIDDLETON, G. V., AND HAMPTON, M. A., 1973, Sediment gravity flows: Mechanics of flow and deposition: *In* G. V. Middleton and A. H. Bouma (eds.), Turbidities and deep water sedimentation: Soc. Econ.

Paleontologists and Mineralogists, Pacific Sec., Los Angeles, California, p. 1-38.
MORGENSTERN, N. R., 1967, Submarine slumping and initiation of turbidity currents: *In* A. F. Richards (ed.), Marine geotechnique: Univ. Illinois Press, Urbana, Illinois, p. 189-220.
MOUNTJOY, E. W., COOK, H. E., PRAY, L. C., AND MCDANIEL, P. N., 1972, Allochthonous carbonate debris flows—Worldwide indicators of reef complexes, banks, or shelf margins: 24th Internat. Geol. Congress Proc., Sec. 6, p. 172-189.
NEUMANN, A. C., AND BALL, M. M., 1970, Submersible observations in the Straits of Florida: Geology and bottom currents: Geol. Soc. America Bull., v. 81, p. 2861-2874.
NOWLAN, G. S., 1973, Conodonts from the Cow Head Group, western Newfoundland: M. S. Thesis, Memorial Univ., St. John's, Newfoundland, 78 p.
OXLEY, PHILIP, 1953, Geology of Parsons Pond-St. Paul's area, west coast of Newfoundland: Newfoundland Dep. Mines and Resources, Geol. Survey Rept. 5, 53 p.
RABINOWITZ, P. H., AND EITTREIM, S. L., 1974, Bottom current measurements in the Labrador Sea: Jour. Geophys. Research, v. 79, p. 4085-4090.
RODGERS, JOHN, AND NEALE, E. R. W., 1963, Possible "Taconic" klippen in western Newfoundland: Am. Jour. Sci., v. 261, p. 713-730.
SMITH, M. E., 1973, Structural geology of Cow Head, Newfoundland (research report): Univ. Massachusetts, Amherst, Massachusetts, 54 p.
SRIVASTAVA, P. N., STEARN, C. W., AND MOUNTJOY, E. W., 1972, A Devonian megabreccia at the margin of the Ancient Wall carbonate complex, Alberta: Bull. Canadian Petroleum Geology, v. 20, p. 412-438.
STEVENS, R. K., 1970, Cambro-Ordovician flysch sedimentation and tectonics in west Newfoundland and their possible bearing on a proto-Atlantic Ocean: *In* Jean Lajoie (ed.), Flysch sedimentology in North America: Geol. Assoc. Canada, Spec. Paper 7, p. 165-177.
SUCHECKI, R. K., 1975, Sedimentology, petrology, and structural geology of the Cow Head klippe: Broom Point, St. Paul's Inlet, and Black Brook, western Newfoundland: M.S. Thesis, Univ. Massachusetts, Amherst, Massachusetts, 180 p.
SWETT, KEENE, AND SMIT, D. E., 1972, Paleogeography and depositional environments of the Cambro-Orodovician shallow-marine facies of the North Atlantic: Geol. Soc. America Bull., v. 83, p. 3223-3248.
THOMSON, A. F., AND THOMASSON, M. R., 1969, Shallow to deep water facies development in the Dimple Limestone (Lower Pennsylvanian), Marathon region, Texas: *In* G. M. Friedman (ed.), Depositional environments in carbonate rocks: Soc. Econ. Paleontologists and Mineralogists, Spec. Pub. 14, p. 57-58.
WEBB, G. W., 1969, Paleozoic wrench faults in Canadian Appalachians: *In* Marshall Kay (ed.), North Atlantic—Geology and continental drift: Am. Assoc. Petroleum Geologists, Mem. 12, p. 754-786.
WHITTINGTON, H. B., AND KINDLE, C. H., 1963, Middle Ordovician Table Head Formation, western Newfoundland: Geol. Soc. America Bull., v. 74, p. 745-758.
—— AND ——, 1969, Cambrian and Ordovician stratigraphy of western Newfoundland: *In* Marshall Kay (ed.), North Atlantic—Geology and continental drift: Am. Assoc. Petroleum Geologists, Mem. 12, p. 655-664.
WILLIAMS, ALWYN, 1973, Distribution of brachiopod assemblages in relation to Ordovician palaeogeography: *In* N. F. Hughes (ed.), Organisms and continents through time: Palaeontol. Assoc. Spec. Papers in Palaeontology, v. 12, p. 241-269.
WILLIAMS, HAROLD, KENNEDY, M. J., AND NEALE, E. R. W., 1973, The Appalachian structural province: *In* R. A. Price and R. J. W. Douglas (eds.), Variations in tectonic styles in Canada: Geol. Assoc. Canada, Spec. Paper 11, p. 203-218.
ZIMMERMAN, H. B., 1972, Sediments of the New England continental rise: Geol. Soc. America Bull., v. 83, p. 3709-3724.

SEPM Special Publication No. 25, p. 155-170, November 1977

PRODUCTION OF FORESLOPE BRECCIA BY DIFFERENTIAL SUBMARINE CEMENTATION AND DOWNSLOPE DISPLACEMENT OF CARBONATE SANDS, MIETTE AND ANCIENT WALL BUILDUPS, DEVONIAN, CANADA

JOHN C. HOPKINS
Dept. of Geology, University of Calgary
Calgary, Alberta, Canada T2N1N4

ABSTRACT

Breccia deposited seaward of the margins of Miette and Ancient Wall buildups (Upper Devonian of Alberta) records synsedimentary submarine cementation of carbonate sands. The size, shape and composition of breccia clasts indicates that their source rocks were differentially cemented (nodular), thinly bedded carbonate sands deposited high on the foreslopes, close to the margins of the buildups. Downslope movement of differentially cemented carbonate sand sequences mixed nodules with uncemented carbonate sands and resulted in deposition of breccia beds.

The breccia was transported downslope by hybrid sediment gravity flows in which upward movement of fluids, grain-grain interactions, and possibly fluid turbulence supported the clasts. Increased pore pressures necessary for initiating such flows on low slopes were the result of metastable packing of sand grains in thinly bedded sequences and/or overriding currents carrying material from the buildups. Rapid dissipation of excess pore pressures, coupled with low slopes, did not allow sustained flow and as a consequence deposition of the breccias occurred within a few kilometers of the buildups.

Carbonate breccia formed by differential submarine cementation and downslope displacement of carbonate sands has not previously been described. One important facet of its origin is that it does not record periods of abrasional erosion of the buildups, but formed at times when rate of sediment supply to the foreslope was relatively high.

INTRODUCTION

Carbonate breccia deposited around the seaward margins of carbonate shelves and buildups is known from many places in the geologic record. Key papers describing such breccias have been summarized recently by Heckel (1974) who noted that the presence of breccia near buildup or shelf margins is usually taken as an indication of the relative degree of "wave resistance" of the margin. If the clasts are largely skeletal in composition, an organic frame or reef is generally postulated; if the clasts are largely calcarenite, near-surface penecontemporaneous cementation of the buildup is considered proven.

Little attention has been paid, however, to the manner in which carbonate breccia clasts are eroded and deposited at shelf or buildup margins. Consequently, as Heckel (1974) has noted, few criteria are available to distinguish between breccias which have been mechanically eroded, and others produced by slope failure of coherent or lithified buildup margin material, yet only those breccias which have been mechanically eroded tell anything of the wave resistance of the margin.

Breccias which occur seaward of the Miette and Ancient Wall buildups (Upper Devonian of Alberta, Fig. 1) are composed dominantly of calcarenite fragments. They are well exposed in outcrop and their position with respect to the margins of the buildups can be directly observed. Hence they provide an excellent base from which breccia source and transport mechanisms can be studied.

The Miette and Ancient Wall breccias have been documented by Pray and others (1968), Srivastava, Stearn and Mountjoy (1972) and Cook and others (1972). Attention of these workers, however, has been focused largely or solely on a single bed of "megabreccia" (containing blocks with diameters up to several tens of meters) which crops out at an exposure of the margin of the Ancient Wall buildup. Generally, it is agreed that large blocks in the megabreccia bed comprise cemented portions of the buildup margin which became detached and moved downslope. Several mechanisms of block detachment and transport have been postulated with emphasis on strong earthquake shocks or tsunamis followed by debris flow (Cook and others, 1972).

More numerous, though less spectacular, thinner beds of carbonate breccia are also present at the Miette and Ancient Wall buildup margins, but have not previously been studied in detail. These breccias form an integral part of the foreslope facies and appear to have a different origin from the megabreccia. Although the clasts are largely calcarenite, they were not eroded from the buildups by submarine or subaerial processes, but rather formed as differentially submarine cemented foreslope sands sloughed away down-

FIG. 1.—Distribution of Upper Devonian carbonate complexes (Leduc Formation and equivalents, stippled) in central Alberta. After Belyea (1964).

slope at times of high rates of sediment supply to the buildup margins (Hopkins, 1971).

The purpose of this paper is to record the form and fabric of these breccias, detail the source, and infer the transport mechanism so that similar sediments may be recognized elsewhere in the geologic record.

Regional perspective.—The Miette and Ancient Wall carbonate or reef complexes, defined by Mountjoy (1967), are two of several large, isolated carbonate masses which developed in an extensive shale basin during late Devonian times in central Alberta (Fig. 1). The carbonate complexes have comparatively wide lateral extent (ten to many tens of kilometers) and restricted vertical extent (a few hundreds of meters).

The Miette and Ancient Wall carbonate complexes as defined, include three gross depositional phases from base to top (Fig. 2a): (1) A widespread carbonate platform[1] which thickens locally to provide a foundation for overlying buildups. (2) A carbonate buildup-shale basin couplet which accounts for much of the vertical accumulation phase of the buildups and infilling of the surrounding shale basin. (3) Lime-sand banks of low relief surrounded by carbonate and argillaceous carbonate mudstones. Banks usually developed above, but were not necessarily confined to areas underlain by buildups.

[1]Mountjoy's (1967, p. 399-400) broad usage of the terms "platform" and "bank" to describe these phases of sedimentation within the carbonate complexes is followed here in order to separate them from carbonates of the sedimentation phase (buildup-shale basin couplet) with which this paper is concerned. The term "buildup" for carbonates of this sedimentation phase is used in accord with the definition provided by Heckel (1974, p. 91).

It is the second phase of sedimentation—that of the carbonate buildup-shale basin couplet—with which this paper is concerned. For convenience, sediments deposited during the buildup phase within the carbonate complexes are referred to collectively as the Miette and Ancient Wall buildups.

Stratigraphically (Fig. 2b) the buildup phase incorporates sediments of the "upper Cairn member" of the Cairn Formation and the Peechee Member of the Southesk Formation. The more or less equivalent basin sediments belong to the Perdrix Formation and the lower part of the Mount Hawk Formation.

Regionally (Fig. 2b) buildup carbonates are equivalent to those of the subsurface Leduc Formation, i.e. the Miette and Ancient Wall buildups are two of many of the well known Upper Devonian Leduc "reefs" of Alberta.

Miette and Ancient Wall Buildups.—Buildup to basin transitions at Miette and Ancient Wall are exposed in several Rocky Mountain thrust sheets. Work for this study was carried out in two areas, one at the southeast margin of each buildup, where exposures are generally excellent. For the Miette buildup, the relevant area is in the vicinity of Marmot Cirque (Fig. 3); for the Ancient Wall buildup, the slopes of Mount Haultain (Fig. 4).

The buildups are composed dominantly of calcarenites, carbonate mudstones and concentrations of skeletal organisms. They are generally fringed by a narrow zone of concentrated stromatoporoid and coral growth (Mountjoy, 1967). Basinal rocks deposited seawards of the buildups comprise three interdigitating facies (Hopkins, 1975): *foreslope facies*—calcarenites and breccias which flank the buildups; *nearslope facies*—carbonate and terrigenous mudstones that accumulated at the "toe of the slope" between the buildup flanks and basin floor; *basin facies*—slightly calcareous terrigenous mudstones which fill the remainder of the basin.

The basin was filled more or less contemporaneously with these deposits so that relief between the tops of the buildups and adjacent flanking sediments was never great. Mountjoy (1967, p. 396) has shown that the average slope of the buildup edge in the Marmot Cirque area was between about 7 to 10 degrees, so that onlapping sediments were deposited on lower slopes. At Mount Haultain, maximum slope for deposition of flanking sediments was between 5 and 10 degrees (Mountjoy, 1967, p. 398). Hence, the deposits discussed in this paper, calcarenites and breccias of the foreslope facies, accumulated on rather low slopes (maximum 5-10 degrees) close to the buildup margins. Five kilometers away from the buildup edge (the maximum extent of

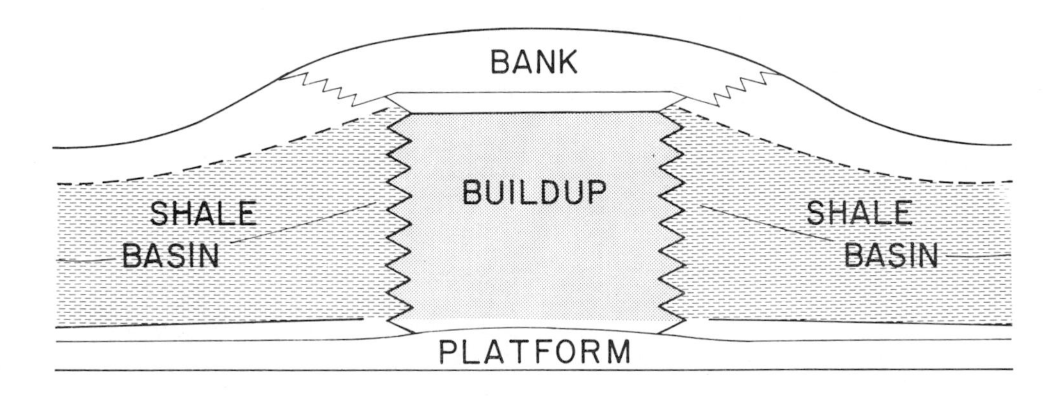

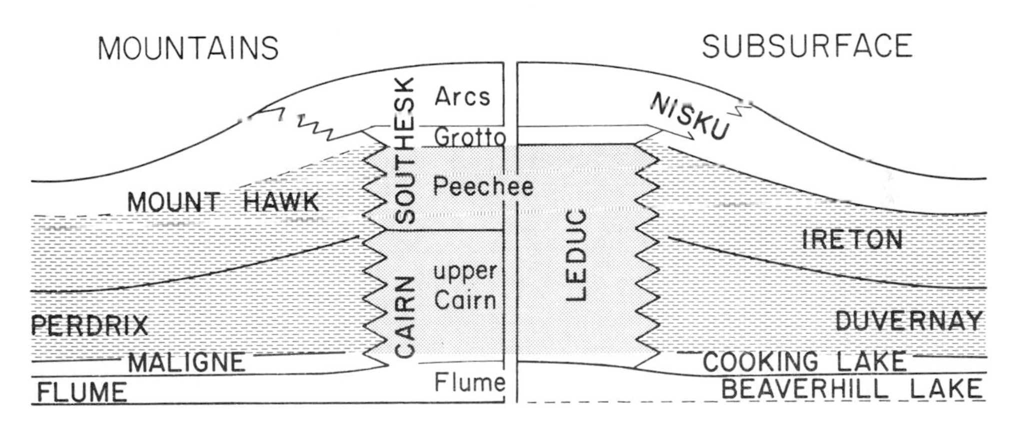

FIG. 2.—Schematic cross section of carbonate complexes (vertical scale greatly exaggerated). *a*—Sedimentological model. *b*—Stratigraphic terminology for mountain and subsurface areas of central Alberta. Correlation after Belyea (1964).

foreslope sediments into the basin) slopes were probably in the order of a degree or two at most.

FORESLOPE DEPOSITS

General Features

The nature of foreslope deposits immediately adjacent to the buildup margin is not well known for the rocks have been dolomitized and are poorly exposed (Marmot Cirque) or are inaccessible (Mount Haultain). The rest of the foreslope is well exposed, however. Deposits relatively close to the buildups (about 100 m to 1 km) comprise three main lithologic and bedding types (Fig. 5): (1) carbonate breccia in beds up to one meter thick, (2) calcarenite beds commonly between 15 and 60 cm thick, and (3) sequences of thinly bedded calcarenite in which individual beds are a few centimeters thick. The breccia and thick calcarenite beds are separated by sequences of thinly bedded calcarenites. Toward the basin, alternating beds in thinly bedded sequences are argillaceous, and ultimately calcarenites give way to interbedded carbonate mudstones and argilla-

FIG. 3.—Foreslope section on southeast side of Marmot Cirque (Miette buildup). View is basinwards (southeast) from the edge of the buildup. Breccia beds are concentrated in three distinct sequences ("B"), each separated by argillaceous and carbonate mudstones.

FIG. 4.—Foreslope section on southeast slopes of Mount Haultain (Ancient Wall buildup). View is towards the buildup (northwest) which crops out behind ridge in background. "B" marks a concentration of breccia beds in the middle of Perdrix Formation. "MB" is the megabreccia bed described by Cook and others (1972).

ceous carbonate mudstones of the nearslope facies.

Up to 40 breccia beds are present in the vertical foreslope sequence (about 150 m). Individual beds are not uniformly distributed throughout the section but are concentrated in certain intervals (Figs. 3, 4). Where such concentrations are present maximum extension of foreslope deposits into the basin, about 5 km, occurs. Elsewhere, argillaceous carbonate mudstones were deposited much closer to the margins of the buildups and the lateral sequence is condensed.

Sediments deposited on the foreslopes adjacent to Miette and Ancient Wall buildups were basically carbonate sands derived from these buildups.

FIG. 5.—Typical foreslope sequence about one kilometer from the edge of Ancient Wall buildup. (For scale; bed "1" is about one meter thick.) "1"—Breccia beds. "2"—Thick calcarenite bed. "3"—Sequences of thinly bedded calcarenite and slightly argillaceous calcarenite.

Carbonate mud was subordinate as mud was generally deposited in deeper, quieter water beyond the foreslopes.

Allochems in foreslope deposits are principally peloids—small (50-200μm), subspherical, hard, cryptocrystalline grains of uncertain origin, but including altered skeletal fragments, carbonate mud aggregates and probably some fecal pellets. Skeletal fragments (stromatoporoid and coral with accessory crinoid, brachiopod and forminifera) are common. Larger carbonate mud intraclasts (up to a few mm) are generally present as minor components.

Breccia Beds

Bedding style.—Breccia beds are commonly between 30 and 100 cm thick. Thickness is not uniform in any one bed which can thicken or thin by a factor of two over a distance of a few hundred meters (the general limit of lateral exposure). More rapid changes in thickness are rare and evidence for channeling is generally lacking. Cook and others (1972, p. 458) concluded that the breccias are "sheet" rather than linear channel deposits.

Sole markings.—Contacts between breccia beds and underlying lithologies are conformable and evidence for erosion or deformation of the underlying strata is lacking, even where mudstones are present. The base of each bed seems to have been a plane of post-depositional solution, and is often knobbly in appearance due to differential solubility between clasts and matrix. Sole markings, if any were present, have been destroyed.

Internal strucutre.—The internal structure of breccia beds is usually chaotic and comprises poorly sorted, randomly oriented clasts supported by a finer matrix (Fig. 6). Where clasts are

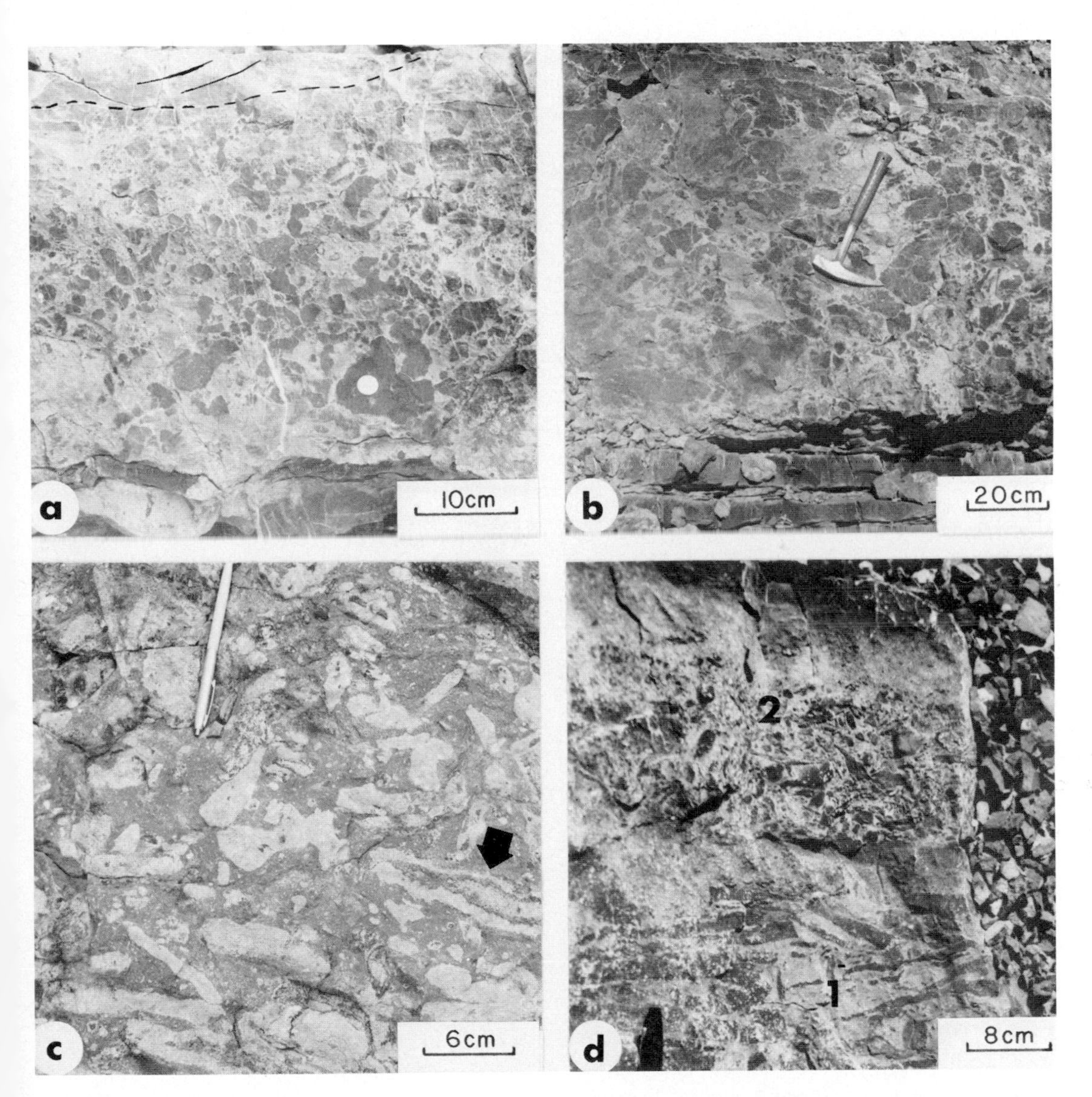

FIG. 6.—Outcrops of breccia beds. *a*—Bed with very irregularly shaped clasts. The position of the large clast near the base (white dot) gives an illusionary appearance of normal grading which was not confirmed on other outcrops of the same bed. Bed is capped by a thin, lensing, cross-laminated calcarenite unit (outlined; see also Fig. 7*a*). *b*—Poorly sorted, randomly distributed subspherical clasts. *c*—Elongate and irregularly shaped clasts crudely aligned parallel with bedding. (Pen is oriented normal to bedding plane.) Most of the clasts are cemented calcarenite, with some tabular stromatoporoids (arrow). *d*—Composite sequence comprising two breccia beds. Lower bed (1) has elongate, wedge-shaped clasts. Upper bed (2) has smaller, angular clasts and is capped by a graded calcarenite unit.

FIG. 7.—Calcarenite units of upper parts of breccia beds. *a*—Cross-laminated calcarenite sharply overlying breccia. Base of photo parallels bedding of breccia bed; sloping black line parallels cross-laminae in carbonate sand unit. Thin section is from the bed shown in Fig. 6a. *b*—Contact between graded calcarenite unit and breccia unit. Polished slab.

elongate, crude alignment with bedding is present (Fig. 6c, d).

In about one-third of breccia bed outcrops examined, an upper calcarenite unit sharply overlies breccia. In one bed (Fig. 7b), the calcarenite unit is graded and a few small breccia clasts are present at its base; in another (Fig. 7a) the calcarenite is cross-laminated.

Not all calcarenite units above breccias are laterally continuous. In a few exposures where beds can be followed laterally, calcarenite is present over some parts of beds, but not others. In the bed shown in Figure 6a, the calcarenite unit comprises lenses of cross-laminated sand (Fig. 7a) filling broad undulations, several meters long and up to 20 cm deep, over the upper surface of the breccia unit.

Breccia clasts.—Clasts within breccia beds are elongate to subspherical (Figs. 6,8). The beds are classified as breccias because the clasts, regardless of their shape, have not been significantly abraded. Clasts are commonly a few centimeters across, although they range from a few millimeters up to about 20 cm.

Most breccia clasts are composed of grain-supported allochems set in calcite cement or, less commonly, mud-supported allochems set in finely crystalline neomorphic calcite.

Single large fossils (stromatoporoids and corals) are often present in breccia beds, but usually account for less than 10 percent of the clasts. A few fossils are encrusted with algae, stromatoporoids, or cemented carbonate sand. However, skeletons are of individual whole organisms rather than broken fragments of any cemented or organically bound mass. An uncommon, but interesting type of clast (Fig. 8c) comprises calcarenite or carbonate mudstone coated with cemented carbonate sand of a different texture.

Unusually large blocks of cemented calcarenite (up to 40 cm thick and one meter long) occur in a few breccia beds. Long axes of blocks are oriented roughly parallel to bedding. No gradation in size is seen between the large blocks and other clasts.

Matrices.—Breccia clasts are supported in a matrix of poorly sorted to well-sorted calcarenite which generally lacks visible cement (Figs. 9a, b). Lithification and porosity reduction in the matrices is the result of compaction accompanied by dissolution of calcium carbonate at allochem boundaries, so that each allochem is in contact with its neighbors in three dimensions. This fabric is referred to as the *compacted fabric.*

In contrast, in calcarenite breccia clasts, even those which had a detrital carbonate mud matrix, lithification and porosity reduction has taken place by the precipitation of calcium carbonate. Sediments which have undergone lithification of this type are said to possess a *cemented fabric.*

Thus a strong distinction is drawn between calcarenites which have a cemented fabric and those with a compacted fabric. The distinction is important, for calcarenites of other foreslope deposits have cemented and compacted fabrics (see below).

At first glance (Fig. 8) it appears that detrital carbonate mud forms a significant proportion of the matrices of breccia beds. This is illusionary, however, and the muddy appearance is due to close packing of numerous cryptocrystalline peloids whose boundaries can be difficult to detect. The "clean-sand" (mud-free) nature of breccia bed-matrices is revealed where dark organic material is concentrated along grain boundaries (Fig. 9a) or, less commonly, where compaction dissolution has not completely reduced porosity and late-stage calcite cement fills remaining pore spaces (Fig. 9d). Rarely, a matrix of cemented calcarenite rather than compacted calcarenite is found (Fig. 8f).

Compacted calcarenites are often partly dolomitized. Euhedral to subhedral dolomite rhombs grow from the points of contact between allochems (Figs. 9a, c) which are replaced as the dolomite rhombs enlarge. In some beds the compacted fabric (assumed) of the matrix has been destroyed and relict peloids are left isolated in euhedral dolomite. Clasts in these beds usually remain undolomitized.

Thick Calcarenite Beds

Calcarenite beds 15–60 cm thick are associated with concentrations of breccia beds although they are less numerous than breccia beds. They also occur within thinly bedded sequences close to the buildups. In these sequences, the thicker calcarenite beds are conspicuous as beds with relatively planar surfaces, in contrast to more irregular bed forms of thinly bedded calcarenites (Fig. 5).

None of the thick calcarenite beds are graded; other internal structures were not seen. As with breccia beds, sole markings, if any, have been obliterated by post-depositional solution along bedding planes.

Petrographically, thicker calcarenite beds usually exhibit cemented fabrics. Compacted fabrics are present in a few beds in more basinal sections.

Thinly Bedded Calcarenite Sequences

Thinly bedded calcarenite sequences that occur between the coarser and thicker calcarenite and breccia beds close to the buildups are basically alternations of cemented calcarenite with compacted calcarenite.

The most common bedding form here is beds of cemented calcarenite ranging from 2 to 10 cm thick and separated by a similar thickness of compacted calcarenite (Fig. 10). The cemented layers range from more or less continuous beds (Fig. 10d) to crudely aligned, discrete nodules (Fig. 10a). Within a single sequence some layers may be entirely cemented and continuous; other layers are nodular. Some nodules, although apparently discrete, are intricately linked in three dimensions (Fig. 11). A cut parallel to the bedding plane through the center of one of these nodular layers revealed an irregular network of random and, for the most part, linked nodules.

SOURCES OF FORESLOPE BRECCIAS

Discussion.—In searching for an origin of the breccias, one is struck by the strong similarity between breccia clasts and cemented nodules of thinly bedded sands. This leads to the proposition that breccia clasts were derived from the thinly bedded cemented carbonate sands. The only alternative explanation is that clasts were eroded (submarine or subaerially) from the buildups. Several points of evidence support the former interpretation.

1. Clasts in breccia beds are rarely larger than about 20 cm across. In a few beds blocks up to one meter long are present; however there is no gradation in size between the clasts and blocks.

The obvious source which would place similar size restrictions on the blocks and clasts is the thinly bedded foreslope sequences which occur close to the buildup margins; cemented nodules and thin beds would yield breccia clasts; cemented thicker beds would produce blocks.

2. Breccia clasts can be subspherical or highly irregular in shape, but none are rounded. Different breccia beds contain clasts of different shapes, but generally the shape is relatively constant within a particular bed (compare clast shape in breccia bed shown in Fig. 6a with that in bed shown in the lower part of Fig. 6d).

It is difficult to conceive the formation of irregularly shaped clasts by normal erosion processes. This problem is minimized for derivation of the fragments from foreslope sequences, for the shape of the clasts would be inherited from

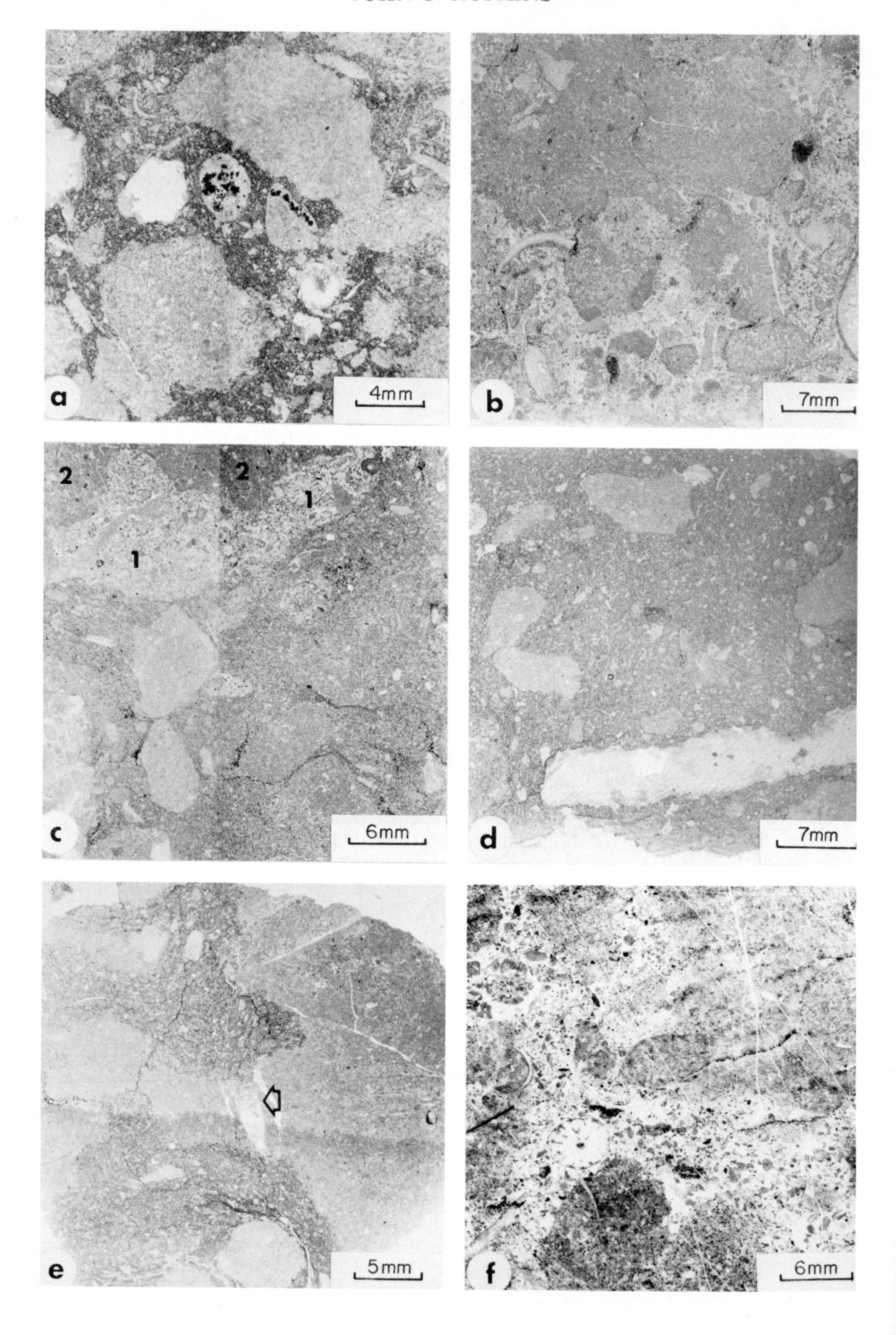
a
4mm
b
7mm
2
2
1
1
c
6mm
d
7mm
e
5mm
f
6mm

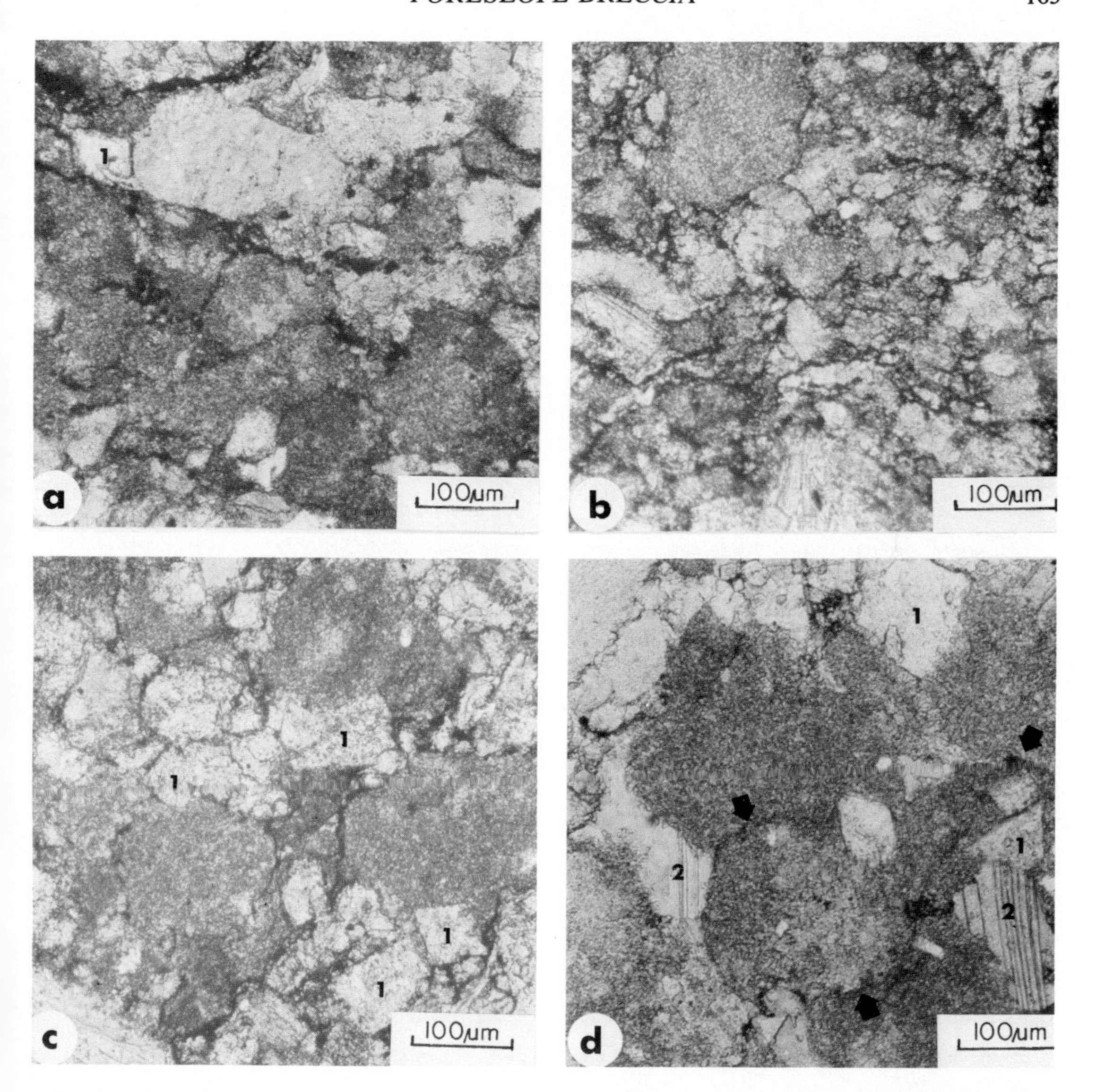

FIG. 9.—Photomicrographs of compacted calcarenite matrices of breccia beds. *a*—Moderately sorted compacted calcarenite composed of peloids (dark) and skeletal fragments (light). Boundaries between allochems are rendered visible by the presence of dark organic material. "1"—dolomite rhombs which have grown between allochems and have partly replaced them. *b*—Poorly sorted compacted calcarenite. *c*, *d*—Compacted and partly compacted calcarenite fabrics from different areas in the same thin section. In c, compaction is complete, the allochems are in three-dimensional contact and dolomite rhombs ("1") have grown at the points of contact between grains. In d, compaction is not complete. Irregular contacts (arrows) formed by dissolution between some allochems indicate some compaction has taken place. Elsewhere, compaction has not completely reduced intergranular spaces, which are now filled by calcite cement ("2").

FIG. 8.—Photomicrographs of breccias. *a*—Typical breccia comprising cemented calcarenite clasts and a few fossils in a compacted calcarenite matrix which appears dark and "muddy" at this magnification. *b*—Similar to a; compacted calcarenite matrix (here light colored) has been largely dolomitized. *c*—Composite clast (upper right) comprising carbonate mudstone (2) coated with cemented calcarenite (1). *d*—Isolated small clasts of cemented calcarenite and a large skeletal fragment in a compacted calcarenite matrix. *e*—Highly irregular clast, a portion of which has been broken during compaction (arrow). The crack has been partly filled with grains from the matrix and later, completely filled with dolomite cement. Orientation unknown. *f*—Rare example of breccia bed in which the calcarenite matrix around clasts has been cemented rather than compacted.

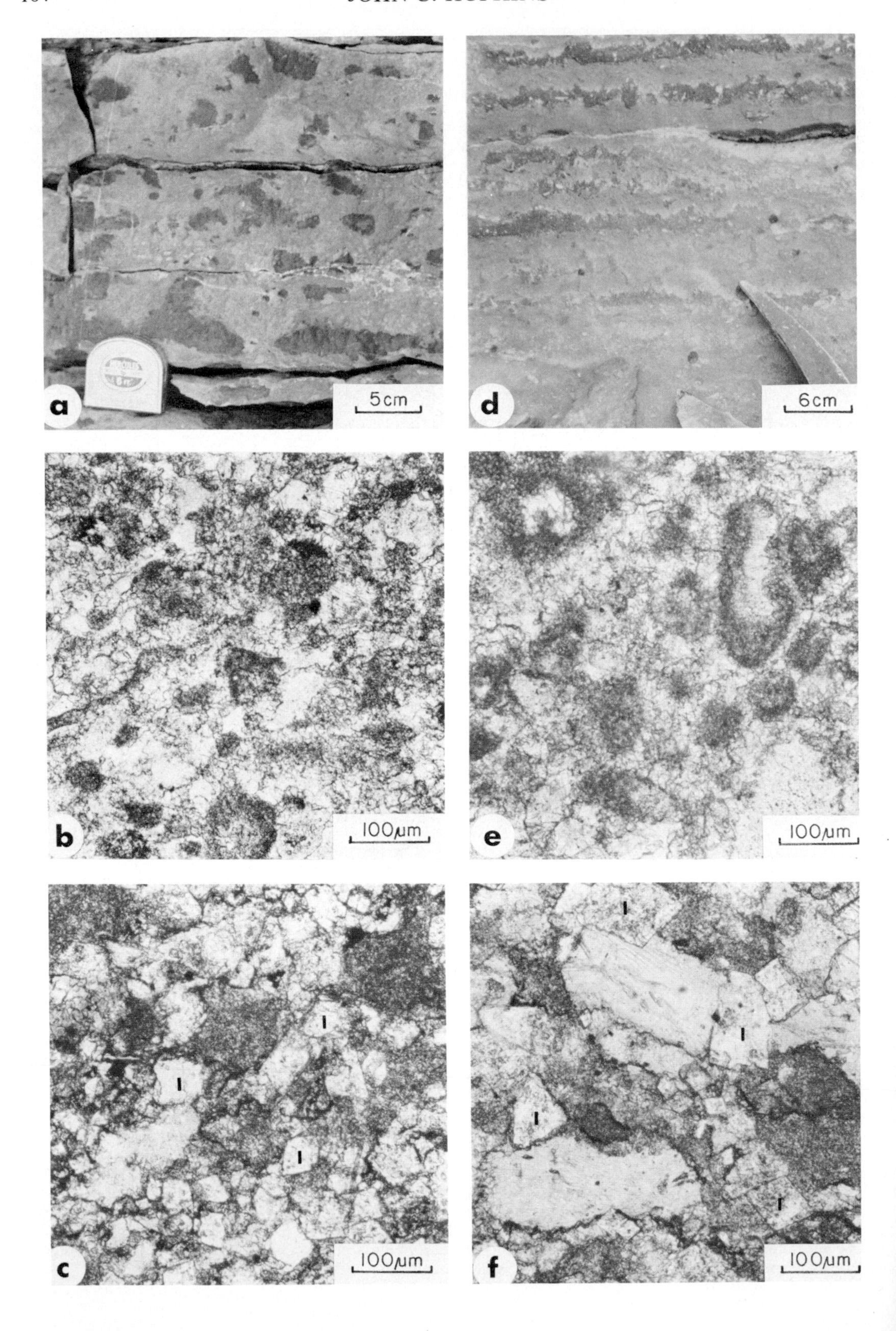
a
5cm
d
6cm
b
100μm
e
100μm
c
100μm
I
I
I
f
100μm
I
I
I
I

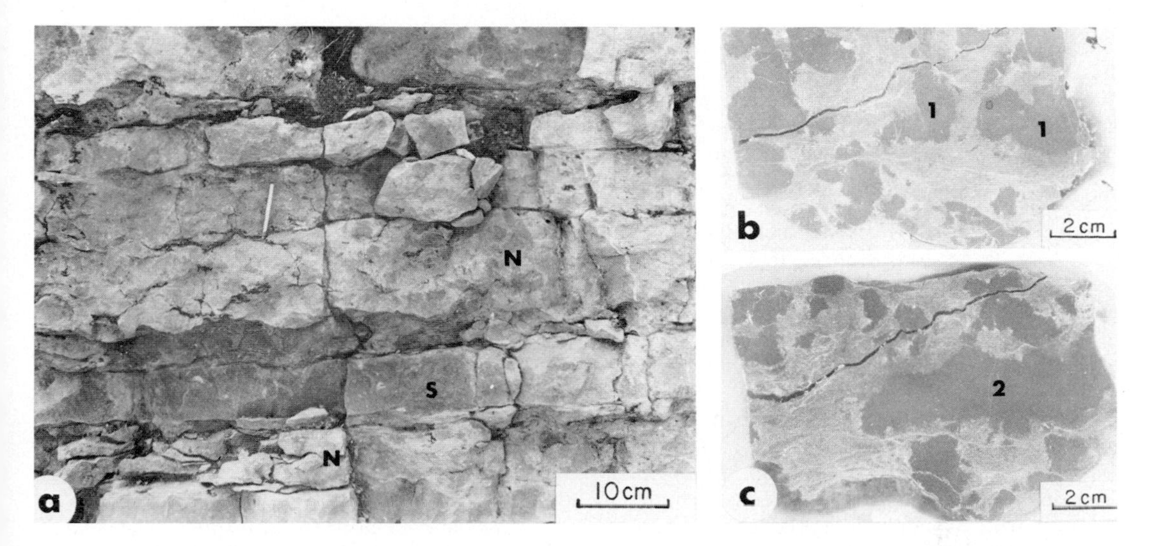

FIG. 11—*a*—Thinly bedded nodular foreslope sequence comprising cemented nodules in compacted calcarenite (N) and a laterally continuous bed of cemented calcarenite (S). *b, c*—Polished surfaces of cuts 2 cm apart from a sample of the upper nodular layer in a. Pronounced three-dimensional irregularity of nodules is shown by the two cemented patches ("1" in b) which are part of a larger cemented nodule ("2") in c).

the cemented nodules. The similarity of clast shape within a particular bed would reflect derivation from a nodular sequence in which the nodules had that shape. Variations in the shape of clasts between different breccia beds can be explained by variations in the shape of the nodules in different thinly bedded, nodular sequences from which each breccia bed was derived.

3. The matrix surrounding the breccia fragments is moderately to poorly sorted, compacted calcarenite.

If the breccias were derived by erosion of the margins of the buildups, somehow matrix material has to be mixed with the clasts prior to deposition of the breccia beds. With derivation from foreslope sequences, however, matrix material is derived from uncemented carbonate sands (now compacted) that occur between cemented layers and nodules which are the source of the breccia clasts.

4. Cemented calcarenites of various textures account for a large proportion of the breccia clasts. The relatively few skeletal forms that occur are corals and tabular stromoatoporoids. Some are encrusted with algae or stromatoporoids; others possess a thin adhering coat of cemented carbonate sand.

Cemented calcarenite nodules and beds in thinly bedded foreslope sequences have very similar textures to the breccia fragments. Corals and tabular stromatoporoids are also present in some samples from thinly bedded foreslope sequences close to the buildups.

The relative scarcity of massive skeletal forms (corals and stromatoporoids) which have been inferred and are known to fringe the buildup at a number of horizons (Mountjoy, 1967) is especially evident. Clasts containing, or composed of massive skeletal fragments should be more common if erosion of the buildup was the dominant mechanism for derivation of clasts. Also absent from breccia beds are clasts composed of fenestral, laminated, and *Amphipora*-bearing carbonate mudstone which are characteristic of the lagoonal facies of buildups.

Interpretation of source.—The size, shape,

FIG. 10.—Outcrop photograph and photomicrographs of a thinly bedded nodular foreslope sequence (a, b, c) and thinly bedded foreslope sequence (d, e, f). *a*—Nodules of cemented calcarenite (dark) in layers parallel to bedding are separated by compacted calcarenite (light). *b*—Cemented calcarenite from a nodule. *c*—Compacted calcarenite adjacent to a nodule. Boundaries between the dark peloids have been almost completely obscured by growth of dolomite rhombs ("1"). *d*—More continuous layers of cemented calcarenite (dark) are separated by compacted calcarenite (light). *e*—Cemented calcarenite from a cemented layer. *f*—Compacted calcarenite between cemented layers. Peloids and skeletal grains form the tight compacted fabric; dolomite rhombs ("1") have grown near junctions between groups of adjacent allochems.

composition and distribution of clasts and the associated matrix of breccia beds, points towards derivation of the breccias from differentially cemented, thinly bedded carbonate sand sequences similar to thinly bedded calcarenite sequences observed in outcrop near the buildup margins. The actual source beds, of course, cannot be observed for they have been displaced downslope. However, it can be postulated that thinly bedded sequences which were displaced to form breccias were composed largely of discrete nodules and uncemented carbonate sand. In this way, any problems concerning breakage of clasts are minimized.

Lithification of foreslope deposits.—Breccia clasts were coherent entities which had obviously become lithified prior to transport down the foreslope, otherwise they would have disintegrated into their respective sand-sized components during transport. Since breccia clasts were derived from thinly bedded sequences, cementation must have been a synsedimentary near-surface process.

Unfortunately, cement fabrics in breccia clasts or thinly bedded sequences are not diagnostic of the type of early cementation—submarine or subaerial. Any explanation involving subaerial cementation, however, would involve exposure of the buildups for which there is no evidence (Cook and others, 1972, p. 471). In any case, subaerial exposure of the upper foreslope in 40 different episodes, to account for the 40 breccia beds present in the foreslope succession, would require extraordinary sea level gymnastics that might even tax the imagination of the most ardent advocate of subaerial cementation.

Hence it seems an inescapable conclusion that differential cementation of the thinly bedded carbonate sands much have been marine and occurred within the sediment pile, close to the sediment-water interface.

EMPLACEMENT OF BRECCIAS

Constraints.—Any explanation of the mechanism of emplacement of Miette and Ancient Wall foreslope breccia beds must take into account the following observations.

1. Petrographic and field evidence indicates that the breccias were derived from differentially cemented carbonate sands which accumulated on the foreslope close to the buildup margins.
2. Evidence for erosion by abrasion is lacking as breccia clasts have not been abraded.
3. Breccia source beds were deposited on relatively low slopes (up to about 10 degrees) close to buildup margins.
4. Breccia beds were deposited on slopes of 5-10 degrees close to the buildups and on lower slopes toward the basin.
5. Breccia beds have a limited lateral extent and to not appear to have been deposited more than about 5 km from the buildup edge.
6. The transporting agent was sufficiently competent to move an assortment of particles from fine sand up to clasts 20 cm across. In a few beds at least, it was capable of moving blocks up to one meter long.
7. Strata beneath breccia beds are not deformed or eroded even where they are comprised of terrigenous and carbonate mud.
8. Breccia beds are devoid of internal structure except for a rude alignment of clasts where clasts are elongate.
9. Some breccias are capped with a carbonate sand unit which can be graded or cross-laminated. Although the contact between breccia and sand is relatively sharp, together they consitute a single bed.

Possible transport mechanisms.—The source, bed form, and fabric of the breccias indicates that they are not the deposits of wave or current action or submarine rock falls. Rather they are the result of gravity-induced, downslope mass movement.

Gravity downslope movement of sediment can involve sliding, slumping, and sediment gravity flows (Middleton and Hampton, 1973). In slides and slumps, movement is along discrete planes with only limited mixing of individual grains or clasts. In sediment gravity flows, on the other hand, particle movement is chaotic, so that the clasts are thoroughly mixed. Four end members can be recognized within the sediment gravity flow spectrum (Middleton and Hampton, 1973), on the basis of the manner in which particles are thought to be supported within the flowing mass: turbidity current flows (clast support by fluid trubulence); debris flows (matrix strength); fluidized flows (upward movement of fluids); grain flows (grain-grain interactions).

Field evidence to clearly distinguish between deposits of these different gravity transport processes is not yet certain. Furthermore, as depicted by Middleton and Hampton (1973, Fig. 10) more than one mechanism may be involved in transportation and emplacement of a particular deposit. Normally, the mechanism which is thought to have been predominant in any one flow, lends its name to the deposit in question.

Discussion.—Evidence for slumping of differentially cemented, thinly bedded foreslope sands is unequivocal in at least one instance (Fig. 12). In this case, movement stopped shortly after initial failure, possibly because the slope was too low, or the cemented layers were continuous rather than nodular and hence did not break up into clasts. Slumping may have been instrumental in initiating mass movement, but slumps must have transformed into some type of sediment

FIG. 12—Slumped sequence of thinly bedded foreslope sands. Deformed cemented layers (dark) are surrounded by dolomitized compacted calcarenite.

gravity flow in order to give the sheet-like form and thorough mixing of clasts of breccia beds.

Determination of the precise nature of the flow mechanism is difficult. Clast support entirely by fluid turbulence seems unlikely; the beds do not exhibit the types of sedimentary structures normally shown by the deposits of turbidity currents. Also strata beneath the breccia beds are not eroded or distrubed, yet it has been shown (Komar, 1970) that turbulent flows of sufficient competence to transport conglomerates (similar in size to the foreslope breccias) would have enormous ability to erode and deform underlying strata.

Clast support by matrix strength as in debris flows also seem unlikely. Some clay-sized material must be present in debris flows to provide matrix strength (Hampton, 1972). With the foreslope breccias, however, there is a lack of clay-sized carbonate in the compacted calcarenite matrices. Elaborate depositional and diagenetic mechanisms which might cause preferential removal of clay-sized carbonate from compacted calcarenites might be appealed to, but we are still left with the problem in those few beds where the matrix is partly cemented.

Fluidized flow has only recently been recognized as a possible important mode of sediment transport. Consequently fluidized-flow deposits are not well known in the geologic record and their principal field features are interpreted in association with dewatering structures (Middleton and Hampton, 1973). Such structures were not observed in the foreslope breccias. Their absence, however, cannot be used as strong evidence against fluidization until more is known of fluidized flow deposits.

Grain-flow deposits are not well known either. Grain flow as a mechanism for emplacement of coarse sands and conglomerates in flysch sequences was popularized by Stauffer (1967). In common with Stauffer's grain-flow deposits, the foreslope breccias typically lack diagnostic sedimentary structures, do not show deformation at the bases of beds, possess a "clean sand" matrix, and the clasts are randomly distributed throughout. Also, where clasts are elongate, they may be crudely aligned parallel to bedding.

The grain-flow mechanism has been questioned by a number of workers and recently Link (1975) reinterpreted the original grain-flow deposits described by Stauffer (1967). Middleton and Hampton (1973) felt that many so called "grain-flow" deposits are probably hybrid sediment gravity flows in which grain-grain collisions, upward movement of fluids, and fluid turbulence all contributed to clast support.

By default, it seems that any sediment transport mechanism for the foreslope breccias must lie within the grain-flow and possibly fluidized-flow fields. Maintenance of these flows on low slopes seems to require the generation of high pore pressures within the interstitial fluid of the sediment mass. Grain flow can occur on low slopes only if high pore pressures reduce internal friction between grains (Middleton, 1970; Stauffer, 1967); by definition, high pore pressures are necessary for fluidized flow.

Excess pore pressures can be generated in sediments under certain conditions. Two important ways are by imposition of shear stress onto a sediment mass by an overlying sediment-laden current or flow (Sanders, 1965: Stauffer, 1967), or by liquefaction failure of an unstable sediment mass (Morgenstern, 1967; Shepard and Dill, 1966). Sediment transport under conditions of excess pore pressures has not generally been favored, for on low slopes deposition will take place shortly after initiation of the flow as the excess pore pressures rapidly dissipate. Middleton (1970) has noted that transport of medium to coarse sands by grain flow under the influence of high pore pressures cannot be effective for more than a few miles.

This observation does not provide a basis for serious objection to transport of the foreslope breccias by a combination of fluidization and grain-grain interactions. In fact, the limited extent of breccia beds (maximum of 5 km into the basin) and their deposition on slopes of 5–10 degrees (close to buildups), requires some mechanism which precludes transformation into flows capable of sustaining themselves for long distances over low slopes.

Interpretation and model for foreslope sedimentation.—The interpretation which best fits the available evidence is that the foreslope breccias are the products of hybrid sediment gravity flows. Clasts were supported by a combination of rising fluids, grain-grain collisions, and possibly fluid turbulence. The flows were not capable of transporting sediment for great distances over low slopes, but came to rest shortly after initiation. A possibility here is that flow was arrested as excess pore pressures dissipated.

The resultant deposits comprise, for the most part, randomly distributed clasts set in a finer matrix. Some beds, however, have graded or cross-laminated calcarenite units above breccia, and these require special explanation. Close juxtaposition of carbonate-sand and breccia units implies genetically related transport and deposition. One possibility is that the carbonate-sand unit records the passage of some overriding sediment-laden current which triggered flow of an underlying differentially cemented sediment mass by increasing pore pressures. Alternatively, sediment gravity flows of differentially cemented sediment were initiated by liquefaction and slumping. Bottom currents, set in motion by displacement of sediment, distributed carbonate sand as a thin tractive blanket over breccia. Turbulent mixing of carbonate sand with overlying waters at the top of a flowing mass may have generated turbidity-current flow leading to deposition of graded carbonate sand units.

Questions concerning the origin of the carbonate-sand units leads to concern about the causes of downslope displacement of foreslope sediments in general. Did breccia flows occur spontaneously, or were they the result of some catastrophic event, e.g. earthquakes or tsunamis, as envisaged by Cook and others (1972) for the megabreccia bed at Mount Haultain? This question can be partly answered from the nature of foreslope deposition.

Thinly bedded foreslope carbonate sands close to buildup margins are sands derived from the buildups by wave and current action and possibly downslope creep. As shown by the thin, even, and relatively undisturbed bedding, and by the presence of a few scattered in-place fossils, the environment was generally one of moderate energy and was not subjected to pervasive traction and redeposition processes. Thick carbonate sand beds close to the buildup margin record episodic rapid carbonate sand deposition perhaps involving short periods of strong current action. Shear stress from such sediment-laden currents may have triggered sediment gravity flow.

Many breccias, however, are not overlain by calcarenites, and thick calcarenite beds do not predominate at the horizons where breccia beds abound. Rather, breccia beds are concentrated in sequences where thinly bedded foreslope sequences are well developed and protrude farthest into the basin. In other words, breccia beds are concentrated at horizons where the supply of sand to the foreslope was relatively high. Under these conditions of rapid loading, unstable packing of sediments and high fluid pore pressures resulting in liquefaction failure and sediment gravity flow could more readily occur.

By this model, the question whether or not failure was spontaneous or was triggered by some external agent is incidental. The primary controlling factor is the rate and amount of sediment supplied to the foreslope. A breccia bed is the result of metastably packed sediments becoming differentially cemented and then sloughing away downslope by failure in uncemented layers.

CONCLUSIONS

Foreslope breccias adjacent to the margins of the Miette and Ancient Wall carbonate buildups exhibit features related to both source and emplacement mechanism, which distinguish them from other carbonate breccias.

1. The foreslope breccias consist of clasts of cemented carbonate sand set in a matrix of compacted carbonate sand.

2. They were derived from thinly bedded, differentially cemented sequences of carbonate sand which accumulated on the foreslopes, close to

the margins of the buildups.

3. Cementation of the thinly bedded sequences was submarine, occurred close to the sediment-water interface, and resulted in the formation of nodules of cemented carbonate sand separated by uncemented carbonate sand (now seen as a compacted calcarenite).

4. Downslope displacement of thinly bedded sequences mixed cemented nodules with uncemented sands and resulted in deposition of breccia beds.

5. The transport mechanism appears to have been some type of a hybrid sediment gravity flow which occurred under the influence of high pore pressures in the interstitial fluid.

6. Breccia beds were deposited relatively close to source because excess pore pressures dissapated rapidly.

7. Primary control for displacement of a thinly bedded sequence (and hence formation of a breccia bed) was the rate of supply of sand from the buildup to the foreslope.

The foreslope breccias appear to be a rather special type of carbonate breccia and it is possible that similar breccias will be found associated with other carbonate buildups in the geologic record. The presence of such breccias cannot be taken as evidence for abrasional erosion or catastrophic break-up of a buildup margin. They might suggest, but will not prove, that submarine lithification of the buildup edge has occurred.

ACKNOWLEDGEMENTS

This paper was derived from part of my dissertation concerning the Miette and Ancient Wall buildup-to-basin transitions. The work was supported by National Research Council of Canada and Geological Survey of Canada grants to Dr. E. W. Mountjoy and Dr. C. W. Stearn, both of McGill University. Receipt of a National Research Council of Canada Scholarship for personal support is gratefully acknowledged.

Many stimulating hours of discussion about the origin of the breccias were spent with Noel James, fellow student at McGill. Volkmar Schmidt and Dick Walls critically read the manuscript and offered useful suggestions for its improvement.

REFERENCES

BELYEA, H. R., 1964. Upper Devonian, Part II, Woodbend, Winterburn and Wabamum Groups: *In* R. G. McCrossan and R. P. Glaister (eds.), Geological history of western Canada: Alberta Soc. Petroleum Geologists, Calgary, Alberta, p. 66-88.

COOK, H. E., MCDANIEL, P. N., MOUNTJOY, E. W., AND PRAY, L. C., 1972, Allochthonous carbonate debris flows at Devonian bank ('reef') margins, Alberta, Canada: Bull. Canadian Petroleum Geology, v. 20, p. 439-497.

HAMPTON, M. A., 1972, The role of subaqueous debris flow in generating turbidity currents: Jour. Sed. Petrology, v. 42, p. 775-793.

HECKEL, P. H., 1974, Carbonate buildups in the geologic record: A review: *In* L. F. Laporte (ed.), Reefs in space and time: Soc. Econ. Paleontologists and Mineralogists, Spec. Pub. 18, p. 90-154.

HOPKINS, J. C., 1971, Production of reef-margin breccias by submarine cementation and slumping of carbonate sands, Miette reef complex [abs.]: Am. Assoc. Petroleum Geologists Bull., v. 55, p. 344.

——, 1975, Distribution and diagenesis of carbonate muds in deeper waters surrounding Miette and Ancient Wall (Devonian) buildups [abs.]: Am. Assoc. Petroleum Geologists Ann. Meeting Abs., v. 2, p. 36.

KOMAR, P. D., 1970, The competence of turbidity current flow: Geol. Soc. America, Bull, v. 81, p. 1555-1562.

LINK, M. H., 1975, Matilija Sandstone: A transition from deep-water turbidite to shallow-marine deposition in the Eocene of California: Jour. Sed. Petrology, v. 45, p. 63-78.

MIDDLETON, G. V., 1970, Experimental studies related to problems of flysch sedimentation: *In* Jean Lajoie (ed.), Flysch sedimentology in North America: Geol. Assoc. Canada, Spec. Pub. 7, p. 253-272.

—— AND HAMPTON, M. A., 1973, Sediment gravity flows: Mechanics of flow and deposition: *In* G. V. Middleton and A. H. Bouma (eds.), Turbidites and deep-water sedimentation: Soc. Econ. Paleontologists and Mineralogists Pacific Sec., Los Angeles, California, p. 1-38.

MORGENSTERN, N. R., 1967, Submarine slumping and the initiation of turbidity currents: *In* A. F. Richards (ed.), Marine geotechnique: Univ. Illinois Press, Urbana, Illinois, p. 189-220.

MOUNTJOY, E. W., 1967, Factors governing the development of the Frasnian, Miette and Ancient Wall reef complexes (banks and biostromes), Alberta: *In* D. H. Oswald (ed.), International symposium on the Devonian System, v. 2: Alberta Soc. Petroleum Geologists, Calgary, Alberta, p. 387-408.

PRAY, L. C., COOK, H. E., MOUNTJOY, E. W., AND MCDANIEL, P. N., 1968, Allochthonous carbonate debris flows at Devonian bank ('reef') margins, Alberta, Canada [abs.]: Am. Assoc. Petroleum Geologists Bull., v. 52, p. 545-546.

SANDERS, J. E., 1965, Primary sedimentary structures formed by turbidity currents and related resedimentation mechanisms: *In* G. V. Middleton (ed.), Primary sedimentary structures and their hydrodynamic interpretation: Soc. Econ. Paleontologists and Mineralogists, Spec. Pub. 12, p. 192-219.

SHEPARD, F. P., AND DILL, R. F., 1966, Submarine canyons and other sea valleys: Rand McNalley, Chicago, Illinois, 381 p.

Srivastava, P. N., Stearn, C. W., and Mountjoy, E. W., 1972, A Devonian megabreccia at the margin of the Ancient Wall carbonate complex, Alberta: Bull. Canadian Petroleum Geology, v. 20, p. 412–438.

Stauffer, P. H., 1967, Grain-flow deposits and their implications, Santa Ynez Mountains, California: Jour. Sed. Petrology, v. 37, p. 487–508.

SEPM Special Publication No. 25, p. 171-186, November 1977

DEEP-WATER LIMESTONES OF THE GREAT BLUE FORMATION (MISSISSIPPIAN) IN THE EASTERN PART OF THE CORDILLERAN MIOGEOSYNCLINE IN UTAH

H. J. BISSELL and HOLLY K. BARKER
Brigham Young University, Provo, Utah 84602, and
Exxon Company, U.S.A., Houston, Texas 77001

ABSTRACT

The Oquirrh Basin of central and northwestern Utah was a major downwarp in the Cordilleran miogeosyncline during the Pennsylvanian and Permian. During Late Mississippian time its forerunner was the Great Blue depocenter in which accumulated from 800 to 1,400 m of dark, gray to black, dense, siliceous and sparsely fossiliferous limestone and calcareous siltstone with minor dark-gray shale units. The Great Blue basin was subovate, and covered about 7,500 sq km area in the eastern part of the miogeosyncline. The following reasons indicate a deep-water origin for most of the limestone of the Great Blue Formation: (1) areally extensive distribution of mostly homogeneous dark colored, argillaceous and siliceous limestone; (2) within the thick succession of this limestone, a monotonous repetition of laminated to thin- and medium-bedded strata is characteristic; numerous strata range from millimeter- to centimeter-thin units; (3) rocks are siliceous and contain substantial amounts of black iron monosulfide, as well as cryptocrystalline varieties of pyrite and/or marcasite; (4) within all sections studied, flysch-like arrangement of strata is typical; some units display graded bedding; (5) most of the formation in outcrops near Provo, Utah, is sparsely fossilferous to nonfossiliferous; forms include a depauperate fauna of thin-shelled chonetid brachiopods; (6) some sections contain exotic blocks of coral-bearing limestone interpreted to have been gravity-driven downslope into deeper waters of the sedimentary basin; (7) some of the shale units contain the trace fossil *Scalarituba missouriensis* Weller (particularly those shales containing the exotic limestone blocks mentioned above); this trace fossil is a form common in deep to intermediate waters of the Mississippian and Pennsylvanian, particularly in the deep-water part of the Oquirrh Basin.

As herein interpreted, the Great Blue depocenter was the forerunner of the Oquirrh Basin. The latter was a deep water depocenter during parts of the Pennsylvanian and Permian. Accordingly, this downwarped segment of a part of the eastern Cordilleran miogeosyncline subsided locally to possibly bathyal depths during Late Mississippian time. Considerable quantities of terrigenous material were swept from the craton on the east and northeast and the Emery uplift and Uncompahgre highland to the southeast. This material moved downslope into deeper waters to accumulate with carbonate material to form impure argillaceous limestone. Siliceous, clayey, and argillaceous lime mud derived from the northeast Nevada highland of Nevada and Utah, the western Utah highland, and Antler orogenic belt in Nevada entered the western side of the Great Blue depocenter.

INTRODUCTION

A century ago geologists of the Fortieth Parallel Survey under Clarence King visited the Oquirrh Mountains southwest of Salt Lake City, Utah. In the report of this Survey S. F. Emmons discussed the region south of Salt Lake City (1877, p. 443-456); fossils of "Subcarboniferous and Waverly types" were collected. Emmons (1877, p. 447) noted the presence of a thick succession of ". . . dark, compact, fine-grained, more or less siliceous limestones, with some interstratified beds of shales and seams of black cherty material." No formal stratigraphic name was applied to these rocks. Later Spurr (1895, p. 374-376) proposed the name Great Blue Limestone to identify these rocks; the name is not geographic but is descriptive in that miners in the Mercur district of the Oquirrh Mountains referred to the bluish color of the limestone. Accordingly, no type section was designated, although Spurr called attention to the fact that this formation was present in the Mercur basin of mining activities.

Gilluly (1932, p. 7, 29-32) redefined the "Great Blue" Limestone, as he wrote it. He noted that the "Great Blue" Limestone conformably overlies the Humbug Formation, and also (1932, p. 31-34) proposed the new name, Manning Canyon Shale, for an aggregate thickness of 325 m of calcareous shale and interbedded limestone stratigraphically above the "Great Blue" Limestone. The term "Great Blue" Limestone, although nongeographic, is so well known to mining public and other geologists that no attempt has been made to replace it with a geographic name. This formation consists of a lower and an upper limestone separated by shaly beds (Long Trail Shale Member) that are about 30 m thick. Fossils present in the three members are of Late Mississippian age. Bissell and Rigby (Bissell, 1959a) published a detailed geologic map of part of the southern Oquirrh Mountains, also identifying and mapping the three members of the formation; they noted that the lower limestone member measures from 150 to 160 m in thickness, the Long Trail Shale about 30 m; and the upper member in excess of 500 m. Bissell (1959a, p. 57) noted that another, unnamed shale member approximately 35 m thick, occurs in the Great Blue about 100 m stratigraphically above the Long Trail Shale Member locally in the Oquirrh Mountains.

The study of the present writers was engendered by our mutual interest in the carbonates of the Oquirrh Basin. Some geologists have studied trace-fossils of parts of the Pennsylvanian and Permian Oquirrh Formation of the Oquirrh Basin, and have concluded that much of this formation accumulated in deep marine waters, particularly in the eastern part of the basin (Chamberlain and Clark, 1973). In our attempts to decipher some sedimentary environments of the Great Blue Formation (as we identify the Great Blue as a formation), we have directed our main attention to the limestone that accumulated in the eastern part of this basin in central Utah. Accordingly, we identify the Great Blue depocenter as the Late Mississippian forerunner to the Oquirrh Basin. Figure 1 locates the sections we studied or visited in reconnaissance, and also outlines what we infer to have been the geographic position of the Great Blue depocenter. In our discussion that follows (Tectonic Setting), emphasis will be placed on sections in the Oquirrh Mountains and those in the south-central Wasatch Mountains (Fig. 1, localities 1 and 2).

TECTONIC SETTING

A discussion of Mississippian tectonics and sedimentation for eastern Nevada and western Utah has been presented by Bissell (1974, p. 84–86), and need not be repeated here. However, it is germane to our discussion to stress the significance of the Las Vegas-Wasatch line, a major tectonic hinge that separated the Great Blue basin from the cratonic shelf and the Emery uplift southeast of it. It is also highly probable that intra-miogeosynclinal welts and positive regions exercised significant influence on patterns of sedimentation in part of the Great Blue depocenter, somewhat as they did for Pennsylvanian time in the Oquirrh Basin (see Bissell, 1962, Fig. 2, p. 30; 1974, Fig. 1, p. 85, and Fig. 2, p. 87). The Antler orogenic belt was outlined in early Paleozoic times as the Manhattan geanticline, and became the site of orogeny in central to western Nevada during the Late Devonian (see Poole, 1974, Fig. 15, p. 73). This belt was also a significant positive feature for much of Mississippian time along the western part of the Cordilleran miogeosyncline. Poole (1974) suggested that basins which formed as a result of crustal buckling became filled with sediments from the Antler orogenic belt, moving sites of thick sedimentation eastward through Mississippian time. He stated (Poole, 1974, p. 78): "Most likely, continental margin deformation during the Antler orogeny warped the continental crust under the shelf and initiated sites destined for subsequent major subsidence and uplift. . . ." Possibly the Great Blue and Oquirrh depocenters behaved as intracratonic basins upon an unstable shelf; we regard them as sedimentary basins that experienced substantial subsidence in an eastern part of the Cordilleran miogeosyncline. They lie west of the Las Vegas-Wasatch line (see Bissell, 1974, Fig. 1, p. 85, and Fig. 2, p. 87). Regardless, Poole's statement is well taken. We do interpret the Great Blue basin as a major downwarped segment of the earth's crust; sediment was swept into it from more than one provenance. In accounting for copious amounts of mostly silt-size and clay-size quartz materials admixed in carbonates of the Great Blue Formation, we believe that much of this material was derived from the cratonic shelf that lay to the northeast and east of the Great Blue depocenter, was transported by longshore and other currents, then moved into the basin. Similarly, much quartzose and carbonate material was derived from the Emery uplift and the Uncompahgre highland that were significant positive areas to the southeast of the eastern part of the Cordilleran miogeosyncline. As herein interpreted, deltas and distributaries, and possibly submarine fan-deltas, that prograded northwesterly from these provenances transported sediment to the proximal part of the Great Blue depocenter in the eastern part of the miogeosyncline. Much of this fine-textured sediment moved downslope into deeper waters of the basin; much of it formed flysch-like thin layers. Possibly much carbonate precipitated out of marine waters of the depocenter, and much detrital silt-size and clay-size carbonate was swept into the basin. This calcisiltite and calcilutite also was arranged in flysch-like layers with siliceous and iron-bearing clastic sediment.

Figure 2 is adapted from Poole (1974, Fig. 22, p. 77). In referring to his Figure 22 Poole (1974, p. 74) stated: "By very Late Mississippian time, the westerly derived clastic sediments filled the foreland trough and spread eastward across the limestone shelf onto the cratonic platform." Bissell (1974, p. 84) made note of the fact that the Antler orogenic belt exercised some degree of control on patterns of sedimentation as far east as western Utah. This is suggested because eastward-thinning tongues of chert-pebble conglomerate, grit, and sandstone of the Diamond Peak Formation (Mississippian) are present near Wendover, Utah-Nevada, various places in the Gold Hill district of western Utah, and other nearby localities (including part of the western half of the Great Blue depocenter). We do not believe that much medium- and coarse-textured clastic sediment was swept into the Great Blue depocen-

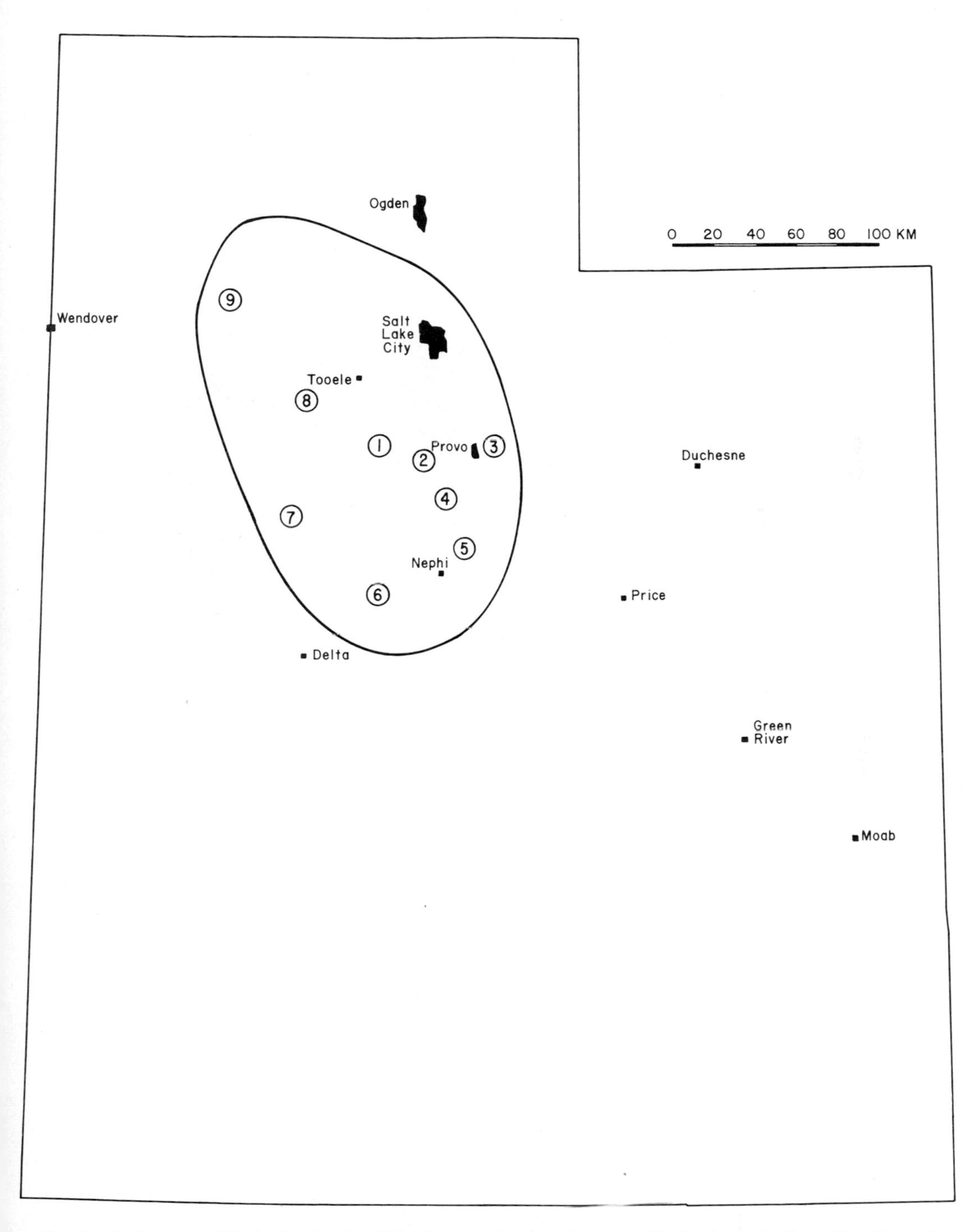

Fig. 1.—Index map of Utah showing localities in central and northwestern Utah where the Great Blue Formation was studied for this report: 1, southern Oquirrh Mountains; 2, Lake Mountain area; 3, Provo area (Provo and American Fork Canyons); 4, West Mountain area; 5, southern Wasatch Mountains; 6, East Tintic district; 7, Sheeprock-Onaqui Mountains area; 8, Stansbury Mountains; 9, southern Lakeside Mountains. Outlined area identifies the Great Blue Basin.

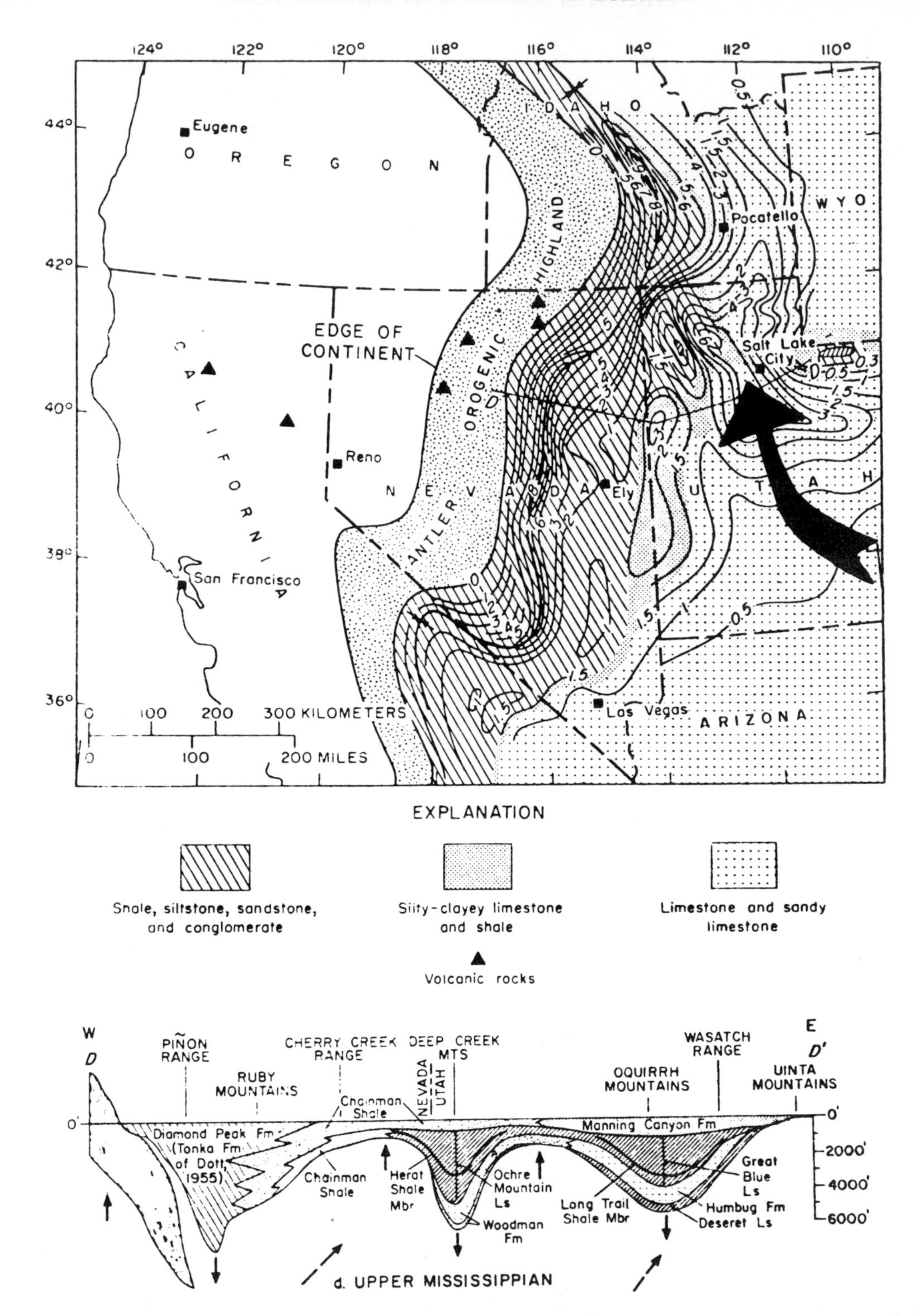

FIG. 2.—Isopachous and lithofacies map of Upper Mississippian rocks (partly restored) for much of the Cordilleran miogeosyncline. Isopachs in thousands of feet. Head of arrow locates central part of the Great Blue depocenter, and shaft indicates sediment transport direction from the Emery uplift and Uncompahgre highland. Modified from Poole (1974, Fig. 22 and top of Fig. 25).

ter from the Antler belt. However, we interpret that some siltstone and shale (including carbonates in the form of calcisiltite and calcilutite), and some lime mud, in the Great Blue Formation could be accounted for by having been transported from the Antler belt by easterly moving currents.

STRATIGRAPHY

Measurements by Gilluly (1932, p. 29) of the lower and upper limestone members and the Long Trail Shale Member (all near Ophir in Ophir Canyon where Long Trail Gulch is tributary to this canyon), give a figure of about 1,100 m for the total Great Blue Limestone at that locality in the Oquirrh Mountains. Bissell (1959a, p. 46, 55–58) reported essentially the same thickness of the formation and indicated mappability of the lower and upper limestone members and the Long Trail Member. As a result of field investigations in 1974 and 1975 by Alan Chamberlain and Michael Metcalf, graduate students at Brigham Young University, the total thickness of the Great Blue Formation proved to be 1,050 m. The lower limestone member is 155 m, Long Trail Shale is 33 m in thickness, and three additional mappable (unnamed) shale units occur within the upper limestone member. The upper limestone member consists of blue-gray and gray-blue micritic limestone with much argillaceous material, with interbedded black and gray silty shale and orthoquartzite. The upper 210 m of the Great Blue Formation consists of cherty and siliceous blue and gray micritic limestone; it is conformably overlain by the Manning Canyon Shale.

In most areas of outcrop of the Great Blue Formation in the Oquirrh Mountains, particularly in the Mercur Basin (see Fig. 3), the lower member is a light to medium gray and blue-gray to gray-blue limestone with fine- to medium-grained orthoquartzite interbeds and lenses. Some limestone weathers with a siliceous rind, and some display liesegang bands. Bedding in the principal limestone ranges from thin- and thick-bedded to massive. However, even massive-appearing units commonly display laminated to thin-bedded layers. The rock is micritic (terminology of Leighton and Pendexter, 1962, as modified from Folk, 1959), and contains considerable amounts of silt size and clay size quartz. Chert blebs, nodules, stringers, and thinly laminated layers are present; the chert is black, gray, and light tan. We interpret that this chert originated essentially contemporaneously with sedimentation, and accordingly formed in deep-water realms. Silica possibly was derived from the same provenances that provided clastics and dissolved materials to the Great Blue depocenter. Much of it likely was derived from the Antler orogenic belt of Nevada, because during Mississippian time that belt was also the site of volcanic activity. Siliceous materials so expelled into the western part of the Cordilleran miogeosyncline were transported in solution and colloidal suspension eastward across the miogeosyncline and then precipitated with carbonate and other materials in the Great Blue depocenter (see Bissell, 1959b, p. 176–182).

Monotonous thick to massive-appearing limestone comprises the Great Blue above the Long Trail Shale in the southern Oquirrh Mountains. However, most units are thinly laminated to

FIG. 3.—View facing north into Mercur Basin and the old townsite of Mercur. Units are as follows: Mlm, lower member of Great Blue Formation; Mlt, Long Trail Shale Member of Great Blue; Mum, upper member of Great Blue; MPmc, Manning Canyon Formation; IPoq, Oquirrh Formation.

thin-bedded but weather to massive-appearing ledges. Limestone in the upper member is medium and dark gray to blue-gray and gray-blue, and is micritic (i.e., less than approximately 0.031 mm diameter; usage of Leighton and Pendexter, 1962, and of Bissell and Chilingar, 1967); many units are argillaceous (quartz silt and clay-size materials), and contain substantial quantities of iron monosulfide. Limestone in uppermost beds of this member is thinly laminated to thin-bedded, contains much admixed silt-size quartz, is very cherty, and contains iron monosulfide and cryptocrystalline varieties of pyrite and/or marcasite. A similar stratigraphic section of the Great Blue Formation crops out along the west side of Lake Mountain, a few kilometers southeast of the Oquirrh Mountains; it measures 780 m in thickness.

Outcrops of the Great Blue Formation in localities of the south-central Wasatch Mountains near Provo, Utah (Figs. 1, 4, and 5), have provided us with much of the data relating to deep-water limestone sedimentation. Baker (1947) reports that total thickness of the Great Blue Limestone is 860 m, as measured at the head of Rock Canyon, a few miles northeast of Provo. It is conformably underlain by the Humbug Formation, and conformably overlain by the Manning Canyon Shale. These formations are exposed along the west side of Mt. Timpanogos, and in American Fork Canyon farther north where a full section of the Great Blue Formation crops out; total thickness there is 860 m (Baker and Crittenden, 1961). A bed of black shale and some beds of rusty-weathering, fine-textured, dark gray, argillaceous (quartz silt and clay-size quartz) limestone are present at the base of the Great Blue in American Fork Canyon. This shale and limestone is in sharp contact with the underlying Humbug Formation. Above the Long Trail Shale Member in that area, the section of the upper member of the Great Blue aggregates about 400 m in thickness and consists of a monotonous, nearly homogeneous sequence of laminated to thin-bedded, dark-gray to black, argillaceous and cherty limestone with thin black shale interbeds. Dense, sublithographic (micritic) limestone containing much admixed silt- and clay-size quartz is typical. Dark-gray to black chert bands and stringers are abundant, arranged parallel to the bedding as distinct beds. Outcrops we studied in American Fork Canyon are best displayed on the north and south side of the canyon, approximately 1 km east of its mouth (Figs. 6 and 7).

Full thickness of the Great Blue Formation is about 860 m in the southern Wasatch Mountains (Brady, 1965, p. 31). Here it is characteristically laminated to thin-bedded, dark-gray to blue and black, argillaceous and cherty limestone, with some orthoquartzite and shale interbeds. Morris

Fig. 4.—Outcropping thin-bedded to laminated argillaceous, impure limestone of the uppermost Great Blue Formation, south of the mouth of Provo Canyon, south-central Wasatch Mountains. Cliff is approximately 45 m high.

FIG. 5.—Outcropping thin-bedded to laminated argillaceous, impure limestone of the uppermost Great Blue Formation, north side of Provo Canyon, approximately one km east of its mouth. Cliff exposes approximately 70 m of strata.

FIG. 6.—View of part of the Great Blue Formation on the south wall of American Fork Canyon approximately one km east of the mouth; approximately 650 m of strata are shown dipping steeply to the south. These are laminated to thin-bedded, argillaceous, impure limestone and minor amounts of shale interbeds.

FIG. 7.—Laminated to thin-bedded, dark gray to blue-black, argillaceous limestone of the Great Blue Formation near road level in American Fork Canyon, directly north of locality shown in Figure 6. Geology hammer in circle.

and Lovering (1961) dropped the name limestone and applied the name Great Blue Formation for outcrops mapped in the East Tintic Mountains (see Fig. 1). They also named four new members (p. 107-113) which are, in ascending order, the Topliff Limestone Member, the Paymaster Member, the Chiulos Member, and the Poker Knoll Limestone Member. Aggregate thickness of the four members, each measured where it is best exposed, is about 750 to 760 m. Lithology of the Great Blue Formation is essentially identical to that in the other sections we studied, in that micritic argillaceous (siliceous) and cherty dark-gray to blue-gray limestone disposed in laminated to thin beds with interbedded shale and orthoquartzite characterizes the formation.

The Sheeprock Mountains and Onaqui Mountains lie northwest of the Tintic mining district (Fig. 1), and were studied by Cohenour (1959), who noted that the Great Blue Formation is approximately 1,350 m thick. Cohenour (1959, p. 90-93) divided this formation into a lower limestone, a middle Chiulos Shale Member, and an upper limestone. Lithology of the Great Blue Formation is very similar to that in the Tintic district with the exception that possibly more sandstone interbeds are present. We interpret this increase in percentage of quartz sand in interbedded units as a function of closer proximity to the source area (see Bissell, 1962, Fig. 2, p. 30).

Northwesternmost outcrops of the Great Blue Formation studied for this project are found in the southern Lakeside Mountains, north of the Sheeprock Mountains and Onaqui Mountains (see Fig. 1). Young (1955) mapped the southern Lakeside Mountains, and estimated the thickness of the Great Blue at 660 to 990 m; he indicated that the formation consists of thin-bedded to massive, cherty, medium to dark gray limestone that weathers blue-gray. We agree, adding that black argillaceous micritic limestone is most typical, being arranged in laminated to thin-bedded units. Minor shale and orthoquartzite interbeds are present.

Within the approximately 7,500 sq km area where outcrops of the Great Blue Formation were studied (Fig. 1), the dominant lithology is fine-textured (commonly micritic) limestone with minor interbeds of shale and orthoquartzite. Limestone is commonly medium to dark gray, blue-gray, gray-blue, or black, and silty or argillaceous (silty size and clay size quartz fragments). Most beds of limestone contain substantial quantities of cryptocrystalline varieties of pyrite and/or marcasite (X-ray studies), and cryptocrystalline material tentatively identified as iron monosulfide (diffractometer studies). Where these iron-containing compounds are abundant in the limestone, the rock is dark gray to black. Chert is commonly present, and is disposed in blebs, nodules, stringers, and thinly laminated to thin-bedded units interstratified in the limestones. Fossils are rare, particularly in sections we studied in the Oquirrh Mountains and in the south-central Wasatch Mountains; where present, thin shelled depauperate forms of chonetid brachiopods are the dominant taxa. Thin-bedded coral-bearing limestone is located approximately midway in the stratigraphic section in American Fork Canyon. However we interpret this coral-bearing limestone as allochthonous, having been formed on shelf areas bordering the Great Blue depocenter to the east and subsequently transported basinward by sediment gravity flows into deeper marine waters. This interpretation is justified because but few isolated exotic blocks were found within shale units that contain the trace fossils *Scalarituba missouriensis* Weller; furthermore, bedding within these shale units was disturbed where the exotic allochthonous limestone blocks were gravity driven into the depocenter. Stratification of authochthonous limestone above and below the shale units varies from thinly laminated to thin- and thick-bedded. Massive-appearing units commonly display fine laminations. Figures 8–11 inclusive illustrate some of the strata in the uppermost 100 m of the Great Blue Formation from outcrops in the south-central Wasatch Mountains.

PETROGRAPHY

Thin sections reveal that carbonate samples we collected from the Great Blue Formation are on the limestone side of the carbonate fence (Bissell and Chilingar, 1967). This limestone is about 65 percent calcite (about the average of all samples); the remaining 35 percent consists of from 25 to 30 percent quartz silt and clay-size quartz and chert, and from 5 to 10 percent of iron sulfide (monosulfide and disulfide, as determined by chemical and X-ray studies). Numerous thin sections were prepared from carbonates of the Great Blue Formation taken in the Oquirrh Mountains, American Fork Canyon, and the south-central Wasatch Mountains. A number of 2 × 3 inch thin sections were prepared from the thinly laminated carbonates (Fig. 12), and under the microscope these thin layers display interlaminated micrite, calcisiltite, quartz siltstone, thin clay layers, and dark gray to black amorphous materal tentatively identified as iron monosulfide and iron disulfide (Fig. 12). Silt-size quartz fragments are subangular to subround; some are arranged in thin laminae, whereas others are scattered at random throughout the micrite, calcisiltite, and iron sulfide-rich lime mud (Fig. 13). Microfossils are rare to absent in the limestone that we regard as deep-water in origin. However, Pinney (1965, p. 38) obtained conodonts from the upper member of the Great Blue at the head of Rock Canyon, northeast of Provo, Utah; he identified *Cavusgnathus unicornis, Gnathodus texanus,* and *Neoprioniodus scitulus* from these samples. They were regarded as of Chesterian age.

Thin sectioned chert contains no radiolarians; this may reflect insufficient sampling rather than total absence of these microfossils. Scanning electron photomicrographs similarly reveal no microfossils in the few samples we tested. Although we were not eclectic in sampling limestone and chert of the Great Blue for laboratory study, we believe a fair representation of lithic types was studied. Our thin sections illustrate the microlithologies which, in turn, reflect lithic features seen in outcrop (Figs. 12 and 13).

Scores of insoluble residues were prepared on samples of the Great Blue Formation; average value of calcium carbonate in these samples is 65 percent. In all instances, after a few days digestion in weak HCl, a dark gray to black scum (resembling soot) rose to the top of the acid. This proved to be partly cryptocrystalline pyrite and/or marcasite (X-ray study), and partly what we tentatively consider to be iron monosulfide (diffractometer study). Insoluble residues in the bottom of the beakers consist of silt size and clay size quartz, chert, and some iron sulfide.

Figure 12a is typical for hand specimens of units of the Great Blue Formation in American Fork Canyon and near the mouth of Provo Canyon (see Figs. 8–11); Figure 12b is a representative slice cut from the area embraced between the 2 cm and 6 cm mark of the hand specimen of Figure 12b. This slice reveals parallel to subparallel laminae and microlaminae of carbonate (mostly calcite), with interlaminated sulfide-rich lime mud (micrite), and admixed silt- and clay-size quartz grains. It contains no fossils; this is the same as the calcareous flysch of the Mississippian marine geosynclinal environment of Nevada as described by Poole (1974), who followed the definition of flysch proposed by Hsü (1970). Our Figure 13 consists of photomicrographs of other representative samples of limestone of the Great Blue Formation from the Oquirrh Mountains and the south-central Wasatch Mountains. Some display poor sorting of the "dumped-in" so-called turbidites, whereas others show lamination and microlamination of sulfide-rich argillaceous limestone. Accordingly, we believe that the microfacies shown in our thin sections and hand specimens support our thesis that the limestone is deep-water marine, possibly to bathyal depth. This latter interpretation refers mostly to the eastern part of the Great Blue basin in central Utah.

Fig. 8.—Thin-bedded, flysch-like carbonate of the upper member of the Great Blue Formation, west side of highway, Provo Canyon (Sec. 5, T. 5 S., R. 3 E.). Height of exposure is approximately 30 m.

Fig. 9.—Close-up of flysch-like carbonates of the upper member of the Great Blue Formation in Provo Canyon (Sec. 5, T. 5 S., R. 3 E.). Slump folds shown in lower right of picture.

Fig. 10.—Details of laminated to thin-bedded, cherty and argillaceous limestone of the upper member of the Great Blue Formation near Olmstead, Provo Canyon (Sec. 7, T. 5 S., R. 3 E.). Geology hammer is at left center of photograph.

Fig. 11.—Close-up of Figure 10.

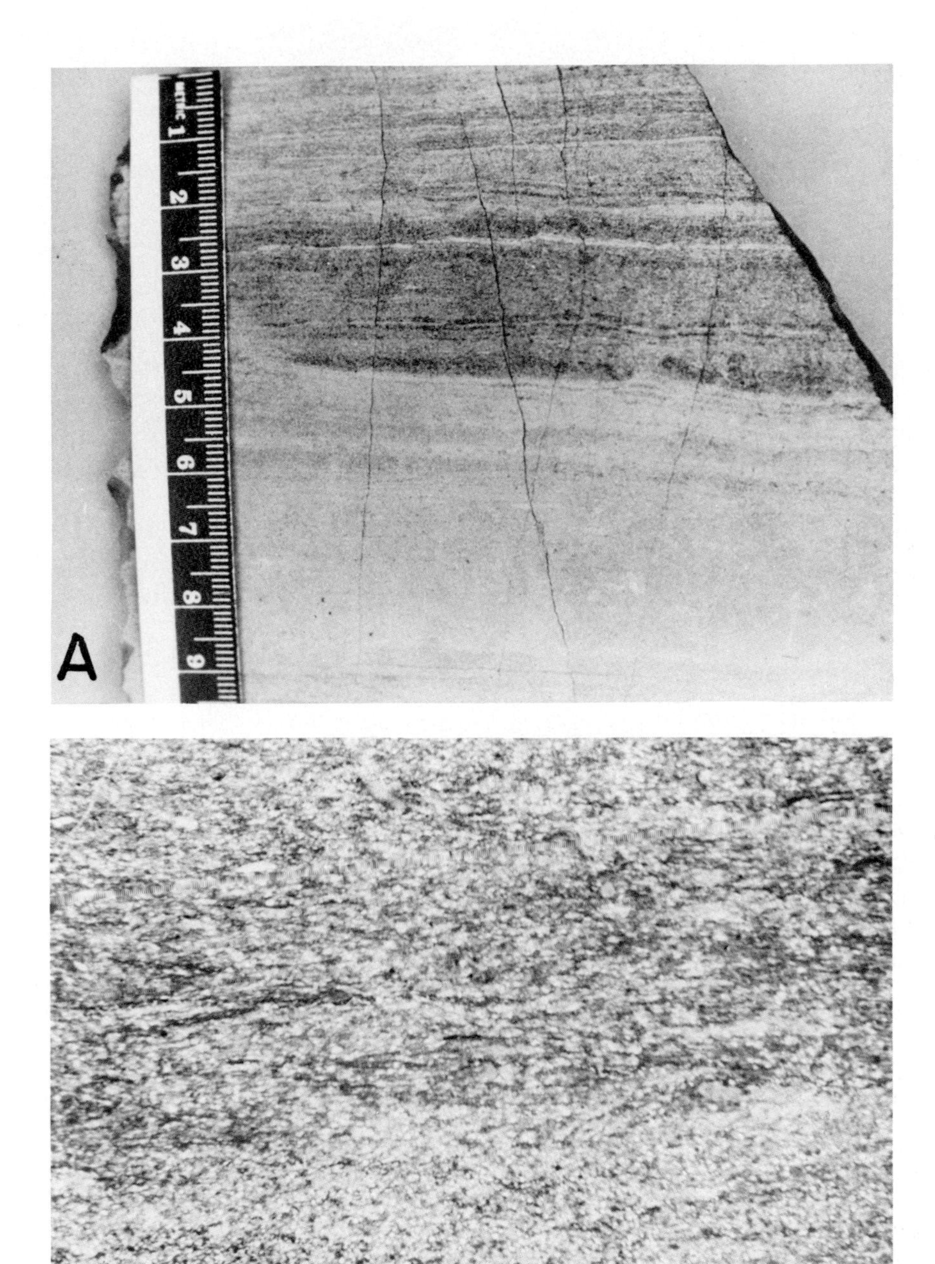

FIG. 12.—*A*—Flysch-like laminae of a sample of the upper member of the Great Blue Formation. *B*—Low-magnification photomicrograph of a thin section taken in the area embraced between the 4 cm and 5 cm bars in Figure 12A. Dark laminae in both illustrations are richer in iron sulfides than light-colored laminae. From outcrops in Provo Canyon near Olmstead (Sec. 7, T. 5 S., R. 3 E.). Bar in photomicrograph is 1 cm long.

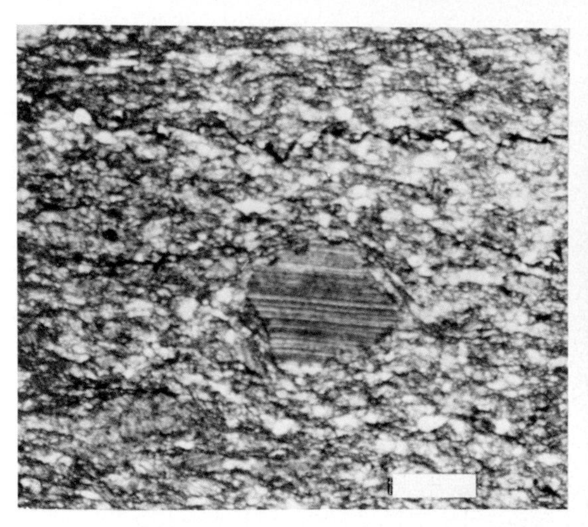

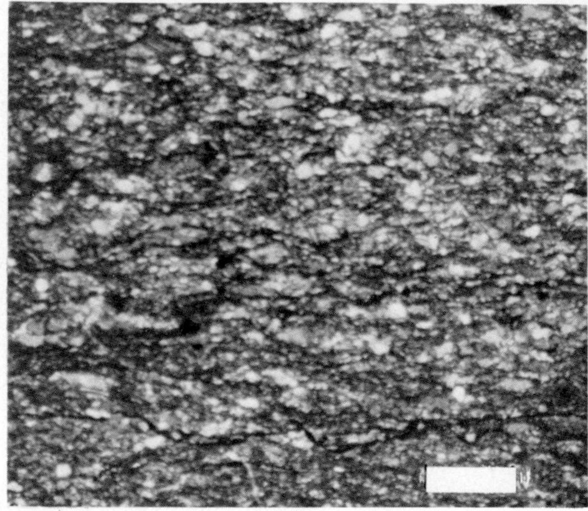

FIG. 13.—Photomicrographs of two representative samples from the Great Blue Formation in central Utah. *Left*—Upper member in the southern Oquirrh Mountains. *Right*—Upper member in American Fork Canyon. Most particles in both thin sections consist of limeclasts and micrite, with darker bands composed of argillaceous (silt- and clay-size) quartz. Large clast in left photomicrograph is washed-in limeclast; stylolite is above it. Length of bar in each photomicrograph is 1 mm.

DEPOSITIONAL ENVIRONMENT

Evidence for deep-water marine deposition of limestone of the Great Blue Formation in the eastern part of the Cordilleran miogeosyncline during Late Mississippian time include the following: (1) areally extensive distribution of mostly homogeneous dark colored (blue to black), argillaceous (silt-size and clay-size quartz) limestone that formed in a sedimentary basin (Great Blue Basin, Chesterian forerunner to the Oquirrh Basin); (2) within this thousand meter-thick limestone, a monotonous repetition of laminated to thin- and medium-bedded strata is characteristic. Most thick-bedded to massive-appearing outcrop units prove to be thinly laminated to thin-bedded when slabbed or sectioned. (3) Limestone contains substantial amounts of iron monosulfide, and cryptocrystalline varieties of pyrite and/or marcasite; (4) most of the formation in outcrops near Provo, Utah, is sparsely fossiliferous to nonfossiliferous; forms present include a depauperate fauna of thin-shelled chonetid brachiopods, and some conodonts. (5) Strata range from millimeter thin to medium-thick beds arranged in flysch-like succession; some strata display graded bedding. The term "flysch" as used herein corresponds to what Poole (1974, p. 58) termed calcareous flysch for impure carbonate rocks interbedded in fine-textured flysch and flysch-like sediments of Late Mississippian age in the Antler foreland basin that lay 300 km west of the Great Blue Basin (see Fig. 2, this paper, also Poole, 1974, Fig. 22, p. 77). (6) Some of the shale members contain the trace fossils *Scalarituba missouriensis* Weller and *Nereites* sp., forms common to flysch and flysch-like sediments of Late Mississippian age of the Antler basin of Nevada, and also common to alleged deep-water accumulations of Mississippian and Pennsylvanian rocks elsewhere. Poole (1974) stressed their significance in suggesting deep-water (possibly bathyal) environment for the Antler foreland basin. They are also found in rocks interpreted as deep-water marine accumulations (Chamberlain and Clark, 1973) of the Pennsylvanian and Permian in the Oquirrh Basin of Utah. (7) Locally, such as in outcrops of the Great Blue Formation near the head of American Fork Canyon of the south-central Wasatch Mountains, coral-bearing limestone occurs as exotic blocks in shale; these are not *in situ* beds, but are interpreted to have originally accumulated in shoal waters along part of the eastern perimeter of the Great Blue basin, and were redeposited as gravity-driven blocks that moved downslope into deeper marine waters. (8) Soft-sediment deformation in the form of a slump folds occur in some outcrops of the Great Blue Formation (Figs. 11, 14); admittedly, these can also form in other environments, but taken together with other evidence cited above, a deep-water origin for these slumps is plausible.

We propose a sedimentary model that involves paleoslopes leading into a bathyal environment to account for deposition of most limestone of the Great Blue Formation (see Fig. 1). It is our interpretation that various positive and shallow

FIG. 14.—Slump folds in limestone of the upper member of the Great Blue Formation near Olmstead, Provo Canyon (Sec. 7, T. 5 S., R. 3 E.). No evidence was found suggesting crustal rather than soft-sediment deformation. If correctly interpreted, these folds formed within the lime muds and other limestone while they were still soft and plastic as they moved downslope from an easterly source into deeper marine realms.

submarine features provided carbonate and siliceous silt and clay detritus to the Great Blue depocenter. These provenances include the Antler orogenic belt on the west in Nevada, northeast Nevada highland in northeastern Nevada and northwestern Utah, the cratonic shelf on the east and northeast, and the Emery uplift and Uncompahgre uplift on the southeast (see Bissell, 1962, Fig. 2, p. 30). No firm data are presently available regarding inclination of paleoslopes leading into the Great Blue Basin. Bissell (1974, Fig. 1, p. 85) presented an isopach map of the Mississippian System in central and western Utah and adjacent eastern Nevada and southern Idaho. Close spacing of contours from this map is noticeable near the Las Vegas-Wasatch line east of the Great Blue depocenter, and slightly more pronounced bunching of contours seemingly is evinced near a highland and/or submarine swell in western Utah. Our Figure 2 (adapted from Poole, 1974, Fig. 22, p. 77, and top of Fig. 25, p. 79) lends credence to the interpretation that these closely space contours represents the position of the paleoslope. On the basis of the outcropping sections of the Great Blue Formation that we studied (Fig. 1), we envision paleoslopes of less than a degree for perhaps 20 km attaining a total depth of roughly 60 m along the western side of the Great Blue depocenter, whereas slightly greater paleoslopes typified the basin to the east and southeast.

Poole (1974, p. 78-79) suggests that ". . . most likely, continental margin deformation during the Antler Orogeny warped the continental crust under the shelf and initiated sites for subsequent major susidence and uplift; continued orogenic compressive stress that was directed continentward resulted in a general eastward shift in sites of thick sedimentation during the Carboniferous." It was probably in such an area of subsidence that the Great Blue depocenter formed. We interpret the Great Blue Basin to have been a downwarped segment along a part of the eastern side of the Cordilleran miogeosyncline, rather than a taphrogenic basin or intracratonic basin. Because of its close proximity to the Las Vegas-Wasatch line to the east, the possibility is considered that the Great Blue depocenter was, in a sense, a foreland basin. This is not to be construed that it behaved like the Antler foreland basin of Nevada that lay east of an orogenic belt (see Poole, 1974) but rather that it experienced marked downwarping directly west of the Las Vegas-Wasatch line, a significant tectonic hinge during Carboniferous and Permian times. This interpretation is interjected as one of various explanations. Because the total thickness of the Mississippian, Pennsylvanian, and Permian Systems of the Great Blue Basin and Oquirrh Basin, aggregates more than 10,000 m, this downwarped part of the Cordilleran miogeosyncline experienced *hypersubsidence* (see Bissell and Chilingarian, 1975). Patterns of depocenters of sediment accumulation and of positive features that were established in the Late Mississippian, continued with slight modification into Pennsylvanian time for the Oquirrh Basin (see Bissell, 1962, Fig. 2, p. 30, Fig. 3, p. 37; 1974, Fig. 2, p. 87; Bissell and Chilingarian, 1975, Fig. 4-6, p. 187).

The Antler orogenic belt was probably an important provenance for much of the argillaceous and some impure calcareous sediment in the Great Blue Basin, as herein interpreted. Fine-grained detritus derived from that source area cascaded downslope into the Great Blue depocenter. In referring to this proto-Oquirrh Basin, Poole (1974, p. 74) stated: "By very late Mississippian time, the westerly derived clastic sediments filled the foreland trough and spread eastward across the limestone shelf onto the cratonic platform." That which Poole alludes to as the limestone shelf (his proto-Oquirrh Basin) is the Great Blue Basin of this report, and is therefore part of the eastern portion of the Cordilleran miogeosyncline (see Fig. 2, this report). We have also noted that the shelf and craton east of this depocenter, and the Emery and Uncompahgre uplifts to the southeast, likely also must have been provenances for tremendous quantities of carbonate and silt-size and clay-size argillaceous and quartzose detritus. It is our belief that the source areas along the eastern side of the Cordilleran miogeosyncline did not account for the iron monosulfide and iron disulfide nor the volcanic material (now chert in the Great Blue Formation); rather they also were derived form the Antler orogenic belt. Iron-bearing compounds derived from various rocks of the Antler belt (see Stewart and Poole, 1974; and Poole, 1974) were carried eastward across the Cordilleran miogeosyncline in solution and colloidal suspension to the Great Blue depocenter where sufficient organic matter was present to aid in deposition of iron compounds. Iron so derived would be precipitated, either directly as black colloidal hydrous ferrous sulfide by reaction with hydrogen sulfide liberated by bacteria from decaying organic matter, or as ferric hydroxide by iron bacteria. We believe that in the decaying, "cesspool-like" conditions of stagnant deep waters of the basin, with iron bacteria in the presence of hydrogen sulfide under reducing conditions, iron-bearing compounds were changed to hydrated ferrous monosulfide, and also to cryptocrystalline varieties of pyrite and/or marcasite. Upon losing water and with the addition of sulfur present in the mud, the monosulfide mineral, melnikovite, may have formed although we have not specifically identified this mineral. Melnikovite or another iron monosulfide could alter to pyrite and/or marcasite.

The presence of about 1,400 m of carbonate, shale, and orthoquartzite in the Great Blue Formation in the western part of the depocenter (Cohenour, 1959), and of a few eastward-thinning tongues of shale and clastics in the formation in the Oquirrh Mountains, suggests an additional sediment source, namely, the western Utah highland (Bissell, 1962, Fig. 2, p. 30). Portions of the Great Blue Basin received substantial thicknesses of orthoquartzite, sandstone, and thin skeletal limestone where proximal to the source areas. Some of the coral-bearing limestone in the Great Blue Formation that formed along edges of submarine escarpments became detached and moved downslope in a westerly direction, becoming resedimented as exotic blocks in dark shale.

It is not our intent to relate the evolution of the Great Blue depocenter to the plate tectonic model as the main thrust of our paper. However, we concur with Poole (1974, p. 78) that: "The geological evolution of western United States can be explained by plate-tectonic models even though many features remain obscure." He continued: "The east-directed overthrusting may have been related to compressive stresses transmitted continentward from the underthrusting oceanic plate which resulted in partial closure of the margin ocean basin. . . ." In conclusion Poole (1974, p. 78) made note that ". . . the origin of the Upper Devonian and Mississippian foreland trough and flysch deposits in the western United States is related to the interaction of oceanic and continental plates along the Paleozoic continental margin." We applaud this conclusion; accordingly, the evolution of the Great Blue depocenter (the proto-Oquirrh Basin) seemingly was related in one way or another with the geologic history of the Antler foreland basin farther west. Both basins, without doubt, were integral parts of the late Paleozoic Cordilleran miogeosyncline. Therefore, any discussion of the history of this major downwarped part of the earth's crust during Mississippian, Pennsylvanian and Permian times must take into account plate tectonic models.

In discussing a late Precambrian continental separation to initiate intitial downwarping and thus formation of the Cordilleran geosyncline, Stewart (1972, p. 1355-1356) stated: "A continental separation appears to be a likely explanation for the origin of the Cordilleran geosyncline in western North America." Later, Stewart and Poole (1974) discussed the evolution of the Cordilleran miogeocline of the Great Basin of the western United States during uppermost Precambrian and lower Paleozoic times. They stated (p. 28): "The continental margin along which the early Paleozoic and latest Precambrian miogeocline was constructed apparently developed by rifting in late Precambrian time (< 850 my). Extensional faulting and flowage related to this rifting continued well into the continent and may have caused major crustal thinning as far east as the Wasatch line, across which the rate of eastward thinning of uppermost Precambrian and Paleozoic increases markedly." In the same publication Bissell (1974, p. 83-87) presented isopachous maps for the Mississippian, Pennsylvanian, and Permian demonstrating the same fundamental pattern of sediment thickening westward from the Las Vegas-Wasatch line into the late Paleozoic miogeosyncline. In the same context Poole (1974, p. 58-82) discussed the Antler foreland basin that was present during Mississippian time between the Great Blue area on the east and the Antler orogenic highland on the west (see Poole, 1974, Fig. 22, p. 77; also, Fig. 2 this paper). Poole (1974, p. 74) stressed the fact that by very late Mississippian time the westerly derived clastic sediments filled the Antler foreland trough and spread eastward into a carbonate depocenter (our Great Blue Basin).

Seemingly, plate separation and plate collision have been the rubic of the Cordilleran geosyncline throughout its history as a miogeocline (late Precambrian through at least late Devonian), and later as a miogeosyncline (Mississippian through early Triassic). Accordingly, rifting, attended by crustal thinning that extended as far east as the Wasatch line, set the stage for the initiation and development of the Great Blue Basin, a deep water sedimentary basin in which flysch-like carbonates of the Great Blue Formation accumulated. Poole (1974, p. 80) illustrates western North America with the zone of subduction for middle Devonian-early Mississippian times along the western margin of the continent, and the Roberts Mountains thrust from the Antler orogenic belt as an easterly translated plate during the late Devonian. Because the Great Blue Basin was directly adjacent on the west of the Wasatch line (a tectonic hinge), downwarping of the earth's crust at this place in the Cordilleran miogeosyncline (possibly initiated by a certain amount of separation) created a deep-water marine sedimentary basin. Through late Mississippian time, this basin progressively filled; at times and places sediment accumulation was that of flysch-like impure limestone. Ultimately, shoal water environments dominated after accumulation of from 750 to 1,400 m of limestone, orthoquartzite, siltstone, and shale. Finally, the Late Mississippain-Early Pennsylvanian Manning Canyon Formation accumulated in neritic environments bordered by estuaries, swamps, and prograding deltas.

ACKNOWLEDGEMENTS

We thank Kenneth Hamblin and Morris Petersen who cirtically read the manuscript and offered valuable suggestions to improve the paper. We are particularly grateful to the co-editors of this volume who devoted much time in an objective critique of our manuscript. Of course, we accept responsibility for our conclusions.

REFERENCES

Baker, A. A., 1947, Stratigraphy of the Wasatch Mountains in the vicinity of Provo, Utah: U.S. Geol. Survey Oil and Gas Inv. Chart 30.

—— and Crittenden, M. D., Jr., 1961, Geology of the Timpanogos Cave Quadrangle, Utah: U.S. Geol. Survey Geol. Quad. Map GQ 132.

Bissell, H. J., 1959a, Stratigraphy of the southern Oquirrh Mountains, Utah—Upper Paleozoic succession: *In* H. J. Bissell (ed.), Geology of the southern Oquirrh Mountains and Fivemile Pass-northern Boulter Mountain area, Tooele and Utah Counties, Utah: Utah Geol. Soc. Guidebook 14, p. 93-127.

——, 1959b, Silica in sediments of the upper Paleozoic of the Cordilleran area; *In* H. A. Ireland (ed.), Silica in sediments: Soc. Econ. Paleontologists and Mineralogists, Spec. Pub. 7, p. 150-185.

——, 1962, Pennsylvanian-Permian Oquirrh basin of Utah: Brigham Young Univ. Geology Studies, v. 9, pt. 1, p. 26-49.

——, 1974, Tectonic control of late Paleozoic and early Mesozoic sedimentation near the hinge line of the Cordilleran miogeosynclinal belt: *In* W. R. Dickinson (ed.), Tectonics and sedimentation: Soc. Econ. Paleontologists and Mineralogists, Spec. Pub. 22, p. 83-97.

—— and Chilingar, G. V., 1967, Classification of sedimentary carbonate rocks: *In* G. V. Chilingar, H. J. Bissell and R. W. Fairbridge (eds.), Carbonate rocks—Origin, occurrence and classification: Elsevier Pub. Co., Amsterdam, The Netherlands, p. 87-168.

—— and Chilingarian, G. V., 1975, Subsidence: *In* G. V. Chilingarian and K. H. Wolf (eds.), Compaction of coarse-grained sediments, I: Elsevier Pub. Co., Amsterdam, The Netherlandsa p. 167-245.

Brady, M. J., 1965, Thrusting in the southern Wasatch Mountains, Utah: Brigham Young Univ. Geol. Studies, v. 12, p. 3-53.

Chamberlain, C. K., and Clark, D. L., 1973, Trace fossils and conodonts as evidence for deep-water deposits in the Oquirrh basin of central Utah: Jour. Paleontology, v. 47, p. 663-682.

Cohenour, R. E., 1959, Sheeprock Mountains, Tooele and Juab Counties, Utah: Utah Geol. and Mineralog. Survey, Bull. 63, 201 p.

Emmons, S. F., 1877, Region south of Salt Lake, Oquirrh Mountains: U.S. Geol. Exploration 40th Parallel, v. 2, p. 443-456.

Folk, R. L., 1959, Practical petrographic classification of limestones: Am. Assoc. Petroleum Geologists Bull., v. 43, p. 1-38.

Gilluly, James, 1932, Geology and ore deposits of the Stockton and Fairfield Quadangles, Utah: U.S. Geol. Survey, Prof. Paper 173, 171 p.

Hsü, K. J., 1970, The meaning of the word flysch—A short historical search: *In* Jean Lajoie (ed.), Flysch sedimentology in North America: Geol. Assoc. Canada, Spec. Paper 7, p. 1-11.

Leighton, M. W., and Pendexter, C., 1962, Carbonate rock types: *In* W. E. Hamm (ed.), Classification of carbonate rocks, a symposium: Am. Assoc. Petroleum Geologists, Tulsa, Oklahoma, p. 33-61.

Morris, H. T., and Lovering, T. S., 1961, Stratigraphy of the East Tintic Mountains, Utah: U.S. Geol. Survey, Prof. Paper 361, 145 p.

Pinney, R. I., 1965, A preliminary survey of Mississippian biostratigraphy (conodonts) in the Oquirrh basin of central Utah: Ph.D. Dissert., Univ. Wisconsin, Madison, Wisconsin, 189 p.

Poole, F. G., 1974, Flysch deposits of the Antler foreland basin, western United States: *In* W. R. Dickinson (ed.), Tectonics and sedimentation: Soc. Econ. Paleontologists and Mineralogists, Spec. Pub. 22, p. 58-82.

Spurr, J. E., 1895, Economic geology of the Mercur Mining District, Utah: U.S. Geol. Survey, Sixteenth Ann. Rept., pt. 2, p. 343-455.

Stewart, J. H., 1972, Initial deposits in the Cordilleran geosyncline: Evidence of a late Precambrian (<850 my) continental separation: Geol. Soc. America Bull., v. 83, p. 1254-1360.

—— and Poole, F. G., 1974, Lower Paleozoic and uppermost Precambrian Cordilleran miogeocline, Great Basin, western United States: *In* W. R. Dickinson (ed.), Tectonics and sedimentation: Soc. Econ. Paleontologists and Mineralogists, Spec. Pub. 22, p. 28-57.

Young, J. C., 1955, Geology of the southern Lakeside Mountains, Utah: Utah Geol. Mineralog. Survey, Bull. 56, 116 p.

SEPM Special Publication No. 25, p. 187-201, November 1977

TRANSITION FROM DEEP- TO SHALLOW-WATER CARBONATES, PAINE MEMBER, LODGEPOLE FORMATION, CENTRAL MONTANA

D. L. SMITH
Montana State University, Bozeman, 59715

ABSTRACT

Limestone and dolomite of the Lodgepole Formation of the northern Rocky Mountains and adjacent plains document a shallow-deep-shallow environmental fluctuation during the initial marine transgression of the Mississippian. Accumulation of this carbonate sequence was strongly influenced by two paleotectonic elements: a relatively inactive shelf in southern Montana and northern Wyoming and a broad, elongate, unstable shelf in central Montana, extending from the western craton margin into the Williston Basin. On the unstable shelf, the Lodgepole Formation unconformably overlies Devonian strata and is subdivided into three members, in ascending order, the Cottonwood Canyon (0-20 m), the Paine (50-70 m), and the Woodhurst (100-200 m). Detailed petrographic, sedimentologic, and stratigraphic studies of these units suggest that deposition occurred in both deep and shallow water.

Dark shale and siltstone of the Cottonwood Canyon Member are interpreted as shallow-water lagoonal sediments, deposited in embayments at the advancing margin of the Lodgepole sea. An abrupt vertical change to carbonate deposition is marked by a thin, widespread, glauconitic and bioclastic wackestone bed at the base of the overlying Paine Member. This bed represents a transitional depth phase of the Lodgepole sea in central Montana and is interpreted to have been deposited in water depths near effective wave base. Following these phases of relatively shallow-water deposition, increased rates of subsidence, coupled with continued transgression, promoted the accumulation of dark, argillaceous, rhythmically thin-bedded strata characteristic of most of the Paine Member. Lime-mud bioherms interbedded with these dark carbonate strata indicate 50 m of depositional relief and suggest water depths of 70-100 m for most of the Paine Member. Overlying the dark Paine carbonates are light-colored grainstone cycles of the Woodhurst Member, products of fluctuation of shallow, agitated environments. These carbonates mark the end of deeper water sedimentation and the beginning of progradation of shallow-water carbonate environments across the central Montana unstable shelf.

INTRODUCTION

Striking stratigraphic and sedimentologic changes in carbonate strata of the Lodgepole Formation in central Montana reflect the alternation of shallow- and deep-water depositional environments. However, unlike many well-documented deep-water continental margin or interior basin carbonates, Lodgepole strata accumulated on a broad, tectonically active, epicratonic shelf, extending from the Cordilleran seaway of eastern Idaho and western Montana into the Williston Basin of northeastern Montana and western North Dakota. The deeper water aspect of these carbonates was originally inferred by Andrichuk (1955) and was later more thoroughly documented by Wilson (1969, p. 14) who noted that Lodgepole carbonate strata "display characteristics so much in common with other limestones whose 'deeper water' origin can be demonstrated stratigraphically that their proper depth of deposition could have been in hundreds rather than in tens of feet." The intent of this paper is (1) to further document the occurrence of deep-water Lodgepole carbonate strata in central Montana, (2) to describe the temporal changes in Lodgepole depositional environments, and (3) to suggest a depositional model for carbonate sedimentation on tectonically active epicratonic shelves.

REGIONAL SETTING, MADISON GROUP

Depositional provinces.—In central Montana, the Lodgepole Formation makes up the lower half of the Madison Group (Sando and Dutro, 1974) which, with its lithic equivalents, constitutes a widespread carbonate blanket, deposited in a great variety of lithotopes during the initial marine transgression of the Mississippian. According to Sando (1967), this blanket accumulated in three depositional provinces: a stable shelf in Wyoming, an unstable shelf in Montana, and a miogeosyncline in Idaho (Fig. 1). On the north side of the Montana unstable shelf, earlier work suggests a second stable shelf province in Canada, similar to that in Wyoming (Edie, 1958). Each of these provinces is characterized by a strikingly different and distinct sedimentary and stratigraphic style. Stratigraphic and sedimentologic evidence suggests that transitions between these provinces were gradual and that, in contrast to many of the better-documented bank-to-basin margins, there were no abrupt shelf edges of sharp hingelines. Thus, the unstable-stable shelf transition in southern Montana conforms more closely to the carbonate ramp model as described by Ahr (1973) and Wilson (1974) than to the shelf-edge model.

Stratigraphic framework.—Stratigraphic rela-

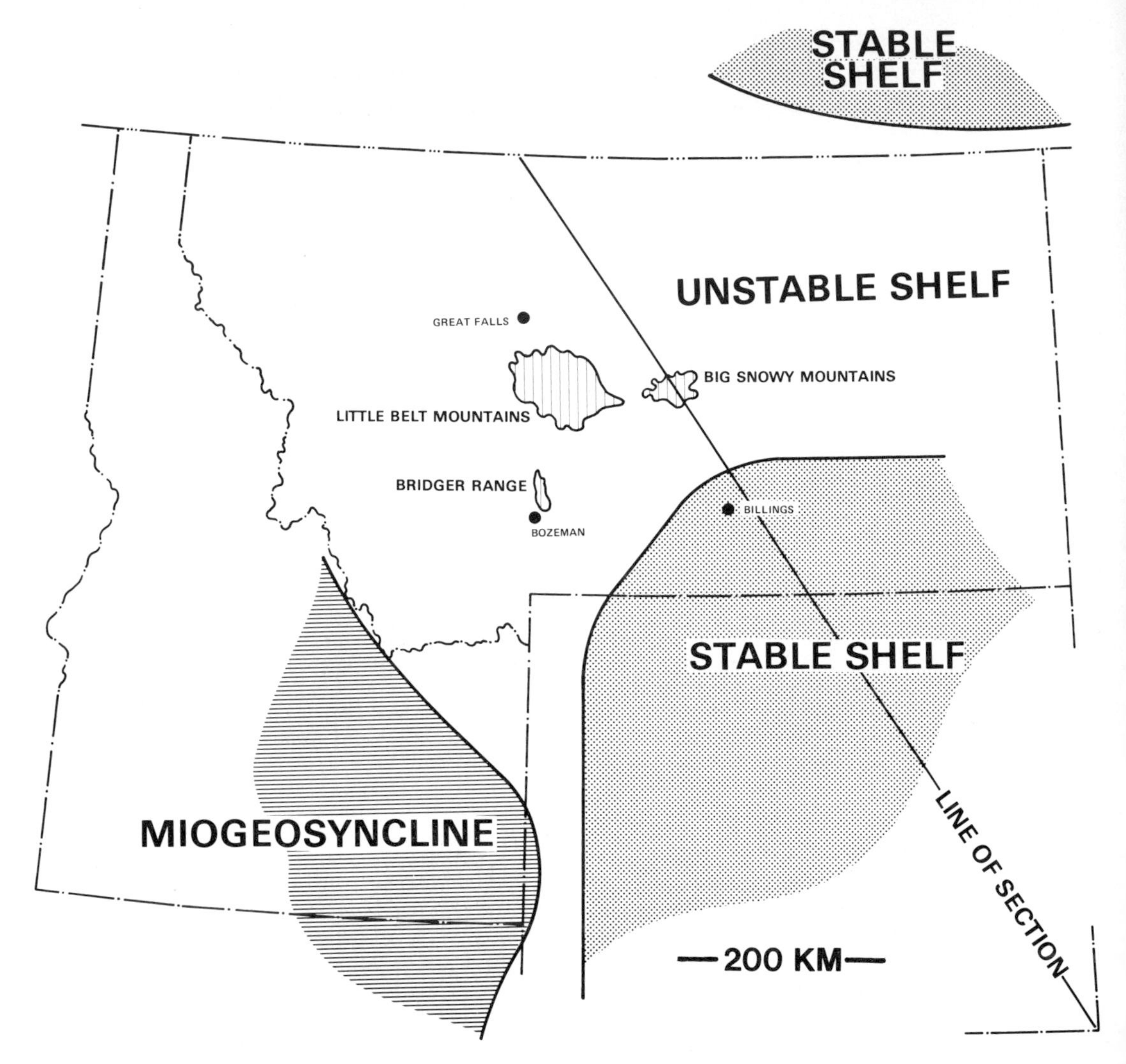

FIG. 1.—Location map, showing depositional provinces of Sando (1967), locations of mountain ranges discussed in text (deformed Precambrian and Paleozoic rocks outlined in vertically ruled pattern), and line of section used in subsequent figures.

tionships between Madison strata of the stable and unstable shelves are shown schematically in Figure 2. The Madison Limestone of Montana and Wyoming is a transgressive-regressive carbonate package, transgression of the sea onto the Montana unstable shelf beginning in Kinderhookian time, resulting in the deposition of the shallow-water Cottonwood Canyon and deeper water Paine members of the Lodgepole Formation. During Osagian time, this transgression continued progressively southward from Montana onto the Wyoming stable shelf. During this transgressive phase, shallow-water carbonates of the Woodhurst Member prograded progressively northward across the Montana unstable shelf, infilling an area that during Kinderhookian time had been the site of deeper water depositional environments. Accumulation of shallow-water carbonates continued during the remainder of Osagian time and into the Meramecian with the deposition of the Mission Canyon and Madison formations on the unstable and stable shelves, respectively. Following this transgressive phase, the sea withdrew from the shelf areas of Montana and Wyoming, and a regional karst topography developed across the exposed Madison surface (Henbest, 1958).

On the stable shelf, the Madison Limestone

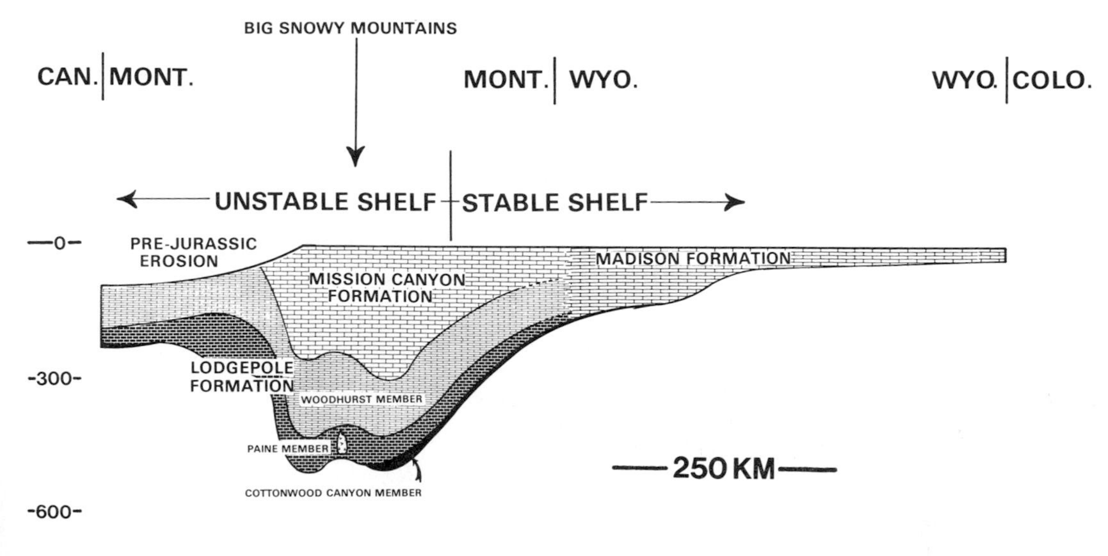

FIG. 2.—Cross section, showing stratigraphic relationships between the Madison Group of the Montana unstable shelf and the Madison Formation of the Wyoming stable shelf. Each rock-stratigraphic unit is designated by a distinctive brick or solid pattern; lime-mud bioherms of the Paine Member are indicated by small "v" pattern. Line of section is shown in Figure 1. Diagram constructed from isopach-facies maps by Craig (1972).

is of formational rank and is characterized by less than 300 m of shallow-water limestone and dolomite. In contrast, the thicker Madison carbonate package of the unstable shelf is of group rank and includes the Lodgepole and Mission Canyon Formations. The Mission Canyon is a thick, massive, cliff-forming carbonate unit, characterized by light-colored, oolitic, bioclastic, micritic, and dolomitic carbonates, generally interpreted to have originated in shallow-water depositional environments of the unstable shelf. The underlying Lodgepole Formation is more thinly bedded and less resistant than the Mission Canyon and is subdivided on the basis of lithology and bedding style into three members, in ascending order: the Cottonwood Canyon, the Paine, and the Woodhurst (Fig. 3).

DEPOSITIONAL HISTORY, LODGEPOLE FORMATION

The initial incursion of the transgressing Madison sea in Kinderhookian time resulted in deposition of siltstone, black shale, and dolomitic shale of the Cottonwood Canyon Member (Fig. 3). These rocks were deposited over an irregular erosion surface in an embayment in southern Montana and northwestern Wyoming and are arranged in two facies tracts: a western siltstone and shale facies and an eastern dolomitic shale and siltstone facies (Baars, 1972, p. 98). Sandberg and Klapper (1967) interpreted the eastern dolomite facies as a shallow marine deposit at the leading edge of the advancing Madison sea; Rodriguez and Gutschick (1970) interpreted the clastic rocks of the western facies as a slightly deeper water deposit.

On the unstable shelf in Montana, the Cottonwood Canyon is abruptly overlain by dark, thin-bedded carbonates of the Paine Member (Fig. 3). This distinctive carbonate sequence represents a dramatic change from clastic to carbonate deposition and is characterized by dark, rhythmically bedded, argillaceous lime mudstone that locally envelops lime-mud bioherms. This dark lime mudstone with its associated bioherms is the principal focus of this paper and is here interpreted as a deep water accumulation, deposited on a downwarped shelf that developed in central Montana during the Kinderhook portion of the Madison transgression.

Overlying the Paine Member are cyclic, shallow-water deposits of the Woodhurst Member (Fig. 3), each cycle featuring a finer grained, nonresistant lower sequence of pellet, bioclastic, and micritic limestone that is capped by thicker, cross-stratified beds of coarse-grained, bioclastic and oolitic grainstone that form distinctive ledges. These cycles are interpreted to be products of the northward progradation and migration of shoal and intershoal depositional environments across the central Montana unstable shelf, filling in the

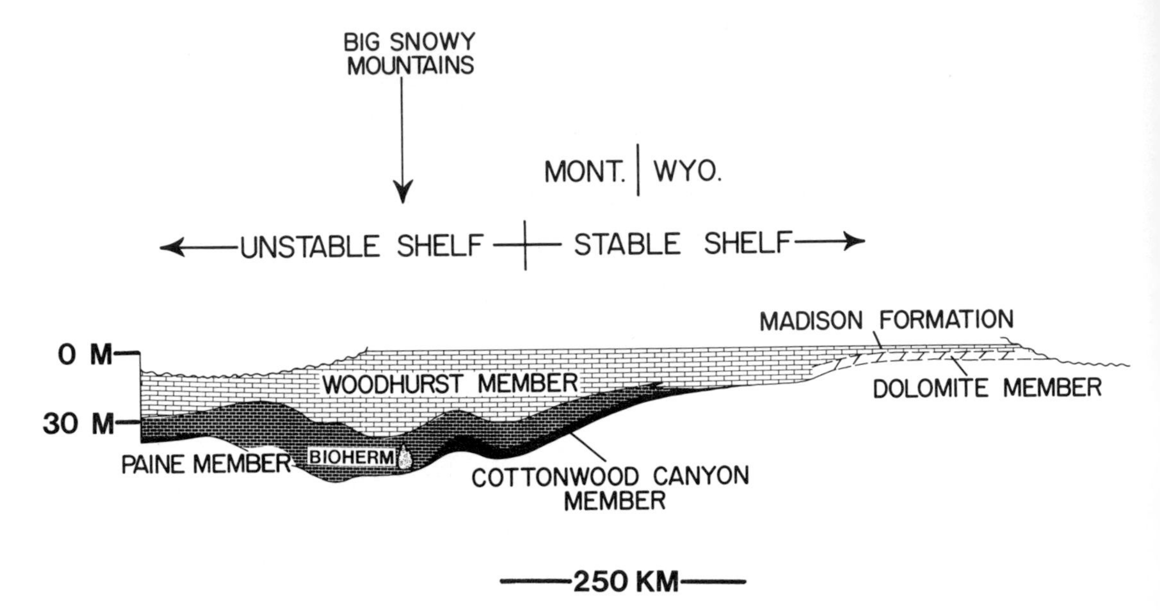

FIG. 3.—Cross section, showing stratigraphic relationships between the Lodgepole Formation of the unstable shelf and approximate age-equivalent strata of the stable shelf. Note near restriction of the Paine Member to the unstable shelf province. Line of section shown in Figure 1. Diagram modified from Craig (1972), Sando (1967), and Smith (1972).

deeper water trough in which the dark lime mudstone of the Paine Member accumulated earlier (Jenks, 1972; Smith, 1972).

PAINE MEMBER, LODGEPOLE FORMATION

Stratigraphy

The Paine Member of the Lodgepole Formation was originally described by Weed (1899) in the Little Belt Mountains of central Montana from outcrops of approximately 70 m of dark, thin-bedded, very fine-grained, argillaceous limestone at the base of the Mississippian carbonate sequence. This unit is recognizable in outcrops and the subsurface throughout most of the Montana unstable shelf but it thins and pinches out on the stable shelf in Wyoming (Fig. 3).

The Paine isopach in south-central Montana is dominated by four elements (Fig. 4): (1) an east-west thinning trend through the Big Snowy Mountains, coincident with the Devonian isopach zero and the Devonian central Montana high (Baars, 1972, p. 94); (2) thickening of the Paine north and south of this trend; (3) a general thickening westward toward the Cordilleran mobile belt; and (4) a gradual thinning toward the Wyoming stable shelf province. The irregular positive and negative behavior of the unstable shelf during a major portion of Kinderhook deposition is reflected in these isopach trends. During later Mississippian and early Pennsylvanian time, the strongly negative Big Snowy trough was to develop across the central Montana area (Craig, 1972; Mallory, 1972), resulting in thick, linear sedimentary sequences of the upper Madison and Big Snowy Groups.

Lithology

Rocks of the Paine Member in central Montana vary both vertically and laterally and, on the basis of lithology, fossil content, primary sedimentary structures, and stratigraphic position in the carbonate sequence, are subdivided into five fundamental lithologic units or facies (Fig. 5): (1) a basal facies of glauconitic wackestone, packstone, and grainstone, (2) a dark lime mudstone facies with rhythmic thin bedding, (3) a bioherm-core facies, (4) a bioherm-flank facies, and (5) a shallow-water facies. The vertical sequence of these lithologies is interpreted to be the product of the change from shallow-water depositional environments of the Cottonwood Canyon Member, to deeper water conditions of Paine deposition, and back to shallow-water lithotopes of Woodhurst deposition. This depth change is postulated to have been in response to the complex interaction of three factors: (1) differential negative epeirogenesis of the unstable shelf, (2) a rapid rise of level of the transgressing Lodgepole sea, and (3) variable rates of carbonate sedimentation.

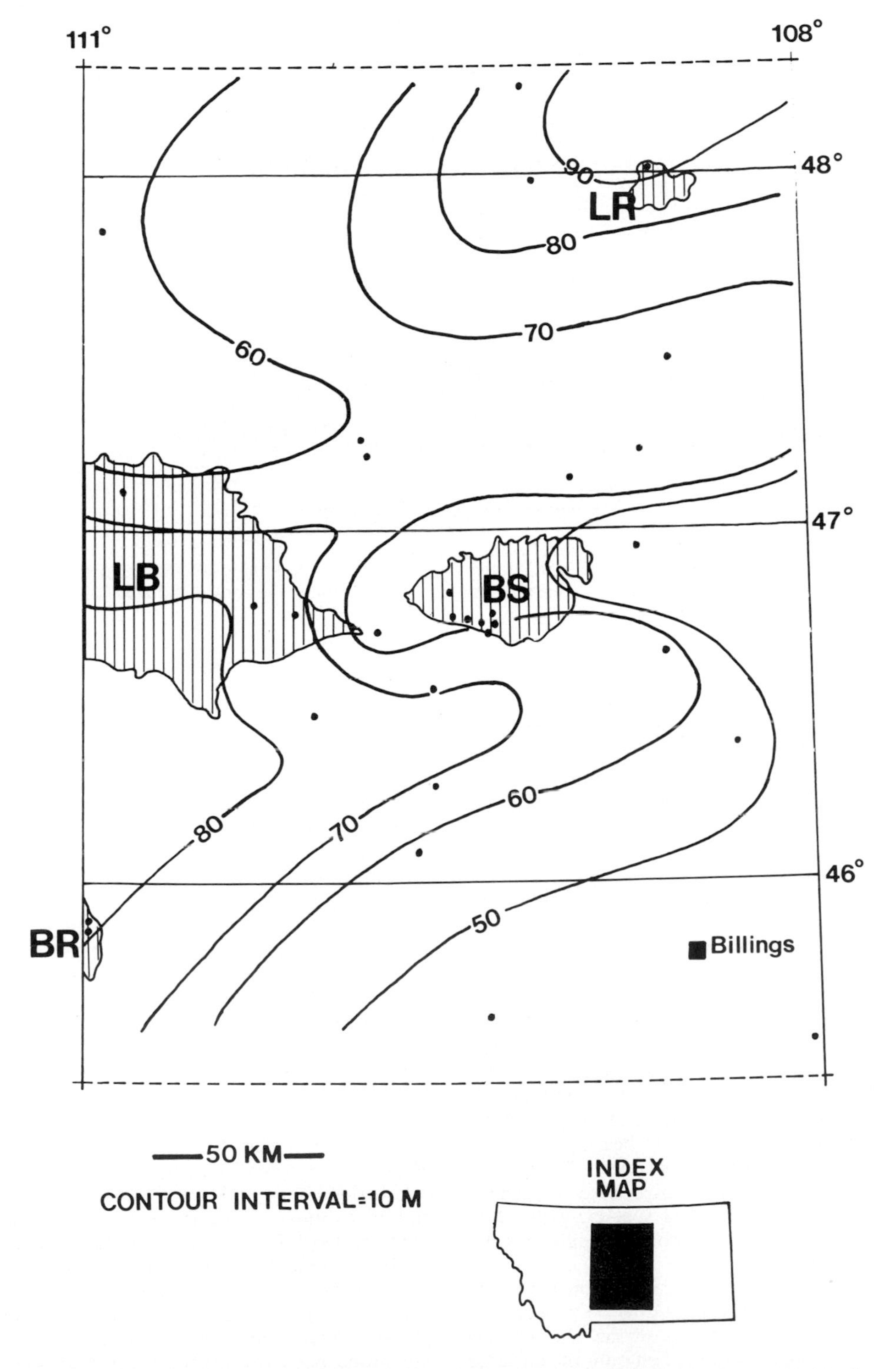

FIG. 4.—Isopach map of Paine Member, Lodgepole Formation. Locations of surface and subsurface control indicated by small dots. Mountain ranges shown with vertically ruled pattern (BR, Bridger Range; LB, Little Belt Mountains; BS, Big Snowy Mountains; LR, Little Rocky Mountains).

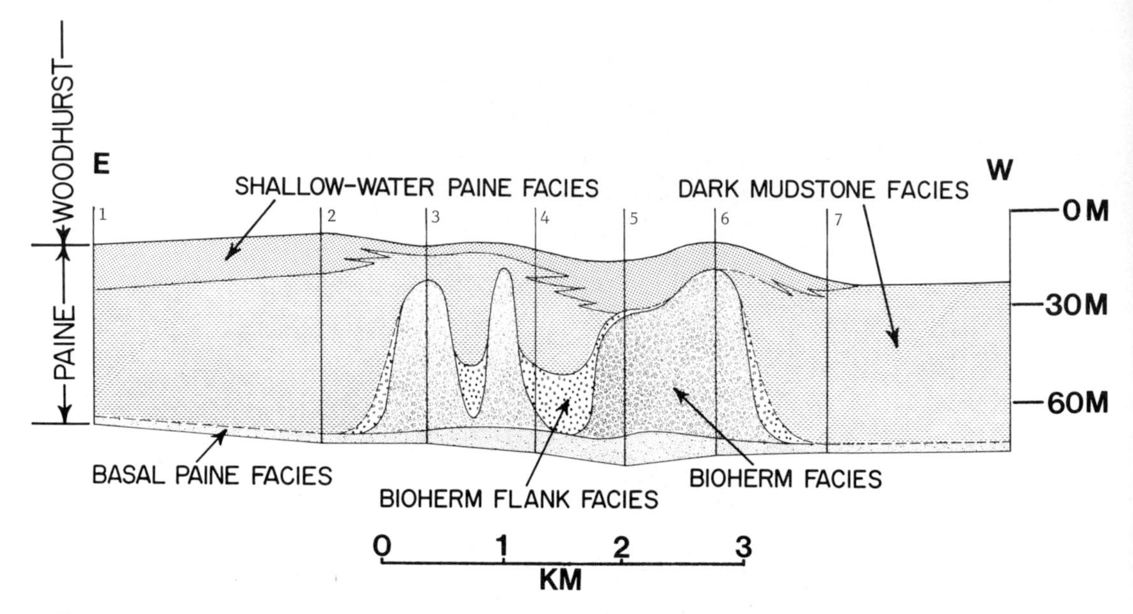

FIG. 5.—West to east facies cross section of the Paine Member, Swimming Woman Canyon, Big Snowy Mountains, Montana. Numbers refer to measured stratigraphic sections. Sections 4 through 6 are shown from right (west) to left (east) in Figure 11a.

Basal Paine Facies

Lithology.—At the base of the Paine Member in central Montana, the Cottonwood Canyon Member is overlain by a discontinuous unit of bioclastic and glauconitic grainstone, packstone, and wackestone 0–17 m thick (Fig. 6). Rocks of this basal facies are medium to thickly bedded. Horizontal and cross-laminations are common but churned, mottled, and burrowed textures predominate. The fauna of this facies is diverse and consists of crinoids, bryozoans, brachiopods, ostracods, corals, and unidentified spicule-like grains. These fossil fragments are generally not abraded or coated but the crinoids are completely disarticulated and the bryozoan fronds are generally fragmented. The remaining clasts in these rocks are sparse pellet-shaped glauconite grains that alter to hematite in weathered samples. Most of these fossil- and glauconite-bearing rocks are wackestone and packstone, the filling and supporting mud being a carbonate-terrigenous mix. However, well-washed spar-cemented bioclastic grainstone is an important minor component of this unit.

Interpretation.—These basal carbonates of the Paine Member are interpreted to be products of relatively rapid transgression of the Madison sea from the Cordilleran mobile belt onto the unstable shelf of central Montana (Fig. 7). Deposits of this transgressive phase resulted in a mantling of terrigenous sediment sources that previously had been available during deposition of the Cottonwood Canyon Member, abruptly terminating local influx of clastic material. The depositional environment was alternately agitated and calm, the mud-rich rocks being products of quiet-water sedimentation below daily wave base, but stirred by storm waves, in areas between grainstone-producing shoals. This was an environment in which a large variety of organisms flourished, lime mud accumulated in quantity, and glauconite formed. Irregular grainstone thickness trends in this unit suggest the buildup of grainstone shoals, decrease of bathymetric relief on the Cottonwood Canyon/Devonian erosion surface by sediment infilling, or most probably, some combination of these factors.

Dark Mudstone Facies

Lithology.—Abruptly overlying the basal unit of the Paine is a 30–70 m sequence of rhythmically thin-bedded dark argillaceous lime mudstone that composes most of the Paine Member in central Montana (Fig. 8). Alternation of thin, horizontally laminated lime mudstone beds with interbeds or partings of more argillaceous carbonate or replacement chert produces the distinctive outcrop pattern of this unit (Fig. 9). Some of these lime mudstone beds are graded (sparse crinoid fragments are the "graded" component), some burrowed and churned, but the sedimentologic norm is horizontal lamination. In addition to these

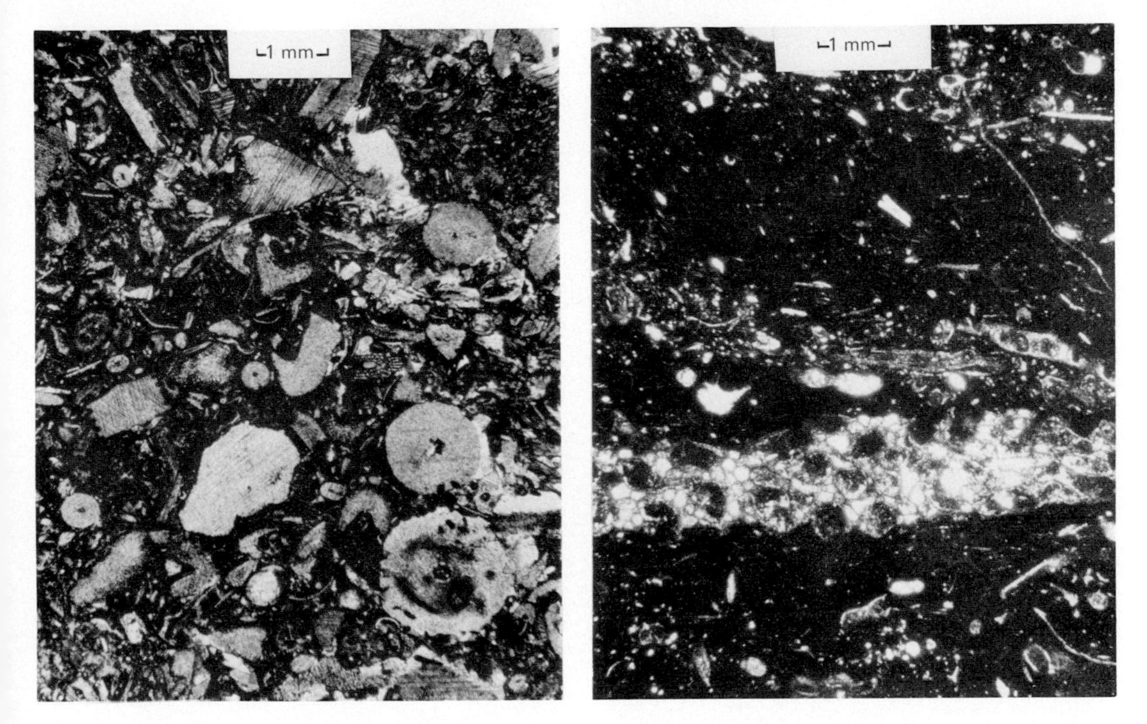

FIG. 6.—Photomicrographs, basal Paine facies. *Left*—Crinoidal packstone/wackestone (plane polarized). *Right*—Bryozoan wackestone/mudstone (plane polarized).

smaller bedding features, large-scale "intraformational truncation surfaces" occur in this facies, similar to those reported from basinal carbonate rocks elsewhere (Wilson, 1969; Davies, this volume; McIlreath, this volume; Yurewicz, this volume). The dark mudstone of the Paine is further characterized by a sparse assemblage of small, commonly articulated crinoid fragments,

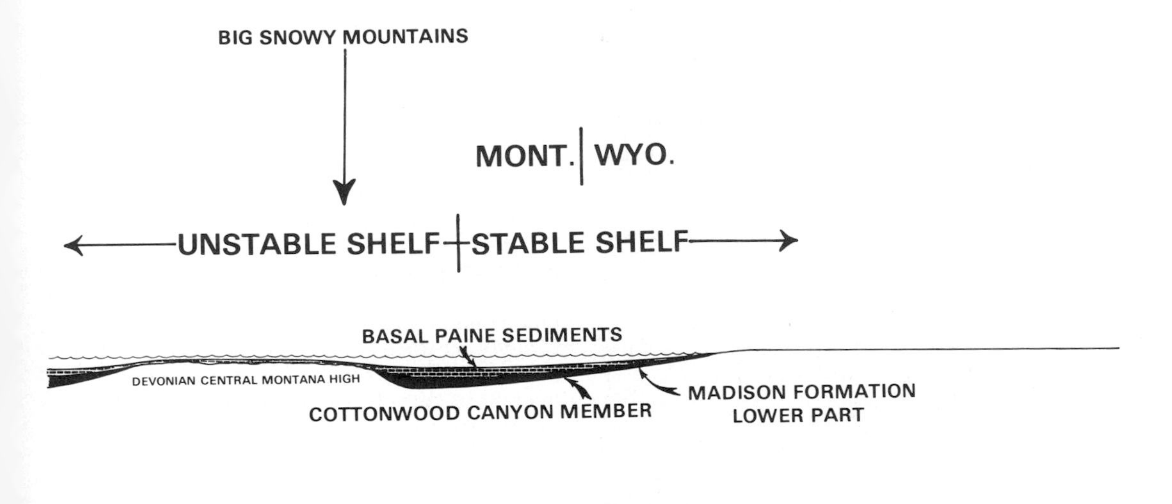

FIG. 7.—Diagrammatic cross section across stable and unstable shelves during deposition of basal Paine Member carbonates, depicting sedimentary mantling of the Cottonwood Canyon Member and central Montana terrigenous source. Line of section shown in Figure 1.

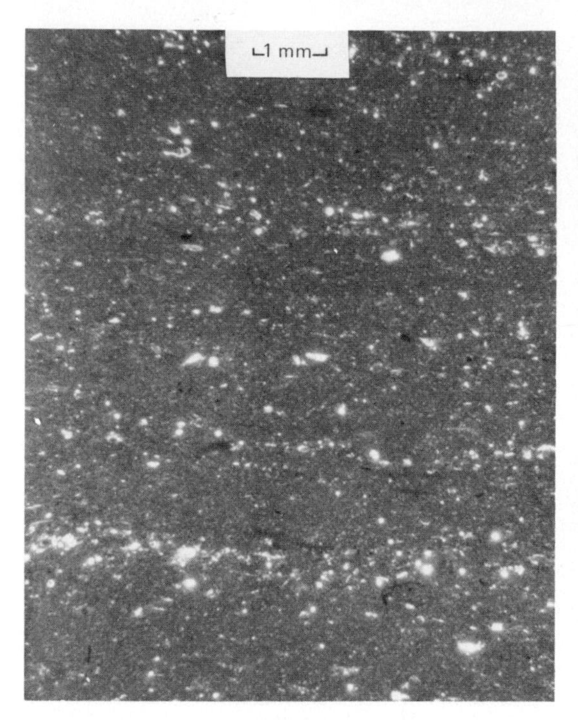

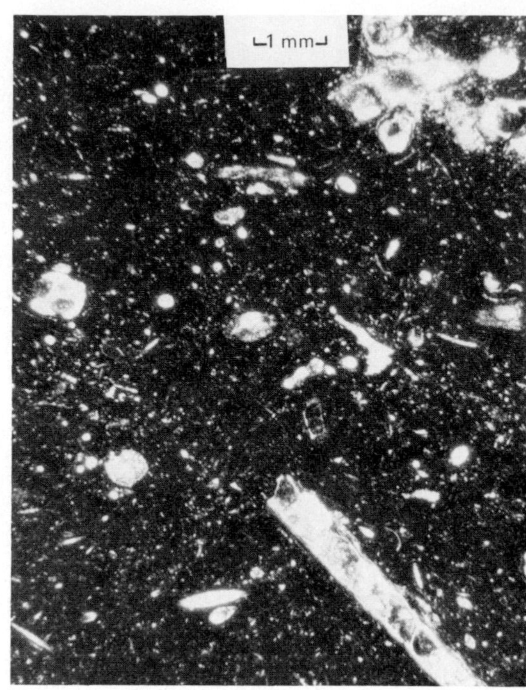

FIG. 8.—Photomicrographs, dark mudstone facies. *Left*—Dark, laminated lime mudstone with sparse bioclasts (plane polarized). *Right*—Dark crinoid/bryozoan lime mudstone (plane polarized).

FIG. 9.—Typical outcrop, dark mudstone facies, Swimming Woman Canyon, Big Snowy Mountains, Montana. Thin lime mudstone beds (dark) alternate with similar thicknesses of replacement chert (light). In other central Montana sections, thin, nonresistant beds and partings of argillaceous limestone and dolomite alternate with the mudstone.

unbroken fenestrate bryozoan fronds, small thin-shelled brachiopods, spicules of unknown origin, and *Zoophycus* trace fossils, suggesting quiet environmental conditions in the biotope. Insoluble residues of these rocks are characterized by small amounts of terrigenous silt and clay and by very small pyrite cubes.

Interpretation.—A deep water interpretation for the dark lime mudstone of the Paine Member is strongly favored, but not proven conclusively, by a composite of the following lines of evidence: (1) dominance of lime mud, (2) even horizontal lamination, (3) large "intraformational truncation surfaces" that may represent submarine mass movements, (4) sparse, specialized fauna of delicate, but commonly unbroken, bryozoans and brachiopods as well as articulated crinoids, (5) *Zoophycus* trace fossils, (6) position of the unit on an unstable shelf with a well-documented tectonic history, (7) overall stratigraphic and sedimentologic similarity to better-documented deep-

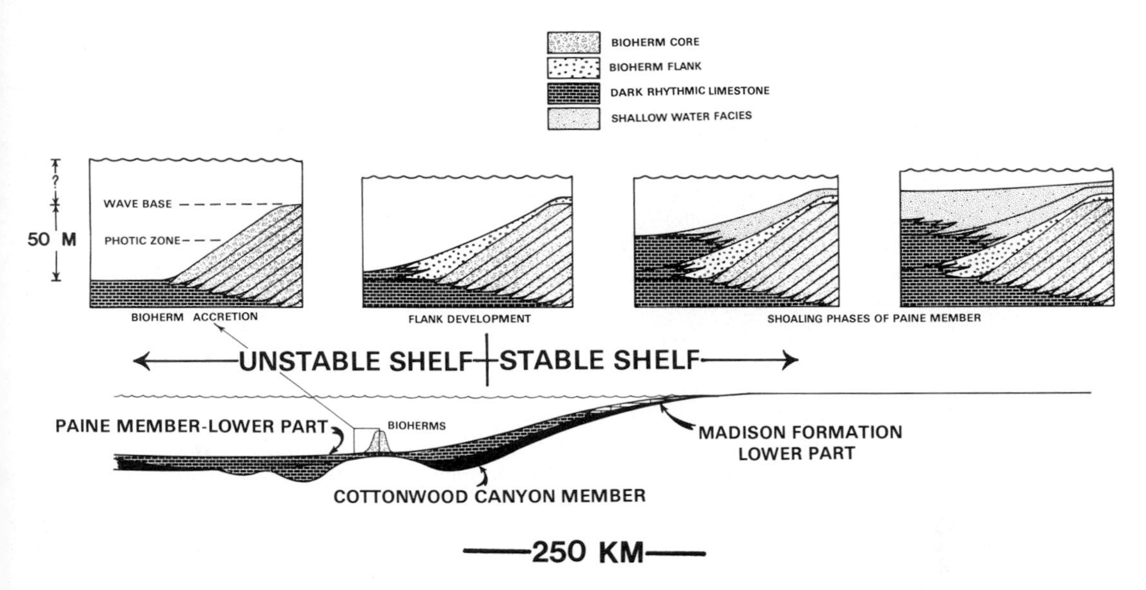

FIG. 10.—Diagrammatic cross section of stable and unstable shelves during deposition of dark mudstone and lime-mud bioherms of the Paine Member, Lodgepole Formation. Upper series of diagrams depicts from left to right; growth of bioherm to effective limiting wave base with accompanying lateral accretion of the core facies; subsequent accumulation of flank facies as a second stage of bioherm development; and eventual cessation of bioherm development and accumulation of enclosing dark mudstone and shallow-water facies. Line of section shown in Figure 1.

water carbonates, and (8) pronounced lack of shallow-water depositional features (compare with Wilson, 1969). Figure 10 schematically depicts the favored hypothesis, featuring the accumulation of the dark mudstone unit on the unstable shelf in water depths below daily wave base and, most probably, below storm wave base. During this phase of Lodgepole deposition, the carbonate source may have been precipitation and sedimentation from the overlying water column, from winnowing and transport of fine-grained carbonate debris from contemporaneous shallow-water lithotopes on the stable shelf, or from some combination of these processes. Terrigenous silt and clay in variable quantities were probably derived from the exposed craton to the east and northeast and were transported to the deeper water lithotope of the unstable shelf by marine surface currents or by prevailing northerly or northeasterly winds.

Bioherm Cores and Flanks

Lithology.—Enclosed within this dark, rhythmically bedded mudstone near the base of the Paine Member are lime-mud bioherms (Fig. 5), similar to other Waulsortian bioherms in the Mississippian of North America (Harbaugh, 1957; Pray, 1958; Troell, 1962; Cotter, 1965; Morgan and Jackson, 1970; Stone, 1972) and northern Europe (Lees, 1964). Bioherm outcrops in the Big Snowy Mountains (Fig. 11) are 70 m thick and up to 700 m in length (Cotter, 1965); those at the crest of the Bridger Range are smaller (Stone, 1972). These are the only well-documented bioherms in the Paine Member of central Montana, although other buildups have been reported. In both areas, the bioherms are subdivided into two facies: a massive, resistant core with a severe and complex diagenetic history (Cotter, 1966; Stone, 1972) and less resistant, layered flanking beds which probably postdate core accumulation.

The Big Snowy Mountains bioherm cores are characterized by alternating, medium to thick layers of lime mudstone and bioclastic wackestone, packstone, and grainstone (Fig. 12), some of which are inclined as much as 35 degrees to the base of the formation and to the stratification of the enclosing dark lime mudstone facies. One continuous distinctive layer was traced 50 m vertically by Cotter (1965), suggesting at least this much depositional relief above the surrounding Lodgepole sea floor, assuming minimal compaction of the thick dark mudstone facies that envelops the bioherms. Organic components of the core facies are principally crinoids and fenestrate bryozoans, along with an abundant and diverse fauna of brachiopods, coelenterates, and mollusks. This fauna is considerably more dense and

FIG. 11.—Paine Member bioherm exposures. *Left*—Big Snowy Mountains bioherms (B) at mouth of Swimming Woman Canyon, view looking southwest. Bioherms form 70 m ledges near base of Paine Member. Ledges above and between bioherm core masses consist of the dark mudstone facies (D) and shallow-water facies (S). Beneath the dark mudstone facies, the bioherm flank facies (F) is poorly exposed. Compare with Figure 5, measured sections 6 through 4 (east and west are reversed in these figures). *Right*—Bridger Range bioherms (B) viewed from north to south along the west flank of the range. Massive ledges of the Woodhurst Member form ridge above the bioherms. Bedding is nearly vertical; the tops of the bioherms are to the east (left).

diverse than that of the enclosing dark Paine carbonates (Merriam, 1958).

Poorly exposed between and on top of the buildups is a flank facies (Fig. 5), composed of large crinoid ossicles in a lime-mud matrix (Fig. 13). In both the Big Snowy Mountains and Bridger Range, rocks of this flank facies appear to overlie and thus postdate, at least in part, both the sides and tops of the core buildups (Fig. 5).

Interpretation.—Origin, maintainence, and ter-

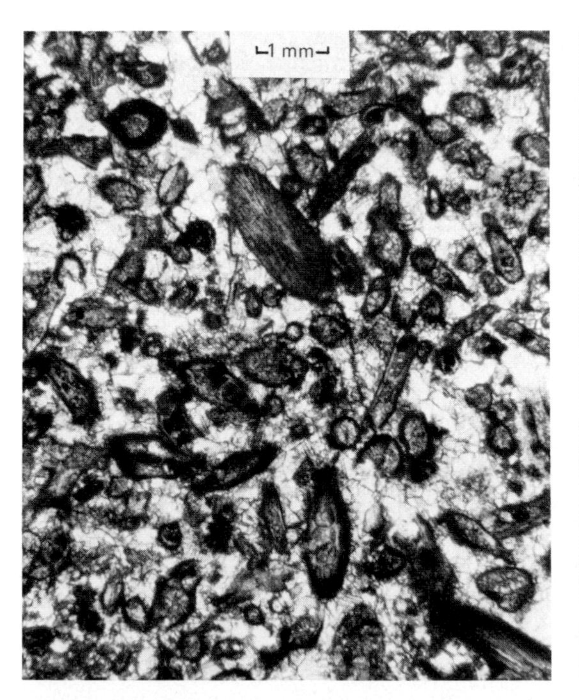

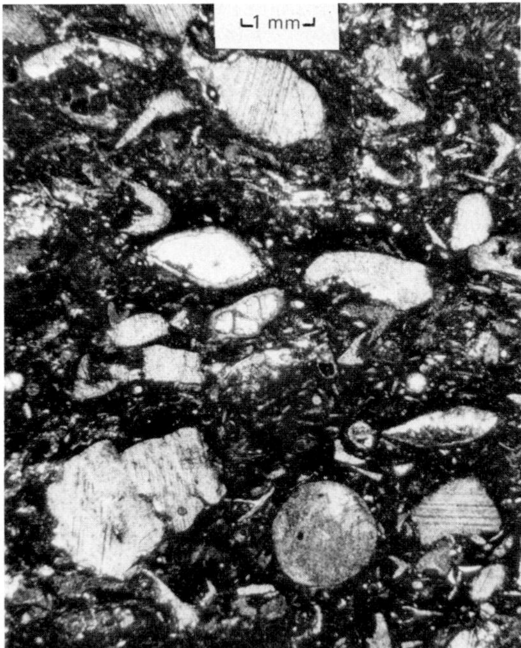

FIG. 12.—Photomicrographs, bioherm facies. *Left*—Crinoid/bryozoan lime grainstone (plane polarized). *Right*—Crinoid/bryozoan lime wackestone (plane polarized).

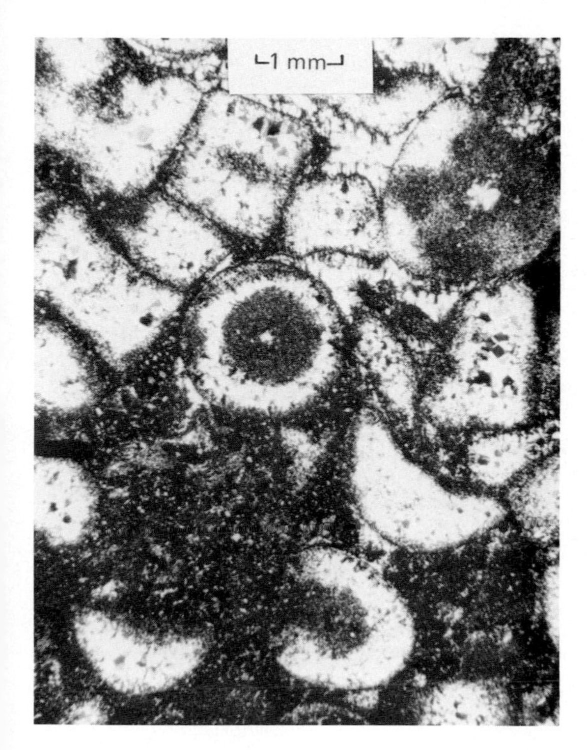

FIG. 13. Photomicrograph, bioherm flank facies. Silicified crinoid packstone with exceptionally large crinoid ossicles (plane polarized).

mination of bioherm growth are somewhat problematic, but the presence of abundant lime mud suggests that the core facies accumulated in calm water, probably below wave base and possibly through sediment baffling and trapping activities of crinoids and bryozoans, as suggested by Pray (1958) for similar bioherms in New Mexico and by Stone (1972) for Paine Member bioherms in the Bridger Range of Montana. Lateral accretion of steeply dipping bioherm core layers occurred after upward growth was limited by effective wave base (Fig. 10a). If, indeed, these bioherms rose more than 50 m above the surrounding sea floor, and the tops were at or slightly below effective wave base in an environment in which a diverse and abundant fauna flourished, then the deeper bottom where the dark mudstone was contemporaneously accumulating may have been somewhat in excess of 70 m below the surface of the Lodgepole sea, assuming a speculative minimum 20 m effective wave base.

At a late stage in bioherm development (Fig. 10b), slightly higher energy conditions produced a rougher water biotope on top of the bioherms. This increase in wave activity may have been due to either a gradual shoaling of the Lodgepole sea or to biohermal accretion into shallow water. Skeletal debris from this biotope accumulated on top and spilled over the sides of the bioherms, forming crinoid packstone and wackestone flanking beds (Fig. 13).

Following cessation of bioherm growth and development of the flank facies, the dark mudstone enveloped the bioherms, interfingering with shallower water deposits developing on top of the bioherms (Fig. 10c).

From the evidence provided by these lime-mud bioherms and the penecontemporaneous dark mudstone, a picture emerges of a widespread deep-shelf environment in central Montana during a major part of Paine Member deposition (Fig. 10). This shelf was below effective wave base. From the quiet shelf sea floor, lime-mud bioherms were built through the collective efforts of bryozoans and crinoids into or near effective wave base where biohermal organisms thrived but where bioherm growth was ultimately terminated in rougher shoal water.

Shallow-water Paine Facies

Lithology.—Gradationally overlying the bioherms and intertonguing with the dark Paine carbonates (Figs. 5, 14) is a unit of bioclastic and pellet mudstone, wackestone, and grainstone in horizontally and cross-stratified, medium to thick beds (Fig. 15). The fauna consists of crinoids, fragments of fenestrate and ramose bryozoans, rugose corals, and *Syringopora* colonies in growth position. Pellets and pelletoids are the only nonskeletal grain components in these rocks. Many of these skeletal and nonskeletal grains are broken, abraded, and coated.

Interpretation.—The characteristics of this unit are in striking contrast to those of the underlying deeper water mudstones and are suggestive of shoal and intershoal depositional environments, slightly above and slightly below wave base, respectively. The cross-stratified grainstone units are interpreted as shoals; the wackestone and mudstone lithologies as protected intershoal deposits. This is the uppermost unit of the Paine Member in central Montana and it grades upward into the shallow-water depositional cycles of the Woodhurst Member (Smith, 1972), thus terminating the deeper water phase of Lodgepole deposition.

PAINE DEPOSITION: SUMMARY

The depositional history of the Paine Member on the central Montana unstable shelf is one of an initial shallow-water, marginal marine phase, during which the dark shale, siltstone, and dolomitic shale of the Cottonwood Canyon Member were deposited in lagoons and embayments of the transgressing Lodgepole sea. As transgression continued, an abrupt change from clastic to car-

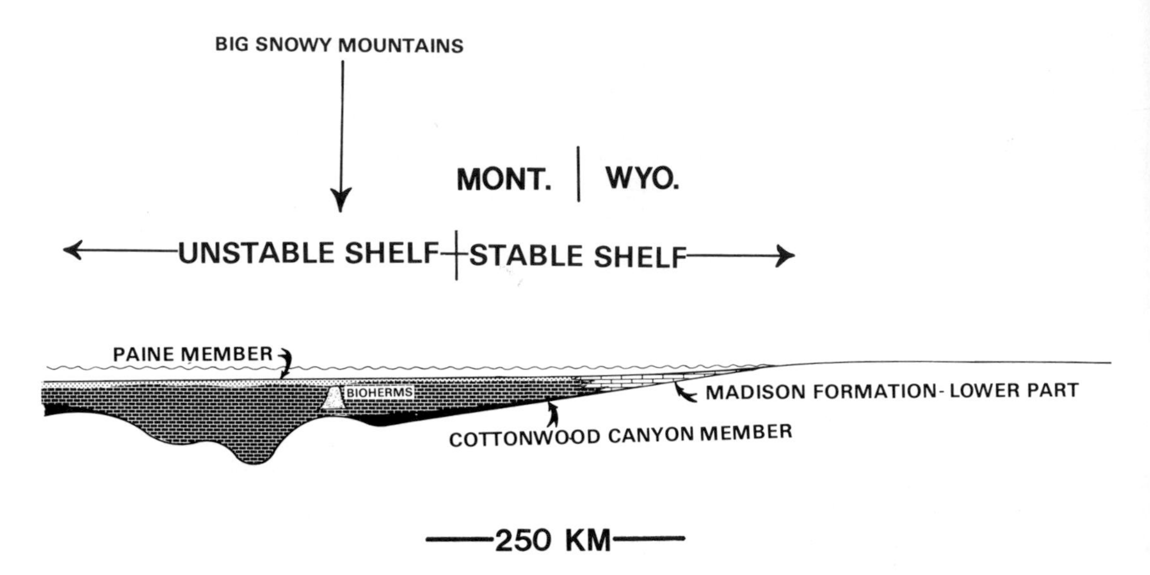

FIG. 14.—Diagrammatic cross section of stable and unstable shelves during deposition of the shallow-water facies (stippled) of the Paine Member. Following deposition of this uppermost Paine unit, oolitic and bioclastic limestone of the Woodhurst Member was deposited in shallow water across the unstable shelf. Line of section shown in Figure 1.

bonate deposition resulted in the accumulation of a thin, discontinuous, bioclastic and glauconitic bed at the base of the Paine Member. This basal unit was deposited near effective wave base and represents a transitional depth phase. Continued transgression, in addition to incipient subsidence of the unstable shelf resulted in deeper water depositional conditions, with the consequent accumulation of dark, argillaceous, rhythmically thin-bedded mudstones of the Paine Member. Lime-mud bioherms enclosed in these mudstones suggest a minimum water depth of 50–70 m. In the later stages of Paine sedimentation, decreased rates of subsidence and/or increased rates of sediment accumulation created shallower water conditions on the unstable shelf and promoted the accumulation of shallow-water grainstones and mud-rich rocks. These shallow-water Paine

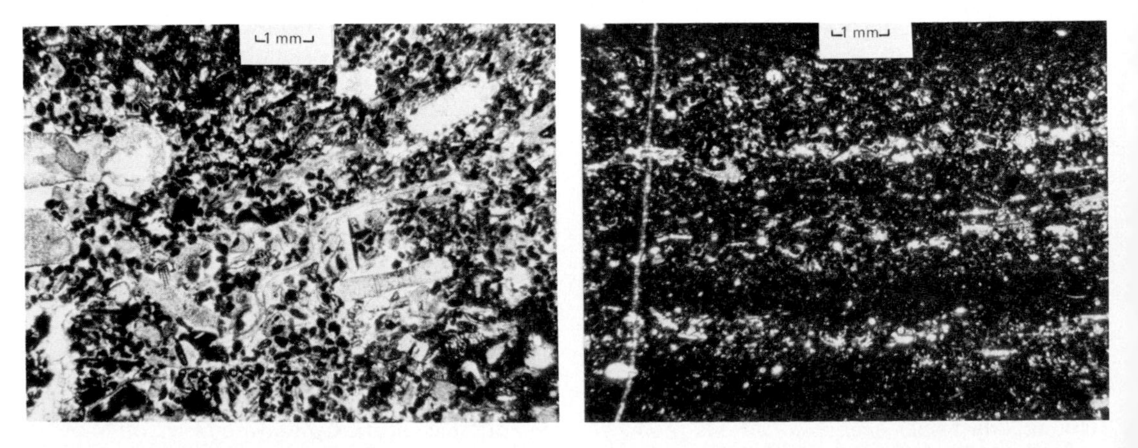

FIG. 15.—Photomicrographs, shallow-water facies, Paine Member. *Left*—Bryozoan/crinoid pellet grainstone (plane polarized). *Right*—Fine-grained, well-sorted, laminated, pellet lime grainstone laminae alternating with laminae of pellet mudstone.

lithologies intertongue with the dark mudstones and are succeeded by shallow-water carbonates of the Woodhurst Member.

DEPOSITIONAL MODEL: PAINE MEMBER

The model tentatively proposed for Paine deposition is depicted in Figure 16. It involves: (1) transgression of the sea, beginning with the inundation of the unstable shelf and later continuing southward, spreading up the ramp onto the stable shelf, (2) epeirogenic downwarping of the unstable shelf, and (3) variable sedimentation rates.

During deposition of the Cottonwood Canyon Member (Figure 16a), sediment influx was predominantly clastic, with sediment sources on the central Montana high and the stable shelf. Sedimentation apparently kept pace with rising sea level and with subsidence, if any, of the unstable shelf.

With continued transgression, terrigenous sediment sources were progressively mantled and carbonate accumulation began with the deposition of the bioclastic-glauconitic facies at the base of the Paine Member (Fig. 16b). Again, sedimentation kept pace with rising sea level, as well as with the incipient downwarping of the unstable shelf. Most of these sediments accumulated at or slightly below wave base.

During the next phase of sedimentation, downwarping of the unstable shelf and/or rise of sea level increased to a point exceeding the sedimentation, resulting in a deeper water, semi-starved unstable shelf in central Montana (Fig. 16c). The dark lime mudstone of the Paine Member accumulated here in deeper water environments along the flanks and axis of this downwarp. Lime-mud bioherms began developing during the early part of this phase of sedimentation and built from the deeper sea floor into less stagnant waters 50–70 m above. On the stable shelf to the south, shallow-water sediments were deposited. However, throughout deposition of the Paine Member, there apparently was no abrupt break in slope

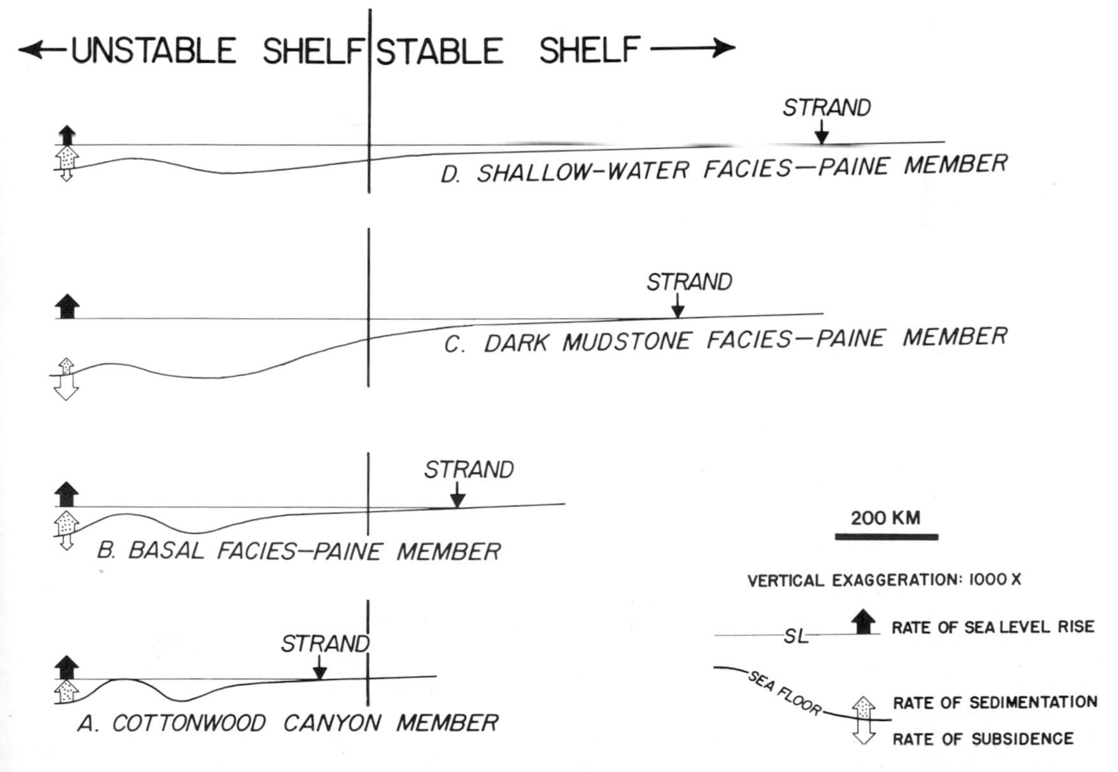

FIG. 16.—Depositional model for Paine Member. Development of the Montana unstable shelf begins with relatively shallow-water depositional environments of the Cottonwood Canyon Member (A), then proceeds through transitional (B) and deeper water phases (C), and returns to a second shallow-water phase with deposition of the shallow-water facies of the Paine Member (D). Alternation of these facies resulted from the complex interaction of rates of sea level rise sedimentation, and subsidence.

between these two provinces with attendant marginal reef buildups.

As transgression of the Lodgepole sea continued, subsidence of the stable shelf reached a maximum, then diminished, and the area took on a more positive aspect. Bioherm growth ceased, followed by accumulation of flank facies and deposition of the shallow-water Paine facies, first on the bioherm highs and later in non-biohermal areas as dark lime mud buried the bioherm flanks and filled the downwarp (Fig. 16d). This phase terminated the deeper water sedimentation that earlier characterized Paine Member deposition.

Widespread shallow-water conditions prevailed on the unstable shelf during the next phase of sedimentation, resulting from rates of shallow-water carbonate production in excess of subsidence and rise of sea level. In response to these conditions, oolite and bioclastic shoals of the Woodhurst Member prograded northward, filling in the unstable shelf.

ACKNOWLEDGEMENTS

Many of the ideas and much of the data presented in this paper were generated during graduate study at the University of Montana and were included in my dissertation. Grateful appreciation is extended to James A. Peterson, University of Montana, who provided advice, counsel, and encouragement during this time and who kindly read this manuscript and offered helpful suggestions for its improvement. Material and financial assistance were generously supplied by the University of Montana, Idaho State University, Montana State University, Esso Production Research Company, and the American Stratigraphic Company.

REFERENCES

Ahr, W. M., 1973, The carbonate ramp: An alternative to the shelf model: Gulf Coast Assoc. Geol. Socs. Trans., v. 23, p. 221-225.

Andrichuk, J. M., 1955, Mississippian Madison Group stratigraphy and sedimentation in Wyoming and southern Montana: Am. Assoc. Petroleum Geologists Bull., v. 59, p. 2170-2210.

Baars, D. L., 1972, Devonian System: *In* W. W. Mallory (ed.), Geologic atlas of the Rocky Mountain region: Rocky Mountain Assoc. Geologists, Denver, Colorado, p. 90-99.

Cotter, E. J., 1965, Waulsortian-type carbonate banks in the Mississippian Lodgepole Formation of central Montana: Jour. Geology, v. 73, p. 881-888.

——, 1966, Limestone diagenesis and dolomitization in Mississippian carbonate banks in Montana: Jour. Petrology, v. 36, p. 764-774.

Craig, L. C., 1972, Mississippian System: *In* W. W. Mallory (ed.), Geologic atlas of the Rocky Mountain region: Rocky Mountain Assoc. Geologists, Denver, Colorado, p. 100-110.

Edie, R. W., 1958, Mississippian sedimentation and oil fields in southwestern Saskatchewan: *In* A. J. Goodman (ed.), Jurassic and Carboniferous of western Canada: Am. Assoc. Petroleum Geologists, Tulsa, Oklahoma: p. 331-363.

Harbaugh, J. W., 1957, Mississippian bioherms in northwest Oklahoma: Am. Assoc. Petroleum Geologists Bull., v. 41, p. 2530-2544.

Henbest, L. G., 1958, Significance of karst terrane and residuum in Upper Mississippian and Lower Pennsylvanian rocks, Rocky Mountain region: Wyoming Geol. Assoc., Guidebook 13th Ann. Field Conf., p. 36-38.

Jenks, Susan, 1972, Environment of deposition and diagenesis of the Lodgepole Formation (Mississippian), central Montana: Montana Geol. Soc., Guidebook 21st Ann. Field Conf., p. 19-28.

Lees, A., 1964, The structure and origin of the Waulsortian (Lower Carboniferous) "reefs" of west-central Eire: Royal Soc. London Philos. Trans., Ser. B, v. 247, p. 483-531.

Mallory, W. W., 1972, Regional synthesis of the Pennsylvanian System: *In* W. W. Mallory, (ed.), Geologic atlas of the Rocky Mountain region: Rocky Mountain Assoc. Geologists, Denver, Colorado, p. 111-127.

Merriam, R. W., 1958, A Madison bioherm, Big Snowy Mountains, Montana: M.S. Thesis, Washington State Univ., Pullman, Washington, 87 p.

Morgan, G. R., and Jackson, D. E., 1970, A probable "Waulsortian" carbonate mound in the Mississippian of northern Alberta: Bull. Canadian Petroleum Geology, v. 18, p. 104-112.

Pray, L. C., 1958, Fenestrate bryozoan core facies, Mississippian bioherms, southwestern United States: Jour. Sed. Petrology, v. 28, p. 261-273.

Rodriguez, Joaquin, and Gutschick, R. C., 1970, Late Devonian-Early Mississippian ichnofossils from western Montana and northern Utah: *In* T. P. Crimes and J. C. Harper (eds.), Trace fossils: Geol. Jour., Spec. Issue 3, p. 407-438.

Sandberg, C. A., and Klapper, Gilbert, 1967, Stratigraphy, age, and paleotectonic significance of the Cottonwood Canyon Member of the Madison Limestone in Wyoming and Montana: U.S. Geol. Survey, Bull. 1251-B, 70 p.

Sando, W. J., 1967, Mississippian depositional provinces in northern Cordilleran region: U.S. Geol. Survey, Prof. Paper 575-D, p. 29-38.

—— and Dutro, J. T., Jr., 1974, Type sections of the Madison Group (Mississippian) and its subdivisions in Montana: U.S. Geol. Survey, Prof. Paper 842, 22 p.

SMITH, D. L., 1972, Depositional cycles of the Lodgepole Formation (Mississippian) in central Montana: Montana Geol. Soc., Guidebook 21st Ann. Field Conf., p. 29-35.

STONE, R. A., 1972, Waulsortian-type bioherms (reefs) of Mississippian age, central Bridger Range, Montana: Montana Geol. Soc., Guidebook 21st Ann. Field Conf., p. 37-55.

TROELL, A. R., 1962, Lower Mississippian bioherms of southwestern Missouri and northwestern Arkansas: Jour. Sed. Petrology, v. 32, p. 629-644.

WEED, W. H., 1899, Description of the Little Belt Mountains Quadangle (Montana): U.S. Geol. Survey Geol. Atlas, Folio 56, 9 p.

WILSON, J. L., 1969, Microfacies and sedimentary structures in "deeper water" lime mudstones: *In* G. M. Friedman (ed.), Depositional environments in carbonate rocks: Soc. Econ. Paleontologists and Mineralogists, Spec. Pub. 14, p. 4-19.

——, 1974, Characteristics of carbonate-platform margins: Am. Assoc. Petroleum Geologists Bull., v. 58, p. 810-824.

SEPM Special Publication No. 25, p. 203–219, November 1977

SEDIMENTOLOGY OF MISSISSIPPIAN BASIN-FACIES CARBONATES, NEW MEXICO AND WEST TEXAS—THE RANCHERIA FORMATION

DONALD A. YUREWICZ
Exxon Production Research Co.
Box 2189
Houston, Texas 77001

ABSTRACT

The Rancheria Formation is redefined to include the Rancheria and Las Cruces Formations of Laudon and Bowsher (1949). The Rancheria Formation (Meramecian) represents deposition in a relatively deep, low-oxygen basin (depths probably less than 250 meters) in south-central New Mexico and West Texas. It generally consists of 60–150 meters of dark gray, cherty, sparsely fossiliferous, fine-grained limestone.

Four major lithofacies can be recognized in the Rancheria Formation. One consists of deep water lime mudstone and spiculitic wackestone deposited from hemipelagic suspension and possibly low density, low velocity turbidity currents (45% of the Rancheria). A second lithofacies consists of lime-silt grainstone that represents deep-water turbidity- and traction-current deposition (22% of the Rancheria). These two lithofacies reflect deposition below wave base under partially euxinic conditions. A third, but minor lithofacies, occurring predominantly along the edge of the basin, consists of coarse skeletal grainstone that represents allodapic sands deposited by turbidity currents and possibly traction currents or sand flows along the edge of the basin (5% of the Rancheria). A fourth lithofacies, composed of lime-silt grainstone, represents shallower water traction-current deposition during the final phases of basin filling (22% of the Rancheria). The sedimentary structures and trace fossils in this lithofacies suggest that it was deposited above wave base in more agitated and better oxygenated water. Fluctuations in benthonic fossil densities and bioturbation indicate that conditions ranged from anaerobic to aerobic during Rancheria deposition.

Intraformational erosion surfaces, interpreted as submarine features, are present locally within the Rancheria. They are most significant along the basin margin in the Sacramento Mountains of New Mexico. Slumping is rare in the Rancheria Formation indicating a relatively stable basin floor. A deep-water origin for most of the Rancheria is suggested by its dark color, the predominance of mud and silt-size sediment, its very low faunal density, the paucity of wave- and surf-formed structures, the absence of typical shallow-water features, the amount of relief along the basin margin, and its similarity to other known deep-water carbonates.

INTRODUCTION

The Rancheria Formation (Meramecian) is a distinctive carbonate unit representing deposition in a relatively deep basin (depths probably less than 250 m) in south-central New Mexico and West Texas. It typically consists of 60–150 m of dark gray, sparsely fossiliferous, cherty, fine-grained limestone. This formation contains many features described by Wilson (1969) as typical of deep water carbonate deposition. It affords an excellent glimpse of basinal carbonate features. Although the overall characteristics of the Rancheria are fairly uniform, subtle differences exist which mark changes in sediment supply, oxygen levels, basin stratification, modes of deposition, and depth. In addition, outcrops of the Rancheria in the Sacramento Mountains represent the northern edge of the Rancheria basin and provide a view of some unique basin-margin features. These include surfaces of discontinuity resulting from submarine erosion, slump features, and fans or sheets of allochthonous shelf debris. This paper will discuss overall features of the Rancheria and some of the unique features associated with the basin margin. A more detailed discussion of Rancheria outcrops in the Sacramento Mountains has been presented in a guidebook prepared for the 1975 Meeting of the Society of Economic Paleontologists and Mineralogists (Yurewicz, 1975). The guidebook includes analysis of regional Mississippian lithofacies, detailed discussions of basin-margin features of the Rancheria, diagenesis of shelf and shelf-edge carbonates, and analysis of critical Mississippian stratigraphic relationships in this region. This paper will provide a broader overview of Rancheria deposits, their distribution, and their genesis.

The Rancheria crops out in four major areas, the Sacramento and San Andres Mountains in southern New Mexico and the Franklin and Hueco Mountains in West Texas (Fig. 1). There also are possible Rancheria outcrops in the Florida Mountains in southwestern New Mexico (Kottlowski, 1963, Fig. 8) and in the Placer de Guadalupe Range, Mexico (Fig. 2; Wilson, 1971, 1975). Data is limited but the Rancheria may extend several hundred kilometers to the east of these exposures in the subsurface as the "Mississippi Limestone" (Lloyd, 1949). These outcrop and subsurface units define the existence of a relatively deep, euxinic, Meramecian basin that extended across West Texas, southeastern New Mexico, and northern Mexico, and received carbonate sediment. Meramecian deposits north and west of the Rancheria are represented by the Hachita and Terrero Formations, both composed of shallow-water carbonates (Armstrong, 1962, 1965; Armstrong and Mamet, 1974). To the south and southeast the Meramecian is represented by basinal shale (Barnett Shale) and basinal sandstone (Tesnus Formation).

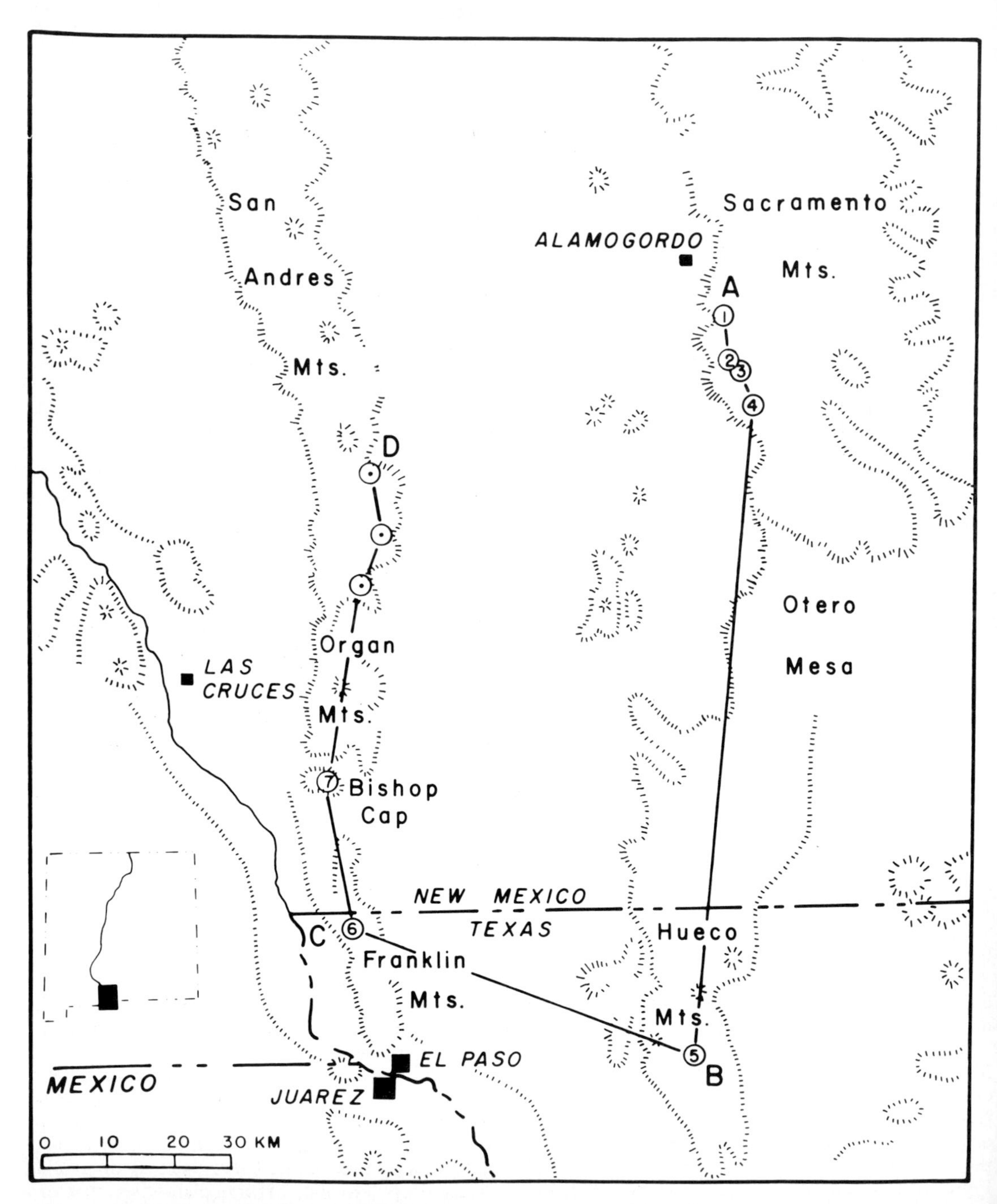

FIG. 1.—Index map of Rancheria outcrop area. Numbered sections from Yurewicz (1973b). Sections in the San Andres Mountains from Laudon and Bowsher (1949).

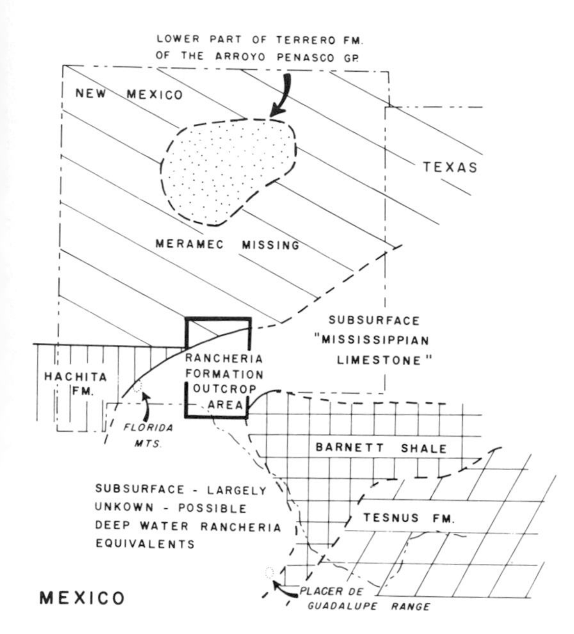

FIG. 2.—Meramecian lithofacies. Data from Armstrong (1962), Armstrong and Mamet (1974), King (1965), Kottlowski (1963), Lloyd (1949) and Wilson (1971, 1975).

STRATIGRAPHY

The Rancheria Formation is redefined in this report to include the Rancheria and Las Cruces Formations as originally defined by Laudon and Bowsher (1949). The Rancheria and Las Cruces Formations consist of dark gray, fine-grained, sparsely fossiliferous limestone that together forms a distinctive basin facies of Meramecian age in south-central New Mexico and West Texas. The two formations were differentiated by Laudon and Bowsher (1949) on the basis of an apparent unconformity between them. This contact has been re-examined and it is interpreted by Yurewicz (1973b) to have formed by submarine rather than subaerial erosion as first implied by Laudon and Bowsher (see discussion under Intraformation Erosion Surfaces, this article). Deep basinal conditions apparently persisted between Las Cruces and Rancheria deposition. The biostratigraphic work of Lane (1974, 1975) does not indicate a significant time break at this contact. The Las Cruces contains a late Osagian to early Meramecian conodont fauna and the basal Rancheria is represented by a Meramecian fauna. Because of these interpretations, the presence of other submarine diastems in the Rancheria, and the limited distribution of the Las Cruces (see Fig. 3), it is proposed here that the two formations be considered as one unit. The Rancheria Formation is elevated here to include both the Las Cruces and Rancheria formations as defined by Laudon and Bowsher (1949). The Las Cruces is redefined as a member of the Rancheria Formation (Fig. 3).

The Rancheria Formation is generally 75 to 90 m thick. It reaches a maximum thickness of 152 m in the Franklin Mountains and thins to a feather edge to the north in the San Andres and Sacramento Mountains. This largely represents depositional onlap thinning across the northern edge of the Rancheria basin (Yurewicz, 1973b, 1975). The Rancheria rests unconformably on the Lake Valley Formation (Osagian) in the north and on Devonian rocks in the south (Laudon and Bowsher, 1949; Pray, 1961; Yurewicz, 1973a, 1973b, 1975; Meyers, 1973, 1974, 1975; and Lane, 1974, 1975). It is overlain conformably by the Helms Formation (Chesterian) in all but the extreme northern exposures where it is overlain unconformably by Pennsylvanian rocks. These relations are summarized in Figure 3. The contact with the Helms has been considered to be an unconformity by Laudon and Bowsher (1949) but no evidence of an unconformity was found by this writer. The contact appears transitional.

RANCHERIA BASIN MARGIN

The northern edge of the Rancheria basin is defined by depositional onlap across the eroded edge of the Lake Valley Formation in the Sacramento and San Andres Mountains. Outcrops in the San Andres Mountains were not examined because of military restrictions, so discussion of the basin margin is directed to exposures in the Sacramento Mountains.

The Rancheria and Lake Valley Formations form a wedge-on-wedge relationship in the Sacramentos, the Lake Valley thinning to the south and the Rancheria thinning to the north (Fig. 4). Thinning of the Lake Valley Formation to the south is predominantly the result of pre-Rancheria erosion (Laudon and Bowsher, 1949; Pray, 1961; Yurewicz, 1973a, 1973b, 1975; Meyers, 1973, 1974, 1975; Lane, 1974, 1975). Yurewicz (1973b, 1975) has shown that complementary thinning of the Rancheria to the north represents depositional onlap thinning across this eroded surface. This eroded surface served as the depositional surface for initial Rancheria deposition. The slope of this surface represents the initial slope of the Rancheria basin. The inclination of this slope is defined by the rate at which the Lake Valley Formation thins from the north to the south (assuming that the top of the underlying Caballero Formation represents a time surface). Since the Rancheria onlapped this surface, eventually filling the basin and eliminating initial basin-margin relief, the slope of the basin edge can also be defined by the complementary rate of Rancheria thicken-

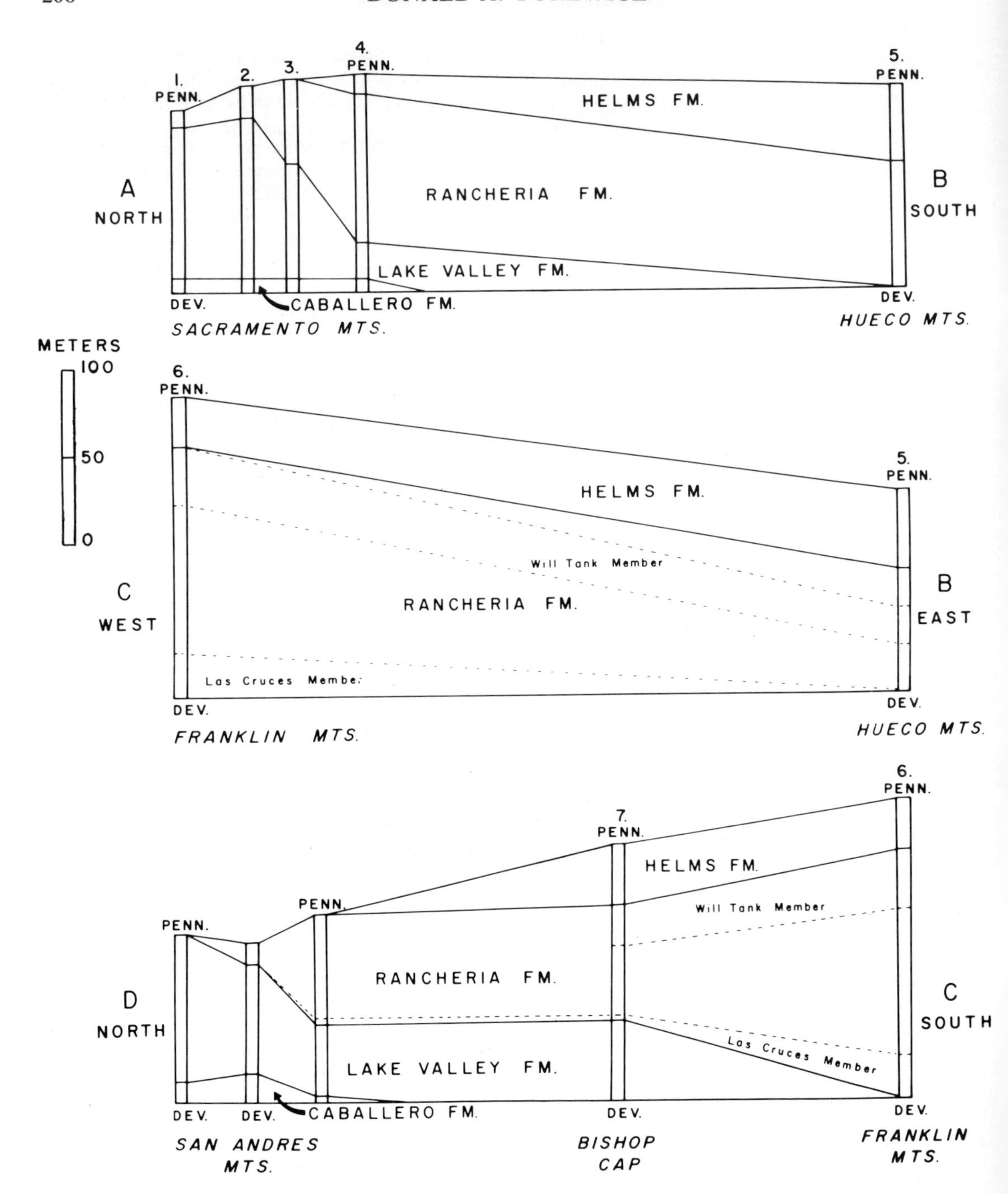

FIG. 3.—Correlation of Mississippian formations in the Rancheria outcrop area. See Figure 1 for location of sections. Sections are not to scale horizontally.

ing from north to south (Fig. 5).

The thin deposits (0–20 m thick) of the Rancheria from Alamo Canyon to Deadman Canyon were deposited on the edge of a submerged shelf platform. These rocks, like those in the thicker sections of the Rancheria to the south, appear to have been deposited in relatively deep water, probably below wave base. Shallow-shelf equivalents of the Rancheria are not preserved in the Sacramento Mountains.

The Rancheria begins to thicken rapidly in Deadman Canyon as it plunges off the edge of the shelf and onto the basin margin. It thickens from 25 to 50 m in Deadman Canyon in less than 0.8 km. The basin margin was steepest here, sloping 2–3 degrees for 0.8 km. It continued to slope southward, although not as steeply (less than one degree), 3.2 km more to Escondido Canyon. The basin floor was nearly level at Agua Chiquita Canyon 2.1 km south of Escondido Canyon. This is based on the slight change in Rancheria thickness from Escondido Canyon south to Agua Chiquita and Nigger Ed Canyons.

Rancheria sections in the Sacramento Mountains are referred to as basin-margin sections in this report and those in the Franklin Mountains, Hueco Mountains and Bishop Cap are referred to as basinal sections.

The Rancheria in the Sacramento Mountains contains many features characteristic of basin-margin sedimentation and erosion. These include submarine channels that dissect the basin margin, slump features, and sheets (fans) of allochthonous skeletal shelf debris. These features are discussed in detail in other parts of this paper.

FOSSILS

The Rancheria Formation has a very sparce biotic assemblage. Body fossils include brachipods, crinoid and other echinoderm fragments, bryozoans, sponge spicules, and tetracorals (all sparse); foraminifera, conodonts, gastropods, and bone or tooth fragments (scarce); and trilobites, bivalves, radiolarians, ophiuroids, and sponge fragments (rare). Although the Rancheria has a low faunal density, it does have a high faunal diversity which probably reflects relatively stable environmental conditions. Faunal lists of identified taxa have been presented for different outcrop areas of the Rancheria by King

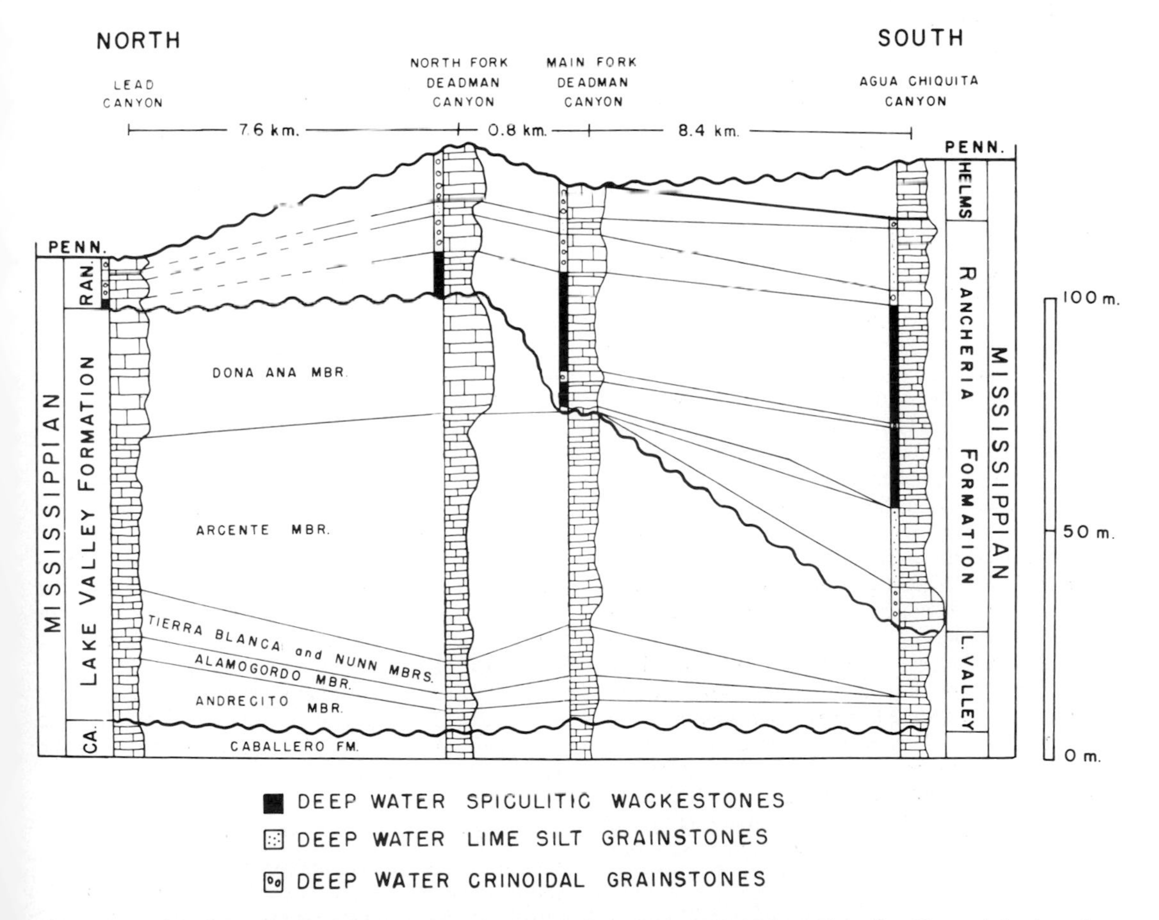

FIG. 4.—Correlation of Mississippian strata in the southern Sacramento Mountains. See Figure 1 for location of sections. Note absence of the shallow-water lithofacies in the Sacramento Mountains.

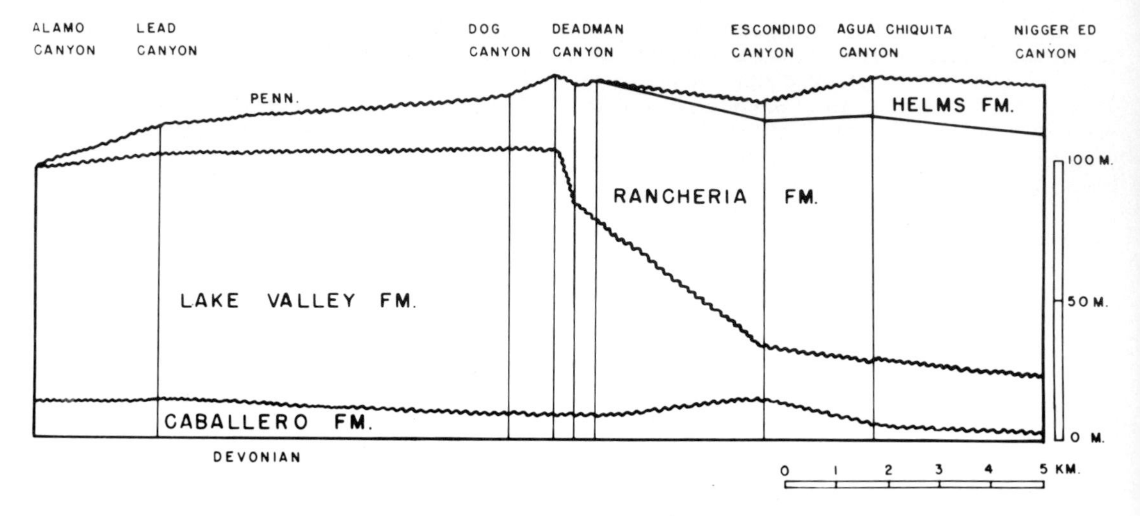

FIG. 5.—Cross-section across the edge of the Rancheria basin, Sacramento Mountains.

and King (1929), King (1934), Laudon and Bowsher (1949), Pray (1952), and Harbour (1972). Burton (1965) and Lane (1974, 1975) have described the conodonts in the Rancheria Formation.

Most of the body fossils are benthic invertebrates and appear to be indigenous to the site of deposition. Only those that are associated with coarse skeletal grainstone (allodapic) show signs of much abrasion and probable transportation. All the strictly shallow-water biotics, such as dasyclad algae fragments, are found within these transported skeletal beds and do not appear to be indigenous. The only body fossil that may provide paleobathymetric information is *Leiorhynchus carboniferum* which is characteristic of the Rancheria. According to Wilson (1969) and others, leiorhynchids in the Devonian and Mississippian are characteristic of basinal euxinic deposits. Sponge spicules are also common in the Rancheria and, according to Wilson (1969), these, too, are common in deep water limestone. Although most sponge spicules are now composed of recrystallized calcite, some retain evidence of a central canal indicating they were siliceous. Others retain their original siliceous composition.

Trace fossils are common in the Rancheria Formation. Most burrows resulted from general bioturbation and have little paleobathymetric significance. The shallow-water lithofacies described in this report, however, contains a very distinctive shallow-water trace-fossil assemblage. It is characterized by the presence of *Zoophycus* (Fig. 9E) and several burrows that indicate the *Skolithos* facies (*Skolithos, Teichichnus*, and *Corophioides*).

LITHOFACIES

The rocks of the Rancheria Formation are differentiated into four major lithofacies on the basis of lithology, fauna, and sedimentary structures. The characteristics of these four lithofacies are outlined in Figure 6. Two minor lithofacies are also present but they account for a very minor portion of the Rancheria and are not critical to a general discussion such as this. The lithofacies represent deposition at slightly different depths, bottom conditions, water conditions, and/or modes of deposition. Three of these lithofacies represent deposition in relatively quiet, deep, euxinic water; they are differentiated primarily on the basis of sediment size and sorting. The fourth lithofacies represents deposition in more agitated, better oxygenated, shallower water, probably above wave base.

The relative proportions of these lithofacies for different parts of the basin are shown in Figure 7. In reference to the distribution of these lithofacies it should be noted that (1) the deep-water spiculitic wackestone and deep-water lime silt grainstone are equally common in the basin and along the basin margin, (2) the deep-water crinoidal grainstone is strongly concentrated along the basin margin, and (3) the shallow-water lime-silt grainstone lithofacies is restricted to the upper one-third of the Rancheria in the basin.

There are several examples of syndepositional massflow failure and resedimentation in the Rancheria Formation. These are discussed at the end of this section.

Deep-Water Spiculitic Wackestone

This facies forms approximately 45 percent of the Rancheria. These rocks consist of lime mudstone, spiculitic wackestone, and minor skeletal wackestone. Sponge spicules comprise 5 to 35 percent of the total rock volume (Fig. 8A). Minor amounts of silt-size skeletal fragments, ostra-

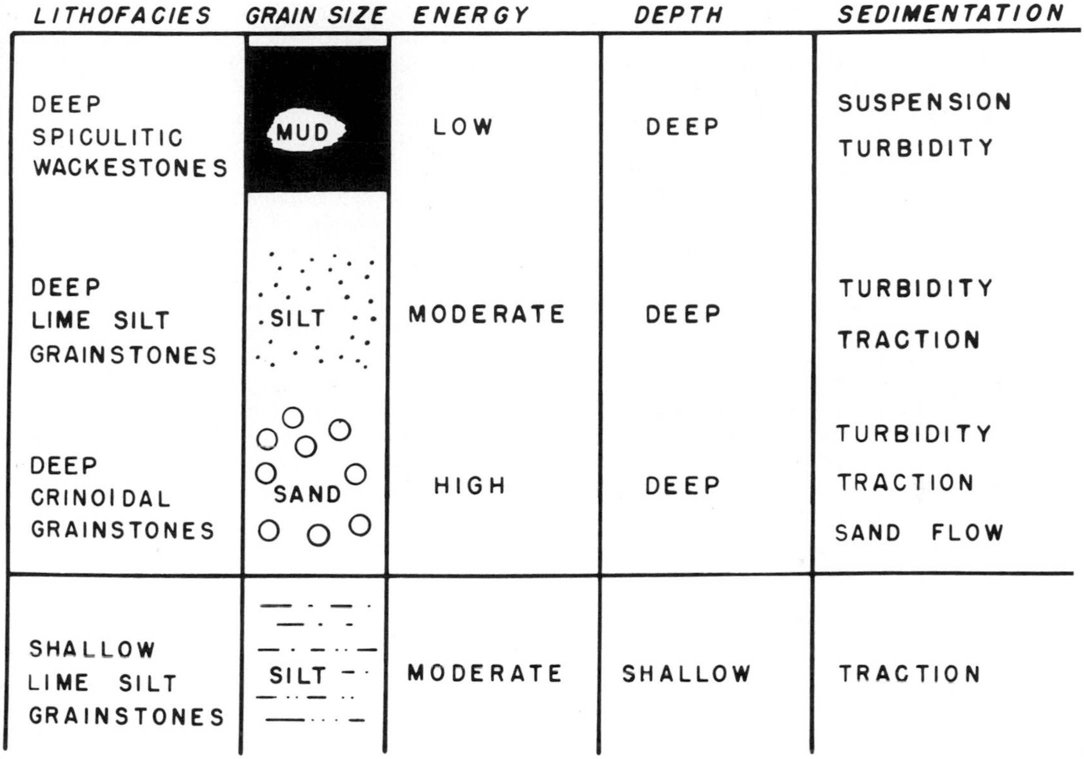

LITHOFACIES
GRAIN SIZE
ENERGY
DEPTH
SEDIMENTATION
DEEP SPICULITIC WACKESTONES
MUD
LOW
DEEP
SUSPENSION
TURBIDITY
DEEP LIME SILT GRAINSTONES
SILT
MODERATE
DEEP
TURBIDITY
TRACTION
DEEP CRINOIDAL GRAINSTONES
SAND
HIGH
DEEP
TURBIDITY
TRACTION
SAND FLOW
SHALLOW LIME SILT GRAINSTONES
SILT
MODERATE
SHALLOW
TRACTION

FIG. 6.—Rancheria lithofacies (in order of description in the text).

DEEP
I
II
III
50 %
40 %
30 %
20 %
10 %
0 %
DEEP
SHALLOW
I
II
III
IV
EARLY
LATE
BASIN MARGIN
BASIN

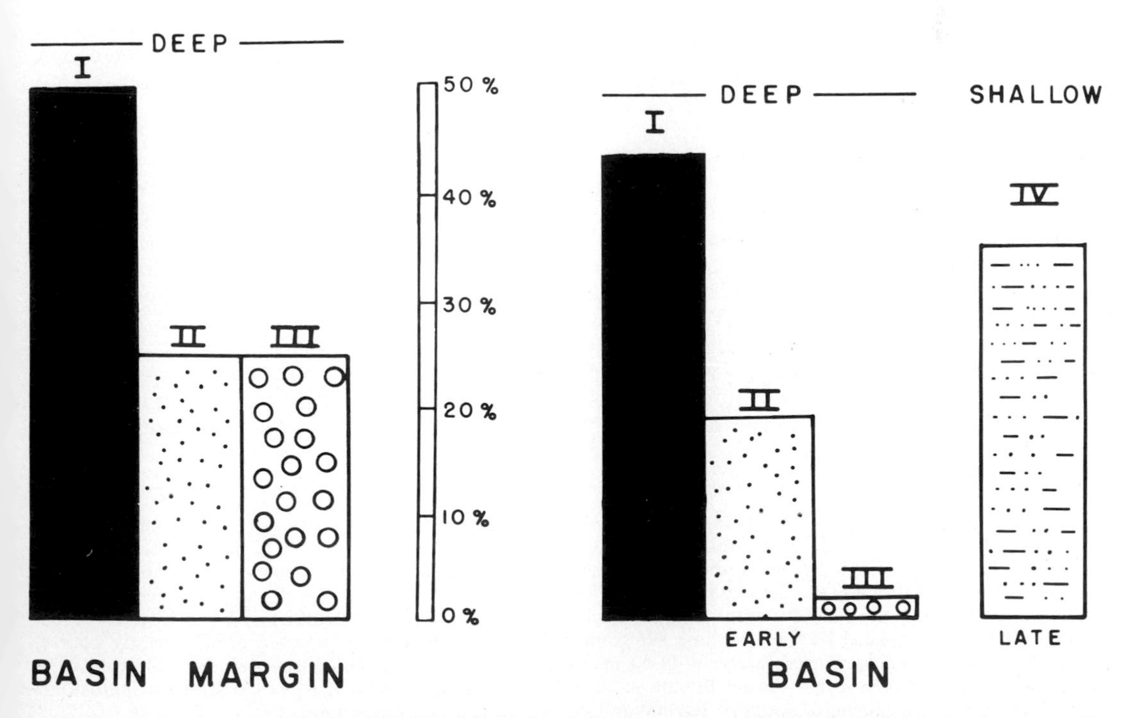

FIG. 7.—Relative proportion of Rancheria lithofacies along the basin margin and in the basin (each area totals 100 percent).

code, foraminifera, and conodonts are also present.

Beds of this lithofacies are dark gray, weather light gray, contain little or no chert, and are remarkably even bedded and continuous (Fig. 8B). Most are medium- to thick-bedded and some are interbedded with much thinner argillaceous limestone beds (bedding thickness is ccording to Ingram, 1954).

Sedimentary and biogenic structures in these rocks vary considerably. These rocks can be grouped, however, into two general categories: (1) nonlaminated beds, and (2) laminated beds.

Nonlaminated beds.—Laminations generally are absent. Beds range from those that are intensely burrowed to those that are nonburrowed and structureless. Burrows are nonspecific and resulted from general bioturbation (Fig. 8C). Body fossils are rare to sparse in these rocks. The nonburrowed or slightly burrowed beds generally contain few or no fossils. The burrowed beds, however, commonly contain a sparse fauna including brachiopods, tetracorals, trilobites, sponge fragments, bryozoans, gastropods, and crinoid fragments. Argillaceous interbeds usually are absent.

Laminated beds.—Beds of this group are characterized by the presence of thin (0.1 to 10 mm) laminae which form 5-25% of the rock. The laminae consist of silt-size material, are discontinuous, and in places, lenticular (Fig. 8D). Some are microcross-laminated. Some laminae appear to have been disrupted by escaping bubbles of trapped water (disruptions on mm scale). Laminated beds commonly alternate with thin argillaceous limestone interbeds. Body fossils are rare except in the argillaceous interbeds.

Origin—Sediment size, color, biota, and the lack of scour features indicate deposition in a quiet, low oxygen, marine environment. Nonlaminated beds appear to be the result of slow pelagic sedimentation. The absence of interbeds and of any indication of bottom traction suggest nearly continuous slow deposition. Variations in the degree of burrowing probably reflect fluctuations in the degree of basin stratification and water circulation at the time of deposition. Nonburrowed beds were deposited under stagnant anaerobic bottom conditions toxic to bottom life whereas the burrowed beds represent slightly better circulation, sufficient at least to replenish nutrients and oxygen to a level capable of supporting a sparse bottom fauna.

The laminated beds can be explained in various ways. They may represent slow deposition from suspension like the nonlaminated beds. In that case the low density of burrows in these rocks reflects low-oxygen, dysaerobic to anaerobic conditions. Laminations are simply due to the introduction of slightly coarser silt-size material. The presence of microcross-laminations indicates intermittent weak bottom currents.

The absence of fossils, except in interbeds, and the presence of microcross-laminations suggest an alternate hypothesis for the deposition of the laminated beds. The laminated beds may represent episodes of more rapid deposition. Sedimentation was too rapid for the sparse bottom fauna to disturb primary structures. Interbeds associated with the laminated beds may represent interim periods of normal slow suspension deposition. Rapid deposition for the laminated beds is supported by the presence of microcross-laminations and the disruption of some laminae by dewatering. The processes by which lime mud is deposited rapidly are uncertain. These beds may have been deposited by low-density, low-velocity turbidity currents. Each bed would represent a series of thin turbidites containing Bouma's C and D horizons. The microcross-laminated silt laminae would correspond to the C horizon and the horizontally laminated lime mud would correspond to the D horizon.

Deep-Water Lime-Silt Grainstone

This facies forms approximately 22 percent of the Rancheria Formation. These rocks are primarily lime-silt grainstone although lime mudstone forms a minor portion of some graded beds. Grains in the lime-silt grainstone are predominantly medium to coarse silt-size lime peloids. Some peloids can be identified as abraded skeletal fragments, but most are cryptocrystalline micrite of uncertain origin (Fig. 8E). Siliceous sponge spicules also are common in the silicified beds. In many beds the peloids appear to be remarkably loosely

FIG. 8.—Rancheria lithofacies. *A*—Thin section of spiculitic wackestone from deep-water spiculitic wackestone lithofacies. Grains are calcified sponge spicules. *B*—Typical outcrop of deep-water spiculitic wackestone strata, Agua Chiquita Canyon, Sacramento Mountains. *C*—Intensely burrowed, nonlaminated beds of the deep-water spiculitic wackestone lithofacies, Deadman Canyon, Sacramento Mountains. *D*—Well laminated bed showing thin, discontinuous laminae—deep-water spiculitic wackestone lithofacies. Note scarcity of burrowing. *E*—Thin section of grainstone typical of the two lime silt grainstone lithofacies. Dark grains are predominantly coarse silt-size lime micrite peloids (approximately 40-60 microns). Quartz silt and neomorphosed skeletal grains are present also (sparry patches). *F*—Partial Bouma sequence (horizons B, C, D) in deep-water lime silt grainstone strata. Two-tenths of a kilometer south of Escondido Canyon, Sacramento Mountains.

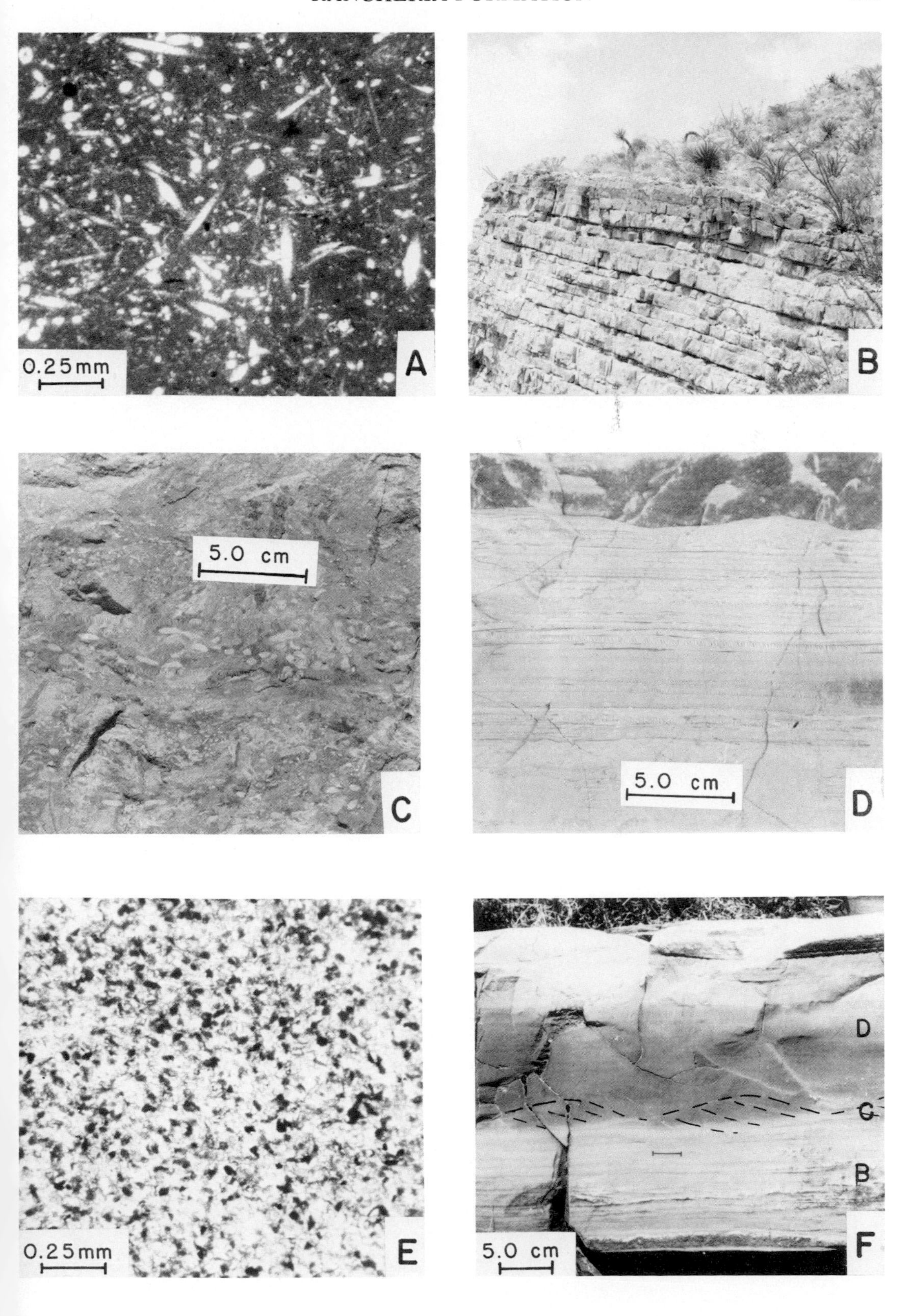
0.25mm
A
B
5.0 cm
C
5.0 cm
D
0.25mm
E
D
C
B
5.0 cm
F

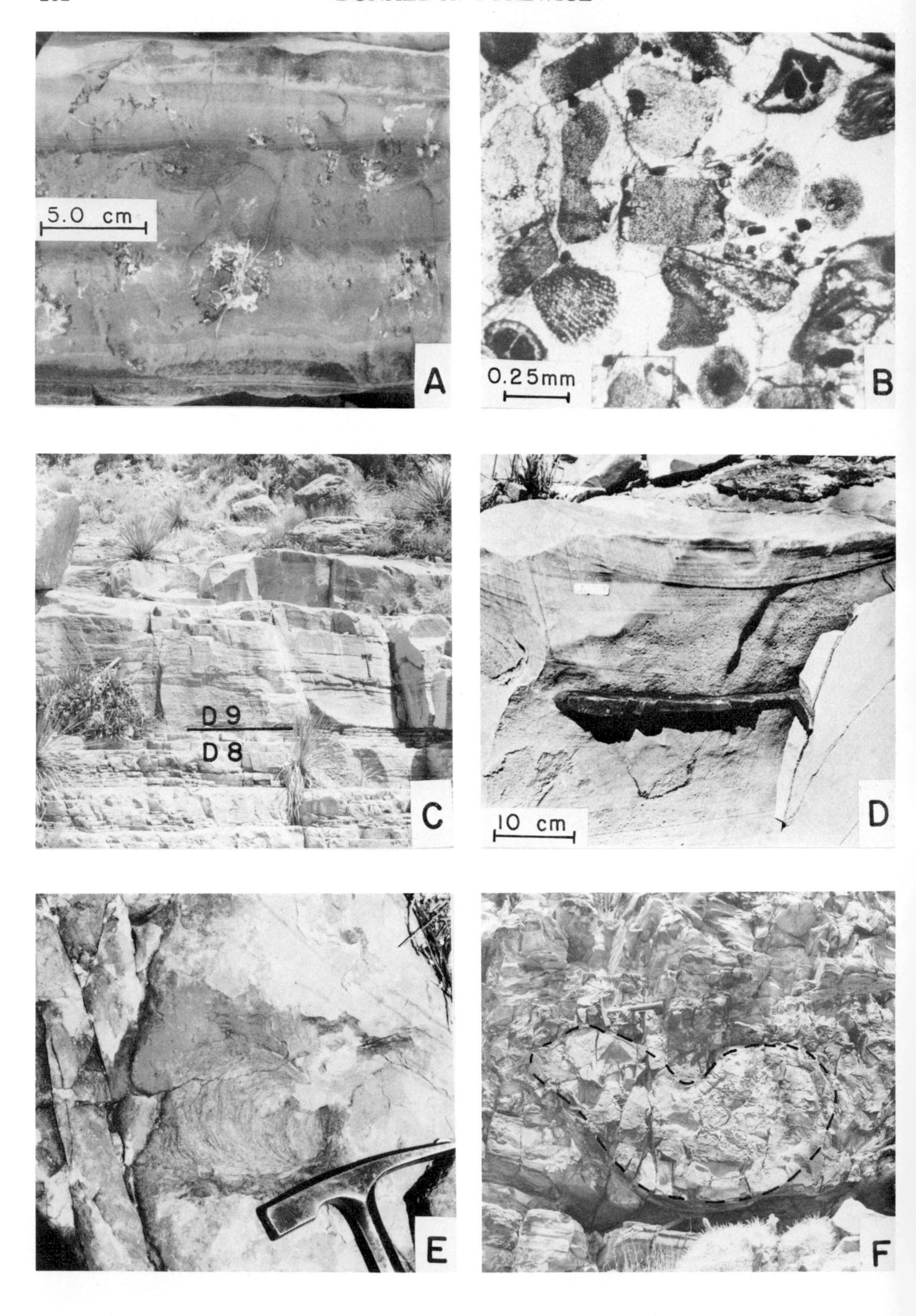
5.0 cm
A
0.25mm
B
D9
D8
C
10 cm
D
E
F

packed; some peloids seem to be floating in cement. The appearance of grains floating in cement is probably due to the presence of some type of neomorphosed skeletal grain which cannot be differentiated from the fine-grained cement.

These rocks are easily recognized in the field. They are dark gray, weather medium dark gray, and contain abundant secondary silica which weathers moderate brown. Siliceous sponge spicules were the probable source of this silica. Beds are thin- to thick-bedded with thin argillaceous limestone interbeds.

These beds of silt-sized grainstone are structureless to finely laminated. Some are cross-laminated and some include a partial Bouma turbidite sequence (horizons B, C, D; Fig. 8F). Some also are normally graded, from coarse silt at the base to lime mud at the top. Graded sequences generally range from 5 to 15 centimeters thick. Load casts and flame structures are common between graded sequences (Fig. 9A). Ripple marks are present but scarce.

Fossils are absent or scarce, and where present they generally are associated with the thinner interbeds. Fossils include crinoid fragments, brachiopods, and, rarely, ophiuroids. There is some bioturbation and some irregular vertical feeding burrows but neither have much paleobathymetric significance. In graded beds burrowing is usually confined to the top of a graded sequence.

Origin.—Low faunal densities and the dark color of these rocks indicate deposition in a low-oxygen marine environment. Beds which contain a partial Bouma sequence or normal grading probably were deposited by turbidity currents. Beds which do not contain a Bouma sequence or grading may have been deposited, or redeposited, by bottom-contouring traction currents (contourites). They contain many features which Bouma (1972) and Bouma and Hollister (1973) suggest are indicative of contourities and differ only in that most are thicker than previously described contourites. This sediment occurs in beds 4 to 30 cm thick and most previously described contourites are less than 5 cm thick. Features shared with described contourites include sharp bottom and top contacts, common cross-laminations and horizontal laminations, very good sorting, very low percentage of matrix, and scarcity of fossils.

Deep-Water Crinoidal Grainstone

This facies forms a minor part of the basin portion of the Rancheria Formation (less than 5 percent of all Rancheria strata) but comprises over 20 percent of the formation along the basin margin (Fig. 7). This rock is predominantly coarse crinoidal grainstone but includes mud-poor packstone as well. Grains include abundant medium to coarse sand-size echinoderm, brachiopod, bryozoan, and mollusc fragments, scarce to common silt-size lime peloids, quartz, foraminifera, ostracodes, phosphatic bone, tooth, and conodont material, abundant to rare oolites, and rare calcareous green algal fragments (Fig. 9B). Most mollusc fragments contain micrite envelopes and, in most cases, the micrite envelope is all that remains of the original shell. All fossils in this lithofacies are fragmented and abraded and indicate current transport. Trace fossils are rare or absent.

These rocks are medium to very thick bedded (Fig. 9C). Bedding plane surfaces are scoured in places with local development of scours as much as four meters wide and one meter deep. Some units of this facies are continuous and uniform for several kilometers while others are discontinuous and lenticular and may pinch out abruptly within 15 meters or gradually over one kilometer. Most units of this facies lie on channelized surfaces.

Origin.—This coarse crinoidal grainstone is probably allochthonous as suggested by the minor volume of this rock in the Rancheria, the abundance of fossil debris in normally dark, fine-grained, nonfossiliferous deposits, and the association with channeled surfaces. The presence of oolites, micrite envelopes, and green algal fragments in some of these units further suggests that these sediments originated in shallow water. These features plus the lateral discontinuity of these units and their variable thickness, indicate that these beds may represent tongues or fans of shelf-derived skeletal debris. This is consistent

FIG. 9.—Rancheria lithofacies. *A*—Load casts between successive graded sequences in deep-water lime silt grainstone strata, Escondido Canyon, Sacramento Mountains. Four graded sequences in this bed. *B*—Thin section of crinoidal grainstone typical of the deep-water crinoidal grainstone lithofacies. Grains are medium to coarse sand-size echinoderm and bryozoan fragments, and minor silt-size lime peloids. Agua Chiquita Canyon, Sacramento Mountains. *C*—Typical outcrop of thick-bedded deep-water crinoidal grainstone strata. Unit D9, Deadman Canyon, Sacramento Mountains. *D*—Medium scale trough cross-stratification in shallow-water lime silt grainstone strata, Hueco Mountains. *E*—*Zoophycus* trace fossil on bedding plane, shallow-water lime silt grainstone strata, Vinton Canyon, Franklin Mountains. *F*—Mass gravity feature in Sacramento Mountains showing lower contorted zone. Note large rolled sedimentary ball. Hammer for scale.

with the concentration of these sediments along the basin margin.

Current structures and grain size indicate deposition by high energy bottom currents. The presence of grading in some units suggests that turbidity currents were at least a partial agent of deposition although the lack of turbidite features in most beds makes it questionable that they are the sole or major agent of deposition. Important depositional agents probably included nonturbid bottom traction currents and possibly nonturbid sand flows.

Shallow-Water Lime-Silt Grainstone

This lithofacies is restricted laterally and vertically in the Rancheria. It forms the upper one-third of the Rancheria in the basin sections (Franklin and Hueco Mountains and Bishop Cap). This rock is entirely lime-silt grainstone, the grains being composed of silt-size lime peloids 20 to 60 microns in diameter. As in the Deep-Water lime-silt grainstone, these grains also appear to be remarkably loosely packed.

This rock is distinguished from the deep-water lime silt grainstone in being very thick to medium bedded. It is dark gray and weathers medium light gray. It contains a moderate amount of secondary chert which weathers moderate brown, forming distinctive brown-weathering slopes.

Beds of this lithofacies are structureless to finely laminated. The laminations are closely spaced, 0.1 mm thick, planar and continuous. Small- to medium-scale cross-stratification, although not abundant, is characteristic of this facies and further distinguishes it from the deep-water lime silt grainstone. Cross-stratified sets are primarily trough structures, 30 to 60 cm wide and 2 to 8 cm deep (Fig. 9D). Some beds are slightly contorted and suggest soft-sediment deformation.

Fossils are scarce except in argillaceous interbeds. These include brachiopods, crinoid fragments, bryozoans, trilobites, and tetracorals. Trace fossils are abundant and include *Zoophycus* (Fig. 9E), a trace fossil of the *Zoophycus* facies (Seilacher, 1967), and numerous trace fossils of the *Skolithos* facies (*Skolithos, Teichichnus*, and *Corophioides*).

Origin.—The sediment size and the type and abundance of current structures in this lithofacies indicate that these rocks were deposited by low- to moderate-energy bottom currents (strong enough to carry silt-size grains in traction and sometimes scour the bottom). There are no structures to indicate turbidity current deposition. Sediments were probably deposited by traction currents.

The presence of a *Skolithos* trace-fossil assemblage suggests moderately shallow water, at least above wave base level. Trace fossils of the *Skolithos* facies are characterized by vertical domichnial burrows and are considered by Seilacher (1967) to represent very shallow, marginal marine conditions. *Zoophycus* is also present in this lithofacies and Seilacher (1967) defined another trace-fossil facies on the basis of its presence, suggesting that it represents fairly deep, outer shelf to upper slope, quiet water environments. Plicka (1968), however, has shown that *Zoophycus* is also present in shallow-water, rapid sedimentation environments. It now appears that *Zoophycus* ranges from deep-slope to shallow-shelf environments. The presence of *Zoophycus* with trace fossils of the *Skolithos* facies is not inconsistent with a shallow-water origin for this lithofacies.

The restriction of this lithofacies to the top or near the top of the Rancheria and the simultaneous appearance of abundant shallow-water trace fossils and medium-scale current structures suggest that this lithofacies represents filling and shoaling of the Rancheria Basin. This lithofacies is succeeded by a relatively minor lithofacies in the Hueco Mountains which suggests even shallower water conditions. This minor lithofacies consists of thin- to medium-bedded lime-silt grainstone. It contains a rich trace-fossil assemblage of the *Skolithos* facies and abundant ripple marks and ripple cross-laminations.

Subaqueous Mass-Flow Deposits

Examples of mass-flow sedimentation are rare in the Rancheria. Only three have been described (Yurewicz, 1973b), two in the Sacramento Mountains and one at Bishop Cap. The two in the Sacramentos occur together, one above the other, in a zone 0.5 to 2.5 m thick that can be traced for one kilometer south from the south side of Escondido Canyon. They occur on the low-angle toe of the basin margin (slopes probably less than one degree) in a deep-water lime silt grainstone unit. The lower mass-flow deposit consists of rolled sediment balls (0.3 to 1.0 m diameter) of lime silt grainstone embedded in a contorted to structureless matrix of the same lithology (Fig. 9F). The base of the lower flow generally is sharp and the underlying beds usually are undeformed. In some places, however, the beds below are slightly folded. The upper mass-flow deposit is a completely homogenous, structureless, closely packed lime-silt grainstone bed 0.3 to 1.0 m thick. Its upper and lower contacts generally appear sharp.

These features appear to represent two slightly separated episodes of subaqueous mass failure and resedimentation. The lower flow was arrested at Dott's (1963) mass-flow stage of deformation in which the yield limit of cohesion is exceeded and the original stratification is preserved but contorted. It represents plastic flow. No stratification is preserved in the upper flow and it was not arrested until it became a mudflow. The liquid limit was exceeded and deformation was by viscous fluid flow.

The third example of mass failure is present in the upper part of the Rancheria at Bishop Cap in New Mexico. It is a homogeneous, structureless, lime mudstone bed, 2 m thick, within a section of deep-water lime-silt grainstones. It is exposed throughout the Rancheria exposures at Bishop Cap (approximately one kilometer). The lower and upper contacts appear to be sharp and even, although pervasive silicification in the surrounding lime-silt grainstone beds makes the contact obscure in places. In several places there appears to have been some deformation of the beds just below the contact. This example of mass failure is interpreted as a subaqueous mudflow resulting from complete loss of cohesion and subsequent viscous fluid flow.

The ultimate cause of failure is not obvious. As Dott (1963) points out there are several conditions that could induce mass failure including overloading, earthquakes, abnormal waves or tidal currents, unusually high pore-fluid pressure, and undercutting. What is perhaps most significant is the scarcity of such features in the Sacramento Mountains where the Rancheria laps onto the basin margin. This suggests that this mild basin slope (less than 2 or 3 degrees; see discussion of basin slope) was relatively stable. It is also pertinent that these slump features are stratigraphically lower than the angular diastems discussed in the next section and are in no way related to them.

INTRAFORMATION EROSION SURFACES

Sharp erosion surfaces within the Rancheria Formation, interpreted here as submarine diastems, represent an important part of the Rancheria record. Submarine intraformation erosion surfaces are common to many deeper water carbonate deposits and their origin is a source of debate (see Wilson, 1969; Davies, McIlreath, this volume). Knowledge of the processes which operate in their formation would aid our understanding of carbonate basin and basin margin processes (sedimentation, erosion, tectonism). Two types of intraformation erosion surfaces are recognized in the Rancheria: angular diastems, and nonangular diastems.

Angular diastems.—Eight large-scale angular discontinuities, that superficially resemble unconformities (Fig. 10A), are present in the Rancheria. These features rarely are exposed for more than several tens of meters along the Sacramento escarpment so it is impossible to establish their overall geometry. Individual truncated surfaces can be traced laterally for 30 to 90 m and contain 2 to 9 m of relief. The beds below the diastem generally are deep-water spiculitic wackestone that appears to be horizontal. Some truncation surfaces are mantled by thin (0.3 to 1.0 m) deep-water crinoidal grainstone units. The crinoidal beds immediately above the diastem are parallel to the diastem and do not onlap it within the outcrop area. The thin units of crinoidal grainstone are followed upwards by additional deep-water spiculitic wackestone. In some diastems the crinoidal grainstone is absent and deep-water spiculitic wackestone is present on both sides of the diastem. Lime mudstone intraclasts up to 15 cm in diameter are present along these diastems in places.

All angular diastems are restricted to the lower portion of the Rancheria in the Sacramento Mountains between Deadman and Agua Chiquita Canyons (Fig. 4). This is the area in the Sacramento Mountains where the Rancheria thins most rapidly against the basin margin and probably represents the steepest portion of the basin edge.

Pray (1961) and Yurewicz (1973b, 1975) suggest that these features resulted from submarine erosion. Evidence for a submarine origin include: (1) the deep-water character of the strata above and below the erosional surfaces, (2) the absence of shallow-water or subaerial features along the contact, and (3) the similarity of these features to intraformational features in other carbonate deposits of known deep-water origin. They are somewhat similar to the angular diastems in the Permian basinal strata of west Texas. Pray (1968), McDaniel and Pray (1969), and Harms and Pray (1974) suggest that those in the Permain of Texas resulted from submarine corrasion along the edge of a density stratified basin. Harms (1974) provides a detailed analysis of the channels in the basin sandstone of the Brushy Canyon Formation. Wilson (1969) described similar features in other basinal limestone but interpreted them as large-scale slippage or slump features. Such an explanation seems unlikely for those in the Rancheria because of: (1) the scarcity of deformed strata that would result from slumping and (2) the presence of intraclasts from the underlying spiculitic wackestones in the allochthonous crinoidal beds immediately above the diastem.

The restriction of these features to the Sacramento Mountains and to an area there where the Rancheria thickens rapidly against the basin margin suggests that they are basin-margin features. As they are never completely exposed it is difficult to make a positive interpretation of their geometry and origin. It is my belief that these features represent broad shallow channels (45 to 180 m wide, 3 to 20 m deep) produced by corrasion on a basin margin depositional slope. They probably belong to channel networks which dissect the basin margin. More work and data are clearly needed for a more detailed interpretation of these features.

Nonangular diastems.—Not all surfaces of erosion in the Rancheria are deeply channelized or angular as those just described. There are also irregular, nonangular surfaces of erosion in which the beds on either side of the diastem parallel the diastem and where erosion generally involves no more than 30 or 60 cm of section. Two major diastems of this type have been recognized in the Rancheria. One is the diastem at the top of the Las Cruces Member in Vinton Canyon in the Franklin Mountains, first recognized by Laudon and Bowsher (1949), and the other is in the middle of the Rancheria in the Sacramento Mountains.

The nonangular diastem in the Sacramento Mountains can be traced for 6.5 kilometers along the basin margin. The rock below the diastem is deep-water spiculitic wackestone and that above is deep-water crinoidal grainstone. This diastem changes considerably as it is traced from the toe to the crest of the basin margin. Along the toe and out into the basin relief varies from 1 to 15 cm. Intraclasts along this portion of the diastem range in size from 1 to 30 cm. As it is followed to the crest, relief along the diastem increases dramatically. Along the crest of the basin margin channels 2 m deep and 2.5 m wide are present and intraclasts are nearly one meter in diameter.

What erosional processes created this diastem? The absence of subaerial features along this contact and the presence of *Rhizocorallium* trace fossils and phosphatic nodules in places along the truncated surface suggest submarine erosion. The increase in relief as it is followed upslope can be interpreted in several ways. (1) If erosion resulted from wave action during an unusually heavy storm, the increase in relief may be due to shoaling. (2) If it was scoured by turbidity currents, the increase in relief may reflect closer proximity to the current's source.

The erosional surface at the top of the Las Cruces Member is very well exposed along 60–90 m of nearly continuous exposures in Vinton Canyon in the Franklin Mountains but cannot be traced further owing to poorer outcrops. This surface is extremely sharp and irregular, with relief of only 2 to 5 cm. The Las Cruces Member consists of very even-bedded, sparsely fossiliferous deep-water spiculitic wackestone. The Rancheria above this contact also consists of deep-water spiculitic wackestone. Immediately above the break is a distinctive lithology about 15 cm thick apparently related to the diastem. These beds contain abundant coarse silt-size recrystallized skeletal fragments that include sponge spicules, molluscs, echinoderms, brachiopods, bones, and teeth. Nonfragmented tetracorals, brachiopods, radiolarians, and foraminifera also are present. Numerous finger-size phosphatic nodules are also characteristic of this horizon. Chert pebbles, plant fossils, and beds of shale, siltstone, and sandstone, all recorded by Laudon and Bowsher (1949), were not seen in my work.

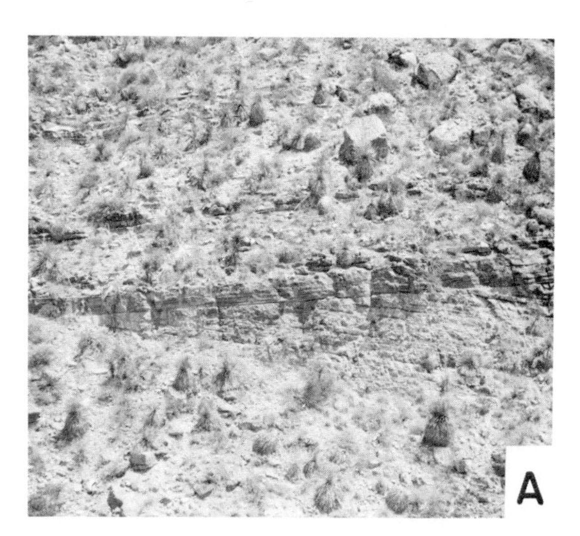

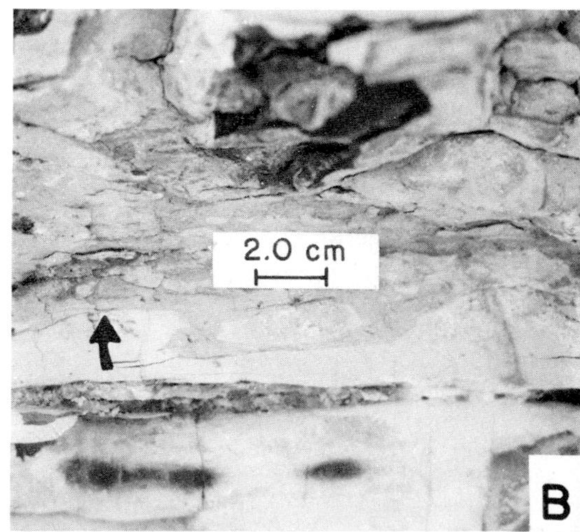

FIG. 10.—Intraformation erosion surfaces. *A*—Angular diastem on the south side of Deadman Canyon, Sacramento Mountains. *B*—Nonangular diastem, upper contact of Las Cruces Member on the south side of Vinton Canyon, Franklin Mountains. Note very low erosional relief and irregularity of contact. Arrow marks the contact.

The overall features of this contact suggest submarine erosion rather than uplift and subaerial erosion, as concluded by Laudon and Bowsher. Evidence for a submarine origin include: (1) the presence of similar marine beds, probably deposited in deep water, below and above this diastem, (2) the presence of phosphatic nodules, (3) the lack of evidence for subaerial exposure, and (4) the absence of shoal-water facies above the contact. The phosphatic nodules also suggest that erosion may have followed or preceded a long period of nondeposition. Finally, the absence of intraclasts along this contact, and the very low relief, the sharpness, and the irregular nature of this contact suggest the possibility of submarine solution rather than submarine corrasion. Unfortunately biostratigraphic control to determine the duration of this break is lacking. The work of Lane (1974) provides the closest control. He collected a sparse conodont fauna from the Las Cruces Member in Vinton Canyon and from a crinoidal unit six meters above the upper contact of the Las Cruces. Conodonts collected from these horizons suggest that the Las Cruces is late Osagion or early Meramecian and that the Rancheria above it is Meramecian.

WATER DEPTH AND ENVIRONMENT OF DEPOSITION

Fossils and general sedimentologic characteristics show that the Rancheria Formation represents marine deposition. The fine-grained texture of the sediment indicates that the bottom was normally quiet and the dark color and sparse biota suggest that water near the bottom was restricted and somewhat euxinic. Variations in density of benthic fossils and bioturbation indicate that conditions ranged from anaerobic to dysaerobic to aerobic. It is almost impossible to determine absolute water depth. Most of the features associated with the Rancheria suggest that it was deposited in relatively deep water (100 to 250 meters), most of it probably below wave base. Evidence for a deep water origin include: (1) absence of conclusive shallow-water sedimentary features, e.g. fenestral fabric, mud cracks, sebkha evaporites, subaerial crusts, etc., (2) low faunal density, (3) absence of any strictly shallow-water fossils except as allochthonous debris, (4) presence of leiorhynchid brachiopods, (5) dark color (due to unoxidized organic material and minor iron sulfide), (6) paucity of wave- or surf-formed sedimentary structures, suggesting deposition below wave base, (7) absence of algae except as rare transported fragments in allochthonous skeletal sand, (8) similarities with other known deep-water basinal limestone, and (9) basin geometry and basin-margin relief.

Although it is nearly impossible to estimate water depths in the Rancheria, it is possible to speculate. There is no appreciable difference in rock types between the toe and the crest of the basin margin in the Sacramento Mountains. Rocks throughout the Rancheria here appear to have been deposited below wave base. This indicates that the shallowest portion of the basin edge must have been deeper than, say, 90 m, to use an arbitrary depth for wave-base. The stratigraphic relief of 75 m along the basin margin indicates that water depths over the deeper part may have been greater than 165 m.

Rancheria outcrops further in the basin expose strata that probably were deposited in somewhat deeper water. The upper one-quarter to one-half of the Rancheria in the basin, however, contains shallow-water structures and trace fossils and was probably deposited near or above wave base. Indicators of very shallow water such as fenestral fabric, mudcracks, evaporites, and algae are lacking. Intertidal and supratidal conditions, at least, were certainly never reached. Shoaling of the Rancheria in the basin outcrops probably represents simple filling of the basin to wave base. This suggests initial depths for basin sections were less than 250 m.

CONCLUSIONS AND SUMMARY

(1) The Rancheria Formation as used here includes the Las Cruces Formation of Laudon and Bowsher (1949). The Las Cruces is herein redesignated as a member of the Rancheria Formation.

(2) The Rancheria contains a very small but diverse assemblage of normal marine to euxinic, benthic body fossils that suggest a deep, stable bottom environment. Trace fossils are not uncommon and can generally be classified as general bioturbation. Trace fossils of the *Zoophycus* and *Skolithos* facies are associated with the shallow-water lime-silt grainstone lithofacies.

(3) Four major lithofacies can be recognized in the Rancheria Formation. Three represent deposition in relatively deep water and the fourth represents somewhat shallower conditions during the final phases of basin filling. The deep-water lithofacies consist primarily of lime mud and lime-silt peloids. They were deposited by a combination of turbidity currents, bottom traction currents, and pelagic suspension deposition. Deposition of this fine-grained sediment was intermittently interrupted by deposition of coarser skeletal sand deposited by turbidity currents, sand flows, and other traction currents. This coarser sediment is concentrated in the Sacramento Mountains along the edge of the basin. The shallower water lithofacies also consists of lime-silt peloids. Shallow-water conditions are suggested by the presence of abundant medium-scale current structures and shallow-water trace fossils. This lithofacies is restricted to the upper one-third

of the Rancheria in the basin sections.

(4) Mass-flow features are present, but rare, in the Rancheria.

(5) Submarine diastems represent an important part of the Rancheria record. Two types of intraformation erosion surfaces are recognized—angular diastems, and nonangular diastems. The angular diastems appear to be broad shallow channels, produced by scour, that dissected the basin margin. The nonangular diastems are low-relief surfaces of erosion and are present in the basin and on the basin margin.

(6) Rancheria deposits in the Sacramento Mountains were deposited along the northern edge of the basin. Evidence for basin-margin sedimentation include: (a) rapid depositional onlap thinning, (b) sheets or fans of shelf-derived allochthonous skeletal debris, (c) submarine channel systems, and (d) evidence of mass gravity failure.

(7) Most of the Rancheria appears to have been deposited in relatively deep water, at least below wave base. Water depths are inferred to have been 100 to 250 m in the basin. Gradual shoaling occurred as the basin filled.

ACKNOWLEDGEMENTS

Gratitude is expressed to Lloyd C. Pray who supervised this study and provided many invaluable suggestions. Thanks also to Frank E. Kottlowski, William Meyers, Richard Lane, and David L. Clark who aided at various stages of this study. Special thanks goes to Mr. Fairchild who kindly allowed access to his ranch. Thanks also to the editors of this publication, Harry Cook and Paul Enos, for their helpful suggestions. This study was supported by the New Mexico State Bureau of Mines and Mineral Resources and the author owes them his deep gratitude. This paper summarizes part of an M.S. thesis at the University of Wisconsin, Madison.

REFERENCES

Armstrong, A. K., 1962, Stratigraphy and paleontology of the Mississippian System in southwestern New Mexico and adjacent southeastern Arizona: New Mexico Bur. Mines and Mineral Resources, Mem. 8, 99 p.

——, 1965, The stratigraphy and facies of the Mississippian strata of southwestern New Mexico: New Mexico Geol. Soc., Guidebook 16th Field Conf., p. 132–140.

—— and Mamet, B. L., 1974, Biostratigraphy of the Arroyo Penasco Group, Lower Carboniferous (Mississippian), north-central New Mexico: New Mexico Geol. Soc., Guidebook 25th Field Conf., p. 145–158.

Bouma, A. H., 1972, Fossil contourites in lower Niesenflysch, Switzerland: Jour. Sed. Petrology, v. 42, p. 917–921.

—— and Hollister, C. D., 1973, Deep ocean basins in sedimentation: *In* G. V. Middleton and A. H. Bouma (eds.), Turbidites and deep-water sedimentation: Soc. Econ. Paleontologists and Mineralogists Pacific Sec., Los Angeles, California, p. 79–118.

Burton, R. C., 1965, Conodonts of the Mississippian System in the Sacramento Mountains, New Mexico: Ph.D. Dissert., Univ. New Mexico, Alburquerque, New Mexico, 214 p.

Dott, R. H., Jr., 1963, Dynamics of subaqueous gravity depositional processes: Am. Assoc. Petroleum Geologists Bull., v. 47, p. 104–128.

Harbour, R. L., 1972, Geology of the northern Franklin Mountains, Texas and New Mexico: U.S. Geol. Survey, Bull. 1298, 129 p.

Harms, J. C., 1974, Brushy Canyon Formation, Guadalupe Mountains, Texas: A deep water density current deposit: Geol. Soc. America Bull., v. 85, p. 1763–1784.

—— and Pray, L. C., 1974, Erosion and deposition along the mid-Permian intracratonic basin margin, Guadalupe Mountains, Texas [abs.]: *In* R. H. Dott, Jr., and R. H. Shaver (eds.), Modern and ancient geosynclinal sedimentation: Soc. Econ. Paleontologists and Mineralogists, Spec. Pub. 19, p. 35.

Ingram, R. L., 1954, Terminology for the thickness of stratification and parting units in sedimentary rocks: Geol. Soc. America Bull., v. 65, p. 937–938.

King, P. B., 1934, Notes on Upper Mississippian rocks in trans-Pecos Texas: Am. Assoc. Petroleum Geologists Bull., v. 18, p. 1537–1543.

——, 1965, Geology of the Sierra Diablo region, Texas: U.S. Geol. Survey, Prof. Paper 480, 185 p.

—— and King, R. E., 1929, Stratigraphy of outcropping Carboniferous and Permian rocks of trans-Pecos Texas: Am. Assoc. Petroleum Geologists Bull., v. 13, p. 907–926.

Kottlowski, F. E., 1963, Paleozoic and Mesozoic strata of southwestern and south-central New Mexico: New Mexico Bur. Mines and Mineral Resources, Bull. 79, 100 p.

Lane, H. R., 1974, The Mississippian of southeastern New Mexico and west Texas—A wedge-on-wedge relation: Am. Assoc. Petroleum Geologists Bull., v. 58, p. 269–282.

——, 1975, Correlation of the Mississippian rocks of southern New Mexico and west Texas utilizing conodonts: *In* Guidebook to Mississippian shelf-edge and basin facies carbonates, Sacramento Mountains and southern New Mexico region: Dallas Geol. Soc., Dallas, Texas, p. 87–97.

Laudon, L. R., and Bowsher, A. L., 1949, Mississippian formations of southwestern New Mexico: Geol. Soc. America Bull., v. 60, p. 1–88.

Lloyd, E. R., 1949, Pre-San Andres stratigraphy and oil-producing zones in southeastern New Mexico: New Mexico Bur. Mines and Mineral Resources, Bull. 29, 79 p.

McDaniel, P. N., and Pray, L. C., 1969, Bank to basin transition in Permain (Leonardian) carbonates, Guadalupe Mountains, Texas [abs.]: *In* G. M. Friedman (ed.), Depositional environments in carbonate rocks: Soc. Econ. Paleontologists and Mineralogists, Spec. Pub. 14, p. 79.

Meyers, W. J., 1973, Chertification and carbonate cementation in the Mississippian Lake Valley Formation, Sacramento Mountains, New Mexico: Ph.D. Dissert., Rice Univ., Houston, Texas, 353 p.

——, 1974, Carbonate cement stratigraphy of the Lake Valley Formation (Mississippian), Sacramento Mountains, New Mexico: Jour. Sed. Petrology, v. 44, p. 837-861.

——, 1975, Stratigraphy and diagenesis of the Lake Valley Formation, Sacramento Mountains, New Mexico: *In* Guidebook to Mississippian shelf-edge and basin facies carbonates, Sacramento Mountains and southern New Mexico region: Dallas Geol. Soc., Dallas, Texas, p. 45-65.

Plicka, M., 1968, *Zoophycus*, and a proposed classification of sabellid worms: Jour. Paleontology, v. 42, p. 838-849.

Pray, L. C., 1952, Stratigraphy of the escarpment of the Sacramento Mountains, Otero County, New Mexico: Ph.D. Dissert., California Inst. Technology, Pasadena, California, 370 p.

——, 1961, Geology of the Sacramento Mountains Escarpment, Otero County, New Mexico: New Mexico Bur. Mines and Mineral Resources, Bull. 35, 144 p.

——, 1968, Basin-sloping submarine (?) unconformities at margins of Paleozoic banks, west Texas and Alberta [abs.]: Geol. Soc. America, Program with Abs., 1968 Ann. Meeting, p. 243.

Seilacher, Adolf, 1967, Bathymetry of trace fossils: Marine Geology, v. 5, p. 413-428.

Wilson, J. L., 1969, Microfacies and sedimentary structures in "deeper water" lime mudstones: *In* G. M. Friedman (ed.), Depositional environments in carbonate rocks: Soc. Econ. Paleontologists and Mineralogists, Spec. Pub. 14, p. 4-19.

——, 1971, Upper Paleozoic history of the western Diablo Platform, west Texas and south-central New Mexico: *In* The geologic framework of the Chihuahua tectonic belt: West Texas Geol. Soc., Midland, Texas, p. 57-64.

——, 1975, Regional Mississippian facies and thickness in southern New Mexico and Chihuahua: *In* Guidebook to Mississippian shelf-edge and basin facies carbonates, Sacramento Mountains and southern New Mexico region: Dallas Geol. Soc., Dallas, Texas, p. 125-128.

Yurewicz, D. A., 1973a, Stratigraphy and sedimentology of basin facies carbonates (Mississippian) of New Mexico and west Texas [abs.]: Geol. Soc. America Abs. with Programs, v. 5, p. 366.

——, 1973b, Genesis of the Rancheria and Las Cruces (?) Formations (Mississippian) of New Mexico and west Texas: M.S. Thesis, Univ. Wisconsin, Madison, Wisconsin, 249 p.

——, 1975, Basin margin sedimentation, Rancheria Formation, Sacramento Mountains, New Mexico: *In* Guidebook to Mississippian shelf-edge and basin facies carbonates, Sacramento Mountains and southern New Mexico region: Dallas Geol. Soc., Dallas, Texas, p. 67-86.

SEPM Special Publication No. 25, p. 221-247, November 1977

TURBIDITES, DEBRIS SHEETS, AND TRUNCATION STRUCTURES IN UPPER PALEOZOIC DEEP-WATER CARBONATES OF THE SVERDRUP BASIN, ARCTIC ARCHIPELAGO

GRAHAM R. DAVIES
Applied Geoscience And Technology (AGAT) Consultants Ltd.,
4515 Varsity Drive NW, Calgary, Alberta Canada T3A028

ABSTRACT

Cherty carbonate rocks and shale of the Middle Pennsylvanian to Lower Permian Hare Fiord Formation in the Sverdrup Basin were deposited in trough and slope environments in water depths estimated to range from 300 m to 1200 m, seaward of a shallow-water shelf constructed of bioclastic and oölitic carbonate. Depositional dips of shelf foreslope beds locally were as high as 40°. Early submarine cementation was pervasive in shelf-edge sediment and in the steeply dipping shelf-derived foreslope sediment tongues. In this high-relief setting, downslope gravitational displacement of sediment and indurated rock was a common occurrence. Different processes or transport mechanisms produced a variety of products, including: (a) bioclastic and lithoclastic turbidites, commonly crinoid-rich, transported and deposited by turbidity currents; (b) flaser breccia, composed of angular, cemented shelf-limestone clasts admixed with plastically deformed lumps of fine-grained sediment, transported by turbidity currents with some interclast shearing during deposition; and (c) debris sheets, composed of angular megaclasts of shelf-derived limestone in discontinuous units, deposited by submarine debris flow.

These displaced carbonate rocks may be related genetically in a depositional progression from debris sheet to flaser breccia to coarse clastic turbidite to distal lime-mudstone turbidite, reflecting decrease in maximum clast size and increase in distance from the source; this progression assumes that a debris flow may evolve downslope into a turbidity current. Individual turbidite beds may contain mixed biotas; one bed contains conodonts from several Pennsylvanian and Permian assemblages.

Gravitational displacement by sliding and subsequent infilling of the slide scar by sediment are invoked as processes forming large intraformational truncation structures in the Hare Fiord Formation in a setting apparently distal to the zone of debris sheets and coarser clastic limestone turbidites. The structures have a smooth, often listric truncation surface that in the largest structures may be traced for at least 1.5 km. Individual structures have truncated and removed up to 150 m of section out of a total Middle Pennsylvanian to Lower Permian formational thickness of about 800 m. Large slump masses, breccias, or rotated blocks have not been observed in the post-slide section. Where the plane of truncation and overlying fill beds lie at a high angle to original ("footwall") bedding, the contact is an angular unconformity; where the plane of discontinuity eventually becomes parallel to bedding, the contact is paraconformable.

Gravity slides and slumps, and displaced sediments and turbidites, are common features of modern continental slopes and rises. These modern examples provide support for the interpretation of structures in the Sverdrup Basin.

INTRODUCTION

The majority of processes controlling deposition and deformation of sediment are governed directly or indirectly by gravitational force. Some of the more spectacular examples of gravitational displacement of marine sediment and rock occur on modern continental slopes and, in ancient rocks, in slope lithofacies adjacent to abrupt shelf margins. This paper describes gravity-displacement products and processes in rocks deposited in slope environments adjacent to the margin of an upper Paleozoic carbonate shelf in the Sverdrup Basin; the rocks are exposed on Ellesmere Island in the high Arctic (Fig. 1).

The transition from shelf-edge to slope environments in the northern Sverdrup Basin in Pennsylvanian to Early Permian time often was abrupt, with high differential relief between the shelf edge and slope floor. In this setting, downslope gravitational displacement of sediment and rock produced a variety of deposits and structures ranging from carbonate turbidites and debris sheets to large-scale truncation surfaces. Recognition of these displacement products is essential to our understanding of processes operating in shelf margin and slope environments, and provides insight into a number of associated problems.

The gravity-displaced rocks and structures described in this paper have not received detailed field study; they are one component of a larger field program covering upper Paleozoic rocks in the Arctic Archipelago conducted by the author in co-operation with W. W. Nassichuk (G.S.C.).

SVERDRUP BASIN

The Sverdrup Basin is an elongate, fault-bounded crustal depression lying off the north-

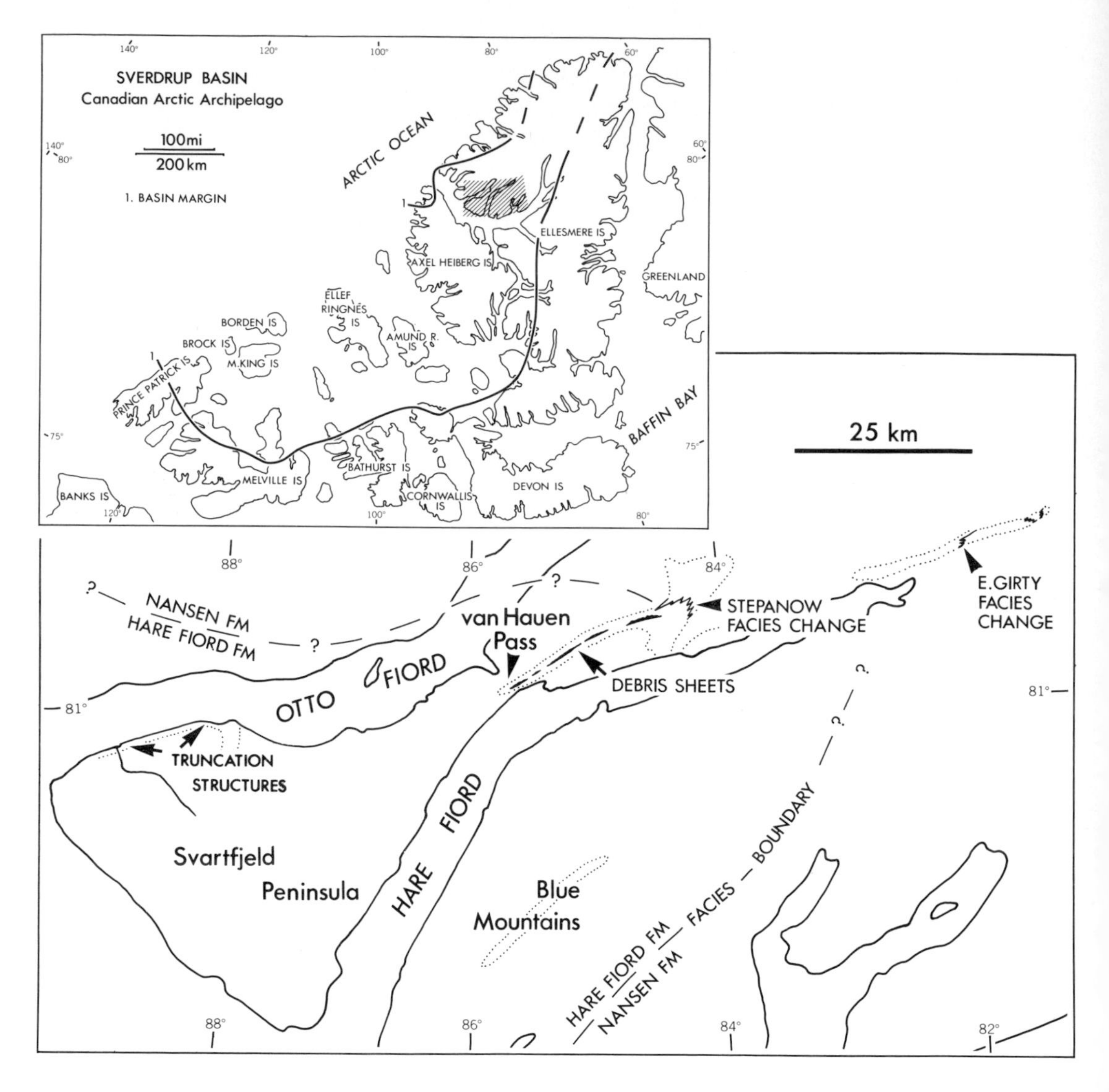

FIG. 1.—Locality map of the Canadian Arctic Archipelago, with detail of northwestern Ellesmere Island. The basin margin boundaries in the regional map outline the erosinal limits of the Sverdrup Basin. Areas outlined by dots on the lower map are outcrop belts of upper Paleozoic rocks containing displaced rocks and truncation structures described in this paper.

western margin of the North American craton in the Canadian Arctic Archipelago (Fig. 1). The basin is filled by about 12,000 m of sedimentary rocks ranging in age inclusively from Late Mississippian to Tertiary. The Mississippian to Permian succession records a sedimentary progression from basal redbeds with minor volcanics, through marine evaporite, into marine carbonate and terrigenous detrital rocks; it thus parallels the succession found in many rifted continental-margin basins[1] (Dickinson, 1974).

The upper Paleozoic rocks in the Sverdrup Basin are exposed best in thrust-controlled outcrop belts on northwestern Ellesmere Island, where they have been described and mapped by Thorsteins-

[1] Editor's note: This succession has similarities to the Late Triassic and Early Jurassic trough sediment in the High Atlas Mountains, Morocco (Evans and Kendall, this volume).

son (1974, and G.S.C. Map Series). The structural geometry of these faults precludes horizontal displacements of more than a kilometer or so; the facies boundaries shown in Figure 1 are not corrected for this displacement.

For the Upper Mississippian to Lower Permian succession in the northern Sverdrup Basin, four major lithologic units are recognized (Fig. 2); these are:

1. The Upper Mississippian Borup Fiord Formation is composed of continental and marine conglomeratic redbeds averaging about 300 m in thickness, and overlies the basal sub-Mississippian unconformity. These redbeds underlie both trough and shelf lithofacies, and probably grade laterally into the lower units of similar redbeds at the margins of the Sverdrup Basin (Canyon Fiord Formation, Fig. 2).

2. The Upper Mississippian to Middle Pennsylvanian Otto Fiord Formation is composed in outcrop of interbedded anhydrite, limestone, and sandstone up to 500 m thick, and confined to the axis (trough) of the Sverdrup Basin. This unit is coeval with thick halite in the subsurface of the southern Sverdrup Basin (Davies, 1975a); the entire evaporitic succession is interpreted to be largely of subaqueous origin (Davies and Nassichuk, 1975a).

3. The Middle Pennsylvanian to Lower Permian Hare Fiord Formation is composed of argillaceous and dolomitic limestone, chert, shale and siltstone up to 1000 m thick. These rocks are of deep-water aspect; together with the Otto Fiord evaporite, they defined the central axis or trough of the Sverdrup Basin in late Paleozoic time. Close to the shelf edge, Middle Pennsylvanian bryozoan mounds (Davies, 1975b) occur at the base of the Hare Fiord Formation (Fig. 2).

4. The Lower Pennsylvanian to Lower Permian Nansen Formation is composed of bedded to reefal limestone and dolostone up to 2000 m thick, and is coeval with the Hare Fiord Formation and uppermost Otto Fiord Formation. These shallow-water carbonates define a broad shelf that bordered the central trough to the east, north and west.

TERMINOLOGY OF DEPOSITIONAL SETTINGS

Terms for depositional settings of upper Paleozoic rocks in the Sverdrup Basin are used in this paper in the following sense.

Shelf.—An areally extensive, relatively low-relief depositional surface covered by shallow water, commonly but not always delineated to landward by a shoreline and to seaward by a marked increase in water depth. Rocks characteristic of this depositional setting are well-bedded, light-colored carbonate of shallow-water aspect.

Shelf edge.—A relatively narrow zone at the outer or seaward margin of the shelf having irregular bathymetric relief over reefs, banks and shoals; some parts may be exposed subaerially. Rocks characteristic of this setting include massive to thick-bedded reefal, bioclastic and oölitic carbonate.

Shelf foreslope.—A narrow zone seaward of the shelf edge characterized by steep depositional slopes and constructed largely of material derived from the shelf and shelf edge. Rocks characteristic of this depositional setting are steeply-dipping tongues of light-colored carbonate interbedded with and transitional into darker-colored rocks.

Slope.—A more extensive depositional environment seaward of the steep shelf foreslope in which water depths increase more gently away from the shelf. Rocks characteristic of this setting include well-bedded, dark-colored carbonate and fine detrital sediment, commonly of turbidite type.

Trough.—The deepest component of the depositional spectrum, transitional from the slope environment by a reduction in bathymetric gra-

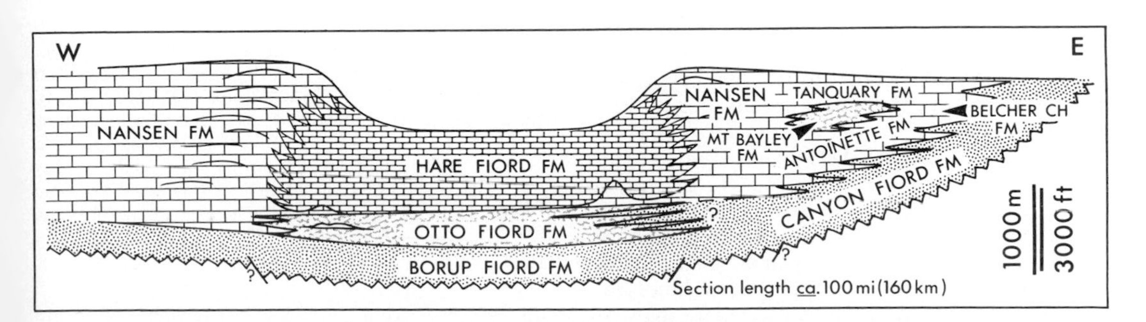

FIG. 2.—Generalized facies relationships and formation units for the Upper Mississippian to Lower Permian section in the northern Sverdrup Basin, based on an east-west section across northwestern Ellesmere Island. Modified from Thorsteinsson (1974).

dient, and defining the axis of the entire depositional "basin." Rocks inferred to be deposited in this setting include dark-colored shale and siltstone with fewer carbonate turbidites.

LITHOLOGY OF THE NANSEN FORMATION—OUTER SHELF SUBFACIES

As sediments and rocks of the outer Nansen shelf contributed to the sediment wedge of the foreslope and slope facies, a brief description of these rocks is warranted here. In overall aspect, the shelf-edge rocks are light-colored, medium-to thick-bedded or massive limestones in which bioclastic wackestone, packstone and grainstone (textural terms of Dunham, 1962) and oöid grainstone are common. Cyclicity is apparent throughout much of the section, particularly in inner shelf settings (Davies, 1975b, Fig. 6). Massive (reefal) and lenticular limestone units generally are more common high in the section (Lower Permian); they are characterized by the skeletal contribution of phylloid algae (Davies and Krouse, 1975, Fig. 2) and the enigmatic organism *Palaeoaplysina* (Davies and Krouse, 1975, Fig. 1; Davies and Nassichuk, 1973) and by radiaxial fibrous calcite cement (Kendall and Tucker, 1973) neomorphic after early submarine cements (Davies, 1974; Davies and Krouse, 1975). The massive units often are capped by fusulinacean and oöid grainstone. Dolomitization is not common in these rocks, but, where found, it preferentially replaces porous oöid grainstone rather than the low-porosity, early cemented biogenic units ("reefs").

LITHOLOGY OF THE HARE FIORD FORMATION

The Hare Fiord Formation typically is a dark-colored, medium-bedded unit composed of argillaceous, cherty and dolomitic limestone (Fig. 3A), siltstone and shale. The percentage of shale generally increases up section (Lower Permian). Individual beds are planar, parallel and continuous for long distances. Burrows, carbonized trails, and other trace fossils of the deep-water *Nereites* facies (Seilacher, 1967) are common in argillaceous partings within the formation. Autochthonous skeletal components include thin-walled brachiopods, crinoid debris, fenestrate bryozoans and siliceous sponge spicules. In thin section (Fig. 3B), many carbonate beds are bioclastic wackestone with partly silicified crinoid columnals, brachiopods and bryozoans. The matrix is argillaceous, sometimes silty (quartzose) lime mudstone that commonly is replaced extensively by microcrystalline authigenic dolomite and by silica. Remobilization of biogenic opal after solution-replacement of sponge spicules by calcite may be a major source of silica in these rocks; the end product is a bedded or lenticular chert unit.

In the slope facies transitional between shelf and trough settings, at least 60 percent of some Hare Fiord carbonate sections is composed of fining-upward size-graded bioclastic (crinoid-rich) limestone beds interpreted to be carbonate turbi-

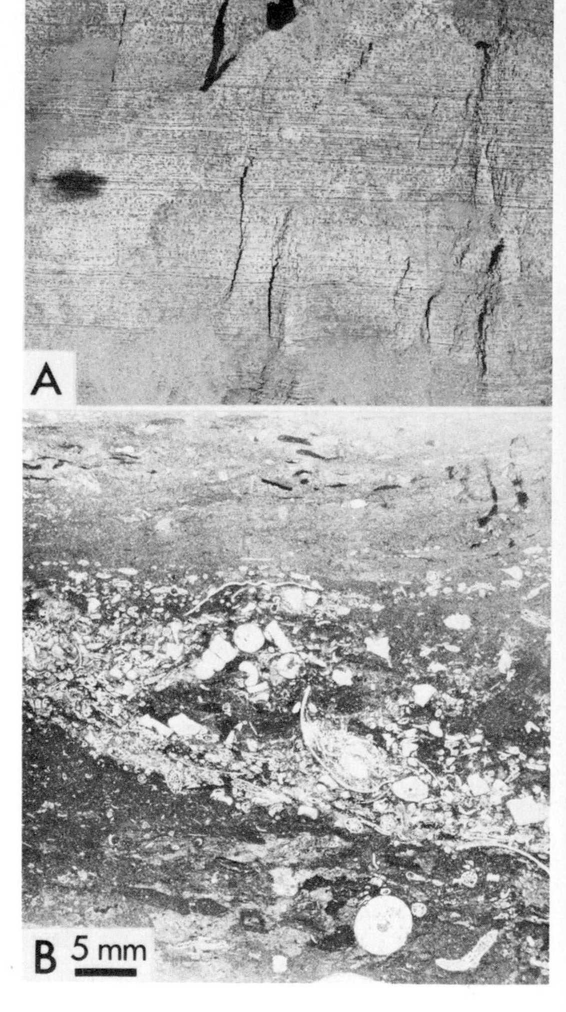

FIG. 3.—Hare Fiord Formation. *A*—Well-bedded argillaceous and cherty limestone typical of the Hare Fiord Formation, exposed on a cliff face on the north side of Svartfjeld Peninsula, Ellesmere Island (Fig. 1). Dark pods and beds are chert (silicified limestone). Note shadow of helicopter (8.5 m long) at left. Width of view about 60 m.

B—Thin section of partly silicified limestone bed from the Hare Fiord Formation, with irregular distribution of bioclasts of crinoids, brachiopods and bryozoans in a partly dolomitized matrix of argillaceous and spicular micrite. Note poorly defined, fining-upward size grading of clasts and burrows in the upper part; beds of this type may be distal equivalents of coarser bioclastic turbidites closer to the shelf edge. Same locality as 3A.

dites. In the same setting, thick resistant beds and isolated blocks of light-colored limestone are found within the formation. This limestone, described more fully in a following section, is composed mainly of shelf-derived sediment and rock.

SHELF EDGE-TO-SLOPE TRANSITION

Where the transition from Nansen shelf carbonate to Hare Fiord slope lithofacies is exposed on northwestern Ellesmere Island, it is always relatively abrupt, with most of the light-colored massive shelf-edge and shelf-foreslope limestone units thinning rapidly over a distance of less than 500 m into dark colored medium bedded argillaceous and cherty limestone of the slope facies (Figs. 2, 4, 7). Pairs of bedding planes (approximating time planes) traced from field photographs (Fig. 4) show a reduction in thickness of rocks between them of more than 90 percent from shelf-edge to slope facies (without allowance for differential compaction). Cook and Taylor (this volume) report a thinning of about 80 percent from shelf to slope sediment in the Cambro-Ordovician of the western United States.

The maximum angle of dip on bedding planes in shelf foreslope rocks, of necessity measured from field photographs and thus actually apparent dip, varies from 15° to 20° for some bedding planes (400 m or more of relief), to between 45° and 50° for the steepest beds. These foreslope dip angles were estimated using the orientation of bedding in rocks of slope facies as a near-approximation to a horizontal datum; obviously, as the depositional surface of the slope facies by definition dips at a low angle away from the base of the shelf foreslope, the values given for foreslope dip are minima. The angle of some foreslope dips exceed the angle of repose of uncemented granular solids in water; significantly, the few accessible, steeply dipping shelf foreslope carbonate tongues that have been sampled are bound inorganically by early submarine cements (Davies, 1974).

Both the relative thinning of units and the angle of dip across the shelf to slope transition may have been modified by postdepositional compaction. However, the degree of differential compaction probably was minimal, as the slope sediment was mainly detrital and bioclastic carbonate with varying amounts of fine terrigenous detritus, rather than relatively pure clay sediment. Thus the figures given here for relative thinning and angle of dip are reasonable approximations to the original depositional geometry.

The shelf-edge section progrades basinward everywhere it is exposed; that is, progressively younger sediments at the shelf edge and shelf foreslope were deposited as a seaward prograding wedge (Figs. 2, 4). This probably is the expected consequence of rapid carbonate sediment production and accumulation on a shelf in near equilibrium with sea level in a subsiding tectonic regime.

Water depth.—Estimates of bathymetric relief from the shelf edge to the floor of the adjacent

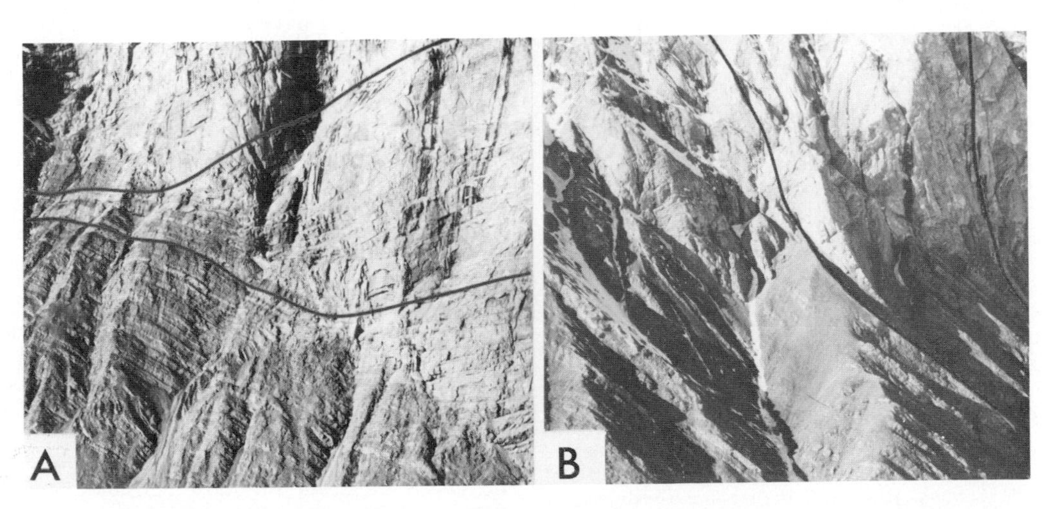

FIG. 4.—Shelf to slope transitions. *A*—Facies transition from light-colored, thick-bedded to massive shelf and foreslope limestone of the Nansen Formation (right) into dark-colored, medium-bedded argillaceous limestone of the Hare Fiord Formation (left). Lines trace two bedding planes across the transition to emphasize depositional slope and thinning. Local structural dip is to the right and away from reader. Locality about 10 km east of East Girty facies change (Fig. 1). Vertical field of view about 500 m.

B—Same as 4A, further east. Local structural dip is steep and to the right.

slope and trough (that is, an approximation to local minimum water depth) may be approached in two ways. First, by tracing bedding planes from the floor of the slope up dip towards the shelf edge (rarely completely exposed), a minimum water depth may be estimated. For one exposure on Hare Fiord (Fig. 7), a minimum water depth of 300 m may be estimated for the onset of Hare Fiord sedimentation in Middle Pennsylvanian time. Water depth increased progressively to at least 500 m, and probably much more, by Early Permian time.

The second approach is based on changes in thickness of biostratigraphic units across the shelf-to-slope transition (Fig. 5). For selected sections of the Hare Fiord and Nansen formations, assuming a more or less horizontal datum near the top of the Otto Fiord evaporite (Fig. 5), a progressive increase in water depth from about 900 m for late Middle Pennsylvanian time to about 1200 m for the end of Asselian (earliest Permian) time may be estimated. Although the latter estimate is more than twice the minimum estimate from bedding plane traces, the combined data certainly support the sedimentological evidence that the Hare Fiord sediments were deposited in a deep, and progressively deepening (albeit perhaps fluctuating) trough.

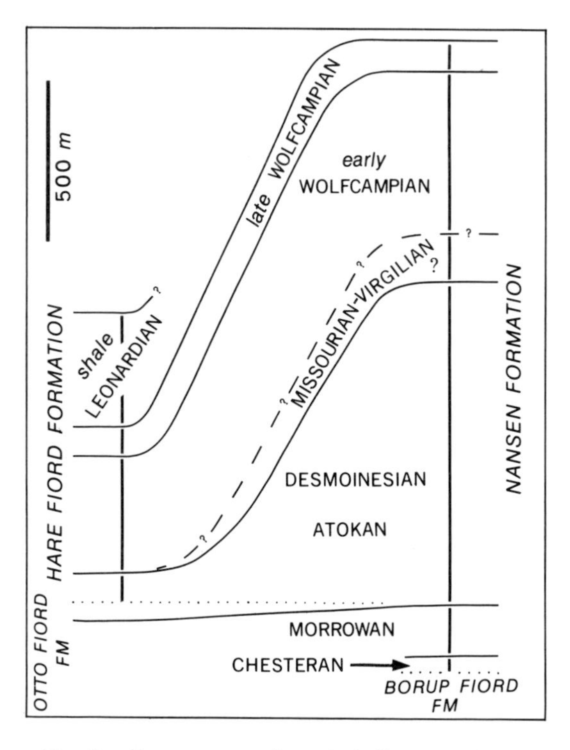

FIG. 5.—Reconstructuction of shelf-to-trough bathymetric relief using biostratigraphic units to define thickness differential between age-equivalent shelf and trough sections. Horizontal datum (not shown) is set within the Otto Fiord Formation at the base of the sections. By this method, estimated relief was 900 m at the end of Desmoinesian time, and 1200 m at the end of early Wolfcampian time (earliest Permian). Paleontological data, mainly foraminiferal, are incomplete. Based on type sections of the Nansen and Hare Fiord Formations, Ellesmere Island.

TERMINOLOGY OF DISPLACED CARBONATES

The increase in the last decade of literature on gravity-displaced carbonate sediment in deep-water sequences has been accompanied by a proliferation of descriptive and genetic terms. Some of the names applied to displaced carbonate (and other rock types), in crude order of increasing clast size, are: allodapic limestone (Meischner, 1964), limestone or bioclastic turbidite (Meischner, 1964; Tucker, 1969; Engel, 1970), debris flow or megabreccia (Cook and others, 1972; Mountjoy and others, 1972), and olistostrome (Flores, 1959; von Görler and Reutter, 1968; Abbate and others, 1970; Richter, 1973). A comprehensive discussion of terminology and processes is given by Hoedemaeker (1973); he introduces "delapsion" as the process of independent gravitating of rock material along sloping surfaces, the general term "olisthon" for all materials that have delapsed, and redefines olisthostrome and olistolith. Carter (1975) has reviewed the processes of mass transport (mainly laminar mass flow), and has proposed several new or revised terms for the spectrum of products from sediment mass transport.

In this paper, I have avoided adopting new or unfamiliar terms and, with one exception, have used descriptive rather than genetic terms for different types of gravity-displaced rocks. The exception is "limestone turbidite," adopted for its brevity in place of "size-graded clastic limestone bed." The essentially synonomous term "allodapic" limestone (Meischner, 1964) has not been adopted here.

GRAVITY-DISPLACED ROCKS IN THE HARE FIORD FORMATION

Two major types of submarine gravity-displaced sedimentary rocks are found in the Hare Fiord Formation; they may represent end products of a spectrum of transportation processes. They are: (a) carbonate turbidites—well-bedded, internally size-graded clastic limestone interpreted to be deposited from turbidity currents, and (b) debris sheets—massive, often discontinuous mega-clastic limestone (breccia) deposited by mass flow.

Limestone Turbidites

Beds of fining-upward size-graded lithoclastic and bioclastic limestone (Fig. 6) interpreted as turbidites are common in the slope facies of the Hare Fiord Formation in many exposures on northwestern Ellesmere Island. These clastic carbonate beds are relatively dark colored, although not as dark as the interbedded argillaceous and cherty limestone and shale. They weather as resistant beds with shaly partings or recessive interbeds.

Limestone turbidites form a significant proportion of the Hare Fiord Formation in the type-section at van Hauen Pass (Fig. 1) where graded beds of crinoidal and lithoclastic debris are common. The thickest turbidite beds that have been measured in Hare Fiord sections are 30 cm thick; probably there are many turbidite beds of greater thickness. Individual samples from turbidite beds contain mixed conodont faunas of multiple Pennsylvanian and Permian ages (P. Bender, personal communication, 1973).

In the Blue Mountains area, southeast of Hare Fiord (Fig. 1), lithoclastic and bioclastic limestone turbidites again are an important component of the Hare Fiord Formation. Where the Hare Fiord Formation buries Pennsylvanian bryozoan mounds exposed in the Blue Mountain frontal escarpment (Davies, 1975b, Fig. 7), the lower 180 m of the formation is composed of more than 60 percent by thickness of resistant limestone beds interbedded with less resistant argillaceous rock (K. J. Roy, personal communication, 1973). Of these limestone beds, many are composed of graded crinoidal and lithoclastic debris. Some of the crinoidal turbidites overlie the flank beds of the partly buried Pennsylvanian mounds (Fig. 6C); they may have been related genetically to submarine slopes above the underlying mounds. Elongate, multi-segment crinoid columnals at the base of these graded crinoidal beds show a preferred orientation which may prove useful in determining transport direction, but has not received systematic analysis in these rocks.

Samples of typical graded crinoidal turbidite beds from the Hare Fiord Formation are composed of up to 95 percent by volume of crinoidal debris, mostly disarticulated columnals (Fig. 6B). The sampled beds are up to 13 cm thick. The largest bioclasts are concentrated in the lower part of a bed but often with a thin basal layer of finer clasts. Basal contacts with the underlying argillaceous and calcareous host sediment are variably planar to irregular, with some indication of load-casting and possibly of scour. Upper contacts are gradational, but may show some evidence of small-scale ripple cross-lamination. Bioturbation is not apparent in the graded clastic beds, but the immediately overlying argillaceous siltstone may contain burrows (Fig. 6A). The complete vertical sequence of sedimentary structures considered typical of turbidites in detrital siliciclastic sediments (Bouma cycle) is not found in these graded carbonate beds.

Clasts other than crinoid debris in the graded beds include bioclasts of bryozoans, fusulinaceans, brachiopods, and corals (rare), and a few lithoclasts of bioclastic, pelletoidal wackestone. In the lower, coarser zone of graded crinoidal beds, the clasts commonly show pressure-solution contacts (overpacking) with a sparry calcite or siliceous matrix; many of the crinoid clasts also are silicified. The upper, finer grained zone may include a mixture of smaller bryozoan and echinoid clasts in grain-to-grain contact with a micritic matrix (packstone texture). Replacement of the cement or matrix by zoned euhedral dolomite is common.

Crinoidal and lithoclastic turbidites also occur interbedded with the uppermost anhydrite units of the Otto Fiord Formation and the lowermost rocks of the Hare Fiord Formation at a facies change exposed a few miles east of Girty Creek on easternmost Hare Fiord (Figs. 1, 7; also Davies, 1975b, Figs. 3, 4, 9). The graded beds are up to 30 cm thick, and are associated with much thicker, resistant limestone units with internal breccia fabrics (debris sheets). Both types of displaced carbonate deposits may be walked out or traced laterally towards the facies change into thicker coeval shelf-foreslope carbonate units with steep depositional dips (some in excess of 40°). These steeply dipping shelf-foreslope beds are composed in part of lithoclasts and bioclasts cement in place by early submarine cements (Davies, Ghent and Krouse, in preparation). A direct connection thus may be established between limestone turbidites (and debris sheets) and the shelf-to-trough transition.

The limestone turbidites within the Hare Fiord Formation at the localities cited have not been dated precisely. From associated faunas within the enclosing sections, they fall within the general time span from mid-Pennsylvanian to Early Permian.

Fine-grained (distal?) carbonate turbidites.—In exposures on Ellesmere Island furthest from the known or projected margin of the Nansen shelf, the Hare Fiord Formation commonly is composed of very even-bedded, dark-colored cherty carbonate with thin argillaceous interbeds. Individual beds are planar and continuous over long distances (Fig. 3A). Thin sections of a few of these carbonate beds show them to be composed of ungraded bioclastic wackestone with scattered, partly silicified crinoids, thin-walled brachiopods and fenestrate bryozoans (Fig. 3B). The matrix is replaced extensively by finely crystalline authigenic dolomite and silica. These carbonate beds may be distal equivalents of carbonate turbidites found closer to the Nansen shelf edge.

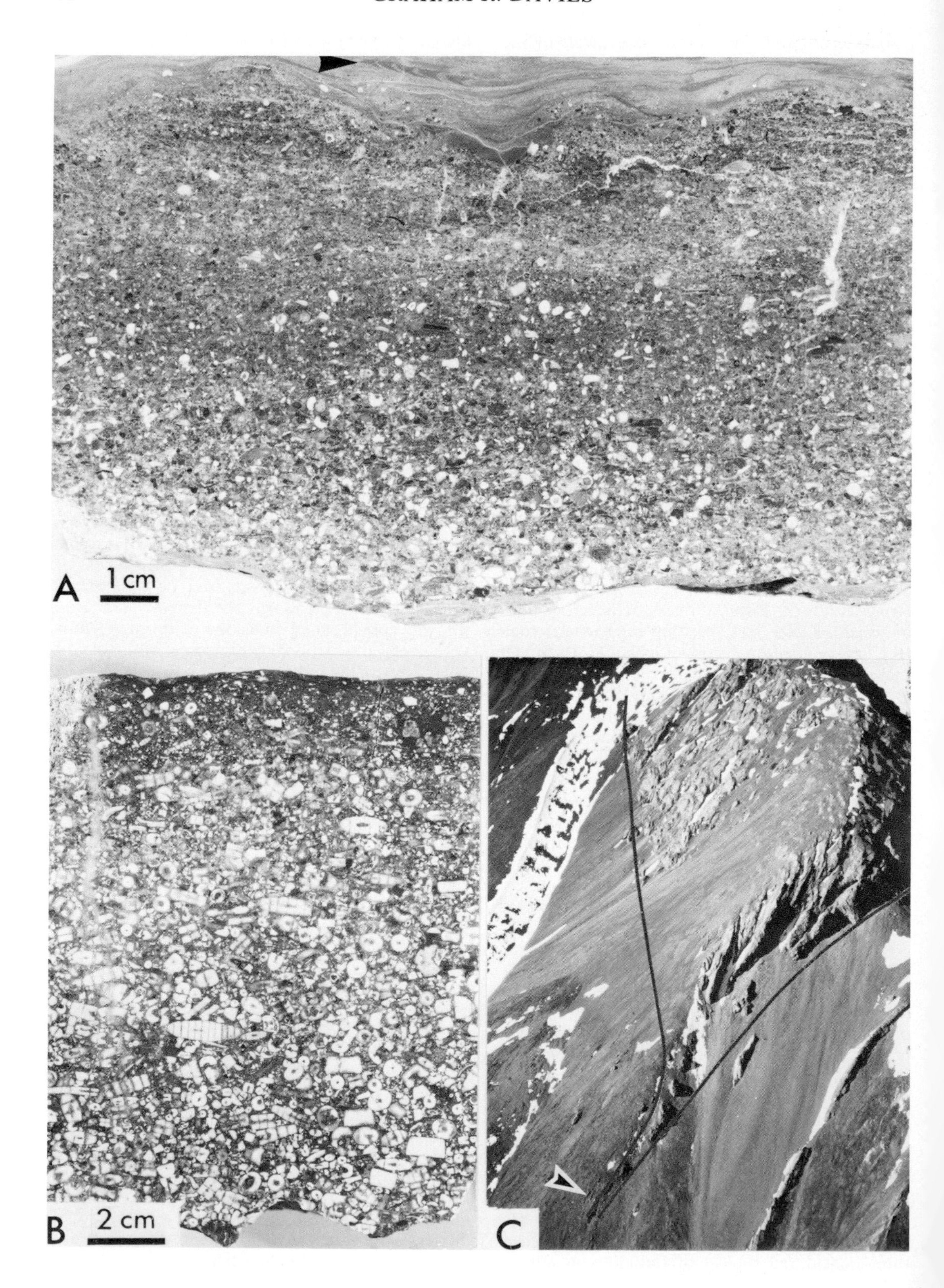
A
1 cm
B
2 cm
C

Limestone flaser breccia (coarse-grained turbidites).—Some clastic limestone beds associated with crinoidal turbidites in the Hare Fiord Formation at the Blue Mountain locality (Fig. 1) are composed in part of lithoclasts of angular, cemented bioclastic limestone mixed with plastically deformed "clasts" of spicular, argillaceous, and micritic limestone (Fig. 8). Texturally, these rocks resemble marl flaser breccias described by Schlager and Schlager (1973); they probably represent a coarse-grained, proximal phase of turbidite deposition. The beds are more poorly size graded (finding upward) than the associated crinoidal turbidities; sampled beds are up to 12 cm thick. The angular lithoclasts include bioclastic grainstone and packstone, some cemented internally with radiaxial fibrous calcite (Fig. 8C), others with sparry calcite. Common bioclasts within the lithoclasts are crinoids, fusulinaceans, bryozoans, encrusting and tubular foraminifers, and algae (Fig. 8C). Poorly preserved clasts of *Palaeoaplysina* also are present. Cementation within the lithoclasts predated clast formation and resedimentation.

The angular carbonate lithoclasts are mixed with irregularly shaped dark-colored clasts of silty, argillaceous, and dolomitic micritic limestone containing sponge spicules. Grain orientations within these irregular clasts show drape, thinning, and other indications of plastic deformation and shearing where they are in contact with the cemented carbonate lithoclasts, forming a flaser-like fabric. The mixture of well-cemented and still-plastic lithoclasts at the time of redeposition provides evidence for the retransported nature and multiple origins of the components.

Although size grading is less obvious in these flaser breccias, the upper parts of individual beds contain finer clasts and more sand-sized bioclasts of crinoids and bryozoans. Silt-sized quartz grains, characteristic of the enclosing Hare Fiord host sediment, also appear in this upper zone.

Debris Sheets

Several types of displaced carbonate rock are found in the Hare Fiord Formation in a belt stretching for 26 km east-northeast of van Hauen Pass, between Hare Fiord and Otto Fiord (Fig. 1). The more obvious and spectacular displaced carbonates in this belt are debris sheets that are exposed as discontinuous masses of resistant-weathering, light-colored, internally brecciated limestone up to 30 m thick enclosed in less resistant dark-colored, medium-bedded argillaceous and cherty limestone and shale (Fig. 9). Some of these prominent carbonate masses are readily confused with in-place mounds or reefs (Fig. 10). Bioclastic and lithoclastic turbidites up to 1 m thick form less obvious yet volumetrically significant components of the regularly bedded Hare Fiord Formation enclosing the debris sheets.

About 12 km east of van Hauen Pass, the normal sequence of well-bedded black shale and cherty limestone of the Hare Fiord Formation is interrupted by several debris sheets 3 m to 10 m thick. The most prominent of these units (Fig. 9) averages about 10 m in thickness, and occurs about 295 m above the base of the formation. This debris sheet may be traced on air photographs for at least 18 km northeastward along strike, to within 8 km of the exposed Stepanow facies change (Fig. 1) from the Hare Fiord Formation to the shelf facies of the Nansen Formation. The facies change north and west of the exposed debris sheets is covered by Mesozoic rocks, but may be less than 10 km distant.

The debris sheets have a planar to undulose basal contact with the underlying recessive, well-bedded cherty limestone and shale, and an irregular upper surface overlain by dark-colored cherty limestone (Fig. 9). Large-scale variations in thickness of the resistant carbonate units result in a lenticular or mound-like profile.

The debris sheets are nonbedded and contain large angular to subangular blocks or clasts of limestone in a chaotic breccia-like fabric. Clast size and clast composition are varied (polymictic). Crinoids, bryozoans, fusulinaceans and ammonoids are visible and abundant on exposed surfaces, particularly in the upper part of the units. Some of these faunal elements, particularly the ammonoids, may have been indigenous, inhabiting the crevassed and irregular "reef-life" surface of the debris sheets. The matrix enclosing the clasts

FIG. 6.—Limestone turbidites, Hare Fiord Formation. *A*—Bioclastic and lithoclastic limestone turbidite. Note the irregular load-casted base, fining-upward size grading of clasts, many small crinoid columnals (white), and burrows (arrow) in bioturbated silty and argillaceous sediment at top. Blue Mountains, Ellesmere Island.

B—Coarse-grained crinoidal turbidite with load-casted base and crude size grading, with larger clasts irregularly distributed rather than confined to the base. Blue Mountains, Ellesmere Island.

C—Oblique aerial photograph of a partly exposed bryozoan mound ("Waulsortian" facies, but Middle Pennsylvanian age) north of the Blue Mountains (Fig. 1). Line overlays trace mound base (right) and top (left); crinoidal turbidite beds similar to 6B overlie the flanks of the mound (arrow) in the basal Hare Fiord Formation.

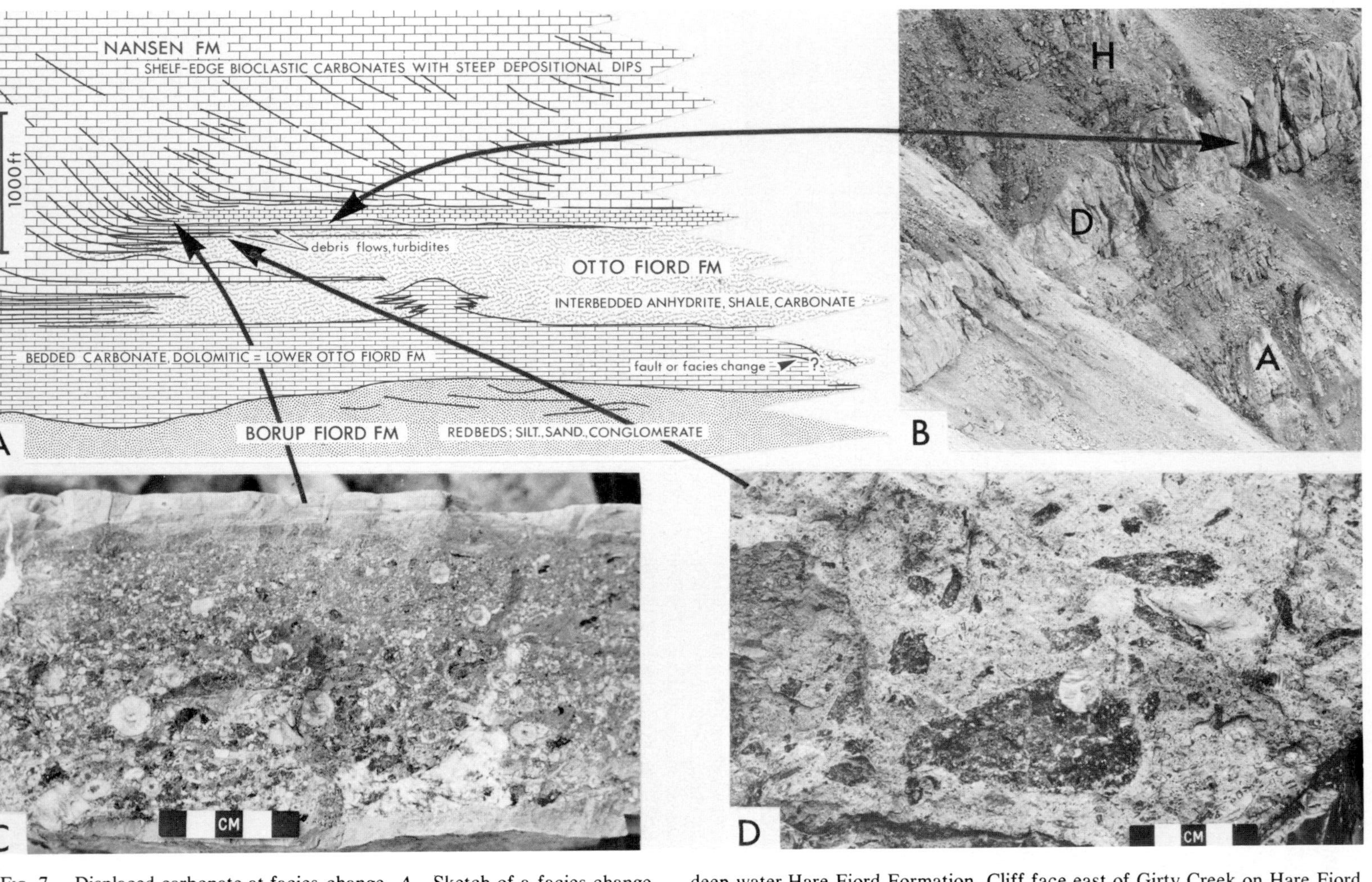

FIG. 7.—Displaced carbonate at facies change. *A*—Sketch of a facies change involving the Otto Fiord, Hare Fiord and Nansen Formations (*see* Davies and Nassichuk, 1975a, Fig. 4; Davies, 1975b, Figs. 3, 4). Steeply dipping tongues of Nansen foreslope limestone plunge downward and thin out into horizontally bedded debris sheets and carbonate turbidites at left center. The thin (50 m) unit of argillaceous limestone immediately below the prograding Nansen limestones (shown by the small limestone block symbols) is a basal unit of the deep-water Hare Fiord Formation. Cliff face east of Girty Creek on Hare Fiord (Fig. 1). *B*—Debris sheets and limestone turbidites in the transition from the Otto Fiord Formation (A = anhydrite) to the Hare Fiord Formation (H); (D) is interpreted to be a debris sheet. *C*—Crinoidal and lithoclastic limestone turbidite from the basal Hare Fiord Formation; this turbidite may be traced into a steeply dipping foreslope tongue. Outcrop photograph. *D*—Part of a debris sheet at the toe of a foreslope limestone tongue. Outcrop photograph.

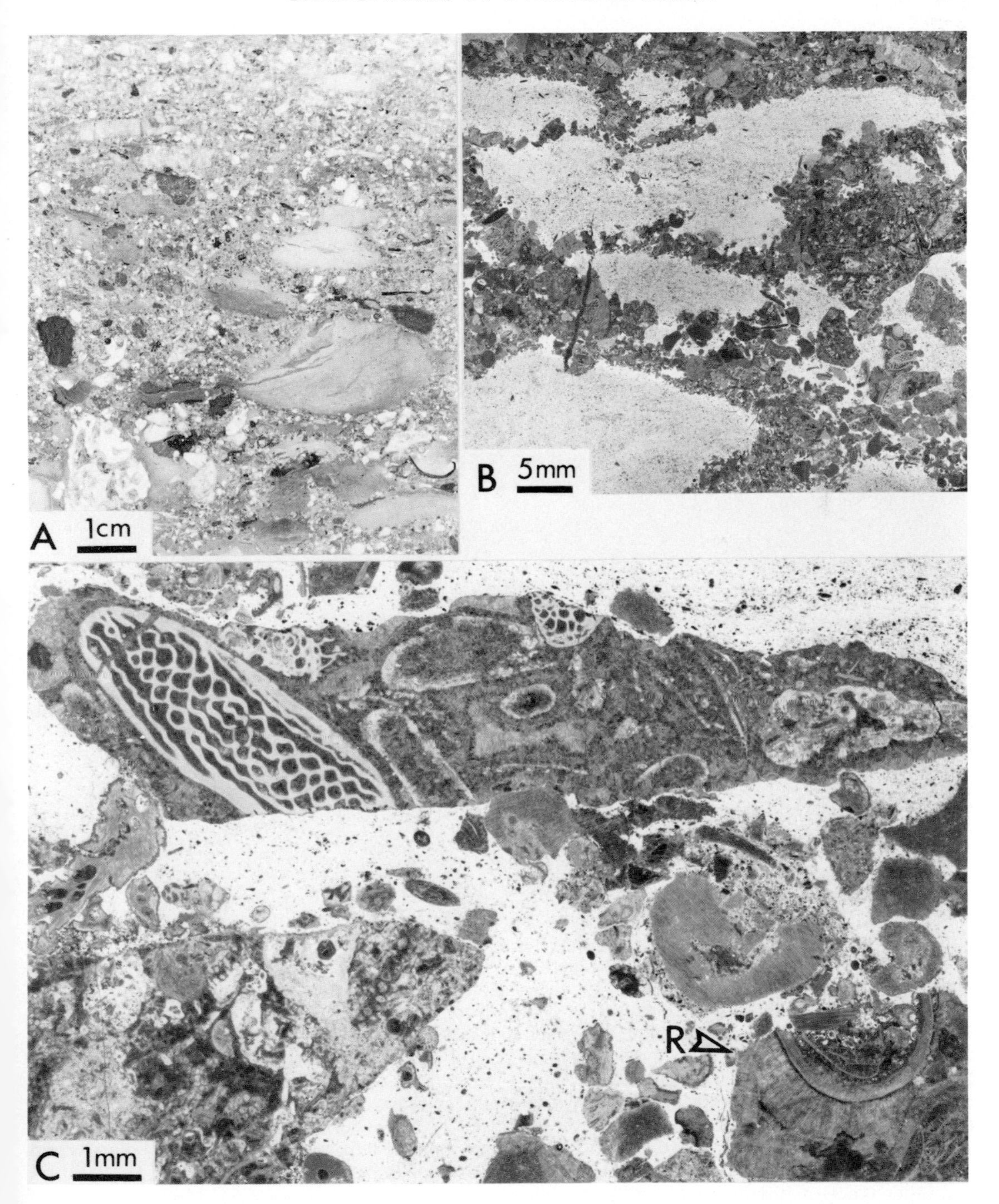

FIG. 8.—Flaser beccias in the Hare Fiord Formation. *A*—Polished slab of a flaser breccia bed from the Blue Mountains area. Note crude size grading, light-colored angular limestone clasts (lower left), and plastically-deformed clasts of darker-colored argillaceous lime mudstone (marl).

B—Thin section of flaser breccia showing irregular lenses of angular limestone clasts embedded in a deformed (sheared) matrix of argillaceous spicular micrite. Negative print.

C—Enlargement of part of 8B, showing angularity of lithoclasts and shallow-water shelf origin of many clast components (fusulinaceans, algae), with scattered crinoids and bryozoans as discrete grains (not in cemented lithoclasts). Radiaxial-fibrous calcite neomorphic after an early marine acicular cement is truncated at the boundary of the clast at lower right (R).

FIG. 9.—Debris sheets in Hare Fiord Formation. *A*—Debris sheets in Hare Fiord Formation, interbedded with well-bedded cherty limestone and thin shales. Note conformable base, irregular upper surface. Figure at lower left (circle) provides scale; debris sheet at left is 10 m thick. Outcrop east of van Hauen Pass, Ellesmere Island.

B—General view of Hare Fiord Formation with enclosed debris sheets of 9A indicated by arrows. Outcrop east of van Hauen Pass, Ellesmere Island, 10 to 15 km west of exposed facies change (Fig. 1).

usually is dark-colored, fine-grained silty and spicular micritic carbonate, often selectively dolomitized, and typical of ''normal'' deep-water rocks of the Hare Fiord Formation. Clasts, particularly bioclasts, are replaced extensively by silica.

Clasts within the debris sheets east of van Hauen Pass are lithoclastic and bioclastic packstone and grainstone (Fig. 11). The most common bioclasts are fusulinacean foraminifers and crinoid columnals. Less abundant but of significance in terms of source environment are clasts of the ?hydrozoan *Palaeoaplysina* (Davies and Nassichuk, 1973), *Tubiphytes, Tetrataxis, Tuberitina,* and other types of encrusting foraminifers, bradyinid and other chambered foramifers, fenestrate and ramose bryozoans, bivalves, gastropods, brachiopods, and phylloid and other types of algae (Fig. 11B, C, D). Cement in angular lithoclasts includes radiaxial fibrous calcite (Fig. 11E) interpreted to be of early submarine origin (Davies, 1974; Davies and Krouse, 1975; Davies, Ghent and Krouse, in preparation); the radiaxial cement is sharply truncated by fracture surfaces bounding the lithoclasts. Clasts of calcareous shale and marl tend to be more rounded than the cemented limestone clasts, and may have undergone some plastic deformation during retransport. Penetration of the argillaceous matrix into skeletal chambers of bryozoans and fusulinaceans suggests that cementation prior to resedimentation in some bioclasts had proceeded only as far as thin fibrous to acicular rims on the inside of skeletal chambers. Most lithoclasts and bioclasts are bounded by irregular surfaces indicative of postdepositional pressure-solution and stylolitization.

Ammonoids, brachiopods, and fusulinaceans from the main debris sheet east of van Hauen Pass yield an Asselian age (Early Permian; Nassichuk and Spinosa, 1972).

Several large but discontinuous limestone masses up to 30 m thick occur at four or more stratigraphic levels in the Hare Fiord Formation about 8 km from the Stepanow facies change (Figs.

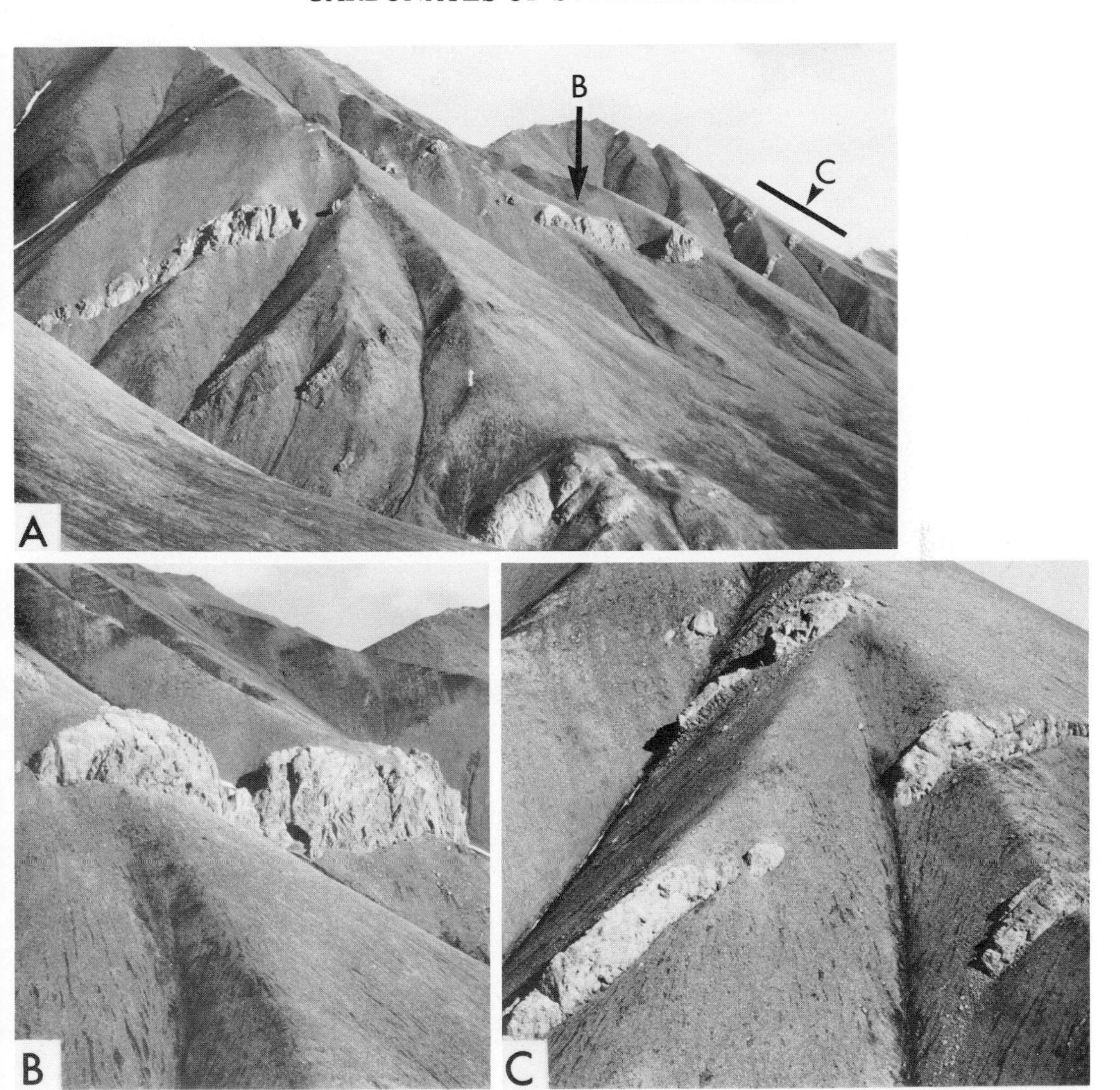

FIG. 10.—Debris sheets, Hare Fiord Formation. *A*—General view of formation with discontinuous or channelized debris sheets enclosed at several stratigraphic levels; debris sheets in this setting, closer to the (exposed) facies change, are more common and thicker than in the area of Figure 9. Outcrop east of van Hauen Pass, 5 to 8 km west of exposed facies change.

B—Massive debris sheet (B in Fig. 10A), 20 to 30 m thick, resembling in-place reef or mound.

C—Discontinuous debris sheets at three or four stratigraphic levels (C in Fig. 10A) in the Hare Fiord Formation.

1, 10). These masses probably are very localized, possibly channelized, debris deposits; they have not been investigated on the ground. Debris sheets also are associated with limestone turbidites of the facies change east of Girty Creek (Fig. 7), described in an earlier section. In that setting, debris sheets may be traced directly into steeply dipping tongues of shelf-foreslope limestone (Fig. 7A, B).

INTRAFORMATIONAL TRUNCATION STRUCTURES

About 45 km west of van Hauen Pass (Fig. 1), a series of spectacular intraformational truncation structures are exposed within the Hare Fiord Formation in a fiord cliff section that is about 800 m high and 13 km long (Davies and Nassichuk, 1975b). The fiord cliffs face Otto Fiord and form part of the northern coastline of Svartfjeld Peninsula. Because of structural complexities and

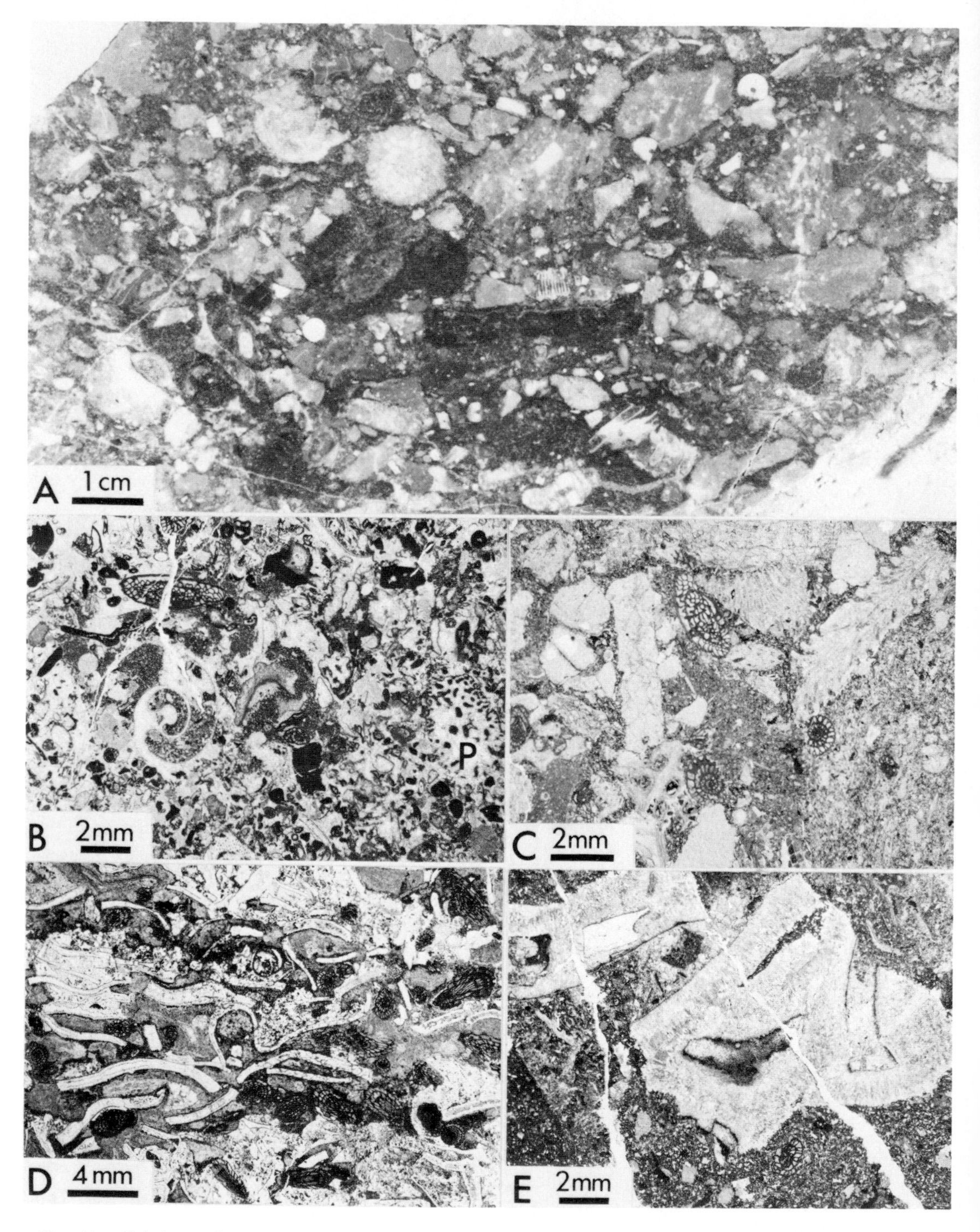

FIG. 11.—Fabrics and composition of debris sheets. *A*—Polished slab of limestone with microbreccia fabric, from the top of a debris sheet exposed east of van Hauen Pass. Angular clasts of cemented limestone, some of shelf origin, are admixed with mudstone clasts and bioclasts of crinoids and bryozoans in a dark-colored argillaceous and micritic matrix.

B—Photomicrograph of a grainstone-textured lithoclast from a debris sheet. Bioclasts characteristic of a shallow-water shelf origin include gastropods, fusulinaceans, algae, a small fragment of *Palaeoaplysina* (P), and *Tubiphytes* (irregular black grains).

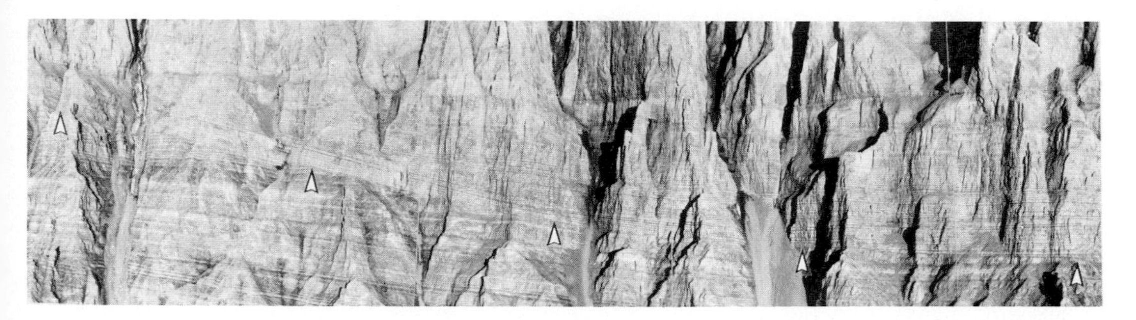

FIG. 12.—Composite photograph of a large truncation structure (arrows) in the Hare Fiord Formation exposed in fiord cliffs on the north side of Svartfjeld Peninsula (Fig. 1). The width of view is 900 m. The structure continues as a paraconformity many hundreds of metres beyond the right-hand western margin. Between 100 m and 120 m of section has been removed by this structure. Note the smooth, curved concave-up (listric) geometry of the truncation surface, the lack of obvious macroscale deformation of beds below or above the truncation surface, and the progressive filling of the slide scar by argillaceous and cherty limestone that thickens down dip into the structure. To the left (east), the contact between "footwall" beds and the sedimentary fill is angular and unconformable. At far right, the overlying beds are parallel and conformable with the footwall beds, yet over 100 m of section is missing; this surface thus is a paraconformity.

cover by younger sediments in the surrounding area, the spatial relationship between the Hare Fiord rocks in the cliff section and the Nansen shelf edge cannot be established with accuracy. A projection (based on field data) of the shelf-edge facies boundary into the subsurface north of the cliff section (Fig. 1) suggests that the Hare Fiord truncation structures may be about 15 km south of the shelf edge.

The top of the Hare Fiord Formation in the cliff section lies about 600 m above the fiord water level; the base of the formation is not exposed. Hare Fiord rocks in the section typically are even-bedded, dark-colored interstratified cherty carbonate and shale (Fig. 3A).

Because of the relative inaccessibility of the cliff section and the limited field time available when the truncation structures were discovered, most of the descriptive data that follow are derived from a continuous mosaic of black-and-white and color photographs taken during several helicopter flights along the cliffs. The low angle of the Arctic sun enabled the shadow of the helicopter on the cliff to be used as a crude scale (Figs. 3A, 13).

About 15 large truncation structures are recognizable in the cliff section. The structures are characterized by a sharp concave-up discontinuity surface that truncates underlying beds and is overlain by a downslope-thickening wedge of sediment in which the lowermost beds lie with an angular relationship on the truncated beds (Figs. 12, 13, 14). As the cliff exposure essentially is a two-dimensional section across three-dimensional structures, various size and geometric parameters discussed here are governed by where the structures are intersected by the cliff face.

The maximum angle of apparent dip of the curved planes of discordance, using the attitude of underlying beds as datum, ranges from 13° to 15° at their up-section ends. Dip decreases progressively down section, with dips on the order of 5° being common in many larger structures. Finally, the plane of truncation flattens out to become parallel with, and indistinguishable from, the underlying undisturbed beds (Fig. 12). The plane of truncation thus is a listric surface, characteristic of landslide and other rotational gravity-slide failure surfaces. The horizontal or parallel-

---

C—Photomicrograph of microbreccia with lithoclasts and bioclasts in an argillaceous and micritic matrix. Muddy matrix has penetrated into pores of bryozoan clast (upper right), but not cemented clasts derived from the shelf. Bryozoans and crinoids apparently were indigenous to deeper water slope environments; they were not cemented before retransport.

D—Photomicrograph of a debris-sheet lithoclast composed mainly of phylloid algal plates and fusulinaceans, both characteristic of the Nansen shelf environment.

E—Radiaxial fibrous calcite cement nucleated on plates of phylloid algae, and truncated at the clast boundary. The fibrous calcite is neomorphic after an early submarine acicular cement, and is characteristic of rocks from the Nansen shelf edge. The algal plates, probably originally aragonitic, underwent partial solution and replacement by calcite spar before retransport.

FIG. 13.—Detail of truncation structure illustrated in Figure 12. Note sharp truncation of footwall beds, lack of local irregularities at truncation surface, down dip thickening of sedimentary fill, with highest beds parallel with original footwall beds. Shadow at lower left center (arrow) is of helicopter; width of view 150 m.

to-bedding component of this surface commonly coincides with a shale bed between beds of cherty carbonate.

There is no evidence in aerial photographs, including close-ups, of any visible deformation of beds below or at the truncation surface, although some down-stepping across cherty limestone beds is apparent. Further, there is no evidence for localized scouring, channeling, or other irregular departures from the relatively smooth, listric plane of truncation.

The sedimentary rocks overlying the truncation surface appear from aerial and photographic inspection to be similar lithologically to the underlying ("footwall") rocks. If there is any trend, it is that the overlying rocks are more argillaceous and cherty (darker colored). Bedding character differs, however, as individual beds and packages of beds are thicker, and thicken progressively down dip (Figs. 12, 13, 14); clearly, the overlying rocks are a sedimentary fill occupying a former depression above the truncation surface. Where dips in the overlying wedge are steepest, the lowermost beds may pinch out on the truncation surface. Depositional dip of overlying beds of the sedimentary "fill" becomes progressively less up section, until the highest beds of the fill are parallel with the truncated "footwall" beds, and the structure is buried.

Perhaps unexpectedly, there is no visual evidence for large-scale breccia, conglomerate, rotated blocks or crumpled or other disturbed bedding structures in the sediment wedge above the truncation surface.

Where the plane of truncation eventually runs parallel with the underlying rocks, the overlying beds also lie parallel with the underlying beds (Fig. 12). The relationship is no longer angular (unconformable), but conformable—yet there may be 100 m or more of stratigraphic truncation (missing section). Thus, this surface is a paraconformity (Dunbar and Rodgers, 1958, p. 119), and may be recognized only by tracing out beds along strike, or possibly by detailed paleontological study.

One of the enigmas of these truncation structures is that they do not occur in facing pairs, and thus do not define trough-shaped structures that one might expect to be formed by localized erosional or gravity slide processes. Only one or two smaller, poorly defined structures that have possible "saucer"-shaped concave profiles have been identified in the cliff section. One of the largest structures possibly may be paired with

FIG. 14.—Eastern end of the Svartfjeld cliff section showing several large, easterly-dipping truncation structures (arrows) in the Hare Fiord Formation. Width of view about 600 m. Large structure at lower left truncates more than 100 m of section. Sedimentary fill of this structure clearly thickens down dip.

a truncation structure that dips toward it 5 km further west; however, precise matching of truncated intervals is not obvious.

The angular or discordant sections of the largest truncation structures are at least 1.5 km long; the structures are much larger if their disconformable, parallel-to-bedding component is added. The most clearly defined truncation structures are 500 to 900 m long. Estimated vertical truncation ranges from a few tens of metres for smaller structures, to 150 m for the largest. The largest individual structures thus have removed nearly one-fifth of the total (preserved) thickness of approximately 800 m of the Hare Fiord Formation in this area.

A plot of all recognizable truncation structures on the photographic mosaic of the Svartfjeld cliff section (not reproduced here) reveals that there is a general reversal in direction of dip of the larger structures on either side of a zone about 4 km from the east end of the cliff section. To the west of this zone, most structures have a major westerly or southwesterly dip component, while to the east of the zone, most structures have a strong easterly dip component. As the cliff exposure is relatively straight, striking west-southwest, these systematic variations in apparent dip are significant. The transition zone of dip reversal from east to west contains in its lower part a complex series of sedimentary structures combining some truncational structures with depositional thinning and thickening of beds (Fig. 15). This zone of reversal of dip components may correspond with the axis of a former bathymetric high or ridge with very gently dipping flanks. Such a ridge may be related to a local southward projection of the Nansen shelf edge. Alternatively, the sedimentary structures in the lower part of the transition zone may have been formed by the coalescing and overlapping of distal lobes of submarine fans deposited downslope from a localized site of turbidity current generation (compare model of Walker, 1975, Fig. 10). If so, the transition zone is an accretionary ridge in its lower part, with relief maintained during later sedimentation.

DISCUSSION

Source and Emplacement Mechanisms For Displaced Carbonate

Limestone turbidities.—The size-graded beds of clastic-textured limestone found in the Hare Fiord Formation are interpreted to be turbidites, the

FIG. 15.—Complex structures in Hare Fiord Formation, Svartfjeld cliffs. *A*—General view of complex structures in cliff face. Truncation structures east and west of this zone of complex structures tend to dip away from it. Structures include truncation surfaces, and apparent depositional thinning and thickening of bedding units. Width of view about 500 m.

B—Detail of thinning and thickening of cherty limestone from upper left (east) side of 15A. Individual beds at right center may be traced laterally as they thin out and dip to the left; some beds pinch out completely. These structures may represent sections through laterally onlapping distal lobes of submarine fans, downslope from a local source of turbidity currents. Alternatively, thinning may result in part from shear along an incipient slide surface in relatively unindurated sediment. Width of view about 100 m.

product of transportation by and deposition from turbidity currents. This interpretation is based on the planar bedded nature of the limestones, continuity of individual beds, internal size grading, clastic texture, granular composition with minor mud component, irregular base (scoured/load casted), and gradational top. Similar bioclastic turbidites (or allodapic limestone) have been described by Meischner (1964), Tucker (1969), Bernoulli and Jenkyns (1970), and Mackenzie (1970). In many ancient basinal or trough successions adjacent to carbonate-shelf transitions, such graded, resedimented carbonates form a major component of the sediment column. Modern examples of displaced carbonate deposits in deep-sea environments, mostly interpreted to be turbidites, have been documented by Rusnak and Nesteroff (1964), Davies (1968), Schlanger and Johnson (1969), Bornhold and Pilkey (1971), and others.

The major source of resedimented clasts in Hare Fiord crinoidal turbidites appears to be shelf-foreslope or slope environments, seaward of shelf-edge organic buildups. This slope source is suggested by the rarity of large indurated lithoclasts or bioclasts diagnostic of the shelf-edge facies, and by the dominance of crinoid and ramose bryozoan clasts. Fossil communities characterized by crinoids and ramose bryozoans appear to be distributed preferentially in deeper-water slope environments seaward or basinward of carbonate shelves. Events such as periodic storms and/or minor earthquakes, acting on uncemented, unstable accumulations of bioclastic sediment on these slopes triggered movement of the loose sediment and generated downslope-moving (gravitational) turbidity currents.

Graded crinoidal limestone beds enclosed in Hare Fiord deep-water rocks overlying steeply dipping flank beds of the Pennsylvanian bryozoan mounds in the Blue Mountains and Krieger Mountains areas may have been deposited by turbidity currents. These turbidity currents were generated on the upper slopes of a submarine topographic high formed above the buried mounds. At least some of the crinoid turbidites that flank these mounds in the Blue Mountain area contain mixed Pennsylvanian and Permian conodont faunas. This suggests that the buried mounds with their overlying sediment persisted as submarine topographic highs into the Permian.

The common occurrence of crinoid-rich turbidites in Paleozoic basinal rocks in the Canadian Arctic and elsewhere suggests that: (a.) there is a strong preference for crinoid community growth in deeper water slope environments where turbidity currents may be generated, or (b.) crinoid (or echinoderm) bioclasts are preferentially remobilized and transported by turbidity currents because of their high internal skeletal porosity and thus low bulk density that affects their behavior as sedimentary particles (Maiklem, 1968). They also may be transported preferentially into distal parts of the turbidite bed as a crinoid concentrate, whereas lithoclastic and other less porous bioclastic components were deposited and bypassed in proximal parts of the bed.

It is possible that the flaser breccia beds described in this paper are proximal equivalents of crinoidal turbidites (Fig. 16); the increase in cri-

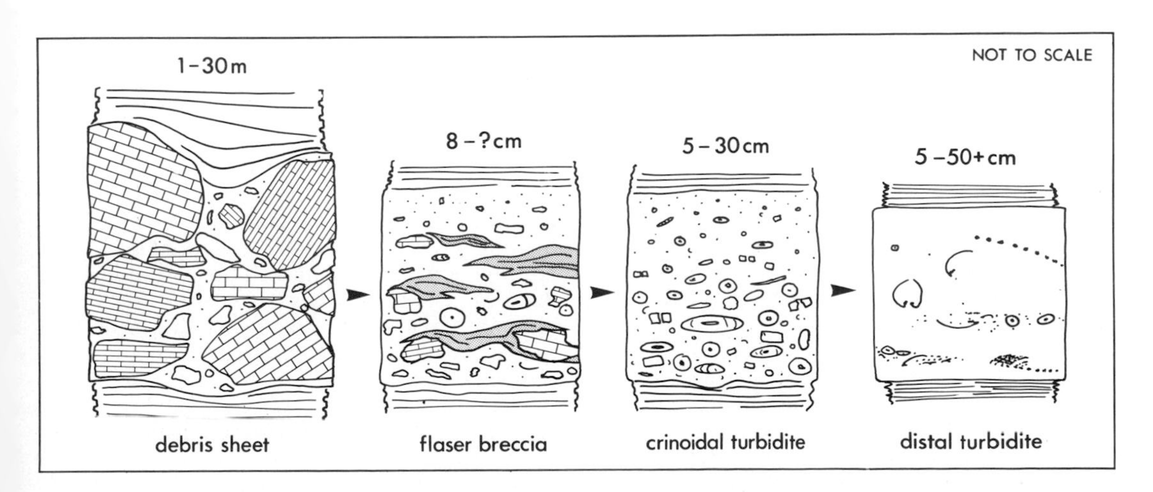

FIG. 16.—Possible genetic relationship among debris sheets, flaser breccia, crinoidal and lithoclastic turbidites, and distal lime mudstone turbidites in the Hare Fiord Formation. Thicknesses given above each type are ranges observed in field. These relationships are inferred from field and compositional data, but have not been established directly by tracing individual beds from one depositional fabric to another. This progression infers a downslope evolution of the transporting mechanism from debris flow into a turbidity current with a mud-rich distal phase.

noid bioclast content towards the top of the flaser breccia bed may support this interpretation. Further, it is possible that the well-bedded, poorly size-graded micritic (wackestone-textured) limestones found elsewhere in the Hare Fiord Formation are distal equivalents of the more coarsely textured crinoidal-lithoclastic turbidites (Fig. 16).

Limestone flaser breccia.—The term marl-flaser breccia was used by Schlager and Schlager (1973, p. 74–76) to describe the basal part of what they termed a proximal fluxoturbidite. Their flaser breccia is characterized by a chaotic fabric of plastically deformed marl lithoclasts separating irregular lenses of subangular limestone clasts and other lithoclasts. The deformed marls form a flaser-like fabric.

As a nongenetic descriptive term, marl-flaser breccia appears most apt for the beds of intermixed plastic and indurated lithoclasts found in the Hare Fiord Formation. However, their proximal character and relationship to limestone turbidites and other displaced sediments cannot be demonstrated clearly. The similarity in composition of cemented angular lithoclasts and associated shaly rocks in the Hare Fiord flaser breccias and in the Hare Fiord debris beds suggests a genetic affinity between these two types of displaced carbonate beds. The thinner bedded, flaser breccia may be the distal equivalent of the thicker, megaclastic debris beds (Fig. 16), a contention supported by the work of Hampton (1972). Flaser breccia in turn may be the proximal equivalent of crinoidal turbidites in which the lighter crinoid clasts remained in suspension and bypassed the coarse lithoclasts to accumulate downslope.

Schlager and Schlager (1973, p. 76) consider marl-flaser breccia to represent an interim stage between grainflow and mudflow transport, with the marly material being absorbed by the moving sediment flow. They assume grain-to-grain shearing during flow, with preferred shearing along the plastic marl flasers allowing differential movement of the lenses of angular clasts.

Deformation of the plastic marl clasts in the Hare Fiord flaser-breccia may be due in part to shearing stresses during deposition. Using Carter's (1975) revised classification of mass-transport processes, deposition of the flaser breccia probably occurred by grain fallout with shear from the basal layer of a turbidity current in which transport of the basal layer immediately prior to deposition was by laminar inertial flow.

Source of material for the flaser breccia must have included indurated shelf-edge carbonate to provide the variety of cemented lithoclasts of shallow-water aspect. The plastically deformed flasers of spicular marl (undated) were derived from the floor of the slope environment.

Debris sheets.—The thick debris sheets in the Hare Fiord Formation northeast of van Hauen Pass have many of the characteristics of carbonate debris beds documented by Cook and others (1972). Field exposures, differential weathering, and gross morphology bear a strong resemblance to the Devonian debris beds in Alberta; compare Cook and others, Plates 2, 7, with Figure 9 of this paper. The light-colored resistant-weathering debris sheets are enclosed in dark-colored cherty, argillaceous lime mudstone and shale of deep-water aspect. Some distance to the northeast, these Hare Fiord rocks grade into Nansen shelf or shelf-edge carbonate. Although the clastic limestone beds within the Hare Fiord Formation in this area have not been traced directly into the Nansen shallow-water carbonate, lithological (grainstone, packstone, fibrous cement) and skeletal (phylloid algae, fusulinaceans, *Palaeoaplysina*) composition of many of the clasts in the debris sheets provides strong evidence for their derivation from the carbonate shelf edge. Additional contributions were made by deeper water, shelf-foreslope and slope sediments. These include lithoclasts of bryozoan and crinoidal packstone and wackestone, silty and argillaceous dolomitic lime mudstone, and other fine-grained, spicular sediments. Discrete bioclasts of crinoids and ramose bryozoans also were caught up in the debris beds; these probably were derived from the crinoid-bryozoan communities typical of shelf-slope environment seaward of shallow-water organic accumulations. Lack of pervasive pre-resedimentation cementation of ramose bryozoan clasts is indicated by the penetration of fine-grained, dark-colored matrix sediment into skeletal apertures (Fig. 11C).

The depositional mechanism responsible for generation of the limestone debris beds probably was submarine debris flow. Hampton (1975, p. 835) defined debris flow as sediment gravity flow in which granular solids such as boulders, pebbles or sand are supported within the flow mainly by the strength of a fluid matrix composed of clay minerals and water. Buoyancy of the fluid matrix also contributes to support (Hampton, 1975). Naturally occurring submarine debris flows probably have a more complex transportational and depositional history; most real debris flows probably are combinations of debris flow and grain flow in the sense of Middleton and Hampton (1973; also see Hampton, 1975, p. 843).

The matrix of the Hare Fiord debris sheets records that lime mud was the dominant contributor to the debris-flow slurry that transported and emplaced the debris. Hampton's (1975) experimental work on debris flow competence utilized clay (kaolin)-water mixtures; the effect of a mixed clay-lime mud or relatively pure lime-mud matrix on debris flow competence apparently has not yet received experimental analysis.

The Hare Fiord debris flows originated in the

vicinity of the shelf edge. Gravitational force acting on unstable sediment on a slope was an obvious prerequisite; the gradient of the slope need only have been a few degrees (Cook and others, 1972), although slopes exceeding 40° may have been present at the Nansen shelf edge. Several other conditions may have contributed to the onset of instability and the subsequent generation of debris flows:

1. Carbonate debris flows of the type described by Cook and others (1972), Mountjoy and others (1972), and in this paper, are characterized by cemented angular lithoclasts of shallow-water-shelf origin that usually bear no evidence of surface boring or organic encrustration. Both of these phenomena are to be expected if indurated clasts were exposed at the sediment-water interface in the vicinity of organic buildups for any period of time (Cook and others, 1972, p. 470). This suggests that much or all of the angular lithoclasts were generated at the time of initiation of the debris flow; it would support a sudden, catastrophic event involving massive fracturing and failure of cemented carbonate rocks.

2. Many lithoclasts in the Hare Fiord debris sheets are cemented by fibrous calcite interpreted to be a neomorphic replacement of an early submarine cement. As early cementation of source rocks for debris flows is very common (Cook and others, 1972; Mountjoy and others, 1972; this paper), perhaps this type of cementation plays a direct rather than incidental role in initiation of the debris flow by producing rigid and brittle carbonate rocks in an unstable setting overlying uncemented, finer grained slope sediment (also see Cook and others, 1972, p. 470; Hopkins, 1972, and this volume).

3. An examination of several sections through Pennsylvanian to Permian carbonate shelf-to-trough transitions on Ellesmere Island indicates that the upper (Permian) part of the shelf carbonate succession progrades towards the trough. Paleontological dating of the debris sheets in the Hare Fiord Formation indicates an Early Permian age. Thus there may be a connection between shelf progradation and generation of debris flows in these deposits. Allied to and supporting this correlation between seaward progradation of the shelf and debris sheets are paleontological data (Fig. 5) suggesting either a hiatus (disconformity) or a reduced section for Late Pennsylvanian time in the Hare Fiord Formation (type section), and a correlative Late Pennsylvanian hiatus or reduced section in several sections of the Nansen shelf carbonate. These anomalies might be produced by a general deepening and transgressive episode resulting in a condensed deep-water section and a reduced rate of shelf sedimentation, yet there is no lithological or other support for this interpretation. More likely, these anomalies are the result of a general lowering of sea level ("regression"), perhaps with exposure and erosion of the shelf, leading to active accretion of clastic carbonate sediments as a shelf-foreslope tongue. This porous and permeable tongue of sediment would be subject to rapid submarine cementation similar to other Nansen shelf-edge rocks, yet would overlie undercompacted and relatively unindurated fine-grained clastic carbonate and argillaceous sediment of the deeper slope lithofacies. In this unstable setting, massive fracturing, slumping and downslope movement of rock would be expected, possibly extending into Early Permian time. General support for a widespread shallowing event (or series of fluctuations) is provided by widespread oöid grainstone sheets capping Nansen shelf cycles from Middle Pennsylvanian time onward, as well as textural evidence from some displaced shelf lithoclasts in debris sheets and flaser breccia suggesting subaerial (fresh-water) diagenesis before retransport.

An examination of facies relationships between the Devonian Southesk Formation (carbonate-buildup facies, the source of debris beds described by Cook and others, 1972) and the Perdrix and Mt. Hawk Formations ("basinal" facies, hosts for the debris beds) as shown diagramatically by Cook and others (1972, Fig. 6), reveals that major debris beds extend basinward only from tongues of carbonate forming seaward prograding phases of Southesk bank deposition.

Pray (in Cook and others, 1972, p. 470) believes that differential compaction of interbedded grain-rich (permeable) and mud-rich (relatively impermeable) deposits in core facies of Mississippian bioherms in New Mexico led to the formation of debris-flow megabreccia at the bioherm flanks, with shearing of massively-cemented skeletal grainstone along basal contacts with relatively uncemented mud-matrix sediment as a probable mechanism. A similar mechanism operating on a larger scale, involving failure by fracturing of submarine-cemented carbonate tongues interbedded with or overlying plastic muddy sediments at shelf-to-trough transitions, particularly where the carbonate tongues are markedly regressive, may be responsible for many debris beds and megabreccia. This mechanism, involving at least initial gravitational downsliding of fractured, indurated limestone blocks into a plastic muddy substrate at inception of downslope flowage, followed by mass flow, explains the intimate admixture of fractured lithoclasts and muddy interclast matrix found in debris beds.

Volumetric Contribution of Displaced Carbonate Sediment to Deep-Water Environments

Displaced carbonate sediment may constitute a major part of the total thickness of deep-water sedimentary deposits. The volumetric contribution of displaced carbonate sediment to the Hare Fiord Formation on Ellesmere Island is variable; in the Blue Mountains area, at least 60 percent of the lower 200 m of the formation is composed of displaced shelf-edge or slope carbonate sediment (turbidite type). Displaced carbonate sediment, again mainly of turbidite type, is a significant component of other Hare Fiord sections on Ellesmere Island, including the type section.

Schlager and Schlager (1973, p. 83-84) calculate that in a total section thickness of about 350 m for the Upper Jurassic deep-sea Tauglboden-Schichten of the Eastern Alps, about 15 m are undisturbed pelagic radiolarite and 335 m are displaced clastic carbonate and resedimented pelagic material deposited mainly by turbidity currents. Much of the Glasenbach basinal sequence of Jurassic age in Austria is composed of displaced carbonate sediment (Bernoulli and Jenkyns, 1970, Fig. 1). Crinoidal and limestone turbidites appear to form at least 50 percent of the thickness of a 50 m section of Devonian sediments deposited in a deep-water basin in North Cornwall (Tucker, 1969, Fig. 2). Thomson and Thomasson (1969) interpret most of the slope and basin lithofacies of the Pennsylvanian Dimple Limestone in Texas as proximal and distal turbidites. Cook and others (1976) describe volcanogenic and carbonate sediment gravity flow deposits which comprise about 50 percent of a 225 m thick Upper Cretaceous basinal section in the Line Islands of the Equatorial Pacific. About one-quarter of the 200 m thick Cambro-Ordovician slope sediment in Nevada is of slump and mass-flow origin (Cook and Taylor, this volume).

The input of displaced carbonate sediment to the basin should decrease with increase in distance from the shelf or slope source. However, displaced carbonate may not be recognized in distal areas where the more obvious grain-size grading, clastic textures and breccia fabrics are replaced by fine-grained sediment without these characteristic features. It is probable that many of these distal deposits of displaced sediment are overshadowed, in terms of recognition and documentation, by the more obvious and spectacular debris flows, coarse turbidites, and other displaced carbonate of proximal aspect. As noted earlier, the very planar and regular-bedded micritic limestone typical of the Hare Fiord Formation farthest from the Nansen shelf edge may be distal carbonate turbidites.

Origin of Intraformational Truncation Structures

In an earlier interim report on the truncation structures within the Hare Fiord Formation, a gravity-slide mechanism was favored over an erosional origin (Davies and Nassichuk, 1975b). This conclusion still is held, although with some reservations. A gravity-slide mechanism is supported *collectively* by the following criteria:

1. The very large size of the structures, with some more than 1.5 m across and with 150 m of vertical truncation, argues more strongly for a slide mechanism than scour by currents. This is particularly so in a deep-water slope environment characterized by fine-grained calcareous and argillaceous sediment.
2. The listric geometry of the truncation surface is typical of landslide and other gravity-slide structures (see, for example, Lewis, 1971, Fig. 2).
3. The sharp and regular truncation of the underlying "footwall" beds, with no obvious local channels or erosional irregularities, is more logically explained by shear than by current scour.
4. The apparent lithologic similarity of the overlying fill to the truncated rocks (both are very fine-grained argillaceous, calcareous and cherty rock) does not suggest the radical change in depositional processes that might be expected to accompany a period of increased current scour.
5. The absence of a coarse breccia or conglomerate at the base of the infill, at least at a scale recognizable by close-up aerial observation, is equivocal; its presence, however, would support a strong channelized traction current.
6. The apparent absence of discontinuous tabular beds confined within the truncation structures argues against deposition by traction currents confined to a channel (*see* Harms, 1974, Fig. 2).
7. The absence of truncation structures with U-shaped cross sections is a strong argument against an erosional channel origin. However, random sections across *small* "spoon"-shaped slide scars also should include some U-shaped structures.
8. Modern analogs provide support for gravity sliding and slumping mechanisms at shelf edges; both processes are common in sediments of slope environments adjacent to modern continental shelves (Stride and others, 1969; Lewis, 1971; Normark, 1974; and many others; Fig. 17, this paper).

Smaller-scale rotational slump structures that bear some resemblance to the Hare Fiord structures have been described from Silurian rocks by Laird (1968). Wilson (1969) recognized similar "cut and fill" or slump structures in lime mudstone at localities in Europe, Montana, and the Guadalupe Mountains of west Texas, and considered them to be characteristic of deep-water carbonate facies.

Harms (1974) has interpreted truncation structures within the Permian Brushy Canyon Formation in west Texas as channels that were eroded and filled by density currents moving basinward from the shallower shelves surrounding the Del-

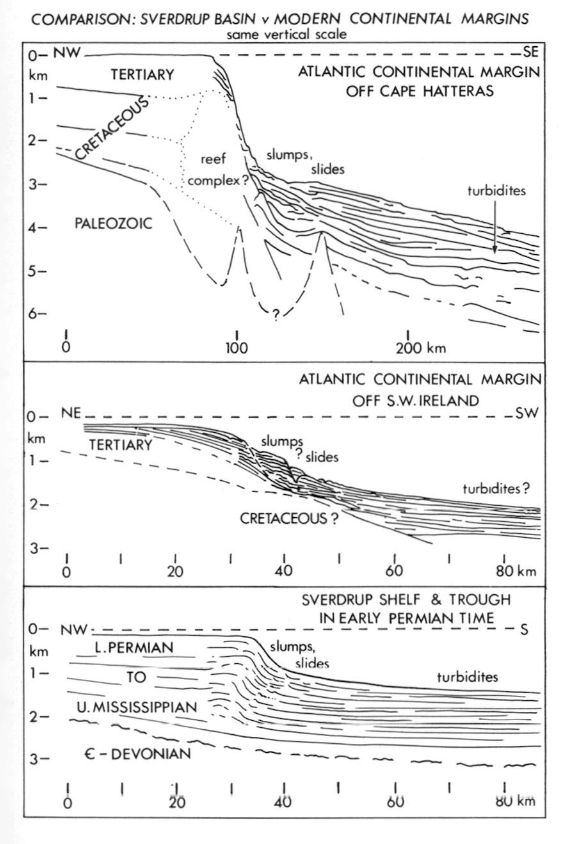

FIG. 17.—Comparison between Sverdrup shelf and trough in Early Permian time with modern continental slopes off Cape Hatteras (from Emery and others, 1970, Fig. 38) and off southwest Ireland (from Stride and others, 1969, Fig. 10), emphasizing the common association of large scale gravity slides and slumps in the slope setting, and turbidites on the floor of the slope. Depth of water in the Sverdrup trough is based on biostratigraphic data (Fig. 5).

aware Basin. Although the general geometry of the Brushy Canyon structures at first appears to resemble that of the Hare Fiord truncation structures, they differ in a number of significant ways: (a) they are cut in and filled by quartzose siltstones and sandstones, rather than fine-grained mudrocks, (b) the channel floors are flat, and the sides commonly slope 20° to 30°, (c) both sides of individual channels are visible in many structures, (d) most channels are only 20 to 30 m deep, (e) the channels are filled in part by breccia, conglomeratic sandstone, and tabular sandstone units confined to the channel structure, and (f) sedimentary structures formed by traction currents are common in associated rocks.

An apparent ambiguity in the gravity slide interpretation is the absence of displaced sediment or rock in any of the truncational structures. In random sections through a number of the structures, it seems logical to expect to see some of the displaced material in the form of rotated blocks, megabreccia, or masses of slumped or distroted beds—yet none have been observed. However, if the cliff section intersects the megastructures in a high up-dip position relative to the original depositional slope, the displaced sediment and rock may have been redeposited south of the section line, closer to the deeper (inferred) axis of the Sverdrup trough. Some support for this is provided by the discovery by Atlantic-Richfield geologists of lenses of limestone microbreccia within normally bedded Hare Fiord rocks about 30 km south of the cliff section containing the truncation structures (Davies and Nassichuk, 1975b). Taking this argument further, if the displaced sediment was relatively unconsolidated and water-saturated, initiation of the slide, perhaps by earthquake shock, might have resulted in liquefaction and mobilization of the fine-grained sediment. This would result in loss of primary bedding character and in transport as a fine sediment flow rather than as large coherent masses. Finally, the lack of coarse conglomerate or displaced blocks in the Hare Fiord megastructures is equally anomalous for an erosional mechanism, as the vast amount of sediment and rocks removed from the structures should have yielded a basal lag-conglomerate in at least some of the structures if they were formed by erosional currents.

Large-scale gravity sliding in the Hare Fiord rocks may have been triggered by earthquakes. The truncation structures are about 50 km east of the northwestern rim of the Sverdrup Basin. This rim was a fault-bounded tectonic element that was intermittently active during Pennsylvanian and Permian time (Meneley and others, 1975), and may have been the locus of numerous earthquakes.

Although the large-scale tectonic setting (continental to oceanic crust transition) of modern continental margins differs from that of the Pennsylvanian-to-Permian shelf edge of the Sverdrup Basin, the general slope morphology and occurrence of slumps, slides and turbidites in the modern setting may be used to support a conceptual sedimentological model for Sverdrup shelf-edge processes. Figure 17 is a comparison between modern Atlantic continental margins in two different areas and the Sverdrup Basin in Early Permian time. The comparison is strengthened by the abrupt nature of the Cretaceous (carbonate?) shelf edge off Cape Hatteras (Emery and others, 1970, Fig. 38), and the comparable water depths for the Atlantic margin off southwest

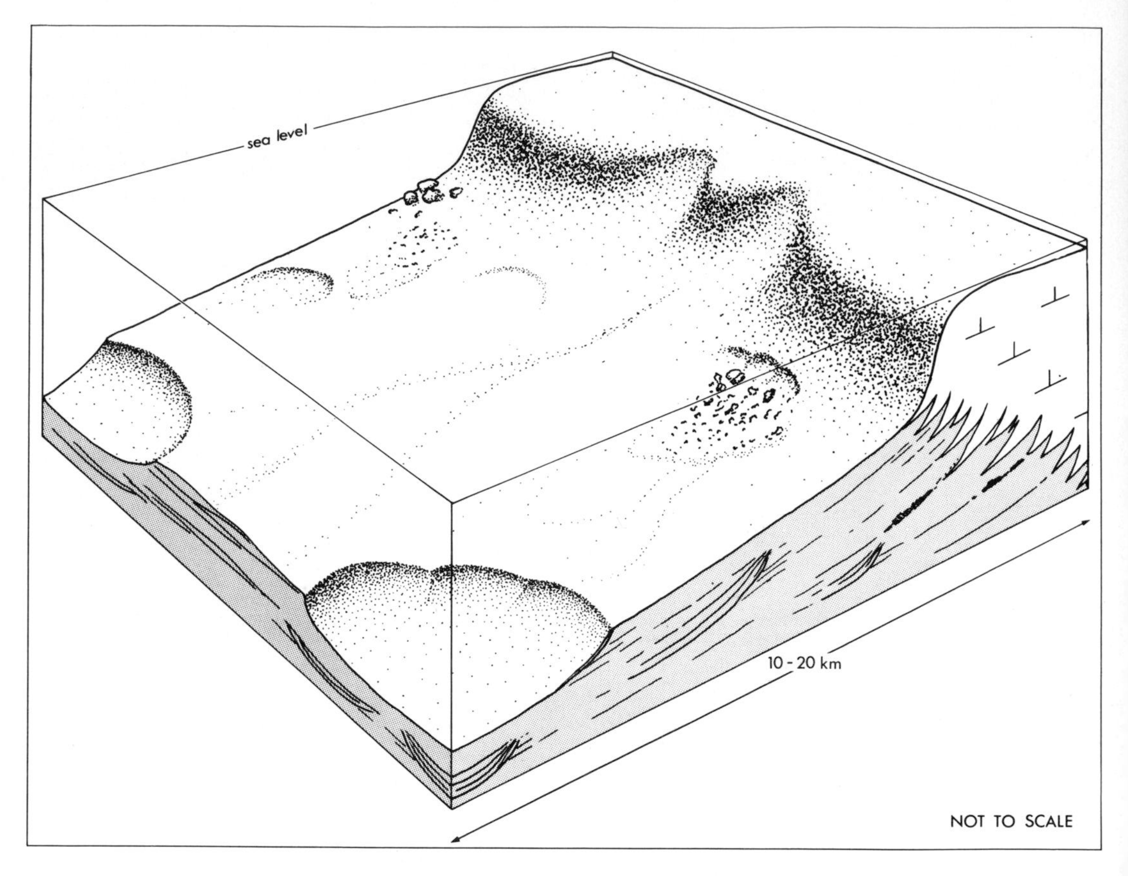

FIG. 18.—Depositional model for gravity-displaced sediment and gravity slides adjacent to the shelf edge of the Sverdrup Basin in Early Permian time. Gravity slides diverge in local dip away from a depositional high formed by distal lobes of a low relief submarine fan system. Debris sheets below high relief shelf foreslopes grade downslope into turbidites.

Ireland and for the Sverdrup Basin (estimated from biostratigraphic data, Fig. 5).

Figure 18 is a schematic diagram of part of the Sverdrup shelf edge and slope as it may have appeared in Early Permian time; it illustrates the possible geometric and spatial relationships between the shelf-edge, shelf-foreslope, and slope environments and debris sheets, gravity slides, and distal lobes of low-relief submarine fans formed in part by turbidites.

CONCLUSIONS

A spectrum of displaced carbonate rocks derived from shelf-edge, foreslope and deeper slope environments are interbedded with deep-water cherty limestone and shale of the Hare Fiord Formation in the northern Sverdrup Basin. The displaced carbonate includes turbidites and flaser breccias emplaced by turbidity currents and debris sheets emplaced by slurrylike debris flows; downslope gravitational movement was the common denominator.

Bioclasts of phylloid algae, fusulinaceans, and the enigmatic organism *Palaeoaplysina*, characteristic of the Nansen shelf and shelf-edge environments, are common components of clasts in the debris sheets and flaser breccias. Fibrous cements of clasts in the debris sheets and flaser breccias. Fibrous cements of marine origin within displaced shelf-edge rocks are truncated at the boundaries of clasts, demonstrating the early, predisplacement origin of the cement. The displaced carbonates may be interrelated in a genetic sequence reflecting decreasing clast size and increasing distance from source, as follows: debris sheet—flaser breccia—bioclastic turbidite—distal lime mudstone turbidite. There may be a genetic link between periods of shelf-edge progradation and the formation of debris sheets. At the other extreme, debris sheets may correlate with periods of shelf erosion and stratigraphic disconformity.

The recognition of displaced carbonate sediments in rocks of deep-water aspect is significant for a number of reasons:

1. It enhances our understanding of processes acting in shelf-edge and slope environments, particularly of processes that are "abnormal" or catastrophic and relatively instantaneous in a geological sense.
2. It allows separation of allochthonous from autochthonous components in basinal rock successions, and thus clarifies their relative contribution and the interpretation of sedimentation rates.
3. Errors in interpretation of paleodepth and paleoenvironment inherent in assuming an autochthonous origin for some displaced carbonates may be avoided.
4. It may provide data for the relative timing of cementation and diagenesis in rocks displaced from the shelf edge; for example, truncated fibrous calcite cements in debris sheet lithoclasts support other evidence for early marine cementation of the shelf edge rocks.
5. The occurrence of mixed biotas, particularly microfaunas, in basinal rocks may be explained by this mechanism; conversely, mixed biotas might be expected in displaced carbonate units, and allowance made for possible errors in biostratigraphic correlations. Some Hare Fiord crinoidal turbidite beds contain mixed conodont faunas representing several Pennsylvanian and Permian stages.
6. The presence and clast size of displaced carbonate may serve as an indicator of proximity to a high-relief shelf edge (or "reef"). Vector information from sedimentary structures, orientation of crinoid columnals, and from other sources, might be used to indicate direction of transport.

At least 15 truncation structures interpreted to be gravity slides occur within the Hare Fiord Formation in a cliff section about 10 km south of the Nansen shelf edge. The largest structures are at least 1.5 km across. Up to 150 m of section were truncated and removed by individual structures. The plane of truncation is listric in shape, decreasing in dip downsection from about 15° to 0°, where it runs parallel to the bedding of underlying rocks. The contact relationship between footwall beds and the sediment fill thus changes from angular unconformity to paraconformity.

For a formational thickness of 800 m spanning Middle Pensylvanian to Early Permian time, the thickness of rock removed by the larger individual truncation structures represents almost one-fifth of the total preserved section, and thus is a significant intraformation discontinuity that may span more than a time-stratigraphic stage.

There are no obvious displaced and rotated blocks of rock or slumped and disturbed beds within any of the truncation structures, suggesting that the removed material has been redeposited elsewhere, downslope from the structures and away from the shelf edge. The structures are comparable with gravity-slide structures in other basinal-rock sequences, and in modern continental-slope settings.

ACKNOWLEDGEMENTS

All fieldwork on Ellesmere Island was conducted jointly with W. W. Nassichuk (G. S. C.), who also provided some additional samples and photographs of debris sheets. C. Johnson and D. Folk of Atlantic Richfield Canada Ltd. provided some field data relevant to the Hare Fiord truncation structures. Discussions with V. Schmidt (Mobil Oil Co.), J. C. Harms (Marathon Oil Co.), and other geologists at the 1975 S. E. P. M. meeting in Dallas helped to clarify some of the ideas in this paper.

REFERENCES

Abbate, Ernesto, Bortolotti, Valeria, and Passerini, Pieto, 1970, Olistostromes and olistoliths: Sed. Geology, v. 4, p. 521–557.

Bernoulli, Daniel, and Jenkyns, H.C., 1970, A Jurassic basin: The Glasenbach Gorge, Salzburg, Austria: Geol. Bundesanstalt Wien Verh., 1970, p. 504–531.

Bornhold, B. D., and Pilkey, O. H., 1971, Bioclastic turbidite sedimentation in Columbus Basin, Bahamas: Geol. Soc. America Bull., v. 82, p. 1341–1354.

Carter, R. M., 1975, A discussion and classification of subaqueous mass-transport with particular application to grain-flow, slurry-flow, and fluxoturbidites: Earth-Sci. Rev., v. 11, p. 145–177.

Cook, H. E., Jenkyns, H. C., and Kelts, K. R., 1976, Redeposited sediments along the Lines Islands, equatorial Pacific: *In* S. O. Schlanger, E. D. Jackson and others, Initial reports of the Deep Sea Drilling Project, v. 33: U.S. Government Printing Office, Washington, D.C., p. 837–847.

——, McDaniel, P. N., Mountjoy, E. W., and Pray, L. C., 1972, Allochthonous carbonate debris flows at Devonian bank ('reef') margins, Alberta, Canada: Bull. Canadian Petroleum Geology, v. 20, p. 439–497.

Davies, D. K., 1968, Carbonate turbidites, Gulf of Mexico: Jour. Sed. Petrology, v. 38, p. 1100–1109.

Davies, G. R., 1974, Submarine cementation, fracturing, and internal sedimentation in Pennsylvanian-Permian carbonate buildups, Arctic Archipelago [abs.]: Am. Assoc. Petroleum Geologists, Ann. Meeting Abs., v. 1, p. 25.

——, 1975a, Hoodoo L-41: Diapiric halite facies of the Otto Fiord Formation in the Sverdrup Basin, Arctic Archipelago: Canada Geol. Survey, Paper 75-1C, p. 23-29.
——, 1975b, Upper Paleozoic carbonates and evaporites in the Sverdrup Basin, Canadian Arctic Archipelago: Canada Geol. Surv., Paper 75-1B, p. 209-214.
——, AND KROUSE, H. R., 1975, Carbon and oxygen isotopic composition of late Paleozoic calcite cements, Canadian Arctic Archipelago—Preliminary results and interpretation: Canada Geol. Survey, Paper 75-1B, p. 215-220.
—— AND NASSICHUK, W. W., 1973, The hydrozoan? *Palaeoaplysina* from the upper Paleozoic of Ellesmere Island: Jour. Paleontology, v. 47, p. 251-265.
—— AND ——, 1975a, Subaqueous evaporites of the Carboniferous Otto Fiord Formation, Canadian Arctic Archipelago: A summary: Geology, v. 3, p. 273-278.
—— AND ——, 1975b, Gravity-slide megastructures in deep-water carbonates of the Pennsylvanian-Permian Hare Fiord Formation of Ellesmere Island: Canada Geol. Survey, Paper 75-1B, p. 227-232.
DICKINSON, W. R., 1974, Plate tectonics and sedimentation: *In* W. R. Dickinson (ed.), Tectonics and sedimentation: Soc. Econ. Paleontologists and Mineralogists, Spec. Pub. 22, p. 1-27.
DUNBAR, C. O., AND RODGERS, JOHN, 1958, Principles of stratigraphy: John Wiley and Sons, Inc., New York, New York, 356 p.
DUNHAM, R. J., 1962, Classification of carbonate rocks according to depositional texture: *In* W. E. Ham (ed.), Classification of carbonate rocks—A symposium: Am. Assoc. Petroleum Geologists, Mem. 1, p. 108-121.
EMERY, K. O., UCHUPI, ELAZAR, PHILLIPS, J. D., BOWIN, C. O., BUNCE, E. T., AND KNOTT, S. T., 1970, Continental rise off eastern North America: Am. Assoc. Petroleum Geologists Bull., v. 54, p. 44-108.
ENGEL, W., 1970, Die Nummliten-Breccien im Flyschbecken von Ajdovscina in Slowenien als Beispiel karbonatischer Turbidite: Geol. Bundesanst. Wien Verh., 1970, p. 570-582.
FLORES, G., 1959, Evidence of slump phenomena (olistostromes) in areas of hydrocarbon exploration in Sicily: Fifth World Petroleum Congress Proc., Sec. 1, p. 259-275.
HAMPTON, M. A., 1972, The role of subaqueous debris flow in generating turbidity currents: Jour. Sed. Petrology, v. 42, p. 775-793.
——, 1975, Competence of fine-grained debris flows: Jour. Sed. Petrology, v. 45, p. 834-844.
HARMS, J. C., 1974, Brushy Canyon Formation, Texas: A deep-water density current deposit: Geol. Soc. America Bull., v. 85, p. 1763-1784.
HOEDEMAEKER, PH. J., 1973, Olisthostromes and other delapsional deposits, and their occurrence in the region of Moratalla (Prov. of Murcia, Spain): Scripta Geol., v. 19, p. 1-207.
HOPKINS, J. C., 1972, Petrography, distribution and diagenesis of foreslope, nearslope and basin sediments, Miette and Ancient Wall carbonate complexes (Devonian), Alberta: Ph.D. Dissert., McGill Univ., Montreal, Quebec, 234 p.
KENDALL, A. C., AND TUCKER, M. E., 1973, Radiaxial fibrous calcite: A replacement after acicular carbonate: Sedimentology, v. 20, p. 365-389.
LAIRD, M. G., 1968, Rotational slumps and slump scars in Silurian rocks, western Ireland: Sedimentology, v. 10, p. 111-120.
LEWIS, K. B., 1971, Slumping on a continental slope inclined at 1°-4°: Sedimentology, v. 16, p. 97-110.
MACKENZIE, W. S., 1970, Allochthonous reef-debris limestone turbidites, Powell Creek, Northwest Territories: Bull. Canadian Petroleum Geology, v. 18, p. 474-492.
MAIKLEM, W. R., 1968, Some hydraulic properties of bioclastic carbonate grains: Sedimentology, v. 10, p. 101-109.
MEISCHNER, K. D., 1964, Allodapische Kalke, Turbidite in riff-nahen Sedimentations-Becken: *In* A. H. Bouma and A. Brouwer (eds.), Turbidites: Elsevier, Amsterdam, The Netherlands, p. 156-191.
MENELEY, R. A., HENAO, DIEGO, AND MERRITT, R. K., 1975, The northwest margin of the Sverdrup Basin: *In* C. J. Yorath, E. R. Parker and D. Glass (eds.), Canada's continental margins and offshore petroleum exploration: Canadian Soc. Petroleum Geologists, Mem. 4, p. 531-544.
MIDDLETON, G. V., AND HAMPTON, M. A., 1973, Sediment gravity flows: Mechanics of flow and deposition: *In* G. V. Middleton and A. H. Bouma (eds.), Turbidites and deep-water sedimentation: Soc. Econ. Paleontologists and Mineralogists Pacific Sec., Los Angeles, California, p. 1-38.
MOUNTJOY, E. W., COOK, H. E., PRAY, L. C., AND MCDANIEL, P. N., 1972, Allochthonous carbonate debris flows—Worldwide indicators of reef complexes, banks, or shelf margins: 24th Internat. Geol. Congress Proc., Sec. 6, p. 172-189.
NASSICHUK, W. W., AND SPINOSA, C., 1972, Lower Permian (Asselian) ammonoids from the Hare Fiord Formation, northern Ellesmere Island: Jour. Paleontology, v. 46, p. 536-544.
NORMARK, W. R., 1974, Ranger submarine slide, northern Sebastian Vizcaino Bay, Baja California, Mexico: Geol. Soc. America Bull., v. 85, p. 781-784.
RICHTER, DIETER, 1973, Olisthostrom, Olistholith, Olisthothrymma und Olisthoplaka als Merkmale von Gleitungs und Resedimentationsvorgängen infolge synsedimentärer tektogenetischer Bewegungen in Geosynklinalbereichen: Neues Jahrb. Geologie u. Paläontologie Abh., v. 143, p. 304-344.
RUSNAK, G. A., AND NESTEROFF, W. D., 1964, Modern turbidites: Terrigenous abyssal plain versus bioclastic basins: *In* R. L. Miller (ed.), Papers in marine geology, Shepard commemorative volume: MacMillan, New York, New York, p. 488-503.

Schlager, Wolfgang, and Schlager, Max, 1973, Clastic sediments associated with radiolarites (Tauglboden-Schichten, Upper Jurassic, eastern Alps): Sedimentology, v. 20, p. 65-89.

Schlanger, S. O., and Johnson, C. J., 1969, Algal banks near La Paz, Baja California—Modern analogues of source areas of transported shallow-water fossils in pre-Alpine flysch deposits: Palaeogeography, Palaeoclimatology, Palaeoecology, v. 6, p. 141-157.

Seilacher, Adolf, 1967, Bathymetry of trace fossils: Marine Geology, v. 5, p. 413-428.

Stride, A. H., Curray, J. R., Moore, D. G., and Belderson, R. H., 1969, Marine geology of the Atlantic continental margin of Europe: Royal Soc. London Philos. Trans., Ser. A, v. 264, p. 31-75.

Thomson, A. F., and Thomasson, M. R., 1969, Shallow to deep-water facies development in the Dimple Limestone (Lower Pennsylvanian), Marathon region, Texas: *In* G. M. Friedman (ed.), Depositional environments in carbonate rocks: Soc. Econ. Paleontologists and Mineralogists, Spec. Pub. 14, p. 57-78.

Thorsteinsson, Ray, 1974, Carboniferous and Permian stratigraphy of Axel Heiberg and western Ellesmere Island, Canadian Arctic Archipelago: Canada Geol. Survey, Bull. 224, 115 p.

Tucker, M. E., 1969, Crinoidal turbidites from the Devonian of Cornwall and their paleogeographic significance: Sedimentology, v. 13, p. 281-290.

von Görler, K., and Reutter, K. J., 1968, Entstehung und Merkmale der Olisthostrome: Geol. Rundschau, v. 57, p. 484-514.

Walker, R. G., 1975, Generalized facies models for resedimented conglomerates of turbidite association; Geol. Soc. America Bull., v. 86, p. 737-748.

Wilson, J. L., 1969, Microfacies and sedimentary structures in "deeper water" lime mudstone: *In* G. M. Friedman (ed.), Depositonal environments in carbonate rocks: Soc. Econ. Paleontologists and Mineralogists, Spec. Pub. 14, p. 4-19.

SEPM SPECIAL PUBLICATION NO. 25, P. 249-261, NOVEMBER 1977

AN INTERPRETATION OF THE DEPOSITIONAL SETTING OF SOME DEEP-WATER JURASSIC CARBONATES OF THE CENTRAL HIGH ATLAS MOUNTAINS, MOROCCO

IAN EVANS AND C. G. ST. C. KENDALL
University of Houston, Houston, Texas 77004, and
University of South Carolina, Columbia, 29208

ABSTRACT

The central High Atlas Mountains of Morocco occupy the site of an Early (Lias) and Middle (Dogger) Jurassic seaway, which was oriented east-west across southern Morocco and filled by carbonate sediments. The axial part of this seaway was the site of deep-water deposition during the Lias. Because the rocks are neither severely deformed nor significantly dolomitized, the facies relations between the axial part of the basin, the slopes and the shelves are easily observed in the excellent outcrop exposures provided by the present-day desert climate.

The Liassic deepwater carbonates were deposited in two major settings, basin and slope.

(1) Basin: With the exception of the lowermost 10–20 metres, the deposits of the trough axis can be considered to be deep-water in origin. The oldest Liassic rocks within the trough axis are laminated micrites that record an early supratidal to subtidal environment of deposition prior to the onset of deeper-water conditions. Upward in the section, this sequence passes fairly abruptly into deeper-water lithoherms, massive micritic buildups that lack the coral and skeletal framework common in shoal-water reefs marginal to the trough. The lithoherms are overlain by an Upper Liassic sequence of rhythmic alternations of bedded marls and micrites that contain coccoliths and ammonites but lack the skeletal remains of benthonic organisms. The micrites contain a monotonous and low-diversity suite of trace fossils that indicate the presence of oxygenated bottom waters at the time of deposition. The rhythms represent the interlayering of pelagic deposits (micrites) and the distal, suspension loads of turbidity currents (marls). These Liassic deep-water deposits are overlain by a progressively upward-shoaling sequence of micrites and marls that are capped by coral reefs (Dogger age) representing the filling of the trough.

(2) Slope: A sequence of Liassic turbidites that are essentially coeval with the basinal deposits represent deposition on the lower portion of the Jurassic slope of the High Atlas trough. Exposed on the southern slope is a sequence of turbidites that is almost one kilometre thick; the base of the sequence is not observed. The turbidite section is made up of materials apparently derived from numerous geographic locations. The oldest observed units are channeled, contain few Bouma cycles, and are interpreted to have been deposited in a mid-fan setting. The younger units contain many well-developed Bouma cycles that are interpreted to represent outer-fan deposition. These strata contain deep-water trace fossils.

INTRODUCTION

The High Atlas Mountains are composed almost wholly of Jurassic limestones, marls and shales. Within the mountains there is a remarkably complete, well-exposed, and structurally simple stratigraphic sequence. The rocks of this sequence form a part of the sedimentary fill of a rift zone that developed during the initial breakup of Africa, Europe and North America (Dewey and others, 1973). This rift became inactive after 20 m.y. in the Late Dogger, and today represents a tectonic scar on the northwest African plate. The ancient rift was characterized by three major depositional settings, the basin (axial), slope (northern and southern) and shelf (northern and southern).

In the Ziz Valley, the study area, the sedimentary facies are symmetrically disposed on either side of the basin axis (Figs. 1 and 2). The threefold facies subdivision of the High Atlas was also recognized by Dubar (1949, Fig. 3; and 1960–1962) and Faure-Muret and others (1971) who used different terminology to describe them (Table 1). These works, together with that of Dubar (1952), provide a basic regional overview of the geology of the central High Atlas area.

The purpose of this paper is to present an interpretation of the depositional setting of the Liassic facies in the axial region of the High Atlas trough. The terminology used to describe the turbidites in this discussion follows Walker and Mutti (1973), while petrologic terminology follows Folk (1962). In order to unravel the geologic history of this portion of the High Atlas Mountains, a transect of approximately 50 km was undertaken between the towns of Midelt and Ksar-es-Souk. Ten sections were measured, totaling more than 4000 metres in cumulative thickness. In addition to the ten sections, additional detailed sections were measured to obtain more specific information concerning the Trias/Lias contact, turbidite units, and deep-water units. In addition, numerous areas were spot checked in order to

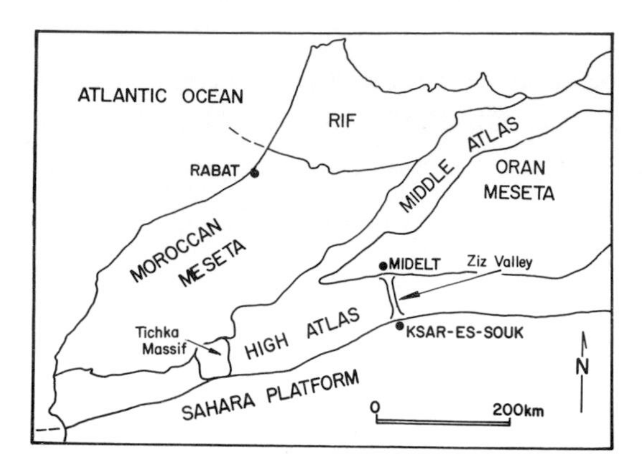

FIG. 1.—Major geologic provinces of Morocco and location of study area (modified after de Sitter, 1964).

determine whether or not the proposed facies distribution patterns were correct. To supplement the field work, approximately 400 thin sections were examined and described. The scanning electron microscope was used to search for nannofossil components in the basinal limestones, and X-ray analyses of insoluble residues were carried out on shelf, slope and basinal samples. Extensive collections of shelf faunas were compiled.

TECTONIC SETTING AND GEOLOGIC HISTORY

During Late Triassic and Early Jurassic time, an elongate E-W trough formed in the region that is today the High Atlas Mountains. This trough, 600 km long and 100–150 km wide, shallowed and terminated against the pre-Mesozoic Tichka Massif in the west, where the sedimentary sequences thin and pinch out. To the east, the trough shallowed near the present Moroccan/Algerian border where the section is thinner and lacks the characteristics of the deep-water facies described in this study (Choubert and Faure-Muret, 1960–1962). The trough existed as a site of marine deposition for approximately 20 m.y. before it was infilled. The Jurassic marine sequences (Lower Lias to Middle Dogger) were subsequently deformed during the Late Eocene-Early Oligocene (Arkell, 1956). This deformation produced a simple style of folding with narrow, steeply dipping anticlines and broad, open synclines (Faure-Muret and others, 1971). These structures were further deformed during the Miocene by a new compressive phase that resulted in the overthrusting of the axial trough limestones both to the north and south over the adjacent shelves. The anticlines within the axial region of the trough were also overturned during this deformational phase. The whole area was then uplifted during the Pliocene, continuing to the present, thus giving the Atlas Mountains their current relief.

The Lower Liassic strata are the initial marine deposits in the Atlas trough; they overlie Upper Triassic continental red beds (marls and fine-grained sandstones) and interbedded basalt flows. Flooding of the rifting trough by waters from the Tethys initiated an episode of carbonate sedimentation that was to continue until Middle Dogger time. Differential subsidence produced an axial depositional basin that was flanked to the north and south by shallow shelf areas which persisted throughout the history of the trough (Figs. 3 and 4). By the close of Lias time, the axial, deeper-water facies were replaced by shallow-water micrites and marls that are capped by a distinctive shallow-water reef facies of Bajocian age.

BASIN DEPTH

Evidence suggesting slope and basinal deep-water settings for some of the Liassic strata includes: (1) stratigraphic and geometric relationships with the shelf facies which indicate deeper-water conditions (Fig. 5); (2) the presence of pelagic faunas and floras of ammonites and coccoliths, and general absence of skeletonized benthonic faunas; (3) the presence of the deep-water trace fossils *Paleodictyon* and *Zoophycus*; (4) no evidence of wave action; (5) the monotonous lithologic sequences of alternating micrites and marls that can be traced laterally over tens of kilometres and attain thicknesses in excess of 500 metres suggest a constant depositional setting below wave base.

Assigning a specific water depth proves to be difficult. However, the Atlas Mountains preserve the basic geometry of the Jurassic Atlas trough. The trough is believed to have been topographically symmetrical. The basic dimensions of the trough (see Fig. 6) represent the present-day dimensions and distribution patterns of the lithologies that are interpreted to represent shelf, slope and basinal environments (Choubert and others, 1956). The early Tertiary deformation of these rocks resulted in an unspecified amount of crustal shortening in this region. In attempting to obtain depth estimates of the basin, the effects of the deformation are not considered because we are presently unable to quantify the deformational effects. In order to calcualte the basin depth it is necessary to know the dip of the Jurassic slope. The average dip value for modern continental slopes is 4° (Gross, 1972), and this figure is used to arrive at the minimum depth value for the basin. Using this 4° value in conjunction with the dimensions of the Atlas trough, the calculated vertical distance between the shelf break and the bottom of the trough would be approximately 315 metres (Fig. 6 and Table 2). To arrive at a minimum value for the total depth of the trough, the depth

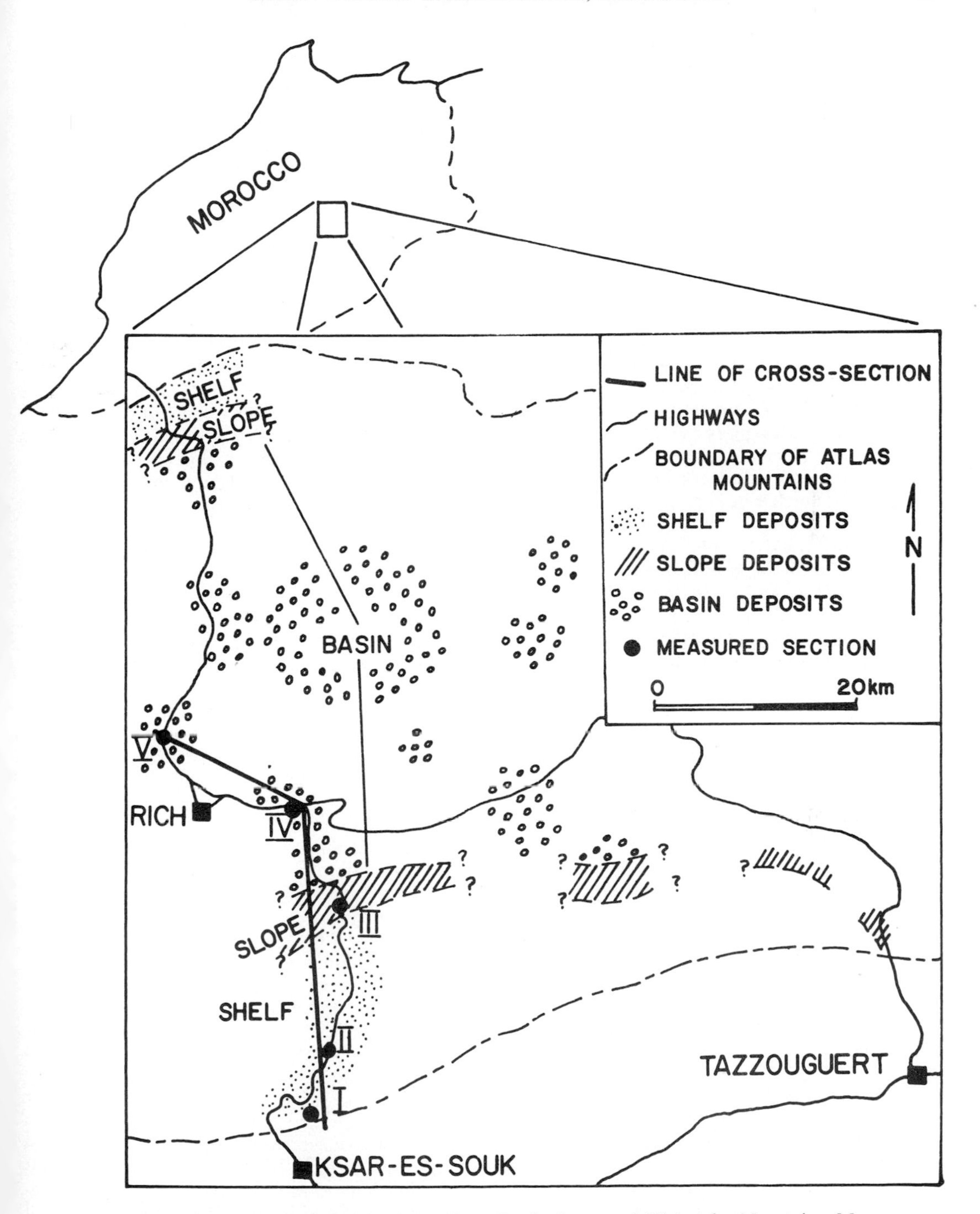

FIG. 2.—Liassic facies distribution map of the Ziz Valley in the central High Atlas Mountains, Morocco.

to the shelf break must be known. In modern oceans this value ranges between 18 and 181 metres (Shepard, 1963). Using the lowest value, a minimum depth estimate for the basin would be 333 metres. We stress that this would be a minimum depth value because if the dip of the slope was higher, not unlikely if in fact the trough was fault bounded, then much greater depths

TABLE 1.—THE RELATIONSHIP BETWEEN PREVIOUS FACIES CLASSIFICATIONS AND THAT OF THIS PAPER

Dubar, 1949	Dubar, 1960–1962	Faure-Muret, 1971	This Paper
Facies du Centre de Haut Atlas	Faune Pelagiques a Ammonites	Fosse de Subsidence	Basin
Facies Mixtes	Facies Mixtes	Regimes Mixtes de Bordures des Fosses	Slope
Calcaires Construits	Calcaires Construits	Frange de Facies Reduits	Shelf

could be attained. If, for example, the dip angle for the slope were doubled and a median shelf break depth chosen (81 metres), a basin depth of approximately 713 metres could be attained. We believe that the figures 333–713 metres represent the low end in the range of estimates, and that when the effects of crustal shortening can be quantified, much larger depth values will result.

FACIES DESCRIPTIONS AND INTERPRETATIONS

The different lithologic sequences of the Atlas Mountains are described in this section, and a hypothetical depositional model is used to explain the lateral and vertical distribution of these sequences. The ages assigned to these sequences are based on the work of Choubert and others (1956).

Slope Facies

This facies bounds the northern and southern platforms (shelves) of the High Atlas trough, but cannot be directly associated with these platforms because thrust faulting and erosion have displaced and removed part of the record. The best developed slope sequence is exposed in the southern part of the trough where two major subfacies are recognized in a measured section (Figs. 2 and 5). Approximately 200 metres of the lower part of this section was structurally disturbed, with parts of the section being repeated. The first subfacies (Sinemurian-Pliensbachian age) consists of a sequence with few Bouma cycles (Bouma, 1962). The second subfacies (Pliensbachian-Toarcian age) overlies the first and is characterized by abundant Bouma cycles. The term "turbidite" used in the naming of the two subfacies is intended to cover a broader spectrum of resedimented deposits in the sense of Walker and Mutti (1973, p. 121–123).

Turbidites with few Bouma cycles.—This subfacies, Sinemurian-Pliensbachian in age, is best developed in the lower part of the section and its base is tectonically, not stratigraphically defined. It is comprised mainly of irregularly bedded black micrites alternating with very thin (1 cm) marly horizons. These beds characteristically thicken from 5 cm to irregularly bedded lenses as much as 3 metres thick over distances up to 100 metres (Fig. 7). Beds with this lenslike geometry appear to be of two main types:

1. Normally graded beds as much as 2 metres thick composed of sand- and pebble-size components including ooids, algally coated particles, intraclasts, and bioclastic debris embedded in a micrite matrix.
2. Beds containing a chaotic arrangement of intraclasts, ooids, and algal-coated grains, with no clear grain-size distribution. The grains, which range from sand- to pebble-size, float in a matrix of dense, black micrite. The beds range in thickness from 20 cm to 3 metres. Boulders are found embedded in the rocks of this subfacies. The boulders contain shallow-water components including coral fragments and algal-coated particles. The enclosing beds are draped over the boulders, indicating depositional relief in the area of the boulder. The boulders range in size from 40 cm to 150 cm.

Turbidites with abundant Bouma cycles.—There is no clear-cut or easily recognizable boundary separating this subfacies of Pliensbachian-Toarcian age from the underlying subfacies. There is, however, a general change in the character of the bedding, and an increase in the number and thickness of marly horizons. More significantly, well-developed Bouma cycles appear. The bases of the Bouma cycles are often scoured to depths of between 20 and 50 cm (Fig. 8). The "a" (lowermost) divisions are conglomeratic, passing up into sand-size carbonates with a micritic matrix. The coarse conglomeratic particles include mud clasts, pisolites, and fragments of corals, bryozoa and brachiopods. The succeeding "b" divisions exhibit characteristic laminations and sand-size carbonate intraclasts in a micritic matrix. The "c" divisions are comprised of finer grained carbonates that exhibit well-developed cross-lamination, flame structures and convolute laminations (Fig. 9). The "d" and "e" divisions are made up of marly micrites and marls, the "d" divisions being laminated. Incomplete "c" through "e" cycles are much more common than

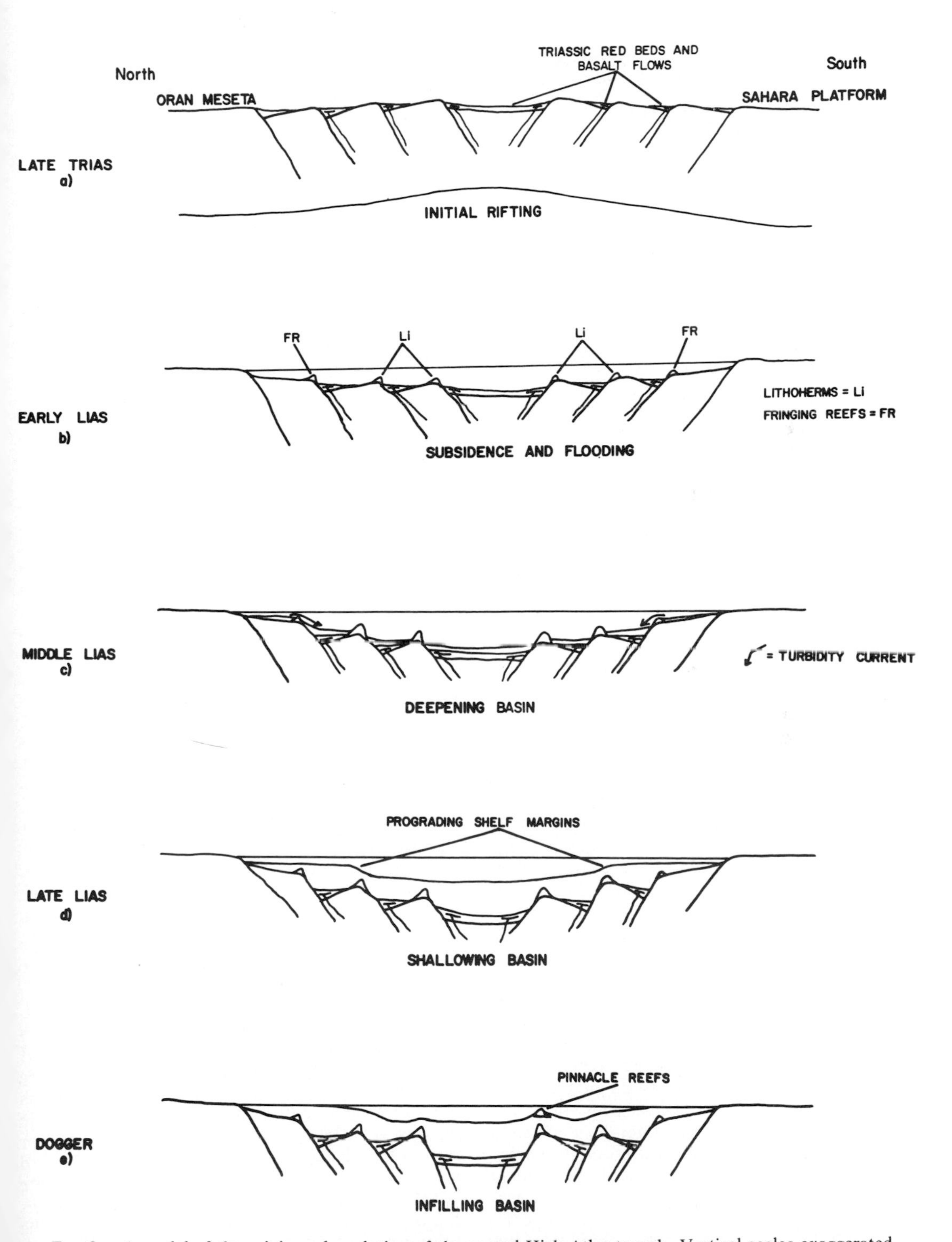

FIG. 3.—A model of the origin and evolution of the central High Atlas trough. Vertical scales exaggerated.

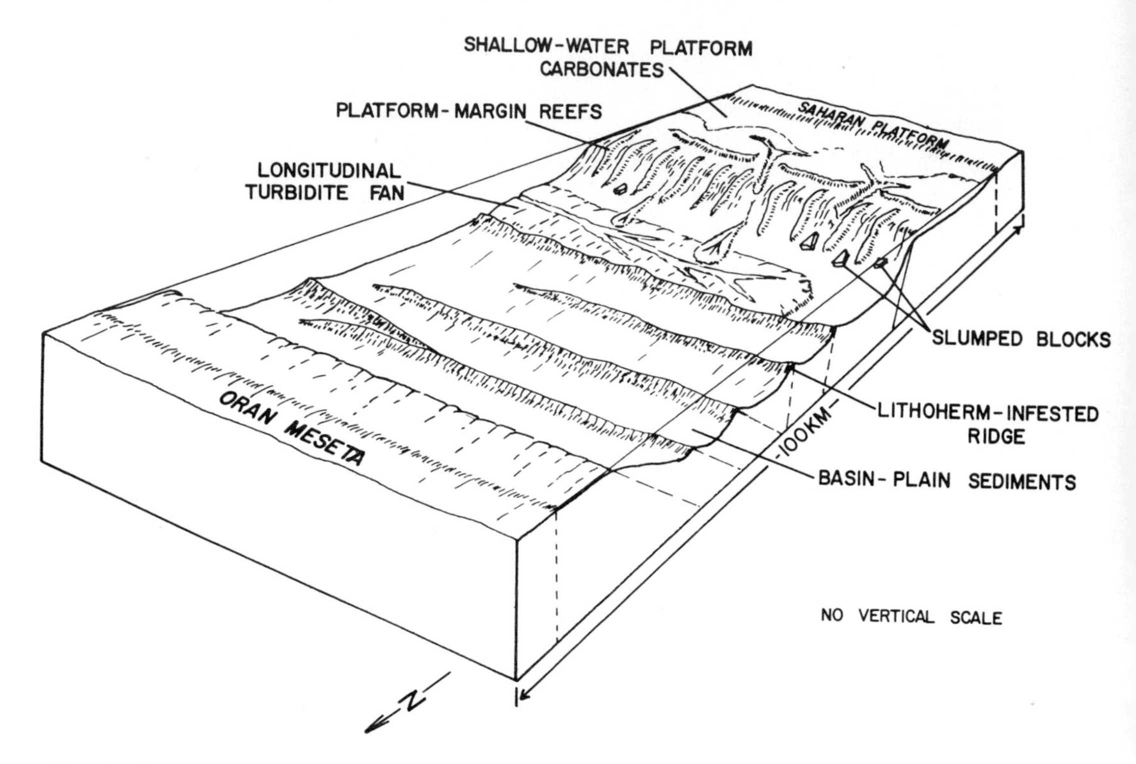

FIG. 4.—Schematic paleoenvironmental reconstruction of the central High Atlas trough during Early Liassic time.

those beginning with "a" divisions. A few flute casts were the only sole markings observed in this subfacies.

In the upper part of this section, a number of unusual features can be seen. Large coral-head boulders (Fig. 10) are found embedded in Bouma "d" and "e" divisions, and at least two horizons exhibit soft-sediment deformation (Fig. 11). These deformed horizons can be traced laterally for distances in excess of 1 kilometre. The upper, irregular surfaces of these deposits, presumably slump sheets, are infilled by subsequent deposits.

Both subfacies are only sparsely fossiliferous (with the exception of transported shelf fauna) with the younger subfacies containing the trace fossils *Paleodictyon* and *Zoophycus*.

Slope facies: An interpretation of depositional setting.—A middle-fan setting is postulated for the rocks with few Bouma cycles in the lower part of the section. The thick beds represent channel-fill deposits, and the laterally equivalent thin-bedded, graded units are intrepreted as being levee and overbank deposits. The orientation of these channels with respect to the shelf margin could not be determined. If, however, the middle-fan interpretation is correct, the channels were probably oriented at a high angle to the shelf margin. The graded, thick beds are equivalent to the A4 units of Walker and Mutti (1973). Thick beds with chaotic arrangement of particles are equivalent to the A3 facies of Walker and Mutti (1973). Both the A3 and A4 facies were assigned to a mid-fan environment by Walker and Mutti (1973).

The upper part of the section is assigned to an outer-fan environment because of the presence of well developed, repeated Bouma cycles (facies C of Walker and Mutti, 1973). The repetition of the Bouma "c", "d" and "e" divisions common in this part of the section is probably the result of the overlapping of feather edges of numerous distal portions of submarine fan lobes. This subfacies is interesting because it contains what Walker and Mutti (1973) consider to be proximal exotic features such as slumped beds and exotic blocks. An explanation for this combination of both proximal and distal features is offered in the following model: the distal parts of submarine fans were probably formed by currents which flowed parallel to the shelf margin (Fig. 12). This mode of transportation and deposition resulted in the distal turbidites being deposited at or near the base of the slope. The source area for the turbidites was located some distance to the east or west of the location examined in this study. Material derived from the slope immediately adja-

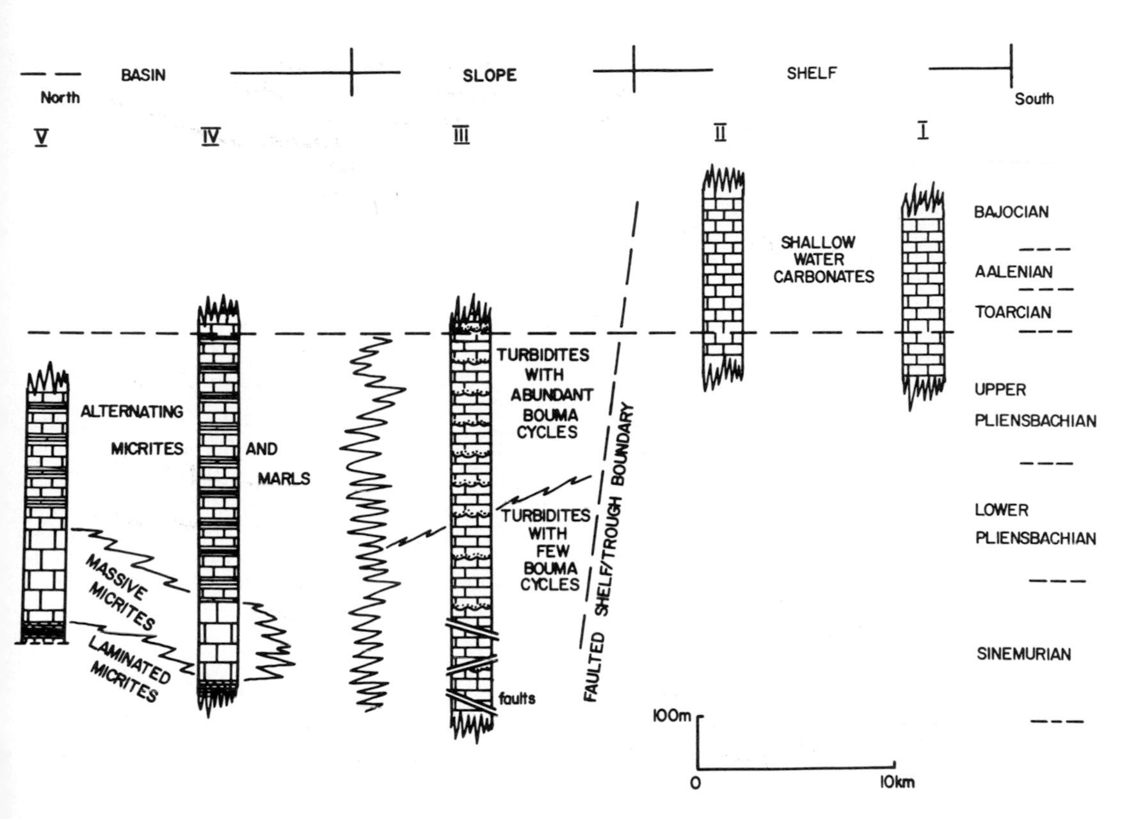

FIG. 5.—Stratigraphic cross section of the southern half of the central High Atlas Mountains. Line of section indicated in Figure 2.

cent to these distal turbidites would then be deposited in association with the distal facies. These proximal exotic features would include the coral-head boulders and the soft-sediment-deformation slump sheets. This hypothesis of flow parallel to the basin margin is further supported by the fact that well-developed "a" and "b" Bouma divisions described from this section do not extend out into the axial part of the basin. In a section measured 8 kilometres basinward from the section which exhibits well-developed "a" and "b" divisions, no such divisions are observed in age equivalent rocks. Instead, rhythmic alternations of micrite and marl, the basin-plain subfacies of the next section of this paper, are found. This suggests that the turbidity currents flowed parallel to the basin margin at or near the base of the slope, confined by a physical barrier (or barriers) that prevented the passage of the bed loads of the turbidites into the central, axial part of the basin. These barriers are hypothesized to be the upturned edges of fault

49km
x
θ
y
40km

49km = Width of Basin + Slopes
40km = Width of Basin only
Measurements based on Choubert (1956)

2y = 9km x = ytanθ

FIG. 6.—Simplified geometry of the central High Atlas trough and the basic dimensions used in the calculation of water depth (Table 2). Width of basin + slopes = 49 km. Width of basin only = 40 km. 2y = 9 km; x = y tan Θ. Measurements based on Choubert (1956).

TABLE 2.—ESTIMATES OF BASIN DEPTH

Θ	x
4°	315 m
5°	394 m
6°	473 m
7°	553 m
8°	632 m
9°	713 m

FIG. 7.—Thickening and thinning of beds in the turbidite subfacies with few Bouma cycles.

FIG. 10.—Coral-head boulder embedded in Bouma "d" and "e" divisions (turbidite subfacies with abundant Bouma cycles). Rubble obscures bedding in photograph. Corals appear as the nodular, rough surface of the boulder, and are embedded in a micrite matrix.

FIG. 8.—Scoured base of Bouma cycle (turbidite subfacies with abundant Bouma cycles) with conglomeratic "a" division and laminated "b" division.

FIG. 11.—Soft-sediment deformation (turbidite subfacies with abundant Bouma cycles). These folds form part of a large scale slump sheet.

FIG. 9.—Convolute lamination in Bouma "c" division (turbidite subfacies with abundant Bouma cycles).

blocks formed during the initial rifting of the crust in this region (Fig. 13). Such faults would have been oriented roughly parallel to the basin axis and are probably represented today by the anticlinal structures within the basin which may represent reactivation of the earlier faults. Two points concerning this stratigraphic section should be emphasized:

1. The stratigraphic section described in this part of the paper probably does not record continuous deposition from a single, geographically fixed, source area. Therefore, attempts to interpret the section in terms of a simple progradational or recessional model are not valid.
2. The fan interpretation as the major environ-

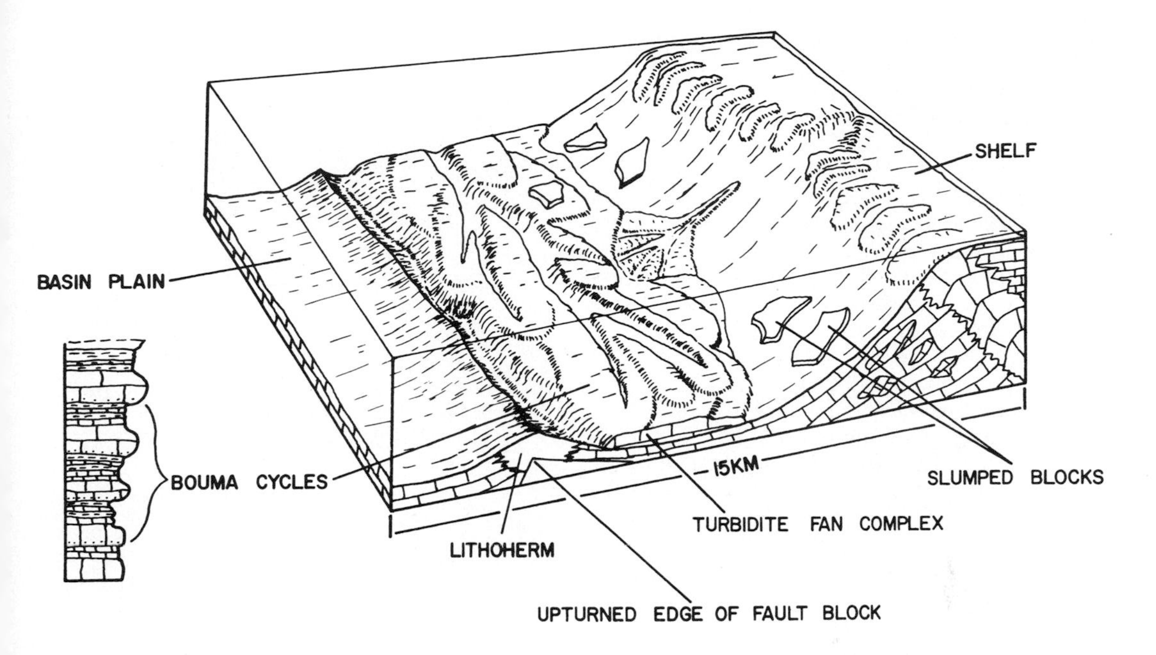

FIG. 12.—Schematic paleoenvironmental reconstruction of the Liassic slope. A longitudinal turbidite-fan complex is shown forming parallel to the shelf margin because of the restricting lithoherm ridge. Flow directions are hypothetical.

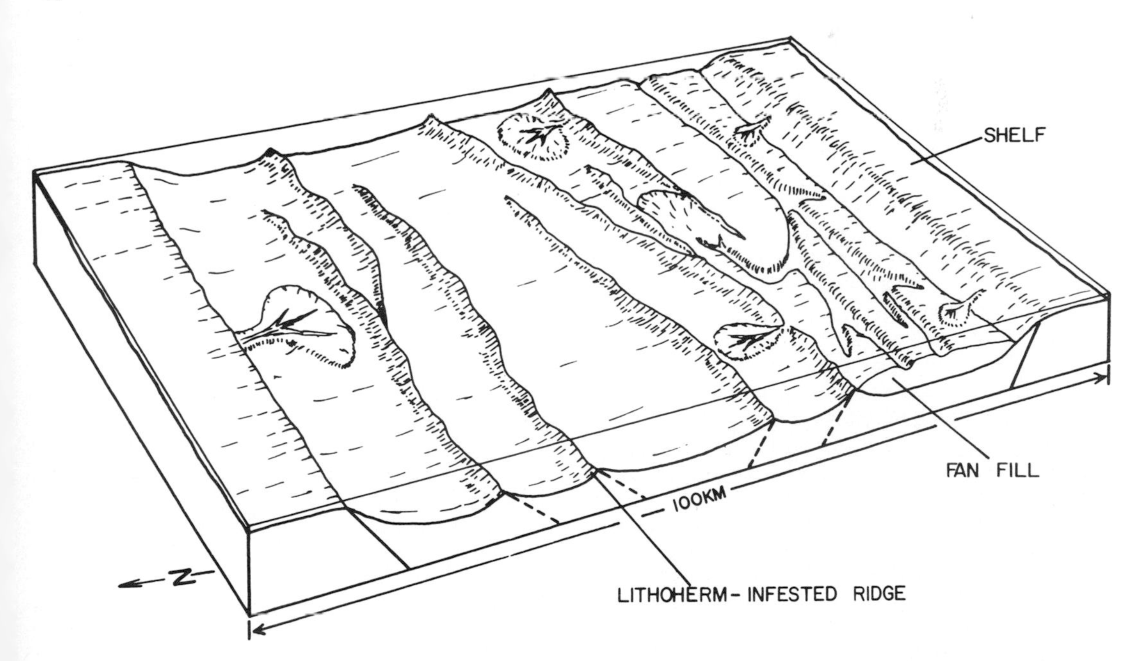

FIG. 13.—Schematic paleoenvironmental reconstruction of the Liassic central High Atlas trough. Turbidite fans are restricted by the presence of lithoherm-infested ridges.

ment of deposition is indicative of deposition on the lowermost part of the slope, or a toe-of-slope setting. The areally larger part of the Liassic slope has been removed by faulting and erosion.

Basin Facies

Basin facies that are the approximate age equivalents of the slope strata occupy a broad area (40 kilometres wide) between the northern and southern slopes of the Atlas trough. Three subfacies are recognized within the axial region of the trough. They are:

1. Laminated micrite subfacies (oldest).
2. Massive micrite subfacies.
3. Alternating micrites and marls (youngest).

The following descriptions are derived from a single measured section, with supplementary information being derived from coeval strata from different geographic locations. Exposures of the underlying Triassic red beds and basalts are observed in the axial part of the Jurassic trough.

Laminated micrite subfacies.—Rocks of this subfacies represent the oldest Liassic units (Sinemurian) observed in the trough; they are in contact with the underlying Triassic red beds and basalts. Characteristically these Liassic carbonates are dark, thinly bedded to irregularly (wavy) laminated micrites interbedded with conglomeratic horizons that range in thickness from 15 cm to 50 cm. The conglomerates are made up of light grey conglomeratic particles ranging in size from sand to pebbles. Many of the larger conglomeratic particles exhibit fenestral structures (Fig. 14A). These conglomeratic particles do not appear to have undergone any extensive transportation as it is possible in many instances to match particles in a "jigsaw" fashion. The matrix is an iron-stained micrite. Within 10–15 metres of the base of the section, the irregularly (wavy) bedded micrites and conglomerates pass gradationally upward into parallel-laminated black micrites. These laminated micrites attain a thickness of 20–25 metres and show little or no disturbance of the laminae (Fig. 14B). This sequence of irregularly bedded micrites, conglomerates and parallel-laminated micrites is widespread throughout the basin, being observed in three measured sections separated over distances of up to 30 kilometres.

Massive micrite subfacies.—Over a distance of 5–10 metres, the parallel-laminated micrites grade upward into a massive micrite facies of Pliensbachian age. Close examination demonstrates that this subfacies is composed of extremely massively bedded units (10–20 metres thick) that commonly interfinger with much thinner bedded (25–75 cm thick) micrites. The massive micrite units are often hundreds of metres long and several hundred metres thick. Locally, sponges, brachiopods and calcareous algae are abundant in the massive micrite. All evidence points to an *in situ* origin for these bodies.

FIG. 14.—Laminated micrite subfacies. *Upper*—Polished slab of conglomerate containing pebbles of laminated micrites with fenestral structures. *Lower*—Laminated black micrites.

Alternating micrites and marls.—This subfacies (Pliensbachian-Toarcian age) is the most widespread (vertically and laterally) of all the basinal facies. These sedimentary rocks consist primarily of monotonous alternations of micrite and marl (Fig. 15). The relative thickness of beds of micrite and marl changes progressively upward. In older sequences, massive micrite beds (75–150 cm) are separated by thin, marly horizons (5–10 cm). In progressively younger sequences, the marls become thicker while the micrites thin until both lithologies have beds of the same thickness (15–25 cm). In the Upper Lias, thin micrites and progressively thicker marls predominate. The rocks of this subfacies lap onto and over the massively bedded micrite units described in the previous

FIG. 15.—Alternating micrites and marls subfacies: Regular, alternating beds of micrites and marls. Length of hammer, 40 cm.

section. Ammonites are locally abundant, and scanning electron microscopy revealed the presence of a few coccoliths in the micrites. The presence of a low-diversity suite of unidentifiable trace fossils indicates oxygenating bottom conditions.

Basin facies: An interpretation of the depositional setting.—The oldest Liassic rocks described in this study could perhaps be best characterized as being axial rather than basinal in terms of their genesis. The presence of irregularly bedded micrites and conglomerates containing particles exhibiting fenestral structures, may suggest the existence of a supratidal environment of deposition in the axial part of the developing Liassic trough prior to extensive flooding. We believe that the conglomeratic horizons formed by solution collapse as there appears to be little evidence to support extensive transportation of the conglomeratic particles. A number of obvious problems arise if we assume that the fenestral micrites were formed essentially *in situ.* Approximately age-equivalent slope rocks described in this paper are turbidites. However, the base of the slope section is never observed, and in the axial part of the basin the proposed supratidal deposits are found only in the lowermost 10 metres of the section. Above the supratidal deposits, the secceeding units represent a fairly rapidly deepening basin. It is possible that a major topographic depression existed and the deepening is merely a reflection of a rapid flooding process. The irregularly bedded micrites and conglomerates are overlain by the well-laminated, black micrites that were probably deposited below wave base. The absence of bioturbation together with the intense black color of these micrites suggests thay they formed under euxinic conditions. The fairly rapid transition from supratidal to deeper-water euxinic conditions with the absence of any significant sedimentary sequence recording this event is not as implausible as it may first appear. Results of the Deep Sea Drilling Project in the Mediterranean indicate that the transition between the evaporitic deposits of the M-horizon and the overlying Pliocene deposits is remarkably sharp and lacks any obvious sedimentary features that may have been expected to mark the flooding of an evaporitic basin (Ryan and others, 1973, cf. Site 132, p. 403; Hšu and others, 1973). Our intention is not to argue that our case is as overwhelming as that of the Mediterranean example, but merely to point out that the record of a catastrophic flooding event need not be accompanied by an extensive sedimentary record.

The deeper-water, euxinic conditions that resulted in the deposition of the thinly laminated black micrites were changed by continued flooding and an increase in the circulation of water in the High Atlas trough. This resulted in oxygenated bottom waters as evidenced by the presence of abundant trace fossils in units younger than the thinly laminated black micrites.

The massively bedded micrites are thought to be lithoherms, similar to the mud baffling systems that occur in 600 metres of water in the Straits of Florida (Neumann and Kofoed, 1972). Although the Jurassic examples did not necessarily form in waters of comparable depth, they are genetically similar. The mud-baffling sponges and algae probably grew along the linear, upturned edges of fault blocks that formed during the initial rifting of this region (Fig. 3). The lithoherms continued to grow along these linear, topographic highs until subsidence outpaced their upward growth and/or until basin-plain sedimentation overwhelmed them. No attempt was made to estimate the number of individual lithoherms within the study area, but they existed in linear groupings throughout the basinal region, being exposed today along the crests of the anticlinal ridges.

The alternating micrites and marls represent basin-plain deposits. The black micrites are interpreted as pelagic sediment containing coccoliths and micrite derived from the bioerosion of reefs that fringed the shelf margin of the basin. The broader shelf area behind the reefs is not believed to have been a contributor of lime mud to the basinal micrites because these micrites are remarkably clean, containing on average only 18.5% insoluble residues. The shelf carbonate sediments contain significantly higher amounts of insoluble residues (55% average), mainly illite and chlorite. Therefore, bioerosion of fringing reefs is proposed as a mechanism that is capable of producing "clean" micrites as there is no reasonable means of segregating lime mud from the clays that are so abundant on the shelf itself (Evans and others, 1977). The marly horizons, regularly interbedded with these micrites, represent the distal deposits of turbidity currents that were generated in the shelf and slope regions of the basin. The coarser materials associated with the slope region were confined to the area parallel to the base of the slope. Lithoherm-capped margins of fault blocks probably acted as barriers to the dispersal of these coarser sediments. The suspension loads of the turbidity currents spilled over the lithoherms and spread evenly over the intervening areas.

POST-LIASSIC DEPOSITION

The facies of the High Atlas trough that have been described in this paper represent those deposits that were laid down in what we consider to be deeper-water conditions. The detailed discussion of the subsequent infilling of the High Atlas trough is beyond the scope of this paper, but the subject is treated in detail by Stanley (1975). Suffice it to say that the alternating micrites and marls (interpreted in this paper as basin-plain deposits) grade upward through a fairly monotonous sequence of marly micrites and marls. Within this sequence is found evidence of shallow-water faunas that increase in abundance toward the upper parts of the section, culminating in an extensive development of coral reefs. The age of these shoaling deposits is Dogger (Bajocian and Bathonian). In the central, axial part of the trough there is what appears to be a single, complete stratigraphic section from the Lias through the Dogger, in which the history of this basin is almost completely documented.

SUMMARY AND CONCLUSIONS

Deep-water Liassic carbonate rocks crop out in the central High Atlas Mountains of Morocco. Two major depositional settings are recognized for these deep-water carbonates; the slope and basin axis. Facies characterizing the slope are turbidite sequences interpreted as forming in mid- and outer-fan locations. Facies of the basin axis include lithoherms, rhythmic alternations of micrites and marls and laminated micrites.

The following geologic history for the High Atlas trough is proposed:

1. Triassic continental sedimentation disturbed by rifting, associated block faulting, and extrusion of basalts.
2. Liassic marine transgression (possibly catastrophic) over the block-faulted Triassic red beds.
3. Deposition of deep-water Liassic carbonate sediments in the rifted basin axis coeval with shallow-water carbonate deposition on the marginal platforms.
4. Infilling of the basin during Dogger time with the deposition of shallow-water deposits and coral-reef formation.

ACKNOWLEDGEMENTS

Financial support for this research was provided by the National Science Foundation, Grant Number GF 32510, by a special foreign currency grant from the Office of International Programs, and from the Earth Science Section of the Division of Environmental Sciences of the National Science Foundation. The authors are grateful for the critical evaluation of the manuscript by Chuck Campbell, Joyce Novitsky-Evans, Martin Schuepbach, John Warme and Jack Wendte and conversations with Emiliano Mutti. Cooperation of the Moroccan B.R.P.M. and the Service Géologique in Rabat is gratefully acknowledged.

REFERENCES

Arkell, W. J., 1956, Jurassic geology of the world: Oliver and Boyd, Edinburgh Scotland, 806 p.

Bouma, A. H., 1962, Sedimentology of some flysch deposits: Elsevier, Amsterdam, The Netherlands, 168 p.

Choubert, G., Dubar, G., and Hindermeyer, J., 1956, Carte géologique du Haut Atlas du nord de Ksar-es-Souk et de Boudenib. Feuilles Rich et Boudenib. Service Géologique du Maroc.

——, and Faure-Muret, A., 1960–1962, Évolution du domaine atlasique marocain depuis les temps Paléozoiques: Soc. Geol. France Mem. Hors Serie, Livre à la mémoire du professeur P. Fallot, v. 1, p. 447–527.

De Sitter, L. U., 1964, Structural geology: McGraw-Hill, New York, New York, 551 p.

Dewey, J. F., Pitman, W. C., III, Ryan, W. B. F., and Bonnin, J., 1973, Plate tectonics and the evolution of the Alpine system: Geol. Soc. America Bull., v. 84, p. 3137–3180.

Dubar, G., 1949, Carte géologique provisoire du Haut-Atlas de Midelt, et notice explicative: Serv. Géol. Maroc Notes et Mem., no. 59, 60 p.

——, 1952, Livret-guide de l'excursion A 34. Haut Atlas Central. 19th Internat. Geol. Congress, Algiers, 1952, 74 p.

——, 1960-1962, Notes sur la paléogeographie du lias marocain (domaine atlasique): Soc. Geol. France Mem. Hors Serie, Livre à la mémoire du professeur P. Fallot, v. 1, p. 529-544.

Evans, I., Kendall, C. G. St. C., and Butler, J. C., 1977, Genesis of Liassic shallow water rhythms Central High Atlas Mountains, Morocco: Jour. Sed. Petrology, v. 47, p. 120-128.

Faure-Muret, A., Choubert, G., and Kornprobst, J., 1971, Le Maroc; Domaine rifain et atlasique: *In* Tectonique de l'Afrique: United Nations Educational, Scientific and Cultural Organization, Earth Science Ser., no. 6, p. 17-46.

Folk, R. L., 1962, Spectral subdivision of limestone types: *In* W. E. Ham (ed.), Classification of carbonate rocks—A symposium: Am. Assoc. Petroleum Geologists, Mem. 1, p. 62-84.

Gross, M. G., 1972, Oceanography, a view of the earth: Prentice Hall, Englewood Cliffs, New Jersey, 581 p.

Hšu, K. J., Cita, M. B., and Ryan, W. B. F., 1973, The origin of the Mediterranean evaporites, DSDP Leg 13: *In* W. B. F. Ryan, K. J. Hsu and others, Initial reports of the Deep Sea Drilling Project, v. 13: U.S. Government Printing Office, Washington, D.C., p. 1203-1231.

Neumann, A. C., and Kofoed, J. W., 1972, "Lithoherms" in the Straits of Florida [abs.]: Geol. Soc. America Abs. with Programs, v. 4, p. 611.

Ryan, W. B. F., and others, 1973, Tyrrhenian Rise—Site 132, DSDP Leg 13: *In* W. B. F. Ryan, K. J. Hsu and others, Initial reports of the Deep Sea Drilling Project, v. 13: U.S. Government Printing Office, Washington, D.C., p. 403-464.

Shepard, F. P., 1963, Submarine geology, 2d ed.: Harper and Row, New York, New York, 557 p.

Stanley, R. G., 1975, A shoaling-upward carbonate sequence in the Dogger (Middle Jurassic) of the central High Atlas of Morocco: M.A. Thesis, Rice Univ., Houston, Texas, 177 p.

Walker, R. G., and Mutti, E., 1973, Turbidite facies and facies associations: *In* G. V. Middleton and A. H. Bouma (eds.), Turbidites and deep-water sedimentation: Soc. Econ. Paleontologists and Mineralogists, Pacific Sec., Los Angeles, California, p. 119-157.

SEPM Special Publication No. 25, p. 263-272, November 1977

ALBIAN SEDIMENTATION OF SUBMARINE AUTOCHTHONOUS AND ALLOCHTHONOUS CARBONATES, EAST EDGE OF THE VALLES-SAN LUIS POTOSI PLATFORM, MEXICO

BALDOMERO CARRASCO-V.
Instituto Mexicano del Petroleo, Mexico 14, D. F.

ABSTRACT

The folded Sierra Madre Oriental permits regional study of Cretaceous contemporaneous carbonate rocks which were part of the large Valles-San Luis Potosi Platform and the Tamaulipas Basin.

Regional patterns of sedimentation in the eastern edge of the platform are characterized by shallow-water carbonate sediment and rudist reefs (El Abra Formation), while on the slope of the platform toward the basin, the forereef and basinal carbonate sediment contains allochthonous breccias ("Tamabra") and fine detrital carbonate (Upper Tamaulipas). These are the product of downdip displacement of sediment from the platform and the mixing or intercalation with autochthonous pelagic calcareous mudstone.

Study of three areas located close to the edge of the platform (Rio Guayalejo, Xilitla and Metztitlan) indicates that at least three different models of contemporaneous (Albian) submarine sedimentation can be inferred.

INTRODUCTION

Part of Mexico during the Cretaceous was a major carbonate province with shallow-water platforms and deep-water basins. The Valles-San Luis Potosi platform and the famous Golden Lane were the largest platforms (Fig. 1). The Valles-San Luis Potosi Platform is spectacularly exposed in many parts of the Sierra Madre Oriental along with part of Tamaulipas Basin which lay to the east. Study of three areas where the basinal, basin-margin, and platform facies are well exposed (Rio Guayalejo, Xilitla, and Metztitlan) allows recognition of different models of contemporaneous (Albian) submarine sedimentation. In the Xilitla area rudist reef facies made its appearance in the Barremian and persisted through the Cenomanian. In the Rio Guayalejo and Metztitlan areas formation of reefs began in the Albian, ending by the Cenomanian (Fig. 2). The existence of reefs provided relief and material for the basins. Sedimentologic and paleontologic data suggest the submarine accumulation of allochthonous (platform-derived) and autochthonous basinal carbonate close to the platform margins.

The canyon of the Rio Guayalejo permits the study of the platform facies, characterized by the massive reefal limestone of the El Abra Formation (Fig. 3), the basin-margin facies, and the basinal autochthonous carbonate rocks with some influx of allochthonous carbonate (Upper Tamaulipas Formation). The Xilitla area has good outcrops along the highway from Xilitla to Laguna Colorada, but away from the highway outcrops are grass covered and it is not possible to observe the rocks in detail. In this area, Albian basinal carbonate facies were incorporated with platform-derived fine and coarse allochthonous material as well as large allochthonous exotic blocks. The stratigraphic section in the Metztitlan area contains the best exposures of basinal and basin-margin carbonate (Fig. 4) consisting mainly of carbonate mudstone, some fine-grained, graded detrital carbonate, and two large, impressive bodies of coarse, allochthonous breccia.

PREVIOUS WORK ON VALLES-SAN LUIS POTOSI AND GOLDEN LANE PLATFORMS

Study of basin-margin and basinal facies in the Sierra Madre Oriental dates from Böse (1906), Baker (1920-1925, published in 1971), and Heim (1926, published in 1940), although these studies dealt primarily with platform facies. At that time, Heim could not explain the origin of the "mixed facies" based on the depth difference between the Tamaulipas Limestone (deep-water facies) and El Abra Limestone (shallow-water facies). Some of Heim's considerations on the "Tamabra limestone" were not very clear which gave rise to some later confusion. Heim (1940, p. 321) stated: "for the entire limestone complex beneath the San Felipe or the Xilitla (Agua Nueva) Formations, the writer propose the name Tamabra Formation, including the different facies," and in another place (p. 325) Heim says: "Thus the Tamaulipas, Taninul and El Abra Limestones are different facies of the same Tamabra Formation, interdigitating laterally and often repeated above one another at the same locality (mixed facies)". Many subsequent authors have cited this passage as the definition of the "Tamabra formation" equating it with "mixed facies."

For many years in Poza Rica and neighboring oil fields located around the Golden Lane platform, the name "Tamabra formation" has been

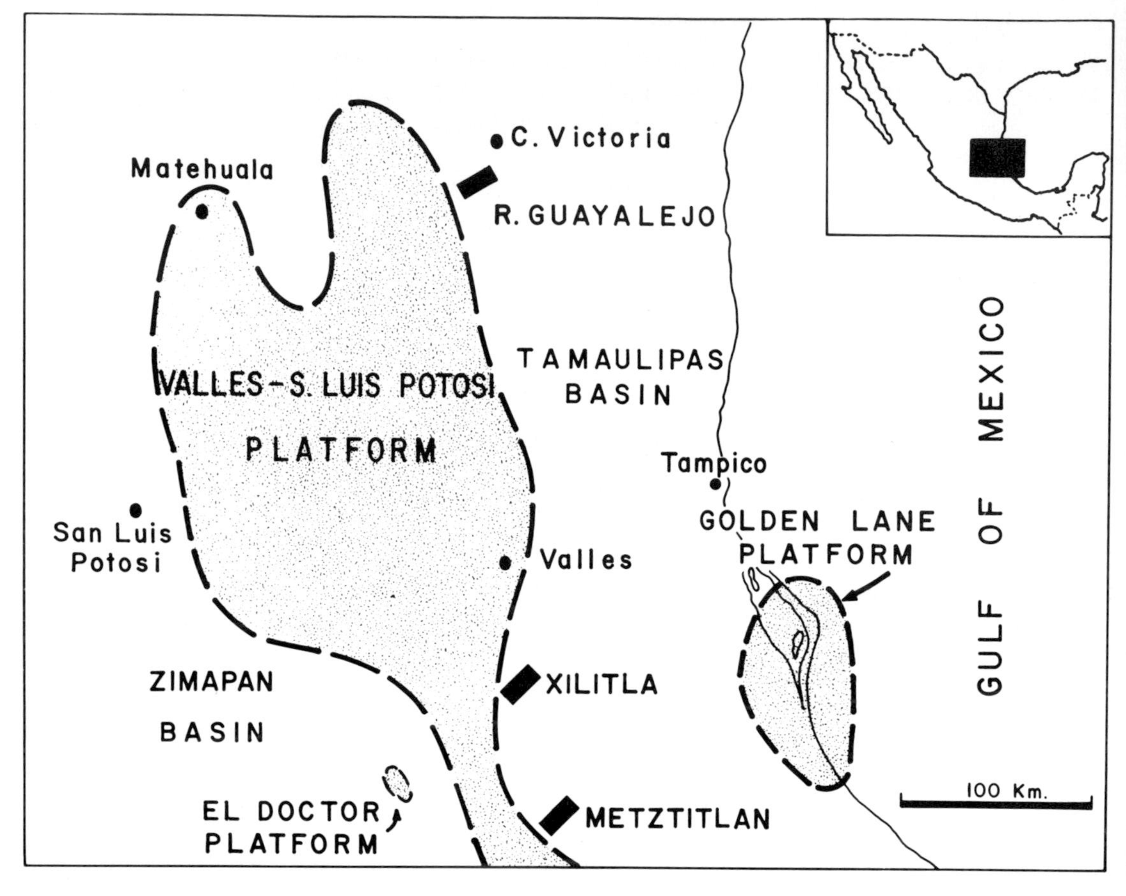

FIG. 1.—Location of Cretaceous carbonate platforms (stippled) and basins. Rectangles are the locations where basinal and basin margin carbonate rocks were studied. Modified from Carrillo (1971).

included in the subsurface geological nomenclature. Many authors used it, for instance: Luna and Muñoz (1949), Suarez (1950), Nigra (1951), López Ramos (1953), Sotomayor (1954), Flores (1955), Mena Rojas (1955), Barnetche and Illing (1956), Acuña (1957), Boyd (1963), Guzmán (1967), Bebout and others (1969), Becerra (1970), Viniegra and Castillo-Tejero (1970), Coogan and others (1972), and Enos (1974, 1975, this volume). Bonet (1963) discussed the misuse of the name "Tamabra formation" in the Poza Rica oil field; he formally proposed the name "Poza Rica formation" as a substitute for "Tamabra." Unfortunately, this proposition was made in a guidebook and is not formally acceptable.

The origin of the sediment composing the "Tamabra formation" which surround the Golden Lane platform has been discussed by many authors, among them Barnetche and Illing (1956), Bebout and others (1969) and Coogan and others (1972) who considered the "Tamabra formation" an *in situ*, shoal-water, mixed reef complex; while Bonet (1963), Viniegra and Castillo-Tejero (1970), Becerra (1970), Carrillo (1971) and Enos (1975, this volume) believe the "Tamabra formation" is a basinal limestone containing sediment-gravity-flow deposits whose constituents were largely derived from a shallow-water calcareous platform.

This conflict and the economic importance of the subsurface "Tamabra" are the reasons for comparing sedimentary models generated from surface data with subsurface occurrences in Poza Rica and surrounding areas. This paper only generates the models, it does not go into comparison with the subsurface Poza Rica area. Criteria for recognizing deep-water carbonates have been reviewed by Wilson (1969) and revised in considering the character of ancient platform margins (Wilson, 1974).

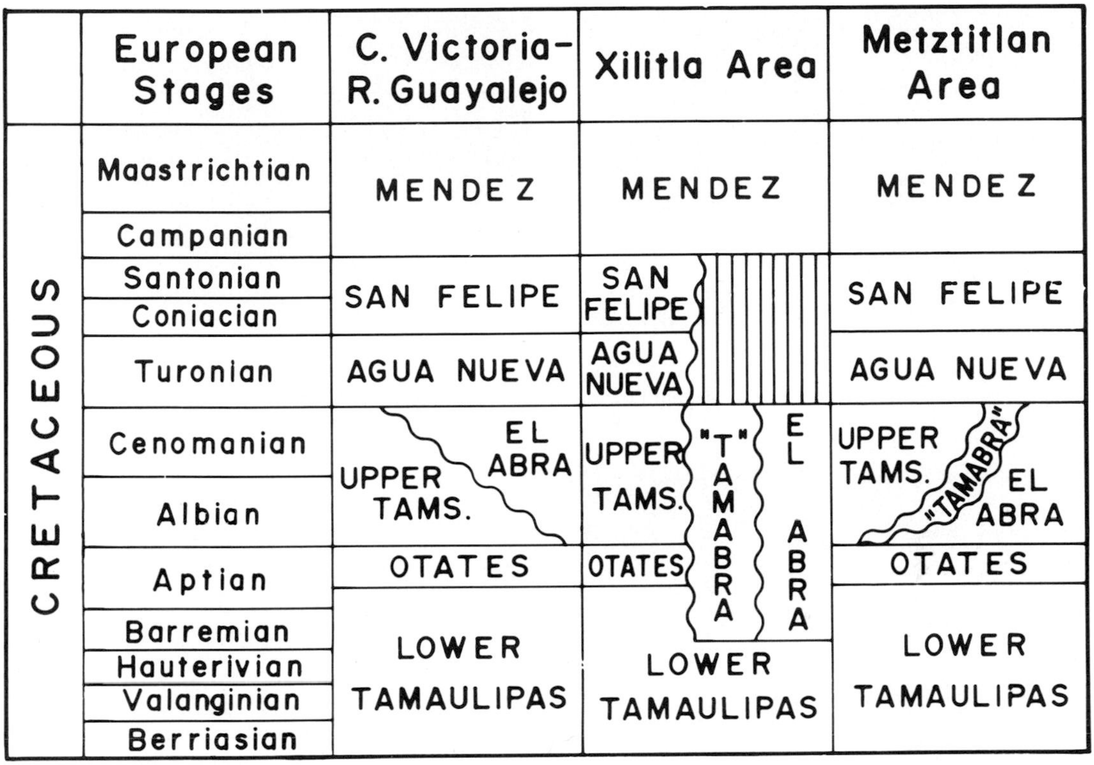

	European Stages	C. Victoria–R. Guayalejo	Xilitla Area	Metztitlan Area
CRETACEOUS	Maastrichtian	MENDEZ	MENDEZ	MENDEZ
	Campanian			
	Santonian	SAN FELIPE	SAN FELIPE	SAN FELIPE
	Coniacian			
	Turonian	AGUA NUEVA	AGUA NUEVA	AGUA NUEVA
	Cenomanian	UPPER TAMS. / EL ABRA	UPPER TAMS. / "TAMABRA" / EL ABRA	UPPER TAMS. / "TAMABRA" / EL ABRA
	Albian			
	Aptian	OTATES	OTATES	OTATES
	Barremian	LOWER TAMAULIPAS	LOWER TAMAULIPAS	LOWER TAMAULIPAS
	Hauterivian			
	Valanginian			
	Berriasian			

FIG. 2.—Correlation of Cretaceous formations of the Valles-San Luis Potosi platform and the Tamaulipas Basin (modified from Carrillo, 1971).

FIG. 3.—View from west to east of the Rio Guayalejo Canyon. *A*—Reefal facies of the El Abra Formation which forms abrupt walls of massive carbonate rocks. *B*—Bedded basinal carbonates of the Upper Tamaulipas Formation. The structure here is an anticline with its axial plane parallel to the photo, and plunging toward the left. Man give scale of foreground (arrow).

FIG. 4.—Partial view of the section in Metztitlan where the basinal facies of the Upper Tamaulipas Formation are exposed.

The present study describes the sedimentology of the Rio Guayalejo, Xilitla, and Metztitlan areas. Although they are essentially coeval, belonging to a common margin of a single platform, they present three very different sedimentary models.

RIO GUAYALEJO AREA

The Rio Guayalejo area differs from the Xilitla and Metztitlan areas in that it lacks abundant breccia. The basinal carbonate in this section is dark-gray, pelletoidal mudstone or wackestone with thin (25–40 cm thick, Fig. 5) to massive (3 m thick) beds. They normally have pelagic

FIG. 5.—Thin-bedded lime mudstone and wackestone. Upper Tamaulipas Limestone, Rio Guayalejo Canyon.

FIG. 7.—Massively bedded breccia with 10 to 30 percent coarse allochthonous clasts derived from the platform. Upper Tamaulipas Formation. Rio Guayalejo Canyon.

FIG. 6.—Convolute bedding in the Upper Tamaulipas Formation showing overturned thin-bedded micrograded lime wackestone. Rio Guayalejo Canyon. Scale bar 25 cm.

microfossils such as globigerinids and calcispheres; there are some impressions of ammonites and brachiopods (*Kingena wacoensis*). Sedimentary structures include thin laminations associated with thin micrograded beds of platform-derived allochthonous carbonate sediment. Convolute bedding is observed in a few beds, which show overturned micrograded sediment (Fig. 6); movement probably occurred early, while the sediment was plastic.

Minor amounts of massively bedded breccia occur at two stratigraphic levels (Fig. 7). The breccia is composed of lithoclasts (10–30 percent by volume) which float in a matrix of autochthonous planktonic foraminiferal lime mudstone. The lithoclasts are of two types: (1) grainstone derived from the platform and (2) clasts typical of the reefal facies, composed mainly of caprinids. The clasts form a random fabric with poor sorting; no grading is obvious. This breccia was probably transported by some type of highly viscous sediment-gravity-flow mechanism. This is suggested by their massive bedding, poor sorting, and clasts up to cobble size floating in a pervasive mud matrix.

XILITLA AREA

Albian carbonate sedimentation of the basin margin in the Xilitla area contrasts with the Rio Guayalejo area in many respects, chiefly in that allochthonous breccia, graded detrital limestone and large exotic blocks predominate over the autochthonous fossiliferous lime mudstone.

Most of the breccia is thick-bedded. It is composed of coarse-grained lithoclastic and bioclastic sediment derived from the platform; the clasts float in a fine-grained matrix of the same composition (Fig. 8). Texture of the rocks suggest transportation as a debris flow (in the sense of Middleton and Hampton, 1973). Some other beds are fine-grained and normally graded and generally pass upward into thin laminations in the same bed.

FIG. 8.—Bedded allochthonous carbonate breccia in the "Tamabra formation." Xilitla area.

Individual exotic blocks have a maximum exposed stratigraphic thickness of 95 m; their true geometry is not known because of vegetative cover or partial exposure in the folds of the Sierra Madre Oriental (Fig. 9). They are composed of caprinids and some other molluscs which were probably part of a reefal environment (Fig. 10). There is no evidence of crushing, convolute bedding or other kinds of deformation within the base of the large blocks. This indicates that the blocks were emplaced at basinal depths when they were already rigid bodies. The basinal host rock shows only small scale contorted bedding in a few places. By palinspastic reconstruction of the platform and basin margins of Albian time (Carrasco, 1973), an approximate horizontal transport distance of 3.5 km from the edge of the platform has been inferred for these exotic blocks. Because well-bedded carbonate mudstone (Fig. 11) with pelagic and deep-water fossils, such as planktonic foraminifers, radiolarians and calcispheres, encloses the exotic blocks and pinches out against them (Fig. 9), there seems little doubt about the deep-water emplacement of these blocks.

METZTITLAN AREA

The Metztitlan area differs from the Rio Guayalejo and Xilitla areas in that typical basinal facies are intertongued with channelized allochthonous breccia.

The typical basinal carbonate rocks are fossiliferous mudstone and pelletoidal mudstone or wackestone. They are well-bedded, with thicknesses ranging from 20 to 60 cm (Fig. 12). A few basinal carbonate beds show boudinage, enclosed above and below by thin beds (Fig. 13). Many beds show millimeter-thick laminations, some of which are micrograded. These beds are interpreted as the product of intrabasin movement of fine calcareous sediment. In thin sections these laminations show no evidences of displaced platform fauna or of clastic material displaced from the platform (Fig. 14). Other beds show normal(?) grading on a major scale and scour attributed to turbidity-current flow. Burrows exist along bottoms of turbidite layers.

FIG. 10.—Core of the allochthonous exotic block shown in Figure 9, composed by caprinids.

FIG. 9.—Partial exposure of an allochthonous exotic block (A); dotted line is the contact with basinal carbonate sediment (B) which pinch out against the block. The rectangle shows the beds in Figure 11. Xilitla area.

FIG. 11.—Typical well-bedded carbonate, which covers and pinches out against the exotic block. This outcrop corresponds to the rectangle in Figure 9. Xilitla area.

FIG. 12.—Basinal thin-bedded carbonate sediment of the upper Tamaulipas Formation. Metztitlan area.

FIG. 13.—Beds of the calcareous basinal sediment of the Upper Tamaulipas Formation showing boudinage. Metzitlan area.

The middle part of the well-bedded sequence of typical basinal carbonate contains 90 m of massive breccia which stands out in outcrop due to its resistance to weathering (Fig. 15). The breccia-basinal facies contact is sharp with no evidence of contorted bedding in the host basinal rocks. The shape of the breccia body suggests it was deposited in a channel which trends away from the platform. In detail the massive allochthonous breccia shows poor sorting, random fabric, and no grading (Fig. 16). Many of the clasts are peloid grainstone or other types of grainstone. The interclast matrix is an organic-rich wackestone with fragments of rudists, echinoids, and a few planktonic foraminifers. This texture is the result of mixing of basinal mudstone with material derived from the platform (Fig. 17). The top of the breccia is in contact with 25 m of well-bedded, typical fossiliferous mudstone with planktonic foraminifers. This evidence is interpreted to show the emplacement of the massive breccia as, perhaps, a submarine debris flow. Another massive body of breccia, probably of

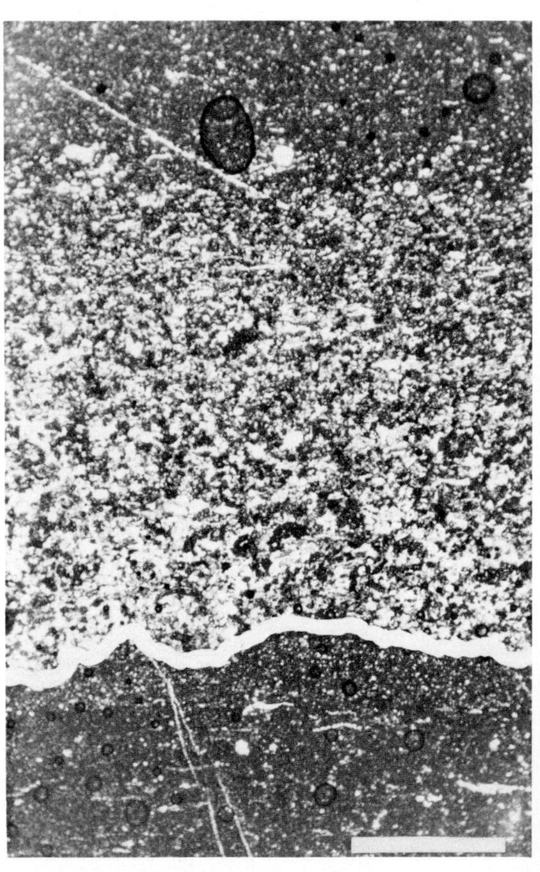

FIG. 14.—Photomicrograph of millimeter-thick laminations which are micrograded in detail. Composed of pelletoidal lime packstone; above and below the micrograded portion is fossiliferous lime mudstone with globigerinids (the white line is the lower contact of the micrograded packstone with the fossiliferous mudstone). Upper Tamaulipas Formation. Metztitlan area. Scale bar 0.110 mm.

FIG. 15.—View of the typical thin-bedded basinal carbonate rocks (A) of the Upper Tamaulipas Formation and thick allochthonous massive breccia (B) of the "Tamabra formation." In the background to the right are reefal facies of the El Abra Formation (C). Metztitlan area.

FIG. 16.—Detail of the allochthonous breccia, "Tamabra formation." Diameter of the coin is 2.5 cm. Metztitlan area.

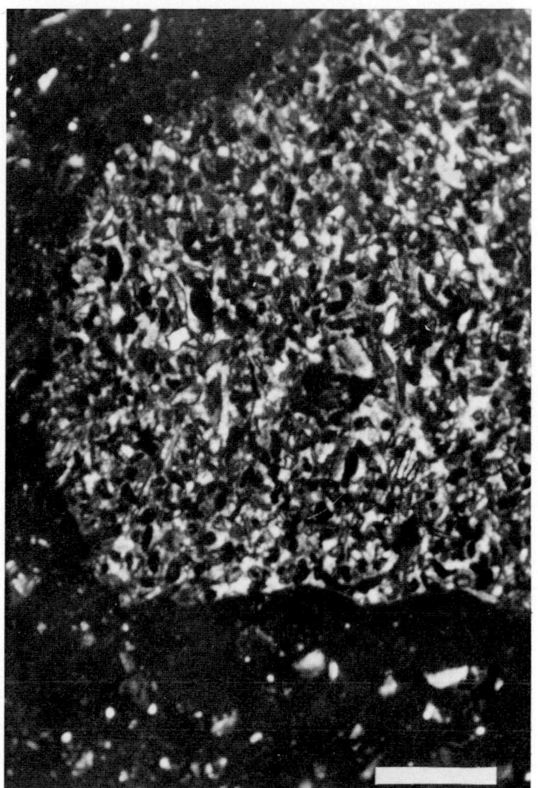

FIG. 17.—Photomicrograph of a portion of the allochthonous breccia ("Tamabra formation"), showing a lithoclast of pelletoid grainstone in a matrix of calcareous wackestone. Scale bar 1 mm.

late Albian-Cenomanian age, overlies the breccia and mudstone just described (Fig. 18).

MODELS OF SEDIMENTATION

Stratigraphic and sedimentologic data of the Rio Guayalejo, Xilitla, and Metztitlan areas are interpretated to show that the autochthonous and allochthonous sedimentary masses were deposited in a deep basinal or basin-margin environment. At least three different models of contemporaneous Albian submarine sedimentation can be inferred from studying the three different areas. Before discussing the different models, let us review some ideas about the morphology of the margins of the Cretaceous platforms and basins.

Many lines of evidence indicate that the Albian-Cenomanian platforms were bounded by steep slopes, probably as the result of rapid vertical accretion of the platform margin. At the eastern edge of the Valles-San Luis Potosi platform prominent escarpments appear to reflect depositional topography as in the Rio Guayalejo Canyon

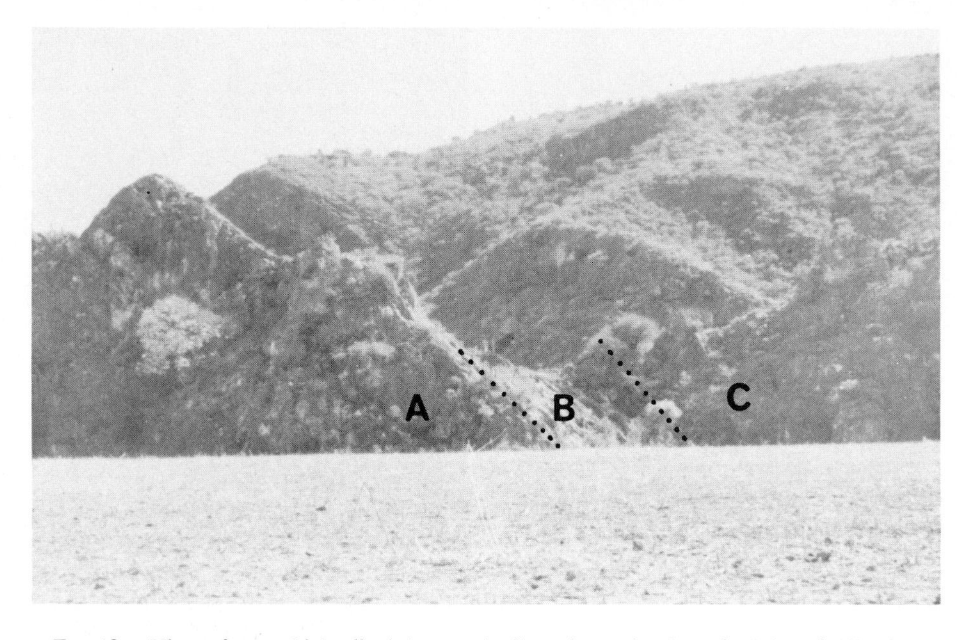

FIG. 18.—View of two thick allochthonous bodies of massive breccia (A) and (C) of the "Tamabra formation." Between the two breccia bodies are 25 m of typical well-bedded basinal mudstone (B) with globigerinids. The lower massive breccia (A) is that shown in Figure 15 (as B). Metztitlan area.

and Metztitlan section. In the Xilitla area the platform edge is faulted and grass covered; thus its original morphology is not known.

The opposite sides of the Guayalejo Canyon (Fig. 3) indicate the topographic expression of the platform and basin: on one side is the El Abra Formation which includes rudist facies; on the other side of the canyon is well-bedded basinal carbonate sediment of the Upper Tamaulipas Formation. At Metztitlan breccias of the "Tamabra formation" (Fig. 15), intervene between the massive cliff-forming limestone of the platform and the basinal facies.

Enos (1974) found the same topographic evidence of steep slopes along the contemporaneous (Albian) El Doctor platform (Fig. 1). Old and recent seismic profiles of the Golden Lane platform indicate that contemporaneous relief of about 1000 m existed on the El Abra Formation adjacent to the Poza Rica area. Recent detailed stratigraphic work by geologists of the Instituto Mexicano del Petróleo (private report for Petróleos Mexicanos) on microfacies and the general data of 87 wells located in the Poza Rica area also indicates relief of about 1000 m by the end of the Albian time. There is no evidence of complicated faulting or folding; if these are present, they are not important compared with the original relief of the platform.

If we now assume that in the Guayalejo Canyon, Xilitla, and Metztitlan areas, relief at any one time was at least a few hundred meters during Albian time, let us consider the models of basin and basin-margin carbonate sedimentation for each area (Fig. 19).

Rio Guayalejo.—Deep-sea sediments of Albian time are characterized by a clear dominance of well-bedded fossiliferous calcareous mudstone with the sporadic influx of some turbidity currents bearing material derived from the platform. Scattered allochthonous lithoclasts probably were transported by some type of highly viscous sediment-gravity-flow mechanism.

Xilitla.—This sedimentary sequence is represented by four types of deposit: (1) calcareous pelagic mudstone which is finely laminated, burrowed, or structureless, (2) allochthonous debris flow deposits which show crude graded bedding or are structureless, (3) exotic blocks with a stratigraphic thickness of 60 and 95 m, and (4) a few platform-derived, fine-grained lithoclastic graded beds emplaced by turbidity currents.

Metztitlan.—This model is distinguished by having allochthonous massive, channelized, debris-flow deposits which locally attain a stratigraphic thickness from 40 to 90 m. They are the product of slumping and downslope movement of poorly sorted clastic material from the platform.

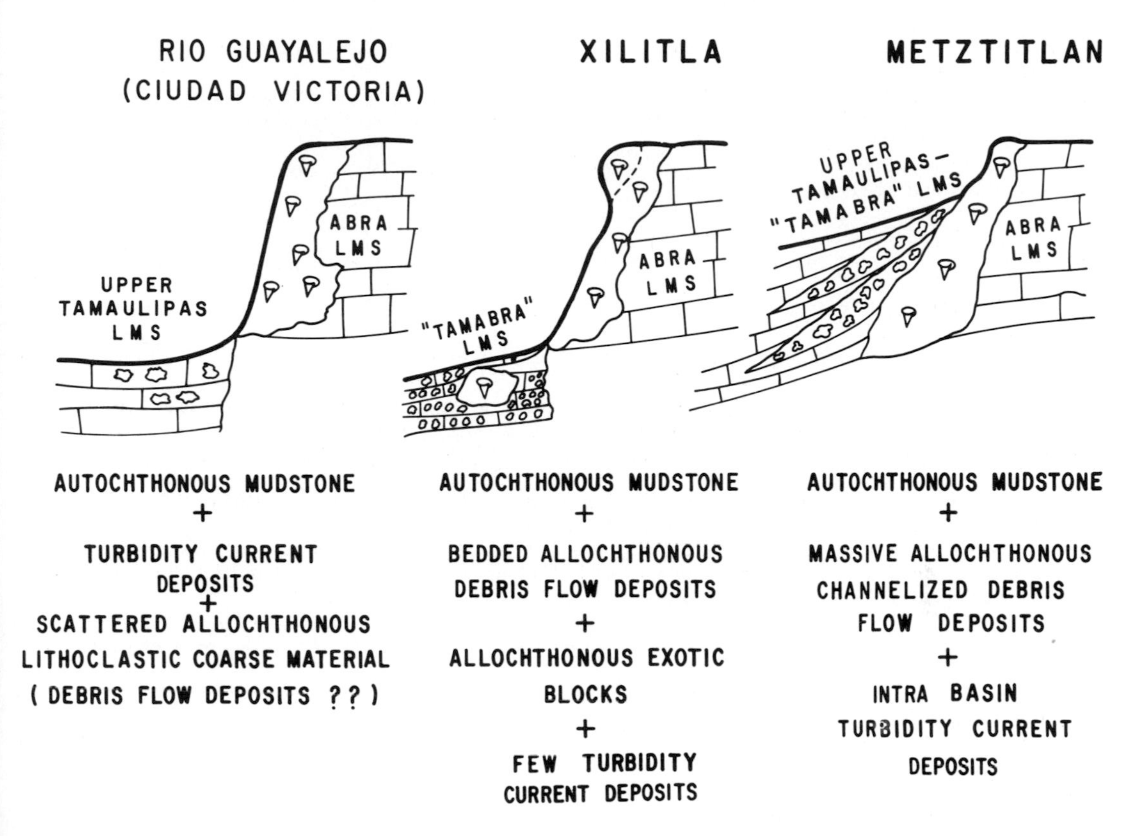

FIG. 19.—Sketch illustrating the different models of Albian sedimentation in Rio Guayalejo, Xilitla and Metztitlan areas. Not to scale.

The breccias are interbedded with typical well-bedded basinal fossiliferous mudstone. It is common in the basinal mudstone to find laminations which are micrograded and the product of intrabasin movement of fine sediment. Grading formed by turbidity currents with clastic sediment derived from the platform is rare.

REFERENCES

ACUÑA, G. A., 1957, El distrito petrolero de Poza Rica: Asoc. Mexicana Geologos Petroleros Bol., v. 9, p. 505-553.

BAKER, C. L., 1971, Geologic reconnaissance in the eastern Cordillera of Mexico: Geol. Soc. America, Spec. Paper 131, 83 p.

BARNETCHE, ALFONSO, AND ILLING, L. V., 1956, The Tamabra Limestone of the Poza Rica oilfield, Veracruz, Mexico: 20th Internat. Geol. Congress, Mexico, D. F., 38 p.

BEBOUT, D. G., COOGAN, A. H., AND MAGGIO, C. M., 1969, Golden Lane-Poza Rica trends, Mexico—An alternate interpretation [abs.]: Am. Assoc. Petroleum Geologists Bull., v. 53, p. 706.

BECERRA-H., A., 1970, Estudio bioestratigráfico de la Formación Tamabra del Cretácico en el distrito de Poza Rica: Inst. Mexicano Petrólio Rev., v. 2, p. 21-25.

BONET-M. FEDERICO, 1963, Biostratigraphic notes on the Cretaceous of eastern Mexico: *In* Geology of Peregrina Canyon and Sierra de El Abra, Mexico: Corpus Christi Geol. Soc., Guidebook Ann. Field Trip, p. 36-48.

BÖSE, EMIL, 1906, De San Luis Potosí à Tampico: 10th Internat. Geol. Congress, Mexico, Guide Excursion 30, 16 p.

BOYD, D. R., 1963, Geology of the Golden Lane trend and related fields of the Tampico Embayment: *In* Geology of Peregrina Canyon and Sierra de El Abra, Mexico: Corpus Christi Geol. Soc., Guidebook Ann. Field Trip, p. 49-56.

CARRASCO-V., B., 1973, Exotic blocks of forereef slope, Cretaceous Valles-San Luis Potosi platform, Mexico [abs.]: Am. Assoc. Petroleum Geologists Bull., v. 57, p. 772.

Carrillo-B., J., 1971, La plataforma Valles-San Luis Potosi: Asoc. Mexicana Geólogos Petroleros Bol., v. 23, p. 1-102.

Coogan, A. H., Bebout, D. G., and Maggio, Carlos, 1972, Depositional environments and geologic history of Golden Lane and Poza Rica trend, Mexico, an alternative view: Am. Assoc. Petroleum Geologists Bull., v. 56, p. 1419-1447.

Enos, Paul, 1974, Reefs, platforms, and basins of Middle Cretaceous in northeast Mexico: Am. Assoc. Petroleum Geologists Bull., v. 58, p. 800-809.

——, 1975, Tamabra Limestone of the Poza Rica trend [abs.]: Am. Assoc. Petroleum Geologists Ann. Meeting Abs., v. 2, p. 91.

Flores, R. J., 1955, Los arrecifes de la Cuenca de Tampico-Tuxpan, México: Asoc. Mexicana Geólogos Petroleros Bol., v. 7, p. 397-500.

Guzmán, E. J., 1967, Reef type stratigraphic traps in Mexico: Proc. 7th World Petroleum Congress, v. 2, p. 461-470.

Heim, Arnold, 1940, The front ranges of the Sierra Madre Oriental, Mexico, from Ciudad Victoria to Tamazunchale: Eclogae Geol. Helvetiae, v. 33, p. 313-362.

López Ramos, Ernesto, 1953, Distribución de la porosidad en las Calizas del Cretácico Medio de la region de Tampico-Poza Rica: Asoc. Mexicana Geólogos Petroleros Bol., v. 5, p. 31-56.

Luna, G. J., and Munoz, F. J., 1949, El campo petrolero de Poza Rica: Asoc. Mexicana Geólogos Petroleros Bol., v. 1, p. 35-46.

Mena Rojas, Enrique, 1955, Estudio geológico-enonómico del Cretácico Superior y Medio el este de la Faja de Oro: Asoc. Mexicana Geólogos Petroleros Bol., v. 7, p. 327-366.

Middleton, G. V., and Hampton, M. A., 1973, Sediment gravity flows: Mechanics of flow and deposition: *In* G. V. Middleton and A. H. Bouma (eds.), Turbidites and deep-water sedimentation: Soc. Econ. Paleontologists and Mineralogists Pacific Sec., Los Angeles, California, p. 1-38.

Nigra, J. O., 1951, El Cretácico Medio de México, con especial referencia a la facies de Caliza Arrecifal del Albiano-Cenomaniano en la Cenobahía de Tampico-Tuxpan: Asoc. Mexicana Geólogos Petroleros Bol., v. 3, p. 107-175.

Sotomayor-C., A., 1950, Estratigrafia y estructura del campo de Moralillo: Asoc. Mexicana Geólogos Petroleros Bol., v. 2, p. 647-677.

Suarez, C. R., 1950, Estratigrafia y estructura del campo de Moralillo: Asoc. Mexicana Geólogos Petroleros Bol., v. 2, p. 647-677.

Viniegra-O., F., and Castillo-T., C., 1970, Golden Lane fields, Veracruz, Mexico: *In* M. T. Halbouty (ed.), Geology of giant petroleum fields: Am. Assoc. Petroleum Geologists, Mem. 14, p. 309-325.

Wilson, J. L., 1969, Microfacies and sedimentary structures in "deeper water" lime mudstones: *In* G. M. Friedman (ed.), Depositional environments in carbonate rocks: Soc. Econ. Paleontologists and Mineralogists, Spec. Pub. 14, p. 4-19.

——, 1974, Characteristics of carbonate-platform margins: Am. Assoc. Petroleum Geologists Bull., v. 58, p. 810-824.

SEPM Special Publication No. 25, p. 273-314, November 1977

TAMABRA LIMESTONE OF THE POZA RICA TREND, CRETACEOUS, MEXICO[1]

PAUL ENOS
State University of New York at Binghamton, Binghamton, 13901

ABSTRACT

The petroleum producing Albian-Cenomanian limestone of the Tampico embayment consists of three regional facies belts: the Tamaulipas (basinal), El Abra (reef, backreef, lagoon), and Tamabra (basin margin). The principle lithologies of the Tamabra Limestone, in order of abundance, are skeletal-fragment grainstone/packstone, breccia, rudist-fragment wackestone, and pelagic-microfossil wackestone. Regional trends are to finer-grained lithologies upward and into the basin. Grainstone/packstone predominates in the greater Poza Rica field; elsewhere breccia is more common to dominant. The skeletal grainstone/packstone was deposited grain-for-grain by sediment gravity flow. The breccia, interpreted as debris flow deposits, contains clasts of basinal pelagic wackestone and cemented, shelf-derived skeletal-fragment limestone. No rudists in growth position were identified in Tamabra rocks although some fossils were mechanically deposited with long axes vertical.

The hypothesis of Coogan and others (1972), that the Poza Rica trend is the true shelf edge and the Golden Lane a faulted backreef-lagoon sequence, is unacceptable because of stratigraphic thickening and evident lack of structure at the Golden Lane escarpment. The hypothesis that the Poza Rica trend is a subsidiary low sea-level shelf margin seaward of the major Golden Lane shelf margin is disfavored, because of the requisite repeated major sea level fluctuations and the facies contrasts between the Golden Lane and the Poza Rica trend.

The geometry and sedimentology of the deposits indicate that the Golden Lane is the true shelf edge and the Tamabra facies is reef-derived debris deposited in the basin. Shelf-edge slope and relief are considered the best guides to basin-margin debris deposits in early stages of exploration for Poza Rica-type reservoirs.

INTRODUCTION

The major oil-producing basin of Mexico is the Tampico embayment[2] with proved reserves of 5.5 billion barrels in 1966. The Tampico embayment extends from the central Mexican platform, near the front of the Sierra Madre Oriental fold belt, to the continental shelf of the Gulf of Mexico. It is separated from the Burgos Basin on the north by the Tamaulipas arch and from the Veracruz Basin on the south by the Jalapa uplift—Teziutlan massif (Fig. 1). Ninety percent of the Tampico embayment production is from Albian-Cenomanian limestone.

This limestone consists of three distinct lateral facies which are given formation rank: (1) El Abra Limestone—shallow-water platform facies, (2) Upper Tamaulipas—fine-grained basinal limestone, (3) Tamabra Limestone—a wedge of coarse skeletal debris and breccia at the basin margin. Upper Tamaulipas, or comparable formations, and El Abra are recognized throughout northeast Mexico (Enos, 1974b). The Tamabra Limestone is best devleoped in the subsurface of the Tampico embayment adjacent to the Golden Lane platform. It is recognized as a facies and locally mapped around other platforms. It is a conclusion of this study that most of the debris of the Tamabra Limestone was derived from reefs at the escarpment of the Golden Lane platform, transported by sediment gravity flows, and deposited at the edge of the basin.

Production and proved reserves from the Tamabra Limestone exceed 2.3 billion barrels; more than 2 billion is from the greater Poza Rica field[3] alone (Table 1, Fig. 2). The Poza Rica field is the northeast flank and nose of a broad anticline which plunges southeast (Fig. 3). The trap is formed by the westward pinchout of the porous Tamabra Limestone into the impermeable basinal Upper Tamaulipas Limestone (Barnetche and Illing, 1956) and by overlying dense basinal limestone of the Upper Cretaceous (Fig. 4).

This report is primarily a study of the rocks from cores in the Poza Rica and other Tamabra fields, from nonproducing wells in the Tamabra facies, and of several wells in the Golden Lane trend (Fig. 5). The regional stratigraphic frame-

[1]Shell Development Company, BRC-EP Release no. 131

[2]A series of oil discoveries in the Isthmus of Tehuantepec since 1972 has greatly altered Mexico's reserve picture (Franco, 1976). Half of Mexico's production may be coming from these fields by 1977. Current production is primarily from Cretaceous limestone with promise of major production from Jurassic limestone.

[3]The amoeboid-shaped greater Poza Rica field (Fig. 2) is subdivided into a number of fields for historical reasons. They are named from NW to SE (Fig. 4): Mecatepec, Poza Rica, Escolín, Presidente Alemán, Papantla, Tolaxca, Cerro del Carbon (Cazuelas).

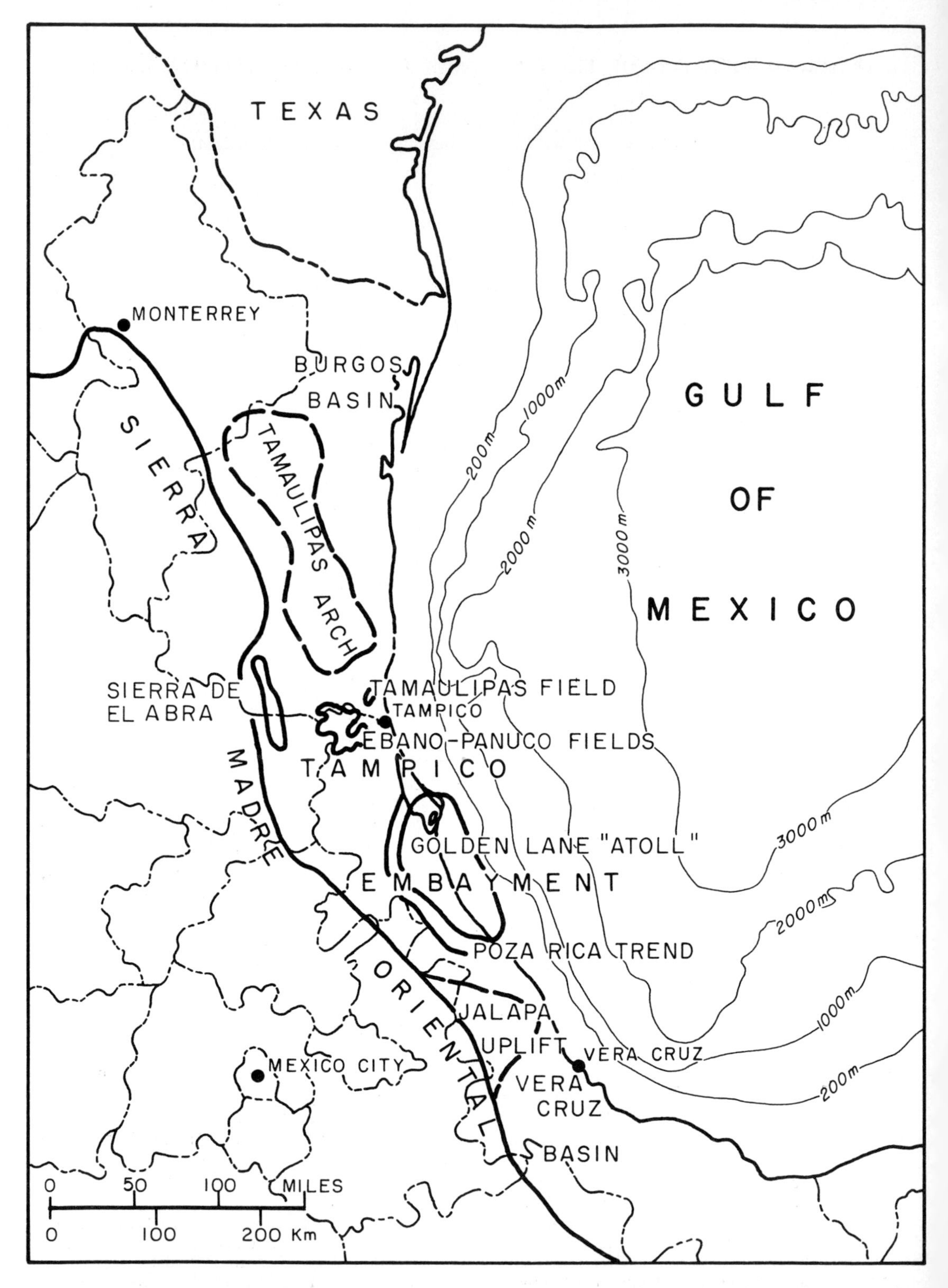

FIG. 1.—Major structural elements of the Gulf Coast of Mexico including the Tampico embayment; after Murray and Krutak (1963).

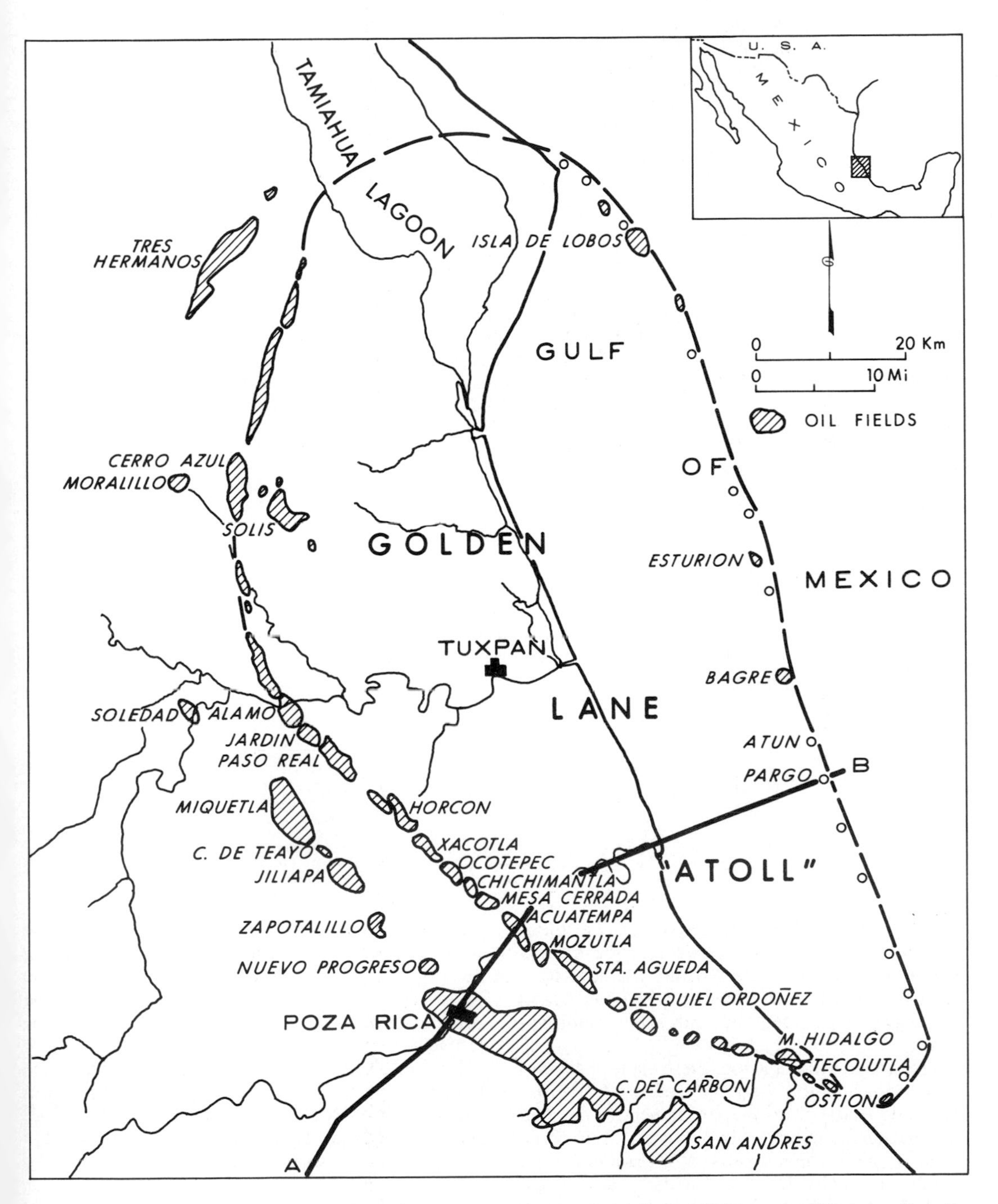

Fig. 2.—Location of Poza Rica and Golden Lane trends showing principal fields and offshore wells; after Guzmán (1967). Cross section is Figure 4.

TABLE 1.—OIL FIELDS OF TAMABRA, ABRA, AND TAMAULIPAS LIMESTONES

Field	Tamabra Fields Production (m.bbls)	Proved Reserves (m.bbls)	API Gravity	Field	Abra Fields (Golden Lane) Production (m.bbls)	Proved Reserves (m.bbls)	API Gravity
				Isla de Lobos	3.3	14.7	40
Tres Hermanos	43.1	96.9	21				
Moralillo	11.2	7.7	21	Naranjos/Cerro Azul	1,164.7	37.8	20
				Solis	6.9	2.7	22
Soledad	4.3	?	33	Alamo, Jardin (w/Naranjos)			
Miquetla	11.6	51.3	35	Paso Real	48.0	?	
Castillo de Teayo	0.3	?	32	Horćon	2.0	1.5	
				Tihuatlan	0.3	—	22
Jiliapa	14.4	23.3	34	Xacotla	0.6	0.5	20.5
				Ocotepec	14.7	7.7	22.5
Zapotalillo	0.1	?		Chichimantla	2.7	1.2	21
				Mesa Cerrada	8.3	4.2	19.5
Nuevo Progresso	4.6	0.4	31	Aguatempa	16.4	24.9	23
Pital y Mozutla	—	0.9	23	Mozutla	7.6	3.3	21.5
Poza Rica	944.0	1,098.3	35	Santa Agueda	73.6	67.5	16
				Ezequiel Ordonez	44.6	14.0	21
Cerro del Carbon	0.5	?	—	Arroyo Grande	—	—	—
San Andrés	(main pay Jurassic)		29	Gutierrez Zamora	0.6	0.2	20
				Vicente Guererro	4.3	0.6	24.5
				Manuel Hidalgo	8.2	1.4	31
				Tecolutla	0.5	0.3	22.5
Totals	1,033.8	1,278.8	29.4		1,407.3	182.5	22.4
	Tamaulipas Fields						
Barcodon	5.7	2.4	17				
Ebano-Panuco	913.1	54.4	12				
Tamaulipas	9.8	56.3	18				
Totals	928.6	113.1	16				

work as developed by geologists of Petróleos Mexicanos (Carrillo, 1971; Guzmán, 1967; Salas 1949) and local correlations within the Poza Rica field established by Barnetche and Illing (1956) were found to be adequate with a few modifications.

Generalized lithologic logs were made of several cores during an initial visit to Poza Rica, Veracruz. Seven additional cores were obtained for detailed examination including thin sections.[4] Ten cores which formed the nucleus of a study of the Tamabra trend by Bebout, Coogan, and Maggio (1969; also see Coogan, Bebout, and Maggio, 1972) were examined briefly. The spacing between the wells was too great to establish correlations using mechanical logs. The logs were useful within the Poza Rica field where isolated wells could be referred to the framework of Barnetche and Illing (1956).

Several early regional seismic lines across the Golden Lane and Poza Rica are available from the literature (Rockwell and Garcia Rojas, 1953; Islas and Equia, 1961); two modern lines were available for cursory examination.

[4]Detailed logs of these cores are available from National Auxillary Publications Service, New York, New York 10017.

STRATIGRAPHIC RELATIONSHIPS OF MIDDLE CRETACEOUS LIMESTONES

Throughout the Albian and Cenomanian the Tampico embayment was the site of almost unin-

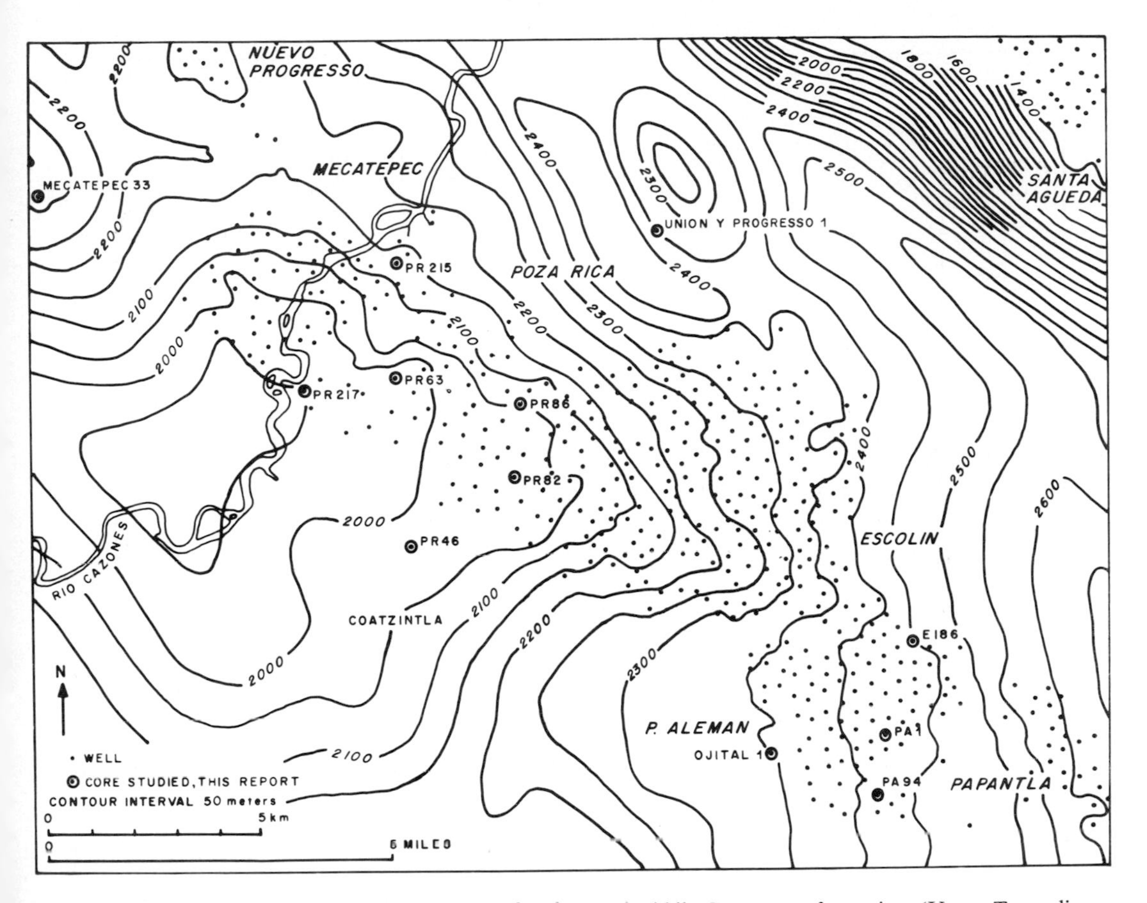

FIG. 3.—Structural contours in meters below sea level, top of middle Cretaceous formations (Upper Tamaulipas, Tamabra, El Abra); courtesy of Petróleos Mexicanos.

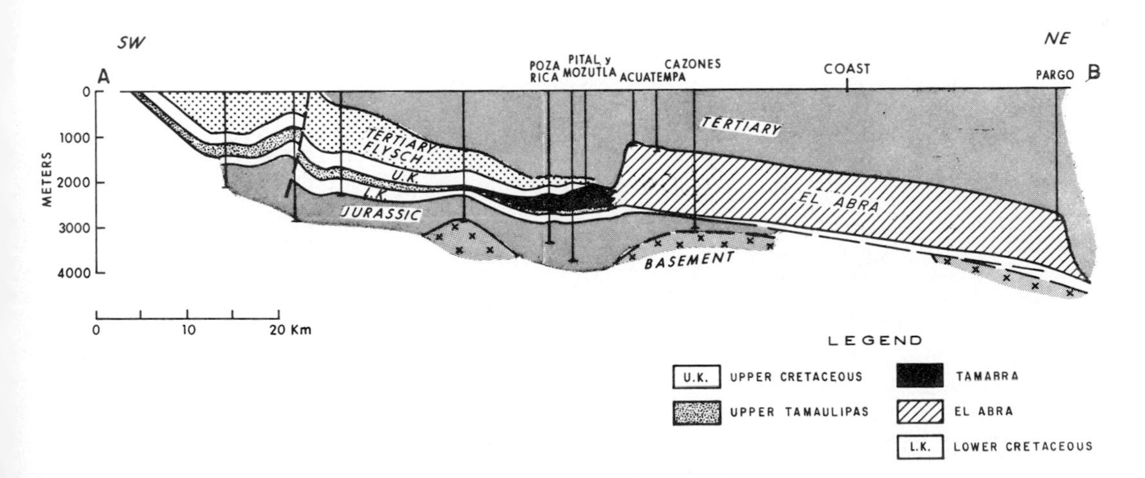

FIG. 4.—Regional cross section (A-B of Fig. 2) of the Tampico embayment showing middle Cretaceous limestone formations; from files of Shell Oil Company.

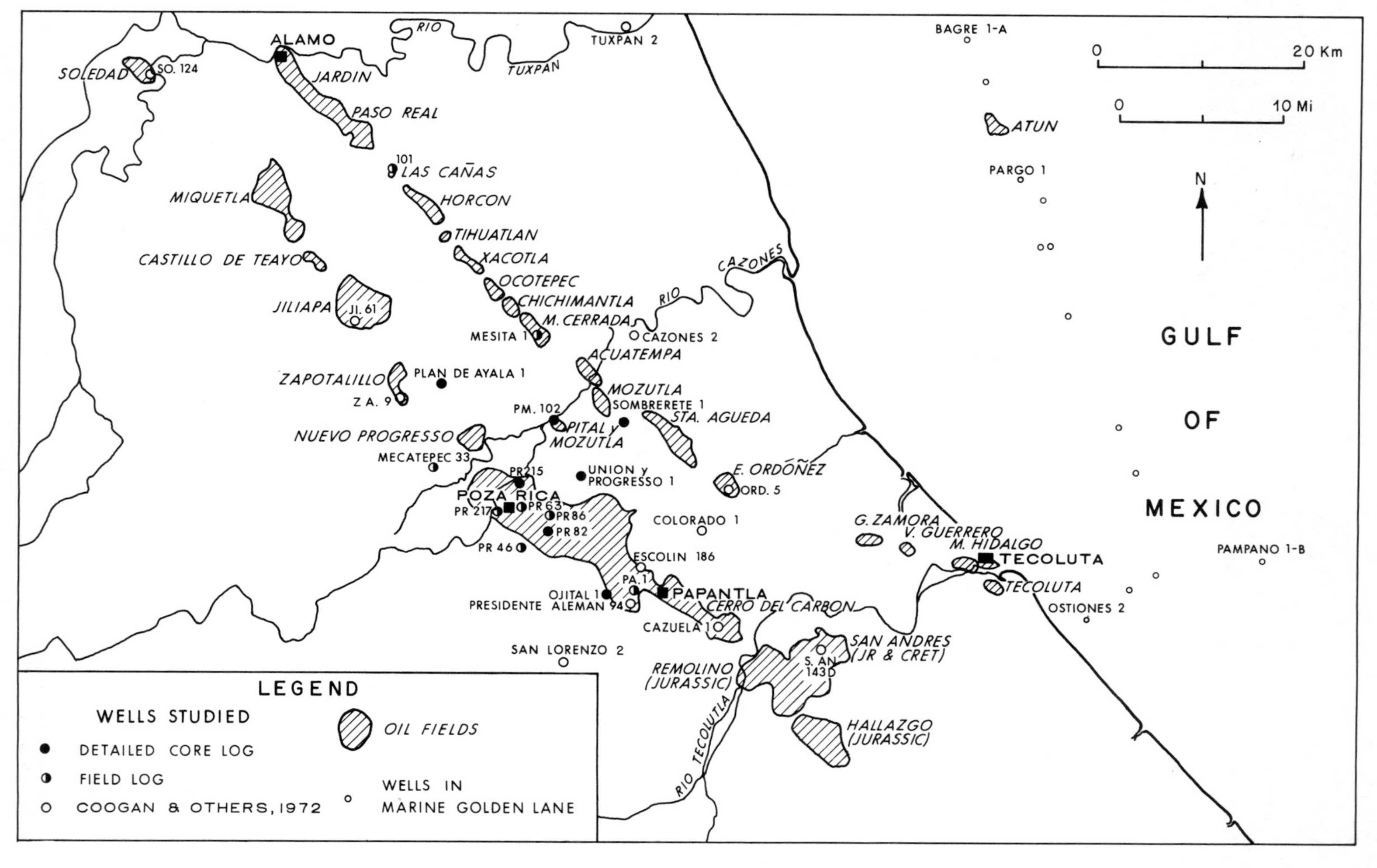

FIG. 5.—Location of studied cores.

terrupted carbonate deposition. The basin extended from the shallow shelf edge of the Valles-San Potosi platform, delineated by rudist reefs exposed in the Sierra de El Abra (Fig. 1; Bonet, 1952, 1963; Griffith, Pitcher, and Rice, 1969), to at least the present shelf of the Gulf of Mexico. The basin was interrupted by the isolated platform or "atoll" of the Golden Lane (Figs. 1, 2). The principal facies within the embayment (Fig. 4) correspond to the basinal Upper Tamaulipas Limestone, El Abra Limestone which generally is interpreted as reef and lagoonal facies, and the Tambra Limestone which is variously described as slope, "mixed facies" (basin and shelf, Heim, 1940, p. 325), forereef (Boyd, 1963), basin margin (Enos, 1974b), debris from erosion (subaerial?) of the reefs (Viniegra and Castillo-Tejero, 1970, p. 317) or even the actual shelf-edge reefs (Coogan and others, 1972).

The Albian and Cenomanian limestone of northeast Mexico forms a unit with no apparent break at the Albian-Cenomanian boundary, that is, the boundary between Upper and Lower Cretaceous series in Europe and the United States. It is convenient to group it informally as "middle Cretaceous" (uncapitalized).

The Upper Tamaulipas Limestone of the middle Cretaceous overlies the argillaceous Otates Formation, (Late Aptian, Early Cretaceous) the approximate age and facies equivalent of the Pearsall and La Peña Formations (Murray, 1961, p. 314). The Otates Formation is only questionably identifiable in the Poza Rica field; the underlying basinal Lower Tamaulipas Limestone (Lower Cretaceous) is distinguished by rather subtle sedimentologic contrasts where it is overlain by the Upper Tamaulipas Limestone rather than the Tamabra.

The Otates Formation is not recognized beneath the Golden Lane where El Albra Limestone passes down into dolomite. Bonet (1952, p. 227) states that the base of El Abra Limestone in the Golden Lane is definitely Albian, a view that is widely accepted (Corpus Christi Geological Society, 1963). Recently R. H. Waite and P. O. O'Neil (unpublished report, 1970) also obtained a date of Late Albian from El Abra in the Golden Lane (Mesita #1, 1920 m) based on the evolutionary stage of *Orbitolina*, a large benthic foraminifer. Temporal correlation between the platform and basinal sequences is still exceedingly difficult because of the dramatic faunal change which accompanies the depositional facies change.

The Upper Tamaulipas and Tamabra Formations are overlain by distinctive thin-bedded basinal limestone of the Agua Nueva Formation. The Agua Nueva is considered Turonian in age (correlation chart, Corpus Christi Geol. Soc. 1963) but Bonet maintains that the base is Cenomanian at Poza Rica (Bonet, 1952, p. 223; 1963, Fig. 1). On the other hand, Turonian fossils have locally been identified from the Tamabra (Becerra, 1970), although the identifications have been questioned (Coogan and others, 1972, p. 1424).

Progressively younger formations, as young as Upper Oligocene, overlie El Abra along the Golden Lane escarpment (Sotomayor, 1954; Viniegra and Castillo-Tejero, 1970). All of El Abra Limestone may be older than the Agua Nueva (Turonian?) which locally overlies it, but this cannot yet be proved by dating of the subsurface El Abra Limestone itself. Viniegra and Castillo-Tejero (1970, p. 314) have very tentatively correlated bentonities near the top of El Abra in Golden Lane wells with some near the base of the Agua Nueva further west in the basin, suggesting that the base of the Agua Nueva may be diachronous. Aguayo (1975) recently found late Turonian pelagic microfossils in a traceable horizon near the top of the El Abra Limestone at its type locality.

Within the basinal sequence, some confusion exists in the well files about the nature of the Agua Nueva-Tamabra contact. The contact is usually picked in practice by a sharp expansion with depth of the electric logs or by deflections on radioactive logs (Fig. 6). This pick does not always correspond to a prominent lithologic break (Barnetche and Illing, 1956, p. 12) from argillaceous limestone with pelagic fossils of the Agua Nueva to rudist-skeletal debris or breccia typical of the Tamabra. Thus, several logs record Agua Nueva lithologies within the Tamabra interval or place breccias in the base of the Agua Nueva. Unfortunately, cores are usually not taken across the contact. However, from limited core material and from well-file descriptions it appears that the Tamabra breccia and rudist limestone interfinger with Agua Nueva lithologies which contain Turonian microfossils. This suggests continuous deposition from the Tamabra into the Agua Nueva rather than an erosional interval immediately following Tamabra deposition as suggested by Barnetche and Illing (1956, p. 26) in the Poza Rica area.

The thickness of the middle Cretaceous varies between 200 and 400 m in most of the basin and between 1000 and 1500 m on the Golden Lane platform (Fig. 7). Middle Cretaceous is absent around the north and south ends of the Golden Lane. This absence may result from one or several periods of erosion (Sotomayor, 1954) or from depositional bypassing, a point of considerable importance in considering the postdepositional history of the middle Cretaceous rocks.

EL ABRA LIMESTONE

El Abra Limestone of the Golden Lane contains most of the facies found in the well-known out-

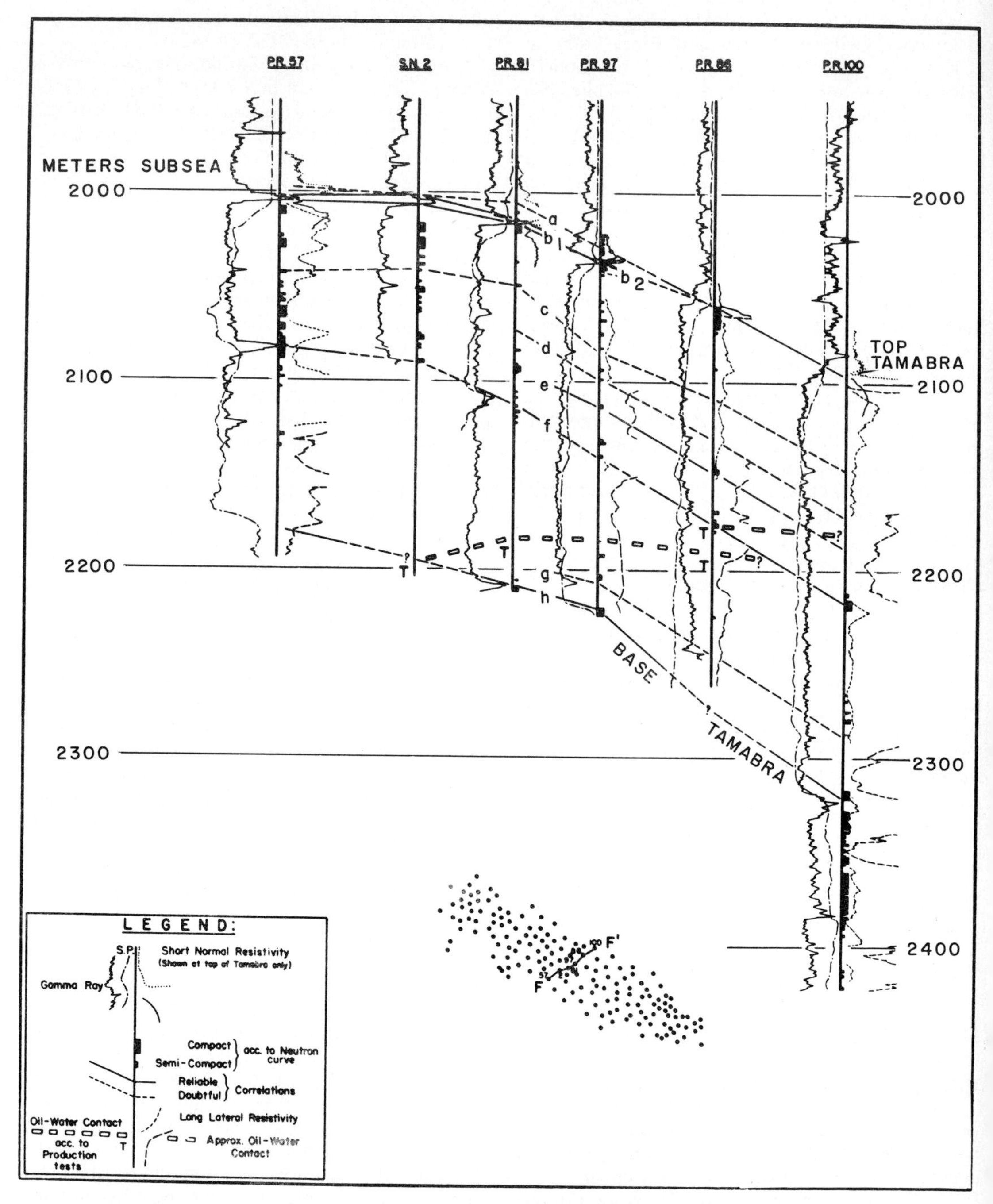

FIG. 6.—Representative cross section of Poza Rica field showing general character of mechanical-log "horizons," some of which are pelagic wackestones, and oil-water contact; from Barnetche and Illing (1956).

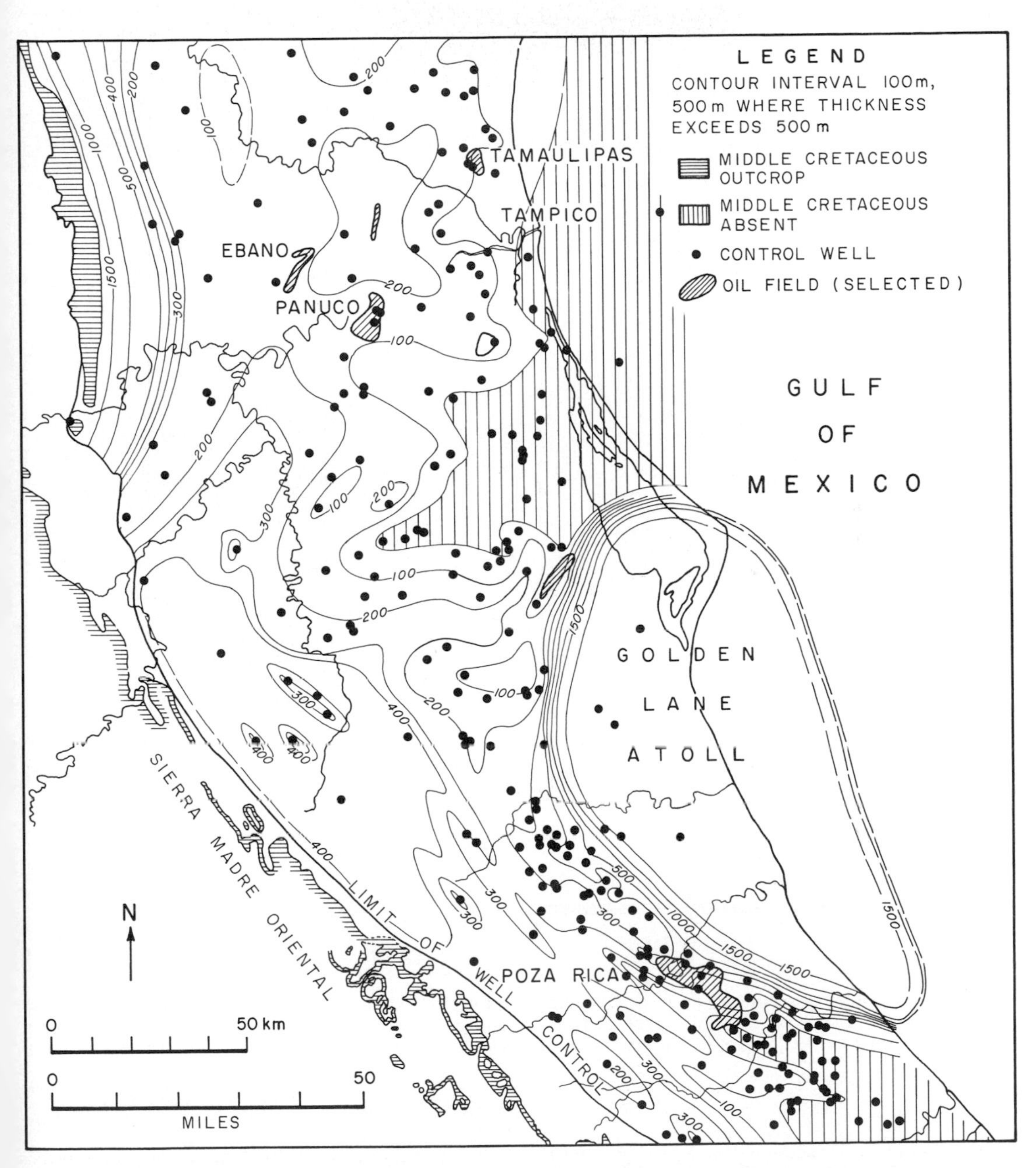

FIG. 7.—Regional isopachous map of middle Cretaceous formations. Note limited control in the Golden Lane where few wells have penetrated the entire El Abra. Data to about 1965; from Shell Oil Company files.

crops of the Sierra de El Abra (Griffith and others, 1969; Aguayo, 1975). In the several cores examined, the prominent lithologies are (1) miliolid-pelletoidal wackestone and packstone, (2) stromatolitic pelletoidal packstone with birds-eye structures (3) coarse rudist-fragment and coralline wackestone and packstone, (4) rounded skeletal-fragment lime grainstone, (5) oolitic grainstone and (6) algal-coated grainstone and packstone. These rocks represent depositional environments which probably ranged from reef through shelf-edge sand shoals to semirestricted shelf and tidal-flat conditions. Evaporite deposits are reported from the interior of the Golden Lane "atoll," indicating considerable restriction (Coogan and others, 1972, p. 1439).

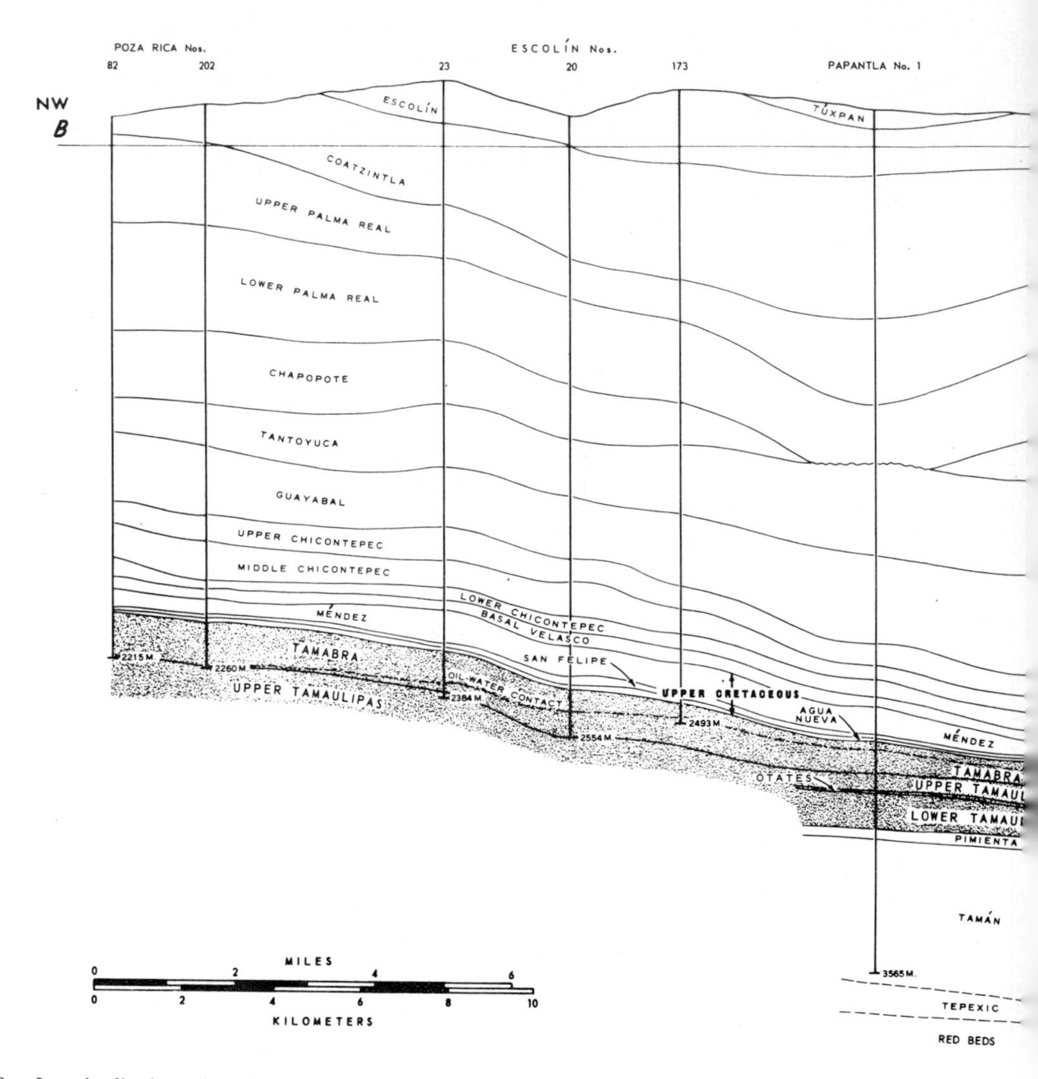

FIG. 8.—Longitudinal section along Poza Rica trend showing truncation of Cretaceous section in San Andrés field. Offlap of Agua Nueva (basal Upper Cretaceous) and consequent unconformity on Tamabra suggested in Remolino #1 may be questioned because of variable lithologic definition of Agua Nueva/Tamabra contact

The sequence of facies within Las Cañas #101 and Mesita #1 wells (Fig. 5) indicates a general trend from reef or perireef conditions at depth (Las Cañas below 1380 m; Mesita below 1690 m) to semirestricted shelf conditions at the top, near the topographic crest of the Golden Lane. This sequence probably results from seaward progradation of the platform margin. The sequence is stratigraphically reversed in E. Ordóñez #5 (Fig. 5). Dolomitization increases with depth in many wells, progressing from dolomitization along fractures to pervasive crystalline dolomite devoid of depositional fabrics.

In the examined Golden Lane wells, which were not good producers, porosity is primarily intergranular in the coarser-grained rocks and "chalky" or leached skeletal and pelletoidal molds in finer-grained rocks. Leached vuggy and cavernous porosity, for which the northern Golden Lane is famous, is encountered in the topographic crest of the escarpment at the top of the sequence. No material is available from fields of the northern Golden Lane, developed between 1904 and 1920. Wells were drilled only into the top of El Abra at which point gushing oil frequently terminated drilling. Only recently has appreciable core mate-

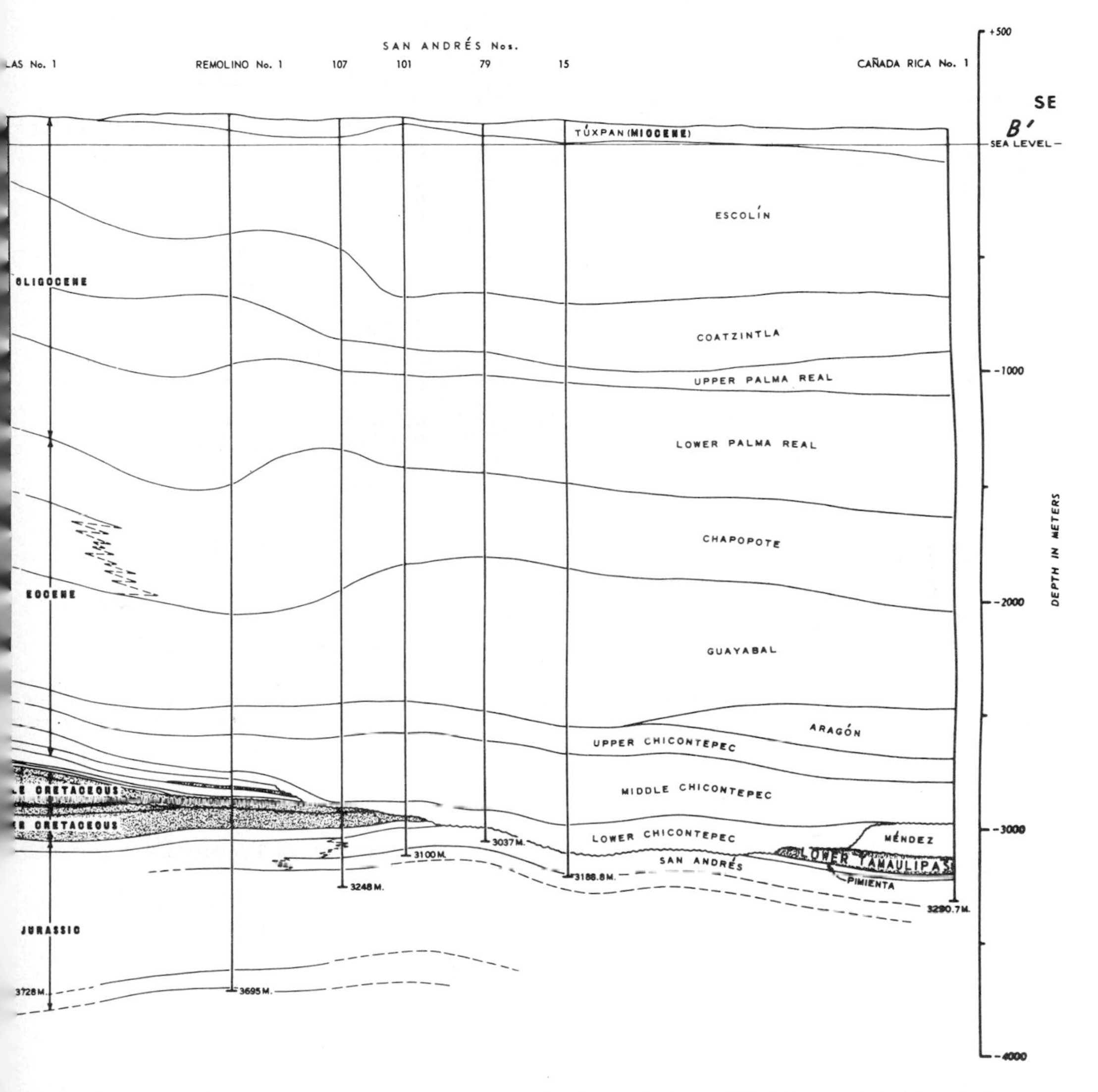

e text). Location in Figure 9; section from Bebout and others (1969), modified from original by A. Ramon, tróleos Mexicanos, 1962.

rial become available elsewhere; previous information on the rocks was confined to pieces blown out of the early wells (Muir, 1936) and to analogy with the outcrops in the Sierra de El Abra.

UPPER TAMAULIPAS LIMESTONE

Extensive intervals of Upper Tamaulipas Limestone were cored in only a few wells. Wackestone containing abundant pelagic foraminifera, radiolarians, and calcispheres is the principal lithology of the Tamaulipas. A few thin beds of skeletal-fragment packstone were encountered. Some wackestone beds are microlaminated; others are structureless or extensively burrowed. Color ranges from nearly white to black. Colors become progressively darker westward into the basin away from the Golden Lane.

Beds of black chert are common where the Upper Tamaulipas is thick. The chert contains silicified ghosts of calcareous pelagic microfossils. Partly silicified laminated wackestone, with well-preserved megascopic details may occur adjacent to the chert. The matrix of wackestones is commonly one to five percent dolomite.

The Tamaulipas Limestone is the main produc-

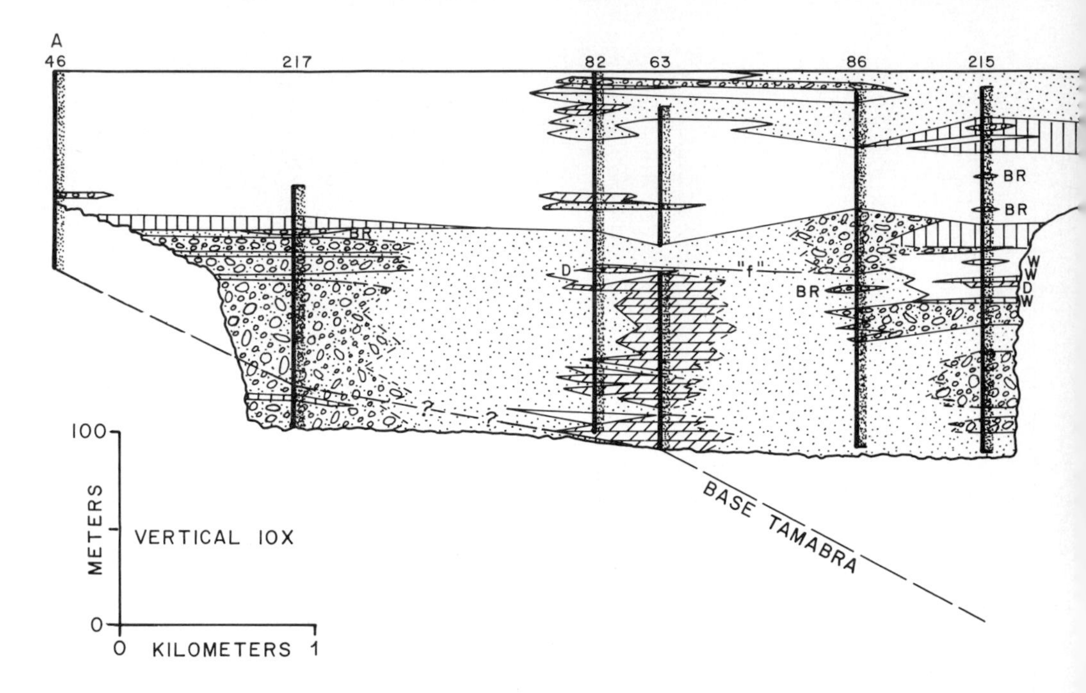

FIG. 9.—Cross section of Tamabra Limestone through Poza Rica field based on wells logged in detail for this study. Datum is top of Tamabra Limestone. Possible interfingering of Tamabra with the overlying Agua

ing horizon in Ebano-Panuco (''northern fields'') where cumulative production is about one billion barrels (Fig. 1, Table 1). The reservoir apparently developed from fracturing.

TAMABRA LIMESTONE

The Tamabra Limestone is used here in the sense of Barnetche and Illing (1956, p. 9) for ''the dominantly bioclastic limestone-dolomite sequence which underlies the San Felipe and Agua Nueva limestone, marl, and shale and overlies and laterally merges southwestwards into the calcilutitic impervious limestone which is similar to the Tamaulipas of the Sierra Madre.'' This is a complete departure from the original definition of Tamabra by Heim (1940, p. 321) as a general formation name to include the Tamaulipas, El Abra, and Taninul (El Abra reef facies) Limestone in the Sierra Madre Oriental. The time-honored, although improper, usage is that followed by Barnetche and Illing.

The Tamabra Limestone forms a wedge 10–15 km wide at the margin of the basin west of the Golden Lane escarpment. The wedge is 200 to 400 m thick within 6 to 10 km of the Golden Lane and thins westward by interfingering with or grading into basinal Upper Tamaulipas Limestone at progressively higher stratigraphic levels (Fig. 4).

The entire middle Cretaceous, including the Tamabra Limestone, is absent at the north and south ends of the Golden Lane (Fig. 7). The hiatus at the south end extends into the San Andrés field where control is available to show that the entire Cretaceous section and part of the Jurassic is truncated (Fig. 8), so that Chicontepec (lower Eocene) rests directly on San Andrés (Upper Jurassic). Truncation by post-Velasco (Paleocene) erosion is a more likely explanation than depositional offlap considering the facies involved, in particular the basinal character of the Tamaulipas and the Upper Cretaceous. A considerable pre-Méndez (Upper Cretaceous) hiatus is also indicated in Cañada Rica #1 (Fig. 8). Nondeposition caused by sediment bypassing in submarine currents, as is presently occurring on the Blake Plateau off the Atlantic shelf, cannot be ruled out from geometry alone.

The possible occurence of a Tamabra facies east of the marine Golden Lane is an open question. A test well, Pampano 1-B (Fig. 5), was drilled more than 3,000 m without penetrating below the Tertiary.

On a more detailed scale, effective mechanical log correlations are possible within the densely

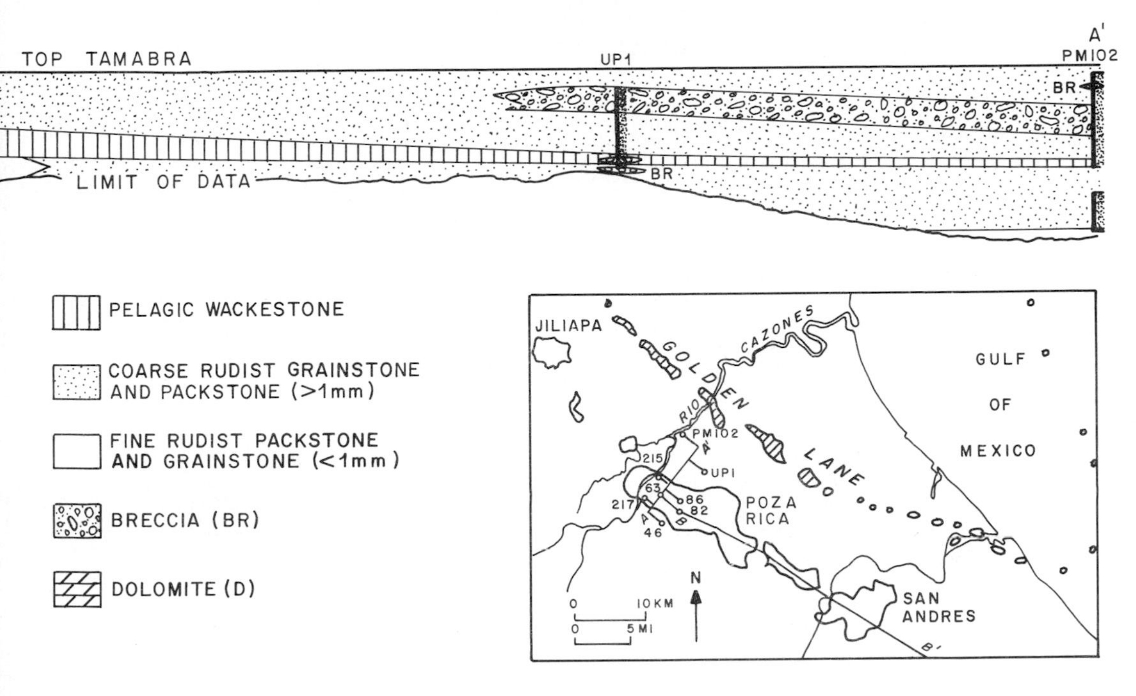

Nueva involves a few tens of meters ambiguity, but would not change interpretations appreciably. Top and base of Tamabra are from well files of Petróleos Mexicanos. B-B′ on location map is Figure 8.

drilled Poza Rica field (Barnetche and Illing, 1956), although the lithologies are not arranged in a layer-cake fashion. The "horizons" correlated by Barnetche and Illing (Fig. 6) are useful in constructing the geometry of the field, but they must be translated into lithologic information with extreme care. Most of the "compact zones," picked by neutron logs, are packstone and grainstone which are relatively less porous because of more extensive cementation (Barnetche and Illing, 1956, p. 14) or less extensive leaching. The radioactive and electric log deflections on which the marker horizons were established cannot generally be tied to a lithology in the core, probably because of poor recovery in just those argillaceous fine-grained intervals which would produce an anomaly. Horizon "f", the most persistent within the Tamabra (Fig. 7), consists of basinal wackestone in some wells and fine-grained, skeletal fragment, dolomitic limestone in others (Barnetche and Illing, 1956; Becerra, 1970). It could be identified in only one of the wells in Figure 9, in part because complete mechanical and lithologic logs were not available. Aside from "b" and "f", the horizons are not very persistent. Thus, it appears that the Tamabra interval is punctuated by a series of dense limestones, in part of pelagic origin, that are discontinuous but might nonetheless influence continuity and fluid transmissibility of the reservoir.

An interpretative cross section of the Tamabra Limestone illustrating the points mentioned above and showing inferred relationships with adjacent units, including interfingering of Tamabra and Agua Nueva, is presented in Figure 10.

Tamabra Petrology

The various lithologies which make up the Tamabra Limestone are complexly interspersed. A lithologic cross section at the latitude of the Poza Rica field (Fig. 9) shows a trend to coarser skeletal fragment rocks with depth in the Tamabra interval and toward the Golden Lane escarpment. Breccia is scattered throughout the Tamabra, but its distribution generally conforms to the trend of coarser grained rocks at the base and nearer the Golden Lane escarpment.

Location of the wells that were studied in detail does not permit construction of a longitudinal section of the Tamabra trend. In general, skeletal grainstone and packstone is most abundant throughout the Poza Rica field. Skeletal rocks are also the dominant lithology in Colorado #1, San Andrés #143D and San Lorenzo #2 wells. Breccia is more abundant than skeletal rocks in the wells studied from the following fields: Sole-

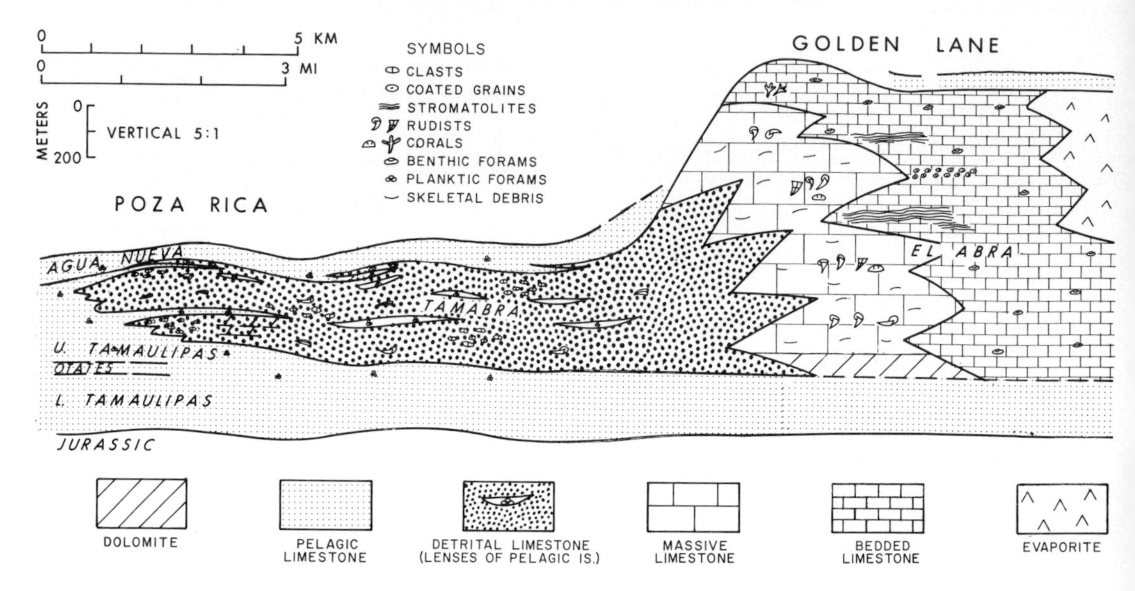

FIG. 10.—Interpretive cross section from Golden Lane to Poza Rica.

dad, Zapotalillo, Jiliapa, Escolín, Presidente Alemán, and Cazuelas.

The principal lithologies of the Tamabra Limestone are (1) skeletal-fragment grainstone and packstone, (2) rudist wackestone, (3) pelagic-microfossil wackestone, (4) breccia, which with few exceptions contains clasts of both skeletal-fragment grainstone and pelagic wackestone in varying proportions, and (5) dolomite. A summary and comparison of the lithologies, excluding dolomite, is in Table 2.

Skeletal-fragment grainstone and packstone.—The most abundant and productive lithology of the Tamabra, at least in the latitude of the Poza Rica field, is skeletal-fragment grainstone and packstone (Fig. 11). Rudists, mainly caprinids, are the dominant to nearly exclusive skeletal fragments.

Bedding character is difficult to determine in cores in which the individual pieces are invariably twisted off and rarely exceed 25 cm in length. The "twist-offs" apparently occur preferentially along stylolites rather than on shale partings or other sedimentary bedding planes, suggesting massive beds. Primary depositional structures, which are rare but significant in environmental interpretation, are interbedded coarse/fine layers (Fig. 12), possible imbrication (Fig. 11B), long-axis orientation (Fig. 13A), and a few graded beds (Fig. 13B).

Grains are generally coarse—many exceed 1 cm—and rounded (Fig. 12). Intraclasts or cemented lumps are virtually absent. The nature and even the amount of original matrix is generally obscured by recrystallization. Inter-particle space filled by coarse, clear, blocky calcite is interpreted as cement-filled pore space in a relatively matrix-free sediment. Intergranular mosaics of equant calcite crystals, 5 to 100 μm in diameter, probably reflect recrystallization of carbonate mud (Fig. 14). No original matrix textures or microfossils have been identified. The close packing of large grains and the abundance of grain-to-grain contacts indicate grain support in the original sediment. Point contacts are not altered by solution; interpenetrating or sutured grain contacts are lacking despite the presence of many stylolites.

Porosity in the skeletal grainstones ranges from less than 5 percent to as high as about 25 percent (Table 3). Skeletal packstone porosity is somewhat lower. Molds of leached fossils are the most abundant pores. Oil staining in molds indicates many are interconnected. Appreciable primary interparticle porosity is preserved and probably contributes significantly to interconnection of the pore network.

Rudist wackestone.—Muddy, rudist (caprinid) limestone (Fig. 15) is rare in the Tamabra (Table 2), although it may form intervals 3 m or more thick. The rudists are fragmented, but external ornamentation is occasionally preserved. A few rudists are encountered with the axis of the shell oriented vertically, suggesting preservation in growth position.[5]

[5]Many rudists, including some caprinids, had recumbent or variable growth positions (Kauffman and Sohl, 1974, p. 403; A. H. Coogan, pers. comm., 1975). Hence absence of vertical orientation cannot be construed as evidence against *in situ* growth, but its presence is generally considered a sufficient, if not necessary, condition.

TABLE 2.—CHARACTERISTICS OF TAMBRA LIMESTONE LITHOLOGIES

Lithology	Skeletal-Fragment Grst/Pkst	Rudist Wackestone	Pelagic-Microfossil Wackestone	Breccia	
Abundance	very common	very rare	rare	common	
Color	dark brown or grey	light brown, cream	white, light grey, rarely dark grey	mottled white, brown and grey, dark matrix	
Oil Staining	common, heavy	rare	lacking	patchy	
Interval Thickness	may exceed 100 m.	generally thin	thin	may exceed 100 m.	
Bedding	vague, massive	vague, massive?	thin	massive	
Sedimentary Structures					
Graded Beds	vr	—	m	vr	
Coarse/Fine Layering	m	vr	r	vr	
m m Lamination	—	—	vr	—	
Imbrication	f	vr	—	vr	
Long-Axis Orientation	m	—	—	—	
Bioturbation	—	—	m	—	
Soft-Sediment Deformation	—	—	vr	f	
Reoriented Geopetals	r	m	vr	f	
Grains					
Whole Skeletons	r	c	c	vr	
Skeletal Fragments	ab	c	f	m	
Pelletoids	vr	f	m	—	
Lithoclasts	vr	r	—	ab	
Intraclasts	vr	f	r	f	
Matrix	r	ab	ab	ab to r	
Grain Size	coarse-fine	coarse-med.	v. finc	coarse-v. coarse	
Roundness	r-subang.	ang.	—	ang.	
Sorting	good to poor	poor	good to poor	poor	
Fossils				matrix	clasts
Caprinids	ab	ab	—	vr	ab
Radiolites	c	c	—	—	c
Other Mollusks	c	c	vr	r	c
Coral	m	f	—	—	f
Red Algae	r	r	—	—	r
Green Algae	f	f	—	—	f
Echinoderms	c	c	f	—	c
Benthonic Forams	—	vr	—	vr	vr
Pelagic Forams	—	vr	ab	ab	—
Calcispheres	—	r	ab	ab	—
Radiolaria	—	—	r	r	—
Cement				matrix	clasts
Micritic	—	r	ab	ab	c
Fibrous (ragged)	r	—	—	—	c
Blocky	ab	c	r	—	c
Silica	r	vr	f	—	vr
Recrystallization	extensive	moderate	slight	slight	moderate
Dolomitization	vr	r	c	ab	r
Pyrite	vr	vr	ab	c	vr
Stylolites	f	f	c	ab	c
Porosity	fair to excellent	poor to good	nil	nil to fair	
Primary Intergr.	c	r	—	r	
Primary Intragr.	c	r	vr	vr	
Moldic	ab	m	—	m	
Vuggy	r	r	r	r	
Fracture	r	f	r	f	
Intercrystalline	—	r	r	c	vr

SYMBOLS: —lacking; vr=very rare; R=rare; f=few; m=many/much; c=common; ab=abundant

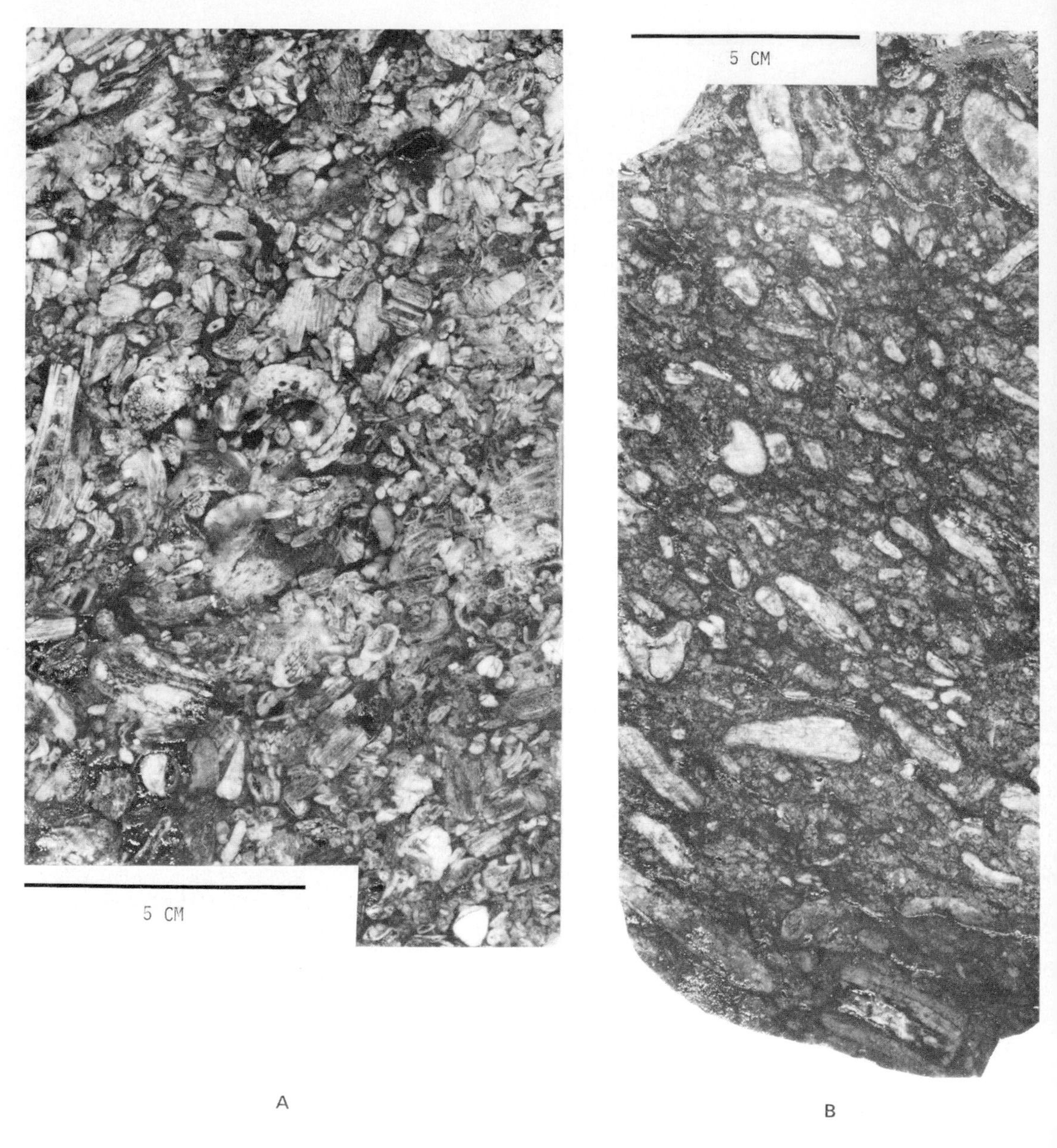

FIG. 11.—Grainstone facies of Tamabra Limestone. *A*—Rudist-fragment lime grainstone; good primary interparticle porosity locally enhanced by leaching. Oil impregnated. Pital y Mozutla #102, 2178 m. *B*—Rudist?-fragment lime grainstone showing considerable rounding, and inclined bedding or imbrication. Extensively recrystallized. Finer-grained parts are oil impregnated. Pitula y Mozutla #102, 2241 m.

The pallial canals of caprinids commonly contain sediment. The attitude of the shell at the time the internal sediment was deposited can be determined in many cases because the fill is graded or is incomplete leaving a flat, originally nearly horizontal, upper surface. The resulting geopetal structures are a reliable, and frequently the only, means of determining the proper orientation of a piece of core if the attitudes of different shells are internally consistent. In a few instances, conflicting attitudes of geopetal structures indicate that final deposition of the shell followed internal sedimentation and lithification or "setting" of the fill in some other position. Several rudist shells which are apparently in upright growth position contain geopetal fills deposited

FIG. 12.—Coarse/fine layering in highly recrystallized skeletal-fragment packstone. Union y Progresso #1, 2579 m.

when the shell was in a different position (Fig. 16). The apparent "growth position" is the chance result of redeposition with the axis upright.

The matrix of the rudist wackestone is micrite containing small, unidentifiable skeletal fragments. No benthic microfossils have been identified,[6] despite their relatively large size, but one rudist wackestone examined in thin section contained abundant pelagic microfossils (Fig. 17).

Porosity is generally less than 5 percent in rudistid wackestone (Table 3). Porosity development is dependent on leaching, principally fossil-mold formation.

[6]Presidente Alemán #1, cores 10-14, contained abundant benthic foraminifera (miliolids, *Dictyconus*) in very limited core recovery of rudist wackestone. However, the depth intervals written on the core boxes and the lithologies do not agree with those recorded in the field description of the core by Petróleos Mexicanos. The top of the Tamabra in the core is too shallow for any part of the southern Poza Rica trend; the depths would be appropriate for the top of El Abra in the adjacent Golden Lane. In most respects, the core resembles a Golden Lane lithologies and is anomalous for the Tamabra Limestone even in adjacent wells (Escolín 186, Presidente Alemán #94). Because of these discrepancies, especially the conflict with the field descriptions, this core is believed to have been mislabelled and so was not considered in this study.

Pelagic-microfossil wackestone.—Beds of light-grey to white wackestone within the Tamabra are generally less than 2 m thick, although mechanical logs and partially recovered cores suggest some beds may be 10 m thick. Contacts with grainstone or breccia, where preserved, are generally stylolitic (Fig. 18A), but originally were probably sharp like that in Figure 18B. Internal sedimentary structures are most commonly a few vague burrows (Fig. 19A), but microlamination is present locally (Fig. 19B). Thin laminae and micrograded beds of pelletoid-benthic microfossil packstone and grainstone occur in a few wackestone intervals (Fig. 19B-D).

Fossils in the wackestone (Fig. 20) are pelagic foraminifera, calcispheres, calcified spicules, calcified radiolaria, echinoid fragments, possible pelagic algae, smooth-shelled ostracods (one occurrence), and mollusc fragments. Vague pelletoids are fairly common. A few intraclasts of microfossil wackestone and lithoclasts of skeletal grainstone have been observed. Microfossils from the wackestone are the best means of dating the Tamabra Limestone (Becerra, 1970), because their preservation is much better than that of the megafossils, problems of reworking are fewer,

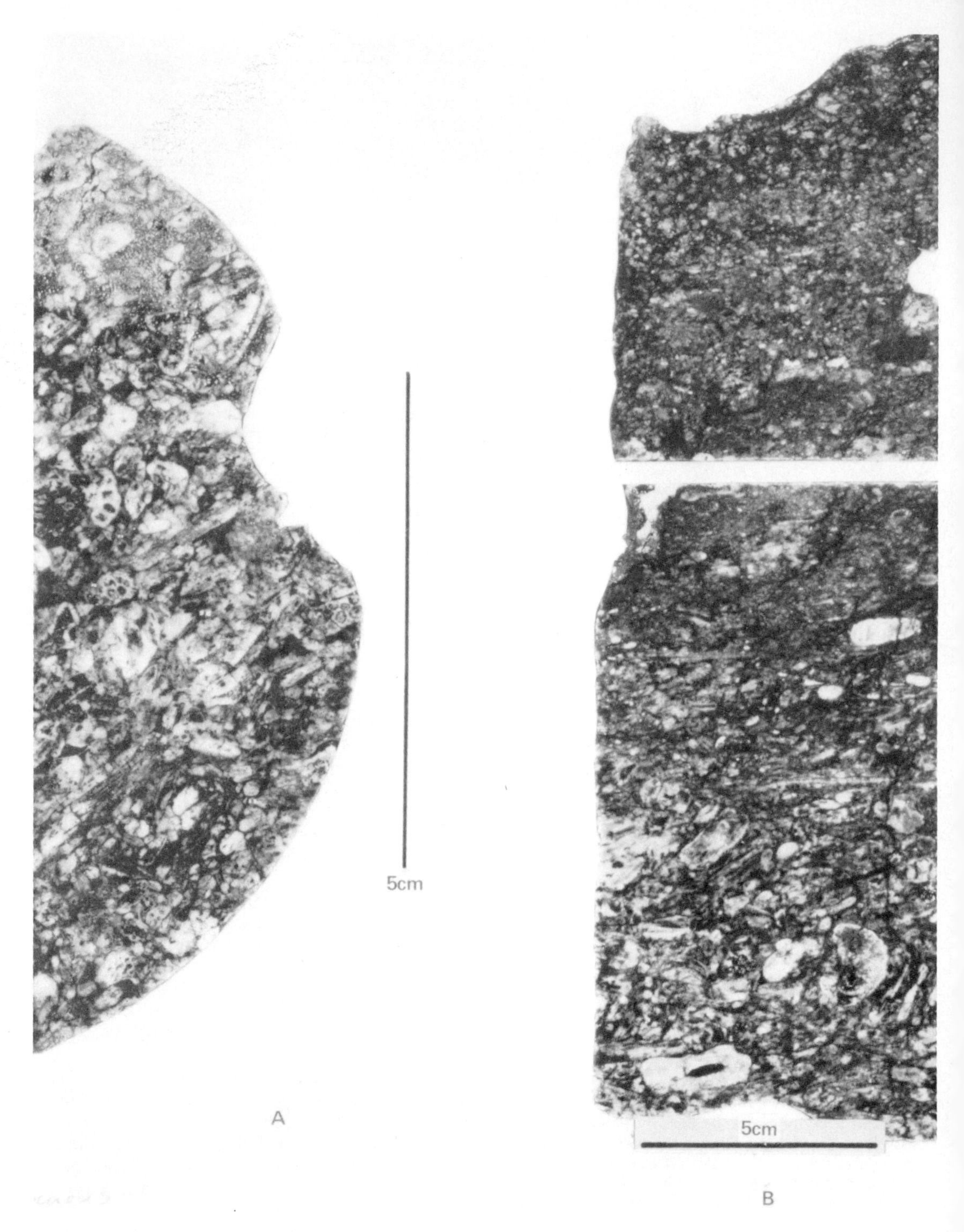

Fig. 13.—Sedimentary structures, grainstone/packstone facies, Tamabra Limestone. *A*—Grain orientation in rudist-fragment lime grainstone; view from top of core shows long axes aligned about 60° clockwise from sawed surface. Pital y Mozutla #102, 2190 m. *B*—Graded bed with large caprinid fragments at base. Faint imbrication. Pital y Mozutla #102, 2211 m.

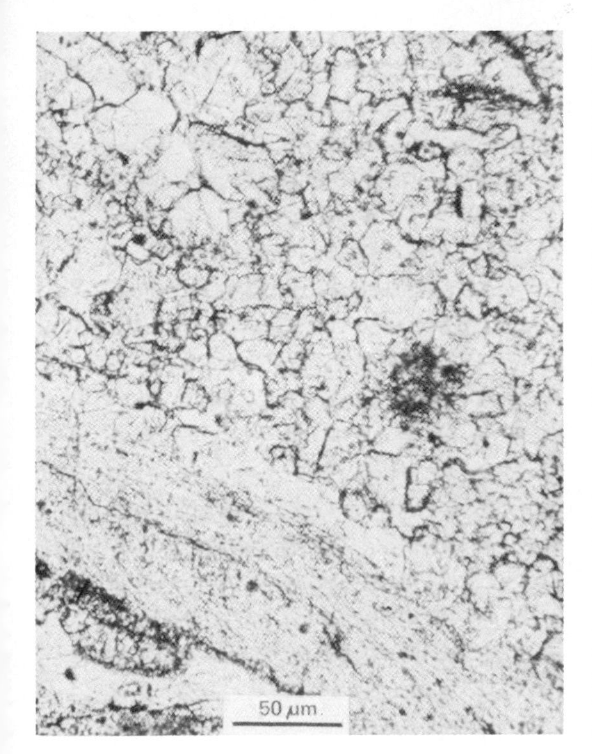

FIG. 14.—Recrystallized matrix of skeletal lime packstone. Note subsequent crystal size, irregular crystal boundaries, and possible relict clots of organic matter. Photomicrograph, plane light, Pital y Mozutla #102, 2180 m.

and better zonation has been established.

A few layers and nodules of chert occur within wackestone intervals in cores from outside the Poza Rica field. The chert is black and translucent with poorly crystalline silica spherulites whose size, shape, and distribution indicate replacement of spherical fossils (calcispheres, radiolaria, foraminiferal chambers). Replacement origin of the chert is further attested by irregular and/or gradational contacts (Fig. 21). Gradational contacts progress within a few centimeters from partial silicification of microfossils in the wackestone, to light-grey calcareous chert with good preservation of sedimentary structures but destruction of microfossil fine structure, to black dense chert (Fig. 21B). Some volume reduction of the wackestone, probably by solution of fine particles, has occurred after chert formation, as shown by compaction around chert lenses. Chert is most common near the top of the Tamabra Formation, especially where Tamabra-type breccia is interlayered with grey, burrowed, slightly argillaceous microfossil wackestone of the Agua Nueva type.

Euhedral rhombs of dolomite replace matrix in many wackestone intervals. The volume of dolomite rarely exceeds a few percent, but some intervals of crystalline dolomite, especially those with a micritic calcite matrix, probably represent more advanced dolomitization of wackestone.

Visible porosity is lacking in microfossil wackestone (Table 3). Light oil stain in a few highly dolomitic wackestones suggests development of fine pores through volume reduction with dolomitization. "Microvugs" less than a millimeter in diameter are occasionally observed in wackestone.

Breccia.—The most spectacular lithology of the Tamabra Limestone is sedimentary breccia (Fig. 22) that occurs at all levels. Breccia appears to be more abundant than skeletal-fragment grainstone and packstone in many areas outside the Poza Rica field, based on single wells from other fields (Fig. 5). Some breccia intervals are less than a meter thick; Poza Rica #217, however, contains about 100 m of breccia (Fig. 9).

Breccia fragments include skeletal-fragment lime grainstone and packstone, pelagic-microfossil wackestone, and, rarely, crystalline dolomite. Clasts of both grain-supported limestone with shallow-water fossils and basinal wackestone are present in nearly all occurrences of breccia, although the proportions vary widely. A few inter-

TABLE 3.—POROSITIES OF TAMABRA LIMESTONE LITHOLOGIES

Lithology	"Average" Porosity*	Maximum Porosity (%)	Principal Porosity Type	Other
Skeletal-fragment grainstone and packstone	Good	25	Moldic	Intergranular
Rudist wackestone	Poor	10	Moldic	Vuggy
Breccia	Poor	8	Intercrystalline	Fracture, Moldic
Pelagic wackestone	Nil	1	Intercrystalline	Fracture
Dolomite	Poor	25	Intercrystalline	Vuggy "Relict moldic"

*Good 10–15%, Fair 5–10%, Poor 1–5%, Nil <1%.

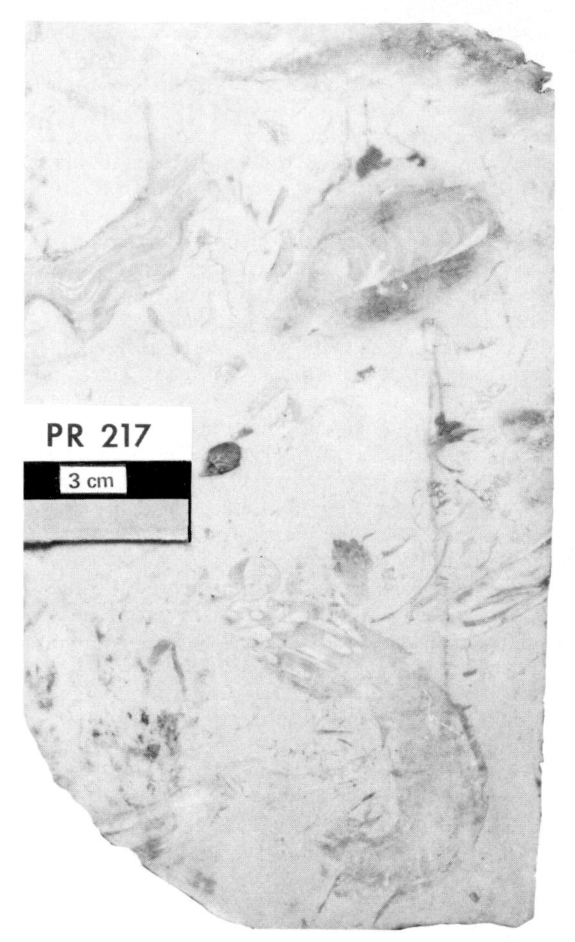

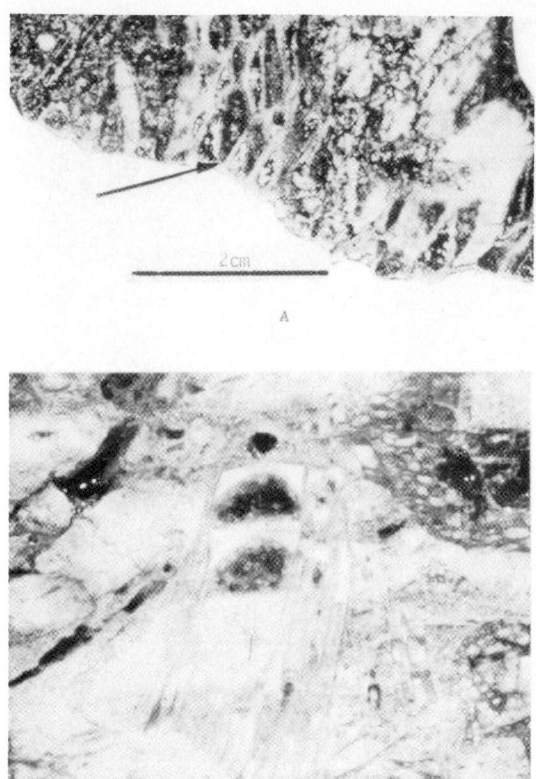

FIG. 15.—Rudist (caprinid and radiolitid) wackestone. See Figure 17 for photomicrographs of matrix. Poza Rica #217, 2106 m.; photo by E. A. Shinn.

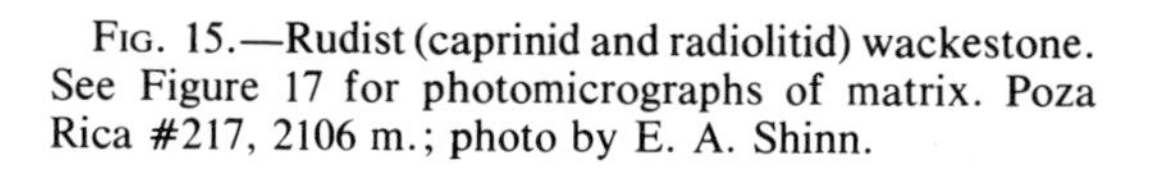

FIG. 16.—Reoriented geopetal fills in caprinids. *A*—Core piece, surface of geopetal in pallial canal oriented vertically (arrow). Cement layer overlies geopetal fill. Poza Rica #86, 2351 m; photo by E. A. Shinn. *B*—Caprinid oriented vertically. Curvature of septa shows that its final position, in which the graded fill was deposited, was upside down. Slab; Escolín #186, 2696 m.

vals of breccia contain predominantely shallow-shelf clasts to the virtual exclusion of basinal clasts, for example San Andrés #143D. In this case, dolomitization has obscured the nature of the matrix, but variations in lithology and bedding attitude between clasts indicates a true breccia.[7]

Breccia clasts are commonly between 1 and 10 cm in maximum dimension, but they range from less than 1 mm to at least 20 cm. Some "beds" of grainstone or wackestone within breccia intervals are confined to one or several core pieces without preserved contacts. These are probably large breccia fragments. Most breccia clasts are angular, but sutured contacts (Fig. 22B) and partial dolomitization obscure the original shapes and sizes.

Preferred orientation of platy clasts in the breccia is not obvious, but inclined fabric is locally evident (Fig. 22B). Many clasts are oriented at high angles, even perpendicular to bedding (Fig. 23). This would be an unusual orientation for any type of particle-by-particle deposition. It is analogous to the mechanical vertical orientation of

[7] "Pseudobreccia" was used without explanation by Barnetche and Illing (1956), apparently for a texture produced by partial dolomitization that isolates nondolomitized patches as apparent clasts. All intervals logged as breccia in this study were true breccia as indicated by polymict composition or rotated layering between clasts.

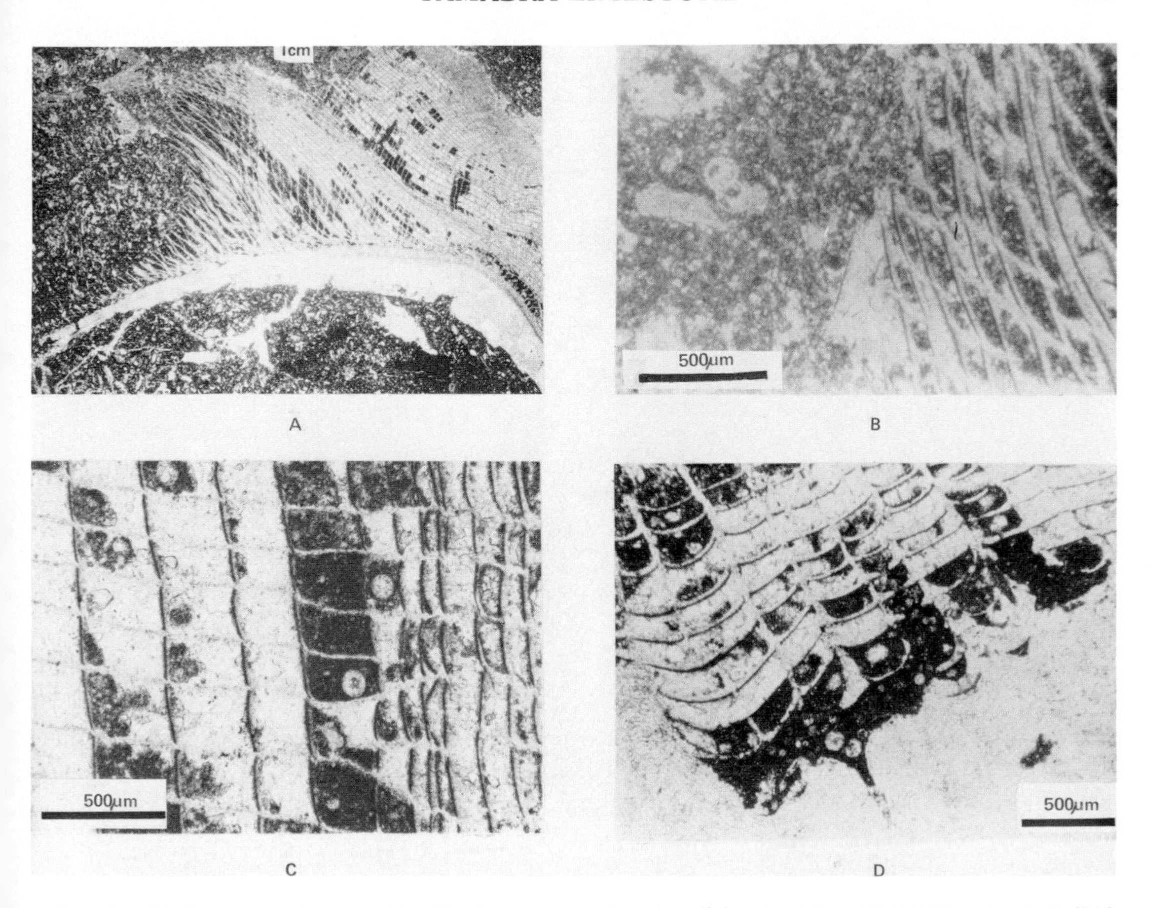

FIG. 17.—Rudist lime wackestone (Fig. 15); photomicrographs, plane light; Poza Rica #217, 2106 m. *A*—Radiolitid fragment in matrix with pelagic microfossils. *B*—Edge of radiolitid shell adjacent to globigerinid foraminifer. *C, D*—Calcispheres within radiolitid.

rudist fragments both in breccia (Fig. 23C) and in wackestone (Fig. 16B).

The amount of matrix in the breccia varies considerably. Clasts may float in fine-grained matrix (Fig. 22A) or be in fitted contact (Fig. 22B). Considerable matrix may have been removed by solution in the development of the sutured contacts; dark residue occupies the remaining interstices. "Open-work" breccia, with primary interparticle pore space and little matrix, was observed only in Zapotalillo #9. Sutured contacts are best developed in the presence of muddy matrix; they are lacking in open-work breccia and grainstone.

Unaltered breccia matrix consistently contains pelagic microfossils (also observed by Becerra, 1970) in micritic and micropelleted carbonate (Fig. 24). Identifiable microfossils are globigerinid foraminifers including *Hedbergella washitensis*, calcispheres, and possible calcified radiolaria. Shell walls of most microfossils are recrystallized sufficiently to make identification difficult; spherical or ovoid calcite rims are all that remain. Mollusc prisms and other benthic remains are rare. They were probably transported along with the coarser clasts.

Dolomite has partially or completely replaced the micritic matrix of much breccia (Figs. 23A, B; 25). Breccia clasts, particularly grainstone, and fossil fragments are commonly not dolomitized or only slightly dolomitized even where the matrix is extensively dolomitized.

Grain-supported clasts in the breccia must have been cemented or they would have been disaggregated during transport. In many clasts an isopachous crust of cement, consisting of clear stubby calcite crystals of uniform length, can be seen around each grain. Grain and cement are broken at the edges of some clasts (Fig. 26A). More typically, the break is around the grain so that

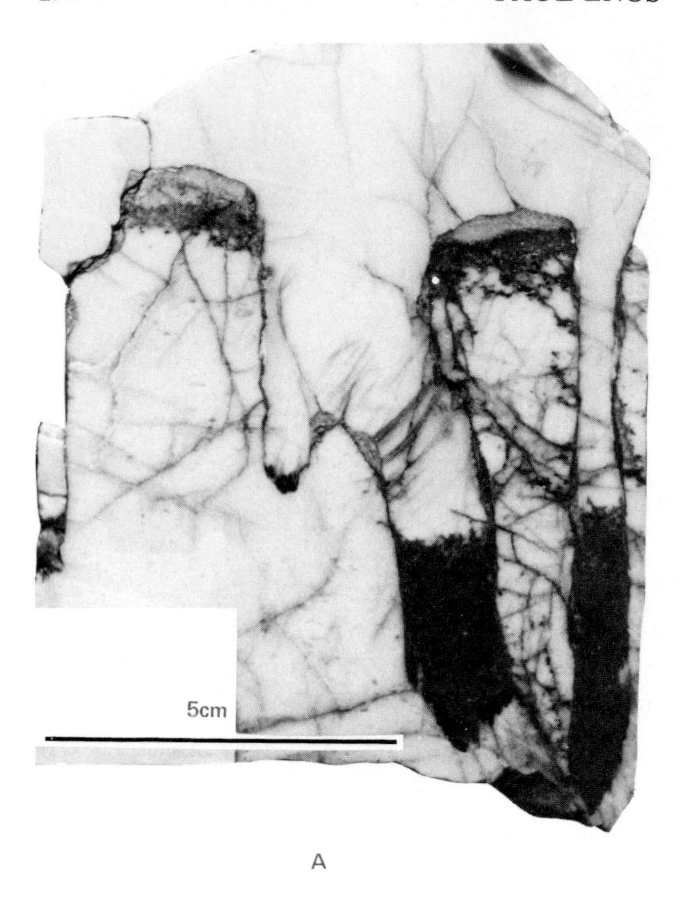

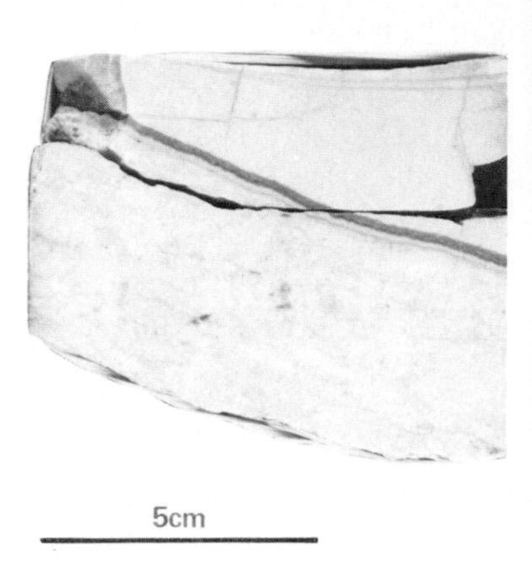

FIG. 18.—Pelagic microfossil lime wackestone. *A*—Stylolitic contact of lime wackestone (top) with coarser grained skeletal-fragment lime wackestone. Top layer contains calcispheres and pelagic foraminifers. The coarser wackestone also contains dominantly pelagic fossils. Note oil stains along stylolites and fractures; light-grey areas on upward projecting columns are dolomite. Poza Rica #217, 2061 m; photo by E. A. Shinn. *B*—Contact of pelagic-microfossil lime wackestone on flat-pebble breccia. Slab; Jiliapa #61, 2251 m.

the grain and a complete or partially broken rim of cement protrude into the matrix (Figs. 24A, 26B). Many fragments of benthic fossils in the breccia matrix have a rim of cement (Fig. 26C). These observations indicate that many of the clasts were lightly cemented but friable prior to transport and redeposition.

In many grain-supported clasts the cement is blocky calcite or a fine calcite mosaic which is not obviously different from the cement of the Tamabra skeletal grainstone and packstone. The clasts, however, are not typical Tamabra lithologies. The grains of the clasts are more angular and finer, the light cream-colored matrix is more abundant, and recrystallization is much less extensive.

FIG. 19.—Pelagic-microfossil lime wackestone. *A*—Burrows, vague horizontal fabric, perhaps from compactive flattening. Shiny areas (left and right center) are pyrite formed in burrow stuffed with organic matter. Near top of Tamabra; Poza Rica #215, 2195 m.; slab; photo by E. A. Shinn. *B*—Argillaceous laminated wackestone with micrograded beds of pelletoidal grainstone (light). Suggestion of ripple forms in grainstone. Poza Rica #215, 2200 m.; slab; photo by E. A. Shinn. *C*—Photomicrograph from B. Interbeds of dark pelagic argillite and pelletoidal grainstone. Suggestion of ripple form near center. Plane light. *D*—Photomicrograph from B. Benthic microfossils (Polymorphinids?) and pelletoids in grainstone layer (bottom); recrystallized planktic microfossils (clear circles) in thin dark pelagic layer. Plane light.

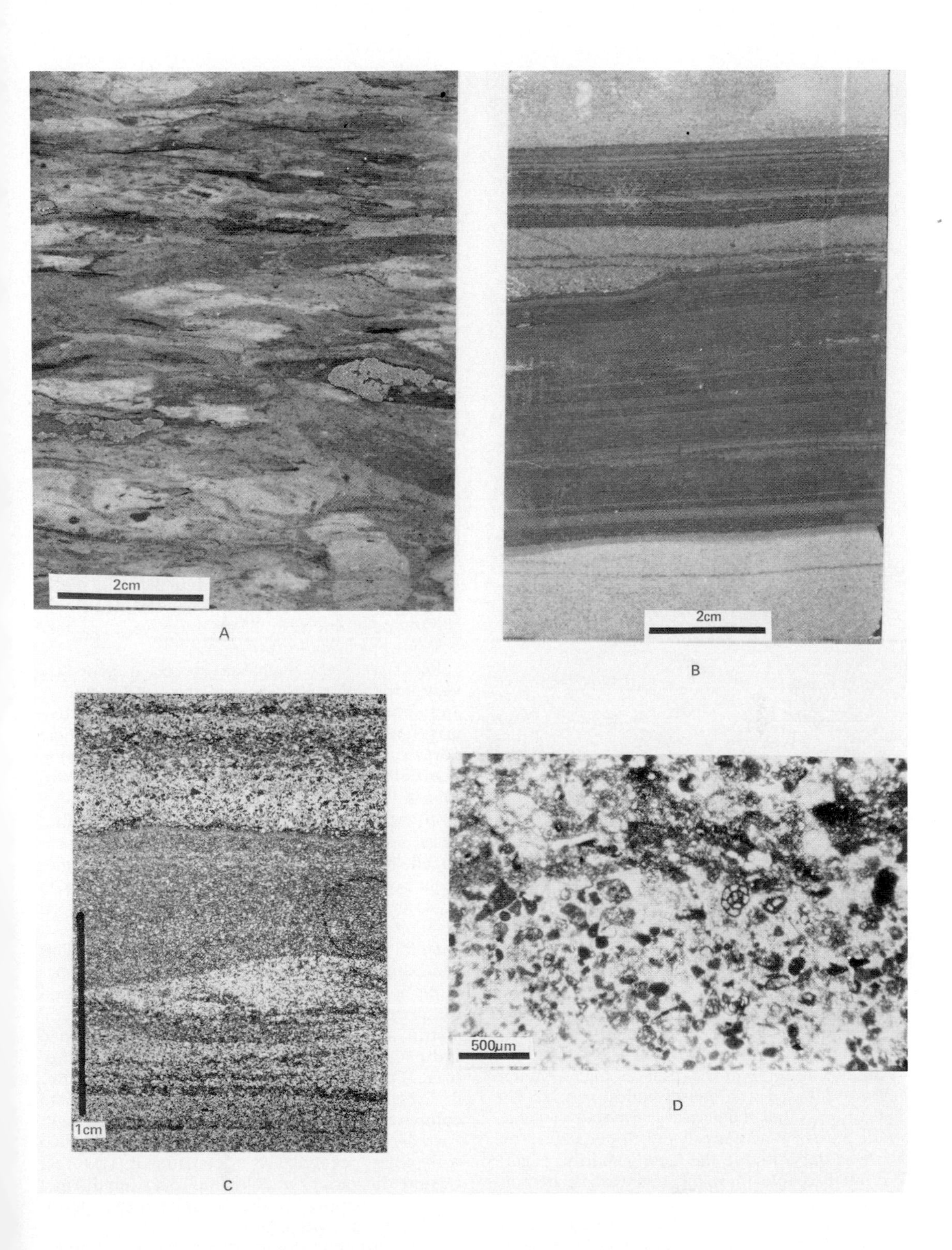
2cm
A
2cm
B
1cm
C
500μm
D

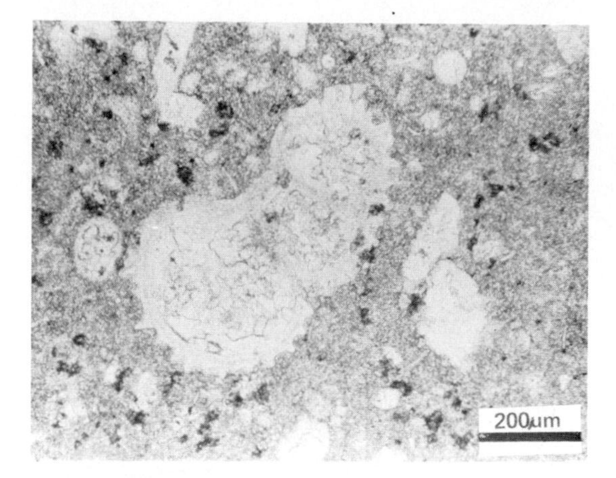

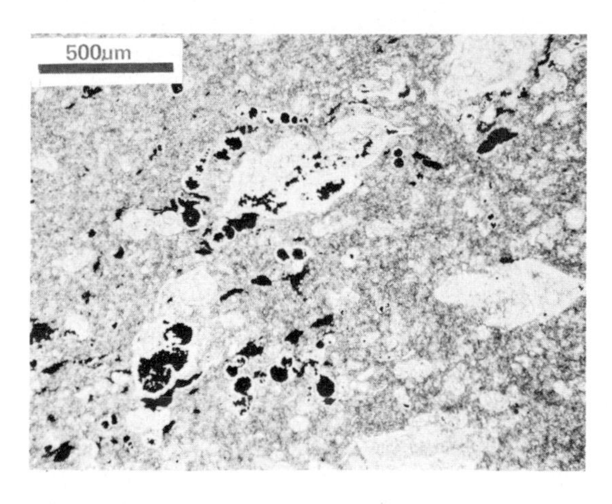

FIG. 20.—Pelagic-microfossil lime wackestone; photomicrographs, plane light. *A*—Globigerinid (*Hedbergella washitensis*?) and fossil fragments in micritic matrix. Poza Rica #217, 2168 m. *B*—Planktic foraminifers, some filled with pyrite spherulites ("framboidal pyrite"). San Felipe or Agua Nueva Formation (Upper Cretaceous) at Tamabra contact, Ojital #1, 2507 m.

Porosity of some of the clasts is high although they are encased in a dense muddy matrix. The implication is that the porosity, generally leached moldic porosity, was largely developed before the clast was deposited in the muddy matrix. Figure 27 is an example in which this can be demonstrated. At the edge of the clast, fossil molds and internal cavities have been infiltrated with the brown mud of the matrix. Toward the center of the clast, where the mud did not penetrate, fossil-mold porosity has been filled by coarse blocky calcite.

Lime mud is cohesive enough to have been transported without cementation. Nearly all of the basinal wackestone clasts, however, whose original shape has not been destroyed by suturing, are angular and undeformed. A few have slightly wavy lamination, suggesting that cementation, if present, was slight. Several very thin, platy clasts also suggest plasticity rather than rigidity (Fig. 23B); cemented clasts of these dimensions would be brittle and probably break during transport. A core-piece of pelagic wackestone with soft-sediment deformation (Fig. 35A) may represent a large uncemented clast. The basinal wackestone clasts differ from typical Tamabra pelagic-microfossil wackestone in that a much higher percentage of the clasts are microlaminated. Microlamination was found to be more prevalent at the basin margin or on the slope than further into the basin, based on an outcrop study of the Tamaulipas Limestone in the Sierra Madre Oriental (W. J. Koch, pers. comm., 1970).

Dolomite clasts, many of which are microlaminated, generally occur with dolomitic matrix. They are probably altered basinal clasts, dolomitized along with the matrix. A few dolomite clasts occur where the matrix and other clasts are only slightly dolomitic (Fig. 23A); these were possibly dolomitized prior to transport.

Porosity of the Tamabra breccia is generally very low (Table 3) because of the muddy matrix and sutured grain boundaries. Many grainstone clasts have extensive moldic and interparticle porosity, but the matrix prevents interconnection between clasts. The producing interval in Poza Rica #217 with 5 to 10 percent porosity and 0.5 to 10 md permeability is a breccia. Vertical fractures, apparently widened by solution, have produced some interconnection between clasts. Fine-scale porosity in the dolomitic matrix and clasts in some breccia is indicated by selective oil staining.

Dolomite.—Dolomitic rocks in the Tamabra Limestone include thin intervals of dense, structureless crystalline dolomite; massive vuggy and fossil-moldic dolomite; breccia with dolomitic matrix and clasts; dolomitic skeletal grainstone and packstone and dolomitic microfossil wackestone. Only the first two were logged as dolomite; in these the original lithology is obscure and dolomite may exceed 75 percent of the rock. Partially dolomitized rocks have been described with their recognizable depositional textures. Despite the variety of dolomitic rocks, no distinct dolomite families could be recognized petrographically or isotopically.

The crystalline dolomite is a mosaic of inter-

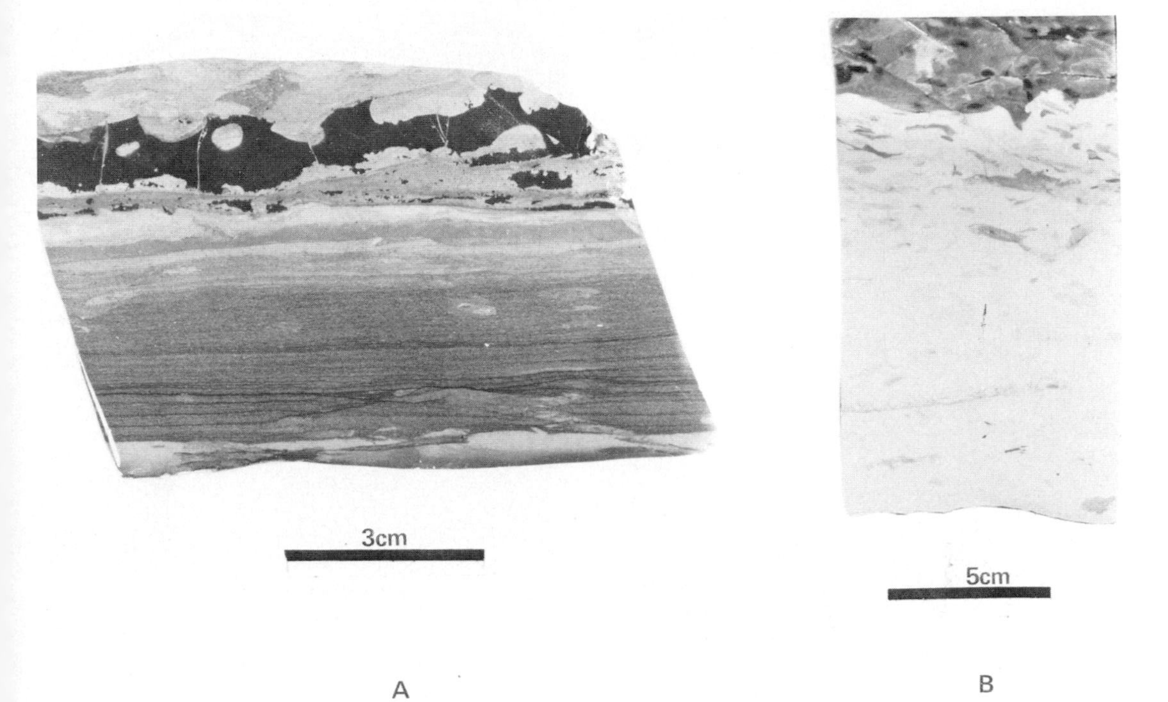

FIG. 21.—Silicification textures. *A*—Irregular replacement textures at edge of chert. Black area is translucent chert with ghosts of microfossils. Patchy silicification extends several centimeters into wackestone. Slab; Zapotalillo #9, 2258 m. *B*—Chert replacement textures (top). Note burrows, especially well-defined one with lateral migration structures. Identical structures appear in Cretaceous cores from the Deep Sea Drilling Project in the Caribbean and western Atlantic. They resemble *Taonurus caudagalli* from the Middle Devonian of New York state and Gaspé, Quebec. Slab; Zapotalillo #9, 2235 m.

locking white or light-grey dolomite crystals with some matrix calcite (Fig. 28A). Recognizable relicts of original grains or sedimentary structures are lacking. Contacts between crystalline dolomite and limestone have not been observed because such dolomite is generally confined to a few isolated core pieces. These occur in generally dolomitic sequences, however, suggesting gradational contacts. Porosity is nil except in a few samples with hairline fractures.

Massive, grey, crystalline, nearly pure dolomite is found near the bottom of Poza Rica #63 and to a lesser extent in Poza Rica #82 (Fig. 9). Some dolomitized fossils and large pores whose shapes suggest fossil molds indicate the original rock contained coarse skeletal fragments (Fig. 28B). These rocks grade into skeletal-fragment lime packstone and grainstone. Porosity is good throughout most of this interval. In addition to the "relict" fossil molds, this massive dolomite contains vugs, practically the only ones observed in the Tamabra Limestone.

ORIGIN OF THE TAMABRA LIMESTONE

The Tamabra Limestone is a wedge of skeletal limestone and breccia containing abundant shallow-water organisms, particularly rudists, interlayered with and enclosed on top, bottom, and the west side by fine-grained limestone with pelagic microfossils (Fig. 4). Eastward it abuts the prominent subsurface escarpment of the Golden Lane (Fig. 29). The several hypotheses advanced to explain the origin of these rocks may be summarized as follows:

1. The Tamabra Limestone "is composed of shallow-water coral-rudistid reefs, debris derived from the reefs and deposited in shoal-water nearby, and forereef talus mixed with basinal muds." El Abra Limestone "was deposited in a shallow-water shelf or lagoon with scattered rudist patch reefs" (Coogan and others, 1972, p. 1419). The prominent Golden Lane escarpment is interpreted as a structural feature, probably a fault (Coogan and others, 1972, p. 1444).

2. The Poza Rica field "represents a zone

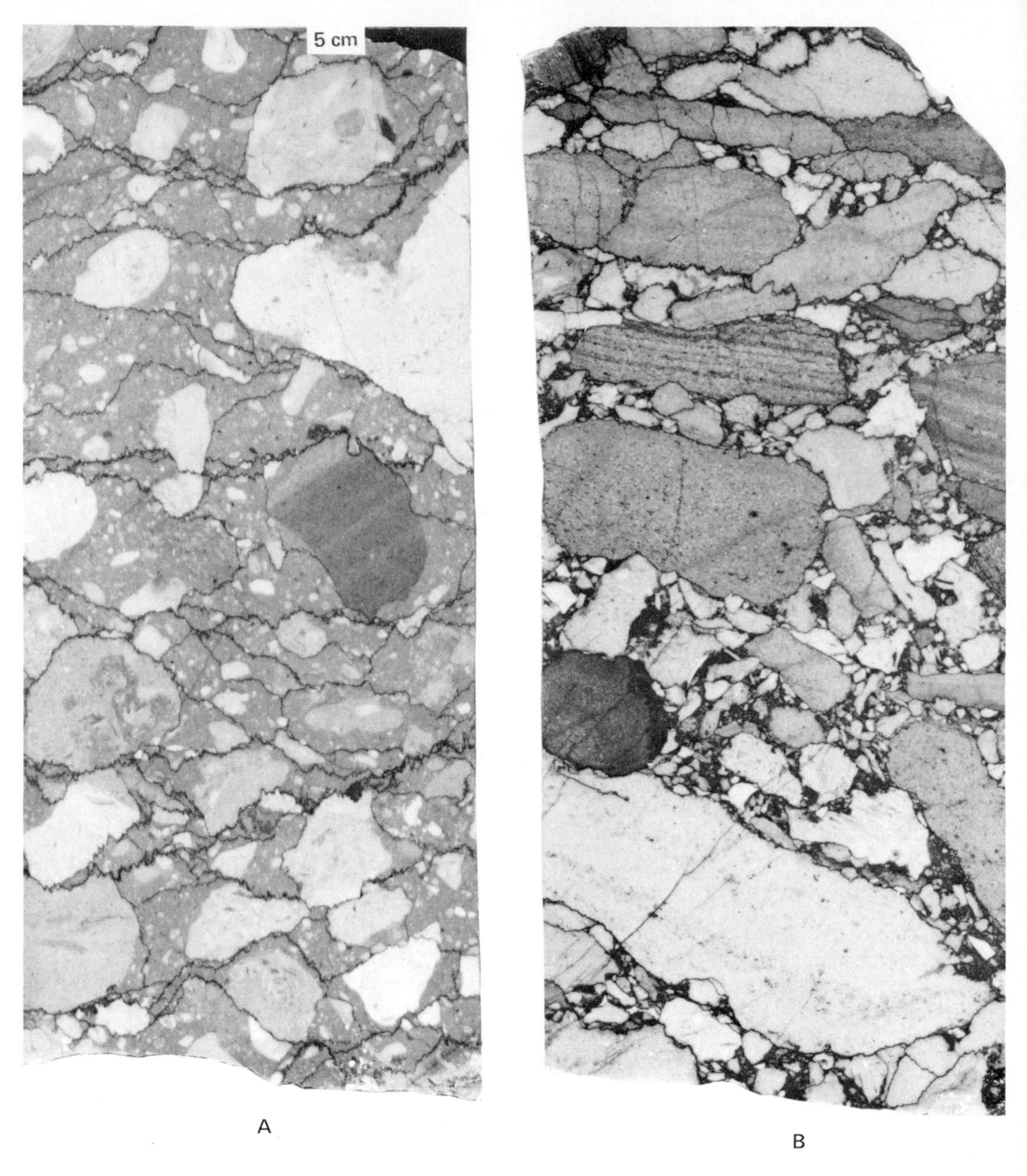

FIG. 22.—Breccia in Tamabra Limestone. *A*—Breccia of rudist-fragment lime grainstone and microfossil lime wackestone in micritic matrix with pelagic microfossils. Numerous stylolites. Poza Rica #217, 2173 m. *B*—Breccia with sparse matrix and sutured clasts of microfossil lime wackestone (laminated) and rudist-fragment lime grainstone. Note inclined fabric of clasts (imbrication?). Same scale as A. Poza Rica #217, 2154 m.

FIG. 23.—Disorientation of fragments in Tamabra breccia. *A*—Oil impregnated, dolomitic breccia; clasts of pelagic-microfossil wackestone, rudist-fragment grainstone, and dolomite (arrow). Note vertical clast, approximately perpendicular to bedding. Poza Rica #217, 2125 m. *B*—Dolomitic breccia, large laminated clast nearly vertical. Escolín #186, 2715 m. *C*—Caprinid oriented as though in growth position; probably in a clast. Light-colored particles inside caprinid are pelagic-microfossil wackestone clasts, thus its environment of growth is not represented even by the clast which contains it. Zapotalillo #9, 2239 m.

5cm
A
B

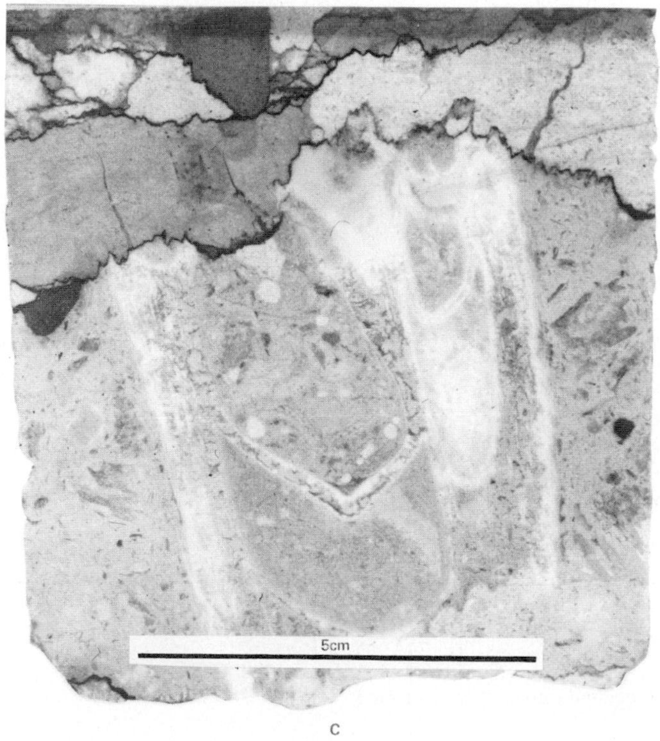
5cm
C

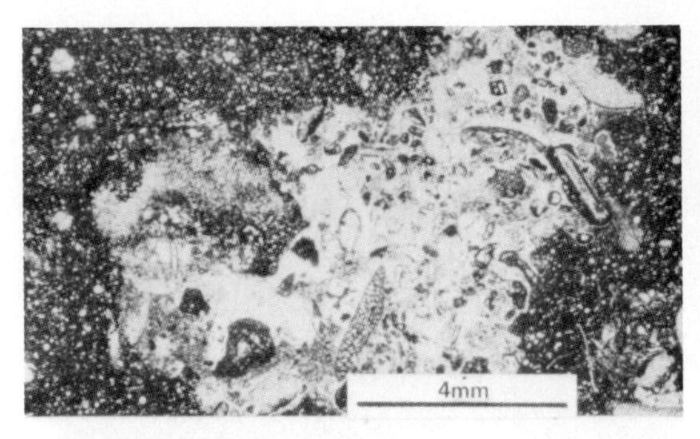

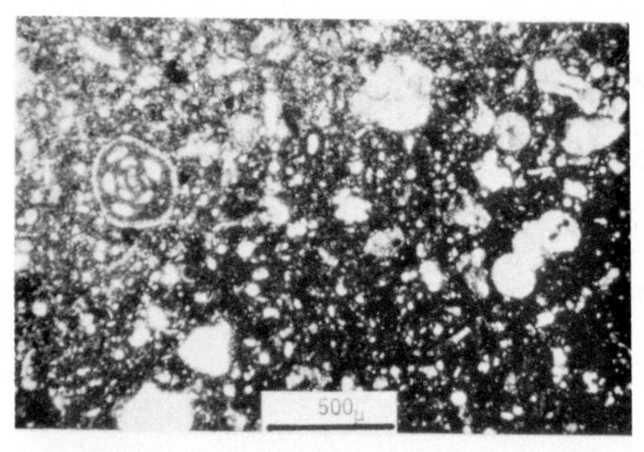

FIG. 24.—Matrix of breccia; photomicrographs in plane light. *A*—Small grainstone clast in pelagic-microfossil matrix. Poza Rica #217, 2168 m. *B*—Matrix of breccia showing globigerinid foraminifera and a miliolid, a unique occurrence in the Tamabra. Poza Rica #217, 2174 m.

FIG. 25.—Dolomitic breccia. Note patchy oil staining. Slab; Plan de Ayala #1, 2264 m.

where elongate rudist beds or banks of rudist shells developed on a shallow, current-swept sea bottom. The whole area was part of a shallow marine shelf . . . formed of bio-clastic lime-sands and lime-silts, lying in front of the Golden Lane reefs'' (Barnetche and Illing, 1956, p. 25). Hypothesis 2 to this point does not differ essentially from 1, although slightly different emphasis is placed on reefs in the Golden Lane. It may be elaborated[8] by specifying that El Abra Limestone developed during high stands of sea level and constructed the Golden Lane escarpment. The Tamabra Limestone developed as a subsidiary shelf margin during less frequent low stands of sea level. High stands of sea level are represented in the Tamabra by pelagic wackestones.

3. The Tamabra Limestone is a deep-water deposit consisting primarily of shallow-water debris derived from the Golden Lane escarpment by turbidity currents (T. F. Grimsdale, unpub-

[8] Barnetche and Illing were primarily concerned with the Poza Rica field; they did not deduce the regional consequences of their hypothesis. The scenario presented here is one possible end member. The other would coincide with hypothesis 1.

FIG. 26.—Cementation of clasts in Tamabra breccia; photomicrographs in plane light. *A*—Grain with early cement rim, broken at edge of clast; same clast as Fig. 24A. Poza Rica #217, 2168 m. *B*—Lightly cemented clast; grains fringed by early cement protrude into matrix. Poza Rica #217, 2173 m. *C*—Benthic foraminifer (*Dictyoconus*) in matrix. Fringe of cement shows it was eroded from lightly cemented rock and redeposited. Poza Rica #217, 2168 m.

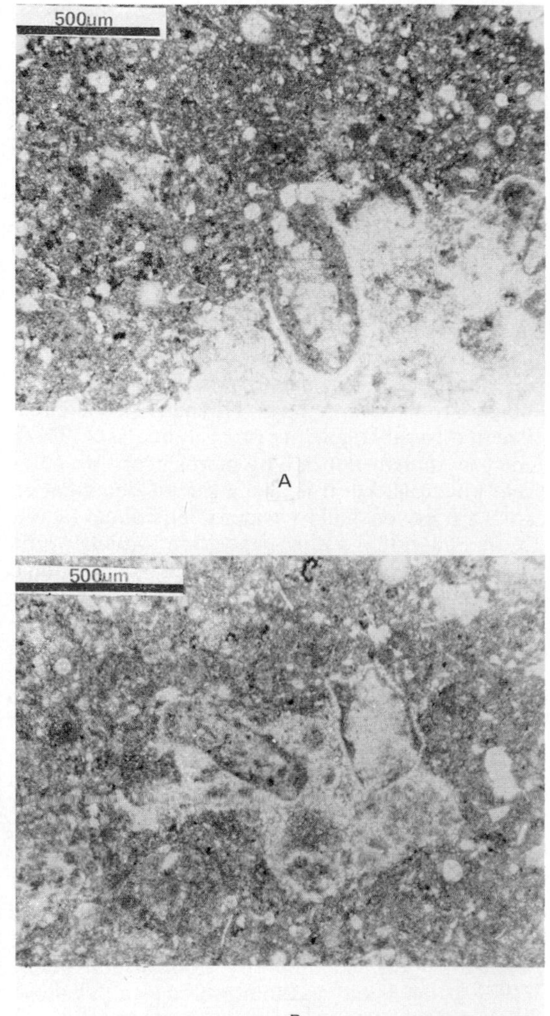

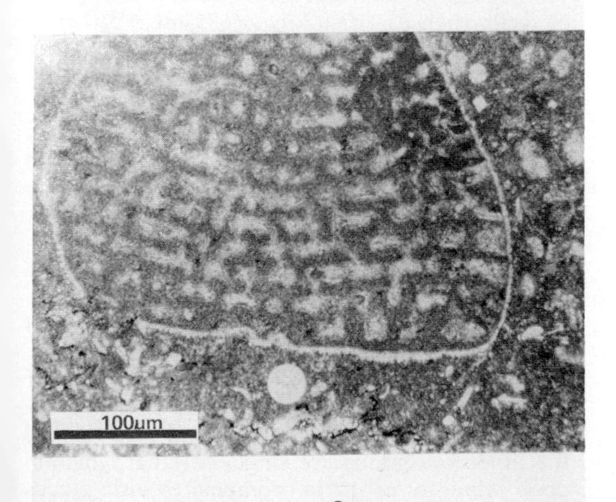

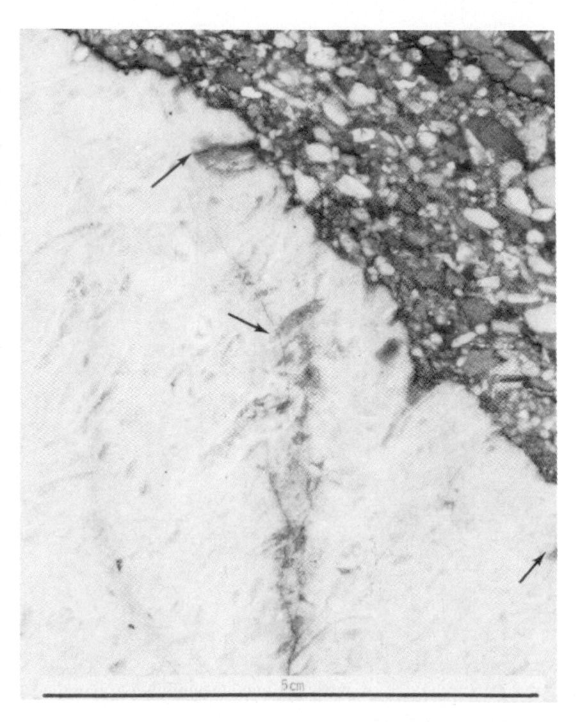

FIG. 27.—Edge of grainstone clast in breccia showing matrix infiltrating molds of fossils (arrows). Elsewhere molds are filled with coarse blocky calcite. Slab; Zapotalillo #9, 2198 m.

lished report, 1953), sliding, and/or slumping.

The hypotheses involve two elements (Fig. 30): the depositional environment of the Tamabra Limestone and the origin of the Golden Lane escarpment. Each element will be considered separately.

Origin of the Golden Lane Escarpment

Published regional seismic lines (Fig. 31, Rockwell and Garcia Rojas, 1953; Islas and Equia, 1961), as well as two lines employing newer seismic techniques, show no offset or folding of pre-middle Cretaceous reflectors at the Golden Lane escarpment. A fault flattening basinward into a bedding plane might be suggested, but this should produce thickening of the middle Cretaceous interval where faulting has repeated the sequence and irregularities in the top of the middle Cretaceous at the adjacent Golden Lane escarpment.

The tremendous increase in thickness in the middle Cretaceous interval at the Golden Lane escarpment (Figs. 3, 4, 31) is a second strong line of evidence for a constructional origin. Although these figures rely heavily on older seismic data, substantiating well control is available in

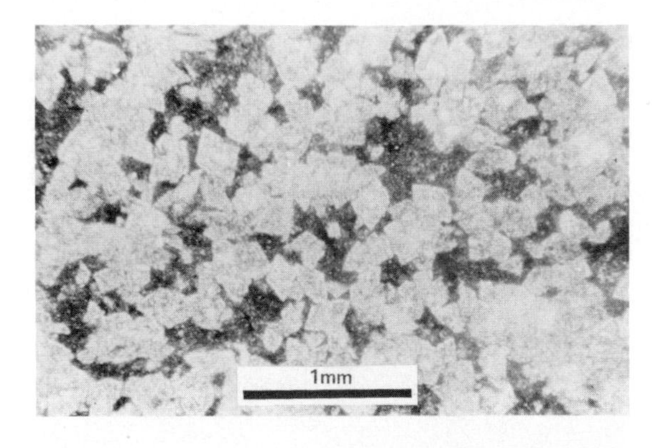

A

B

FIG. 28.—Dolomite. *A*—Zoned dolomite rhombs replacing skeletal wackestone (?). Photomicrograph in plane light; Ojital #1, 2538 m. *B*—"Relict" fossil molds and vugs in dolomitized skeletal fragment grainstone. Residual oil in leached pores. Slab; Poza Rica #63.

the Poza Rica trend (Fig. 7), and the few wells which have penetrated the enitre El Abra sequence in the Golden Lane confirm the essentials of the seismic interpretations (e.g. Arroyo Grande #1, 1323 m of El Abra, Lopez Ramos, 1950; Las Cañas #101, 1174 m; and in the platform interior; Muro #2, 1350 m; Cazones #2, 1425 m; Tuxpan #2, 1395 m; all from Viniegra and Castillo-Tejero, 1970).

The Golden Lane contains intervals of algal stromatolites, birds-eye structures, algal pisolites, miliolid wackestone and grainstone, and oolite. These are indicative of supratidal, restricted lagoonal, and open-shelf environments (cf. Bathurst, 1971). The Golden Lane also contains packstone intervals with abundant caprinids, radiolitids, coral heads, encrusting red algae, and possibly bryozoans and stromatoporids (E. Ordóñez #5, 1400–1430 m, Mesita #1, 1693–1795 m, 2355–2360 m, Las Cañas #101, 1380–1493 m, 1952–1958 m, 2237–2243 m). This assemblage is suggestive of outer-reef conditions. Other coral-rich intervals in the Golden Lane cores contain delicate branching corals in a muddy matrix (Mesita #1, 1480 m). These are interpreted as a patch-reef coral thicket or the coral fringe of a mud shoal, which has many analogs in the Holocene shelf margin of south Florida (Griffith and others, 1969; Enos, in press).

In the Tamabra Limestone basinward of the Golden Lane escarpment, none of these assemblages are represented by more than a few broken, isolated fossil fragments or by lithoclasts. Thus, the escarpment not only contains probable outer reef intervals, but it is also a line of demarcation for a variety of shallow-water depositional facies. It was an entity during deposition, which would not be the case if it were a postdepositional structure.

These points make a structural origin (hypothesis 1) for the Golden Lane escarpment untenable.

Depositional Environment of the Tamabra Limestone

Shallow-water arguments.—Evidence cited in favor of shallow-water reef or bank origin for the Tamabra Limestone follows. These points will be discussed individually before marshalling arguments which favor deep-water origin.

1. The "caprinid-micritic facies" (rudistid wackestone of this report) occurred in 4 of 10 Tamabra wells examined by Coogan and others (1972, p. 1433) and was interpreted as a "shallow-shelf rudistid reef deposit."

2. The cavernous and dolomitic intervals found in wells such as Poza Rica #63 and #82 (Figs. 9, 28) are considered to represent an axial reef zone in the Poza Rica field (Barnetche and Illing, 1956, p. 15).

3. "The delicate ornamentation still seen on many of the rudistid and other fossils, despite recrystallization, could not have withstood long transportation" (Barnetche and Illing, 1956, p. 27).

4. Miliolids are lacking in the Tamabra Limestone. "If . . . derivation from the Golden Lane was the chief source of the sediment, we should expect to find miliolids in it,

5. "as well as lime-sand grains formed of reef rock and previously cemented bio-clastic limestone. . . . Thin section studies indicate that all the lime-sand grains are single fossil fragments: none were found to be composite with pieces of several fossils cemented together as might be

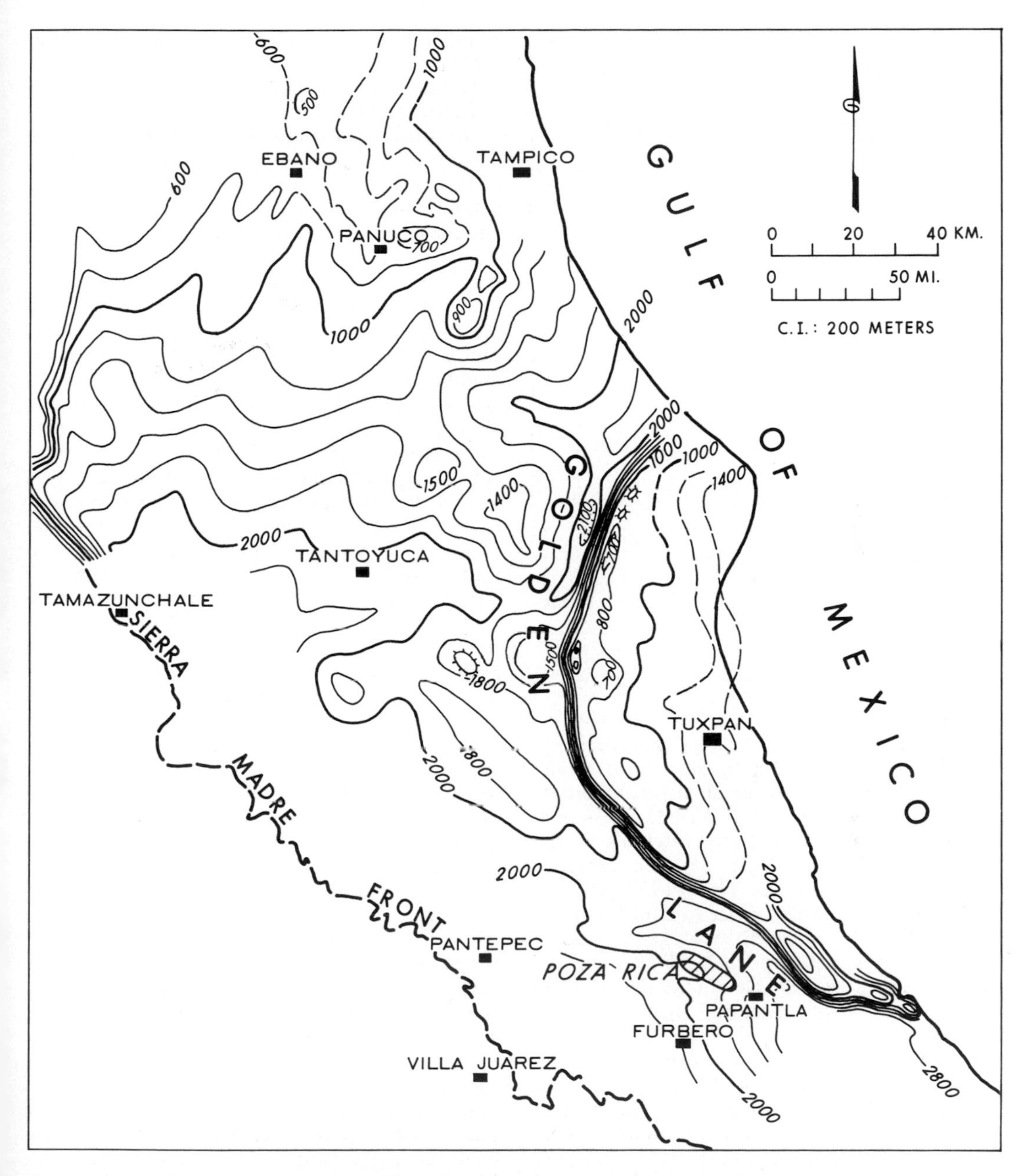

FIG. 29.—Golden Lane escarpment shown by regional structural contours on top of middle Cretaceous formations. Depths are in meters below sea level. Contour interval 200 m; reduced to 100 m where indicated by hachures or odd hundreds. From Barnetche and Illing (1956).

expected on the derived hypotheses'' (Barnetche and Illing, 1956, p. 27).

6. The abundance of leached porosity in the Tamabra Limestone suggests contact with volumes of undersaturated water (Salas, 1949, p. 140) which would be favored in the vadose or shallow phreatic environments. Subaerial exposure would in turn be more likely for shallow-water than for basinal sediment.

Discussion.—The ''caprinid-micritic facies'' or rudistid wackestone (point 1) contain abundant caprinids and radiolitids, the rudists of platform-

		ORIGIN OF GOLDEN LANE ESCARPMENT	
		STRUCTURAL	CONSTRUCTIONAL
TAMABRA DEPOSITIONAL ENVIRONMENT	SHALLOW WATER	1	2
	DEEP WATER		3

FIG. 30.—Hypotheses of origin of Tamabra Limestone.

margin reefs. Some intervals have nothing to suggest that they are not shallow-water reefs. In general, however, they lack the associated corals, coralline algae and occasional stromatoporoids (the exception is San Andrés #143D, 2980–3000 m) which characterize the reef intervals of the Golden Lane (Boyd, 1963, p. 55), the Sierra de El Abra (Griffith and others, 1969, p. 127; Aguayo, 1975), northern Coahuila (lower Albian, Bloxsom, 1972, p. 16) and the Glen Rose of Texas (middle Albian, Perkins, 1974, p. 142). These are the accessory organisms capable of binding reefs into solid wave-resistant structures (Kauffman and Sohl, 1974, p. 401).

It has been shown that the micritic matrix can be a pelagic sediment (Fig. 17) and that some rudistids in "growth position" were redeposited (Fig. 16). Some rudist wackestone contains ovoid or "half-moon" clasts of graded muddy sediment (Fig. 32). These are interpreted as filled central cavities of rudistids (Caprotinids?) which were cemented and then isolated by complete leaching of the shell. They are now packed together in random orientation without any preserved molds, showing that they were leached and then redeposited. Thus the rudist wackestone was in part, at least, detrital sediment.

Clasts of probable basinal origin occur within the "caprinid-micritic facies" in Soledad #124 at 1950 m. Breccia containing some basinal clasts occurs adjacent to the "caprinid-micritic facies" in other wells, including San Andrés #143D at 2980m.

The cavernous and dolomitic intervals from the Poza Rica field (point 2) are probably altered from skeletal-fragment grainstone and packstone which surrounds them (Figs. 9, 28). There is no evidence of *in situ* reef growth unless dolomitization itself is considered indicative of reefs. Dolomitization and development of cavernous porosity are considered primarily diagenetic problems and will be discussed in a later paper (in preparation).

The preservation of delicate ornamentation of fossils (point 3) is rare in the Tamabra Limestone, except within clasts in the breccia and in some muddier rocks. Most of the particles in Tamabra grainstone and packstone are fragmented and abraded (Figs. 11–13). As indicated above, some wackestone is considered detrital, so the significance of preservation may lie in indicating a mode of transport rather than a lack of transport. Some authors have maintained that in detrital deposits, delicate preservation characterizes turbidity-current deposition because suspension transport in a muddy current would be much more gentle than traction transport. Perservation of delicate structures is possible even in coarse breccia deposited at the basin margin (Fig. 33). Fragile architecture was preserved in radiolitid rudists which apparently were reworked into an environment with pelagic microfossils (Fig. 17A). Clearly delicate preservation is ambiguous evidence; here the matter must rest.

The excellent rounding of originally angular fossil fragments in typical Tamabra grainstone is good evidence for extensive reworking, probably by wave action in a shallow marine environment, whether or not that was the environment of final deposition. The experiments of Kuenen (1956) indicate that current transport of a few miles, without constant reworking by waves, would not be adequate for appreciable rounding, even of carbonate grains.

The almost total lack of miliolids in the Tamabra Limestone (point 4) is in addition to the lack of other benthic foraminifera, stromatolites, and other characteristic features of the immediate backreef in the Golden Lane and the Sierra de El Abra. This is a red herring not adequately explained by any of the hypotheses. If the Tamabra Limestone is the shelf margin (hypothesis 1), miliolid grainstone and wackestone would be expected somewhere in the 10 to 15 km wide Tamabra belt; there is no reason for the facies to end abruptly at the Golden Lane escarpment. If the Poza Rica trend and El Abra of the Golden Lane are similar shelf-edge deposits active at different positions of sea level (hypothesis 2), the facies distributions would be expected to be similar. If the Tamabra Limestone is debris

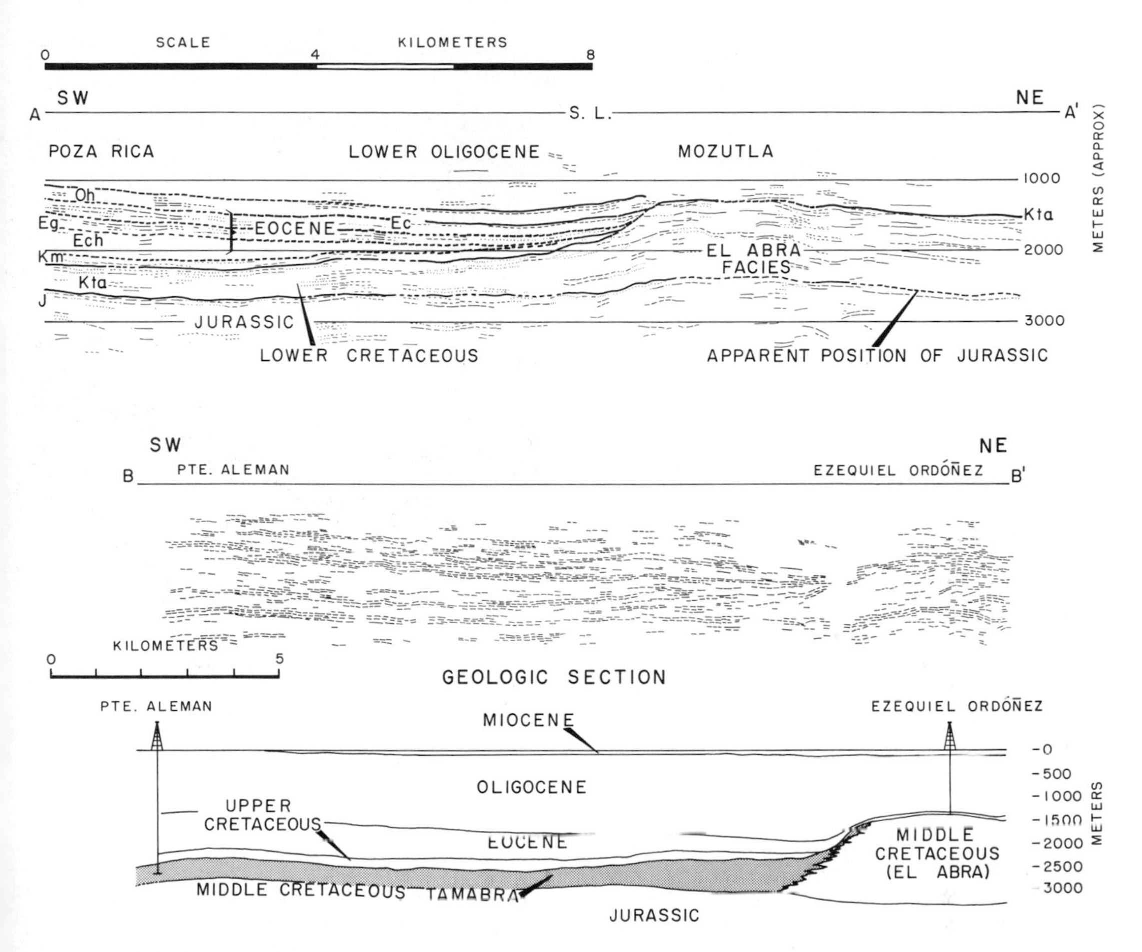

FIG. 31.—Regional seismic cross sections of Golden Lane escarpment and Tamabra interval. Locations can be determined from field names in Figure 5. At bottom is geologic cross section of B-B′ migrated at the Golden Lane escarpment and corrected for "velocity pull-up" under the Golden Lane. A-A′ from Rockwell and Garcia (1953, p. 2559); B-B′ from Islas and Equia (1961, p. 31). Since the appearance of these publications, Golden Lane wells Arroyo Grande #1, Mesita #1, Cazones #2, and Las Canas #101 have penetrated through middle Cretaceous rocks to verify essentials to the geologic interpretation of the seismic events. Note that Lower Cretaceous is omitted.

derived from the Golden Lane (hypothesis 3), all the facies present in El Abra at the crest of the Golden Lane should be represented in the Tamabra.

It is possible, however, to construct a rational, if somewhat contrived, model for the exclusion of miliolids under hypothesis 3. The miliolids are a backreef facies found behind the shelf edge; only the shelf-margin facies are represented by debris in the Tamabra (A. H. Bercerra, pers. comm., September, 1969). Virtual absence of lagoonal clasts from bank-edge debris redeposited basinward of the Miette bank (Devonian, Canada) has been reported by Cook and others, (1972; H. E. Cook, pers. comm., 1976). Debris would be shed most rapidly from those portions of the shelf edge which were accreting or actually over-steepening the most rapidly. These would be the actively growing reefs and sand shoals of reef-derived material, not the semirestricted backreef deposits.

"Previously cemented bioclastic limestone" and "reef-rock" particles (point 5) are indeed present in abundance in the Tamabra limestone, primarily in the breccia but also in lithoclastic, skeletal-fragment packstone and grainstone. They were probably not observed by Barnetche and Illing (1956) because of the great abundance of skeletal-fragment packstone and grainstone and correpsonding reduced importance of breccia in

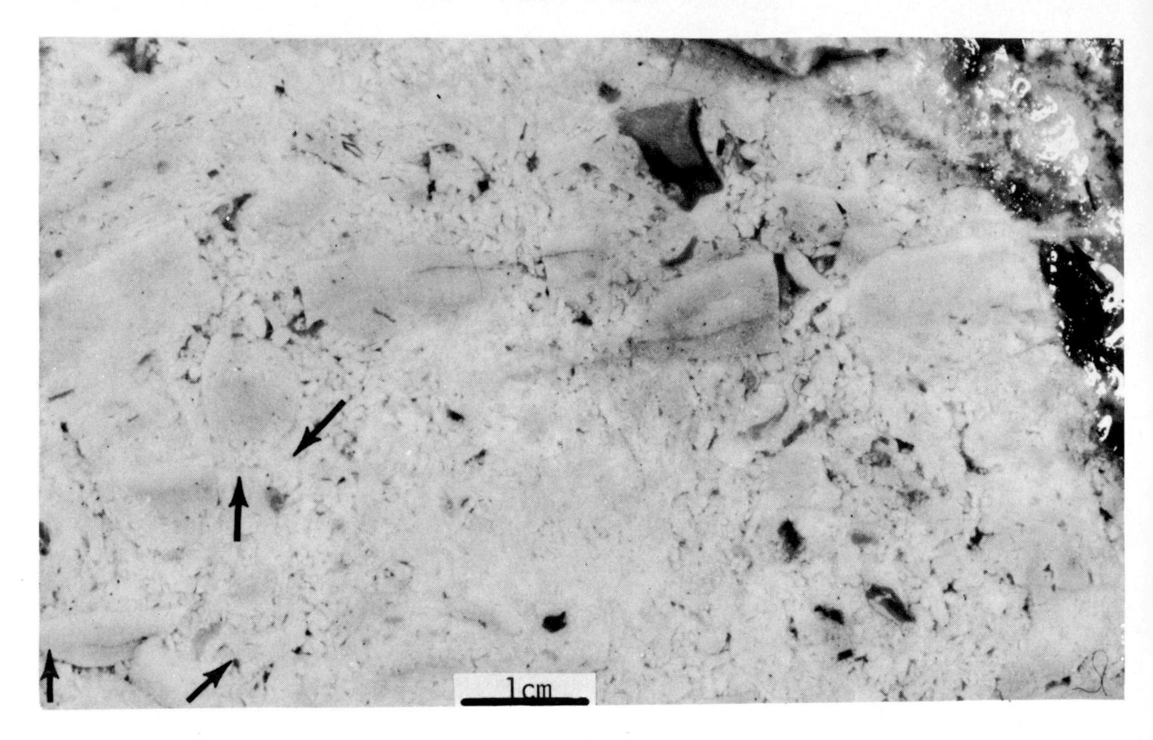

FIG. 32.—Graded beds in clasts within rudist lime packstone/wackestone; arrows show direction of fining. Ovoid and half-moon shapes of clasts indicate that they are geopetal fills of fossil cavities that have been eroded from the skeleton (mold?) which formerly enclosed them. Slab; Plan de Ayala #1, 2283 m.

much of the Poza Rica field.

The abundance of leached porosity (point 6), which reflects a particular diagenetic environment, is a tenuous argument for a particular depositional environment. Large-scale leaching suggests subaerial exposure because the vadose and uppermost phreatic zones are the realms of largely unsaturated waters and of maximum volume of water movement. These are also the environments where leaching is well known because of cave development, a spectacular phenomenon produced by practically observable processes. Other possible environments for leaching, such as deeper phreatic (cf. Back, 1963), hydrothermal, or submarine (cf. Friedman, 1965), are poorly known. A tendency to discount these environments may reflect our ignorance of them.

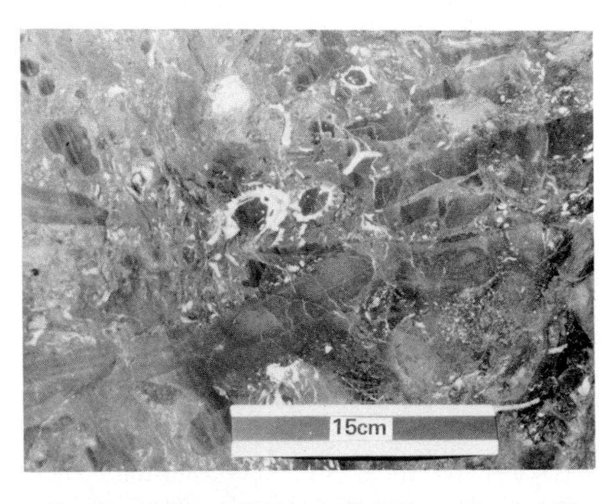

FIG. 33.—Recrystallized radiolitid rudist shells, showing good preservation of shell ornamentation, in coarse basin-margin breccia. Middle Cretaceous outcrop near El Doctor, Querétaro, Mexico.

If it could be documented that leached porosity in the Tamabra Limestone results from subaerial exposure during or soon after deposition, hypotheses 1 and 2 would be strengthened. By hypothesis 2, sea level fluctuations sufficient to expose the basin during the middle Cretaceous would be demonstrated. An alternative hypothesis of massive leaching by deep circulation of meteoric waters through submarine springs, based on present-day limestone hydrology in Florida (Back, 1963; JOIDES, 1965) appears equally viable (Enos, in preparation) and is compatible with hypothesis 3.

A number of positive areas, both islands and platforms, have been delineated in the Upper Jurassic of the Tampico embayment (González, 1969; Stabler and Mejia, 1975). These positive areas are located beneath Tamabra fields, including the southern part of the Poza Rica fields. The possibility arises that development of the Tamabra facies may depend on antecedent banks; this would, in turn, suggest shallow-water deposition for the Tamabra. Carbonate accumulation should be enhanced on a positive feature only in shallow water where organic productivity would be favored. On closer examination, the case becomes less than compelling for four reasons. (1) Subsurface data on the Jurassic is virtually confined to deeper wells through Cretaceous reservoirs, especially thinner Cretaceous intervals in basinal rocks; hence, much of the apparent coincidence is merely the availability of data. (2) The available data show that some Tamabra fields are not coincident with highs; many highs probably also exist outside Cretaceous fields. More data will become available as Jurassic exploration proceeds. (3) The Poza Rica and Golden Lane trends are apparently unrelated to trends of Jurassic shelf edges (González, 1969, Figs. 2, 6; Stabler and Mejia, 1975). (4) Finally, the Lower Tamaulipas Limestone blanketed the entire Tamabra trend, so basinal conditions were established prior to the middle Cretaceous. Thus inheritance of relief would be indirect.

Deep-water argument.—A deep-water or basinal environment of deposition of the Tamabra Limestone is supported by the evidence which follows.

1. Thin interbeds of pelagic wackestone recur at many levels within the Tamabra Limestone.
2. Abundant breccias within the Tamabra Limestone have matrices of pelagic mud and contain clasts of basinal wackestone mixed with clasts of shallow-shelf, skeletal-fragment limestone.
3. The gross geometry of the Tamabra Limestone is more consistent with a wedge of forereef debris than with a shallow shelf margin, especially in the lack of expression of a possible shelf edge.
4. The internal geometry of the Tamabra, although complex, is simpler and more laterally homogeneous than in a typical shelf-margin accumulation. A general trend to finer grain size within the Tamabra from the toe of the Golden Lane escarpment toward the outer edge is the reverse of trends to be expected across a shallow shelf margin (cf. Enos, 1974a).
5. Sedimentary structures consistent with, although not diagnostic of, redeposition of shelf skeletal fragments by sediment gravity flow occur in the Tamabra.

Discussion.—The occurrence of pelagic wackestone beds and the abundance of breccia with pelagic elements are clear evidence that much of the Tamabra Limestone is a basinal deposit. The repeated pelagic intervals require that (a) the interbedded limestone with fragments of shallow-water skeletons were also deposited in deeper water against a background of normal pelagic deposition (hypothesis 3) or (b) sea level fluctuated repeatedly (hypothesis 2).

Near the Golden Lane escarpment, the Tamabra belt is a wedge or prism of fairly uniform thickness (Fig. 7) which tapers toward the outer edge as the Tamabra-Tamaulipas contact climbs in the section to the southwest. If the Tamabra Limestone represents an undeformed shelf margin (hypotheses 1 and 2), some expression of the shelf edge would be expected in isopachous and/or structural contour maps. This is not the case (Figs. 3, 7, 29). The southwestward migration of the Tamabra-Tamaulipas transition with time (Barnetche and Illing, 1956, p. 26) suggests that the transition was of low relief.

The carbonate shelf margin is typically a zone of rapid and complex but, in a general sense, systematic variation of facies. In the Tamabra trend, however, about the same lithofacies are present near the inner edge at the toe of the Golden Lane escarpment as at the outer edge (Fig. 9). The most obvious trend is toward finer grain size at the outer edge of the Tamabra trend. This is contrary to expectation if the trend is a shelf margin with material derived from shelf-edge reefs, but is consistent with derivation of material from the Golden Lane escarpment.

No single sedimentary structure is uniquely diagnostic of deposition by sediment gravity flow (turbidity current, debris flow, grain flow; Middleton and Hampton, 1973) but familiar suites of structures characterize deposits which can be inferred to result from these mechanisms. Many of these structures are sole marks which cannot be seen to advantage in cores, especially in limestone without shale breaks. Perhaps the most characteristic structure is graded beds which are rare but striking in the Tamabra Limestone (Figs. 13B, 34). Other structures that indicate deposition in a current-dominated environment are coarse/fine layering (Fig. 12), imbricated clasts (Fig. 11B), and alignment of elongate fragments (Fig. 13A).

Finer grained rocks show many sedimentary structures to better advantage (Fig. 19B–D). These examples are from thin detrital limestone, with fragments of platform fossils, interbedded with pelagic limestone. Soft-sediment deformation indicating bottom instability is also present, although rare, in wackestone intervals (Fig. 35).

The common orientation of clasts in Tamabra breccia with the long axis at a high angle to bedding (Fig. 23) is noteworthy. A parallel occurrence is the upright orientation of rudist fragments as the result of redeposition (Fig. 16B). Vertical orientation of clasts could not consistently result from ordinary current deposition nor from other means

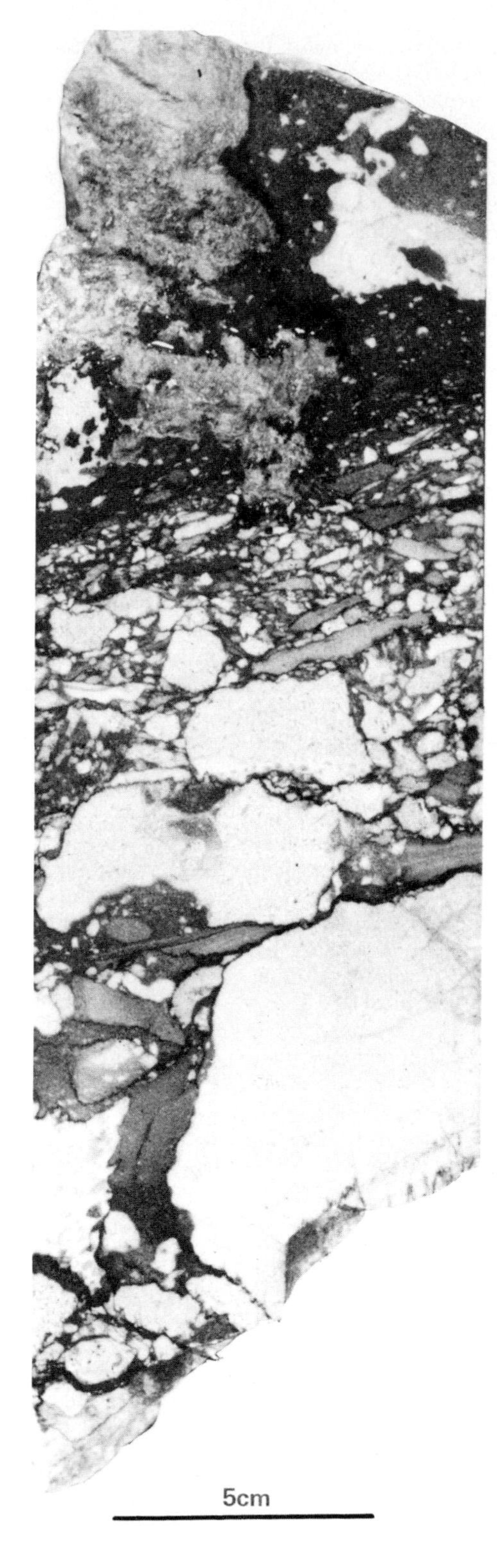

A

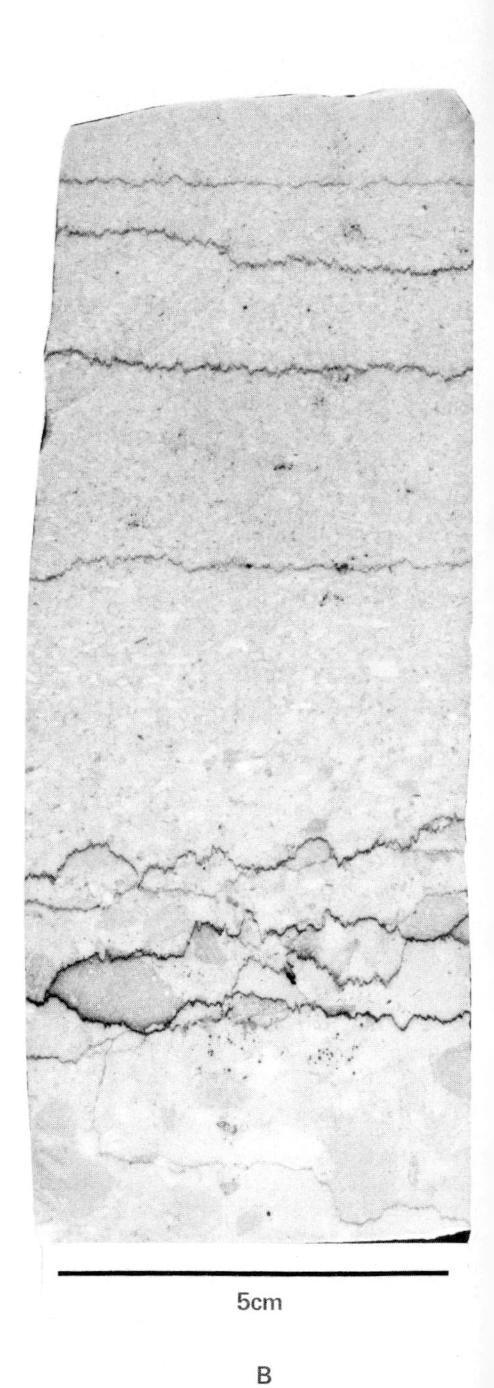

B

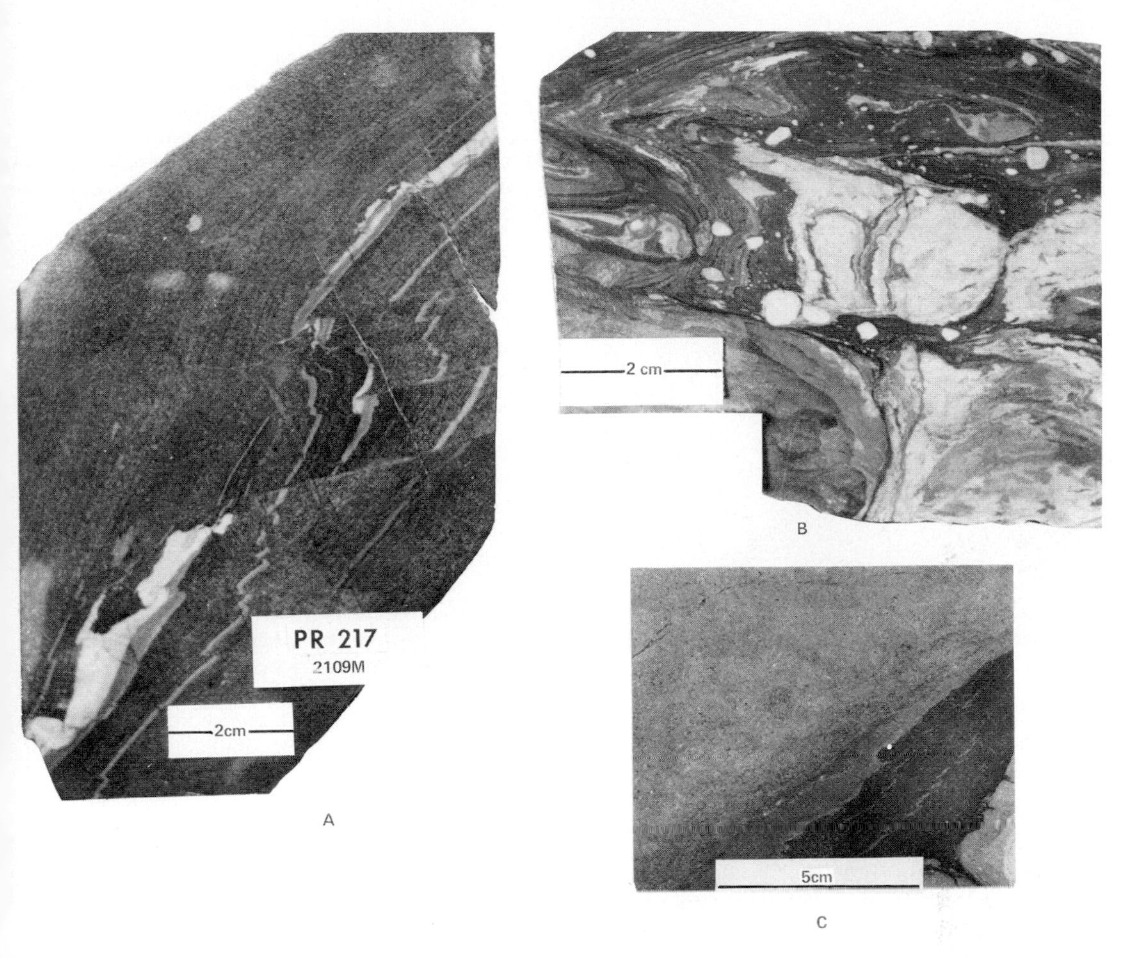

FIG. 35.—Soft-sediment deformation. *A*—Flowage and incipient fractures. Poza Rica #217, 2109 m.; photo by E. A. Shinn. *B*—Very fluid deformation, possibly involving soft clasts. Poza Rica #215, 2196 m.; photo by E. A. Shinn. *C*—Flowage and incipient fractures. Zapotalillo #9, 2300 m.

of particle-by-particle deposition such as rock fall. This orientation is interpreted to reflect deposition from a very viscous or a plastic medium, such as a debris flow. Breccia in which the matrix is slightly argillaceous shows distortion consistent with this mechanism (Fig. 36). The pervasive muddy matrix of many breccias is also consistent with this type of transport. Mud would be winnowed by shallow traction currents capable of transporting large clasts.

Discussion of Hypotheses of Origin

Hypothesis 1 is considered untenable because it requires that the Golden Lane escarpment be structural. The sedimentological arguments presented against shallow-water deposition, of the breccia in particular, contradict both hypotheses 1 and 2. A major objection to hypothesis 2 is the magnitude and number of sea-level fluctuations required to account for the interbeds of pelagic limestone and breccia. The magnitude would not be great early in the middle Cretaceous since of the relief of the Golden Lane escarpment was constructed during the middle Cretaceous (Figs. 3, 31). At the end of the middle Cretaceous, however, it would be in excess of 1000 m. The number of sea-level fluctuations required cannot be precisely stated since it was argued that Illing's mechanical-log marker horizons do not represent persistent beds of pelagic limestone. The maxi-

FIG. 34.—Graded beds of Tamabra breccia. *A*—Zapotalillo #9, 2342 m. *B*—Plan de Ayala #1, 2229 m.

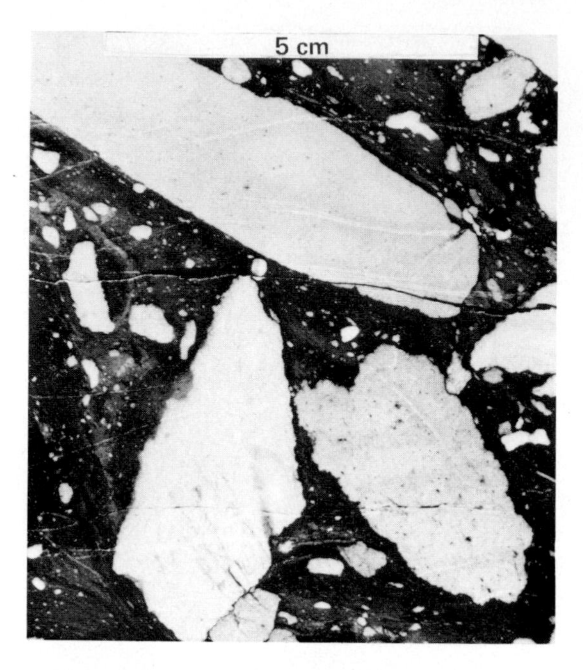

FIG. 36.—Breccia with contorted pelagic matrix (especially lower left). Slab; Poza Rica #217; photo by E. A. Shinn.

mum number of pelagic intervals in any well studied was about five, but it would almost certainly have been more with better core recovery. Furthermore, it was argued that most of the pelagic limestone occurs as discontinuous lenses so that the total number in any well represents only a few of the probable incursions of pelagic sedimentation. Finally, the pelagic matrix of the breccia indicates that this abundant lithology also represents some form of deeper water sedimentation.

Repeated large fluctuations in sea level are recognized during the Pleistocene and the Permian (Dunham, 1972), but these were glacial periods. Other causes of sea level fluctuations are possible, of course, such as changes in the volume of the ocean basins through sea-floor spreading or mountain building. No glaciation nor evidence for major fluctuations in sea level during the middle Cretaceous have been previously reported, although a major transgression occured in the Turonian which covered at least part of the crest of the Golden Lane escarpment with pelagic limestone (Agua Nueva). The weight given the objection to repeated fluctuations in sea level will depend largely on the prejudices of the reader—precisely the problem with this type of argument. Many ridiculed hypotheses in geology have later become "new insights."

Hypothesis 3, that all of the shallow-water skeletal fragments and clasts were derived from the Golden Lane reefs, transported down the escarpment, and deposited in the basin, is favored. It is consistent with the regional geology, the geometry of the Tamabra Limestone, and the detailed stratigraphy and sedimentology.

CONCLUSIONS

Environment and mechanics of deposition.—The Golden Lane escarpment must have had high potential for the development of sediment gravity flows. Disregarding the probability that erosion, karst formation, and differential subsidence have subdued the final depositional topography, the relief at the end of the middle Cretaceous was more than 1000 m and the slope up to 30° (scaled from Fig. 3). Instability in the depositional slope would result and mass movement would be promoted by any steepening or rapid loading of the slope. In a carbonate province, this would occur in the areas of reef growth, where skeletal production is intense and whence much of the sediment of the entire carbonate shelf margin is ultimately derived (Enos, 1974a). The reef itself is a minor part of the sediment package; most of the sediment accumulates in banks of skeletal sand.

Gravity flows of debris into the adjacent basin to produce the Tamabra Limestone sampled both the reef core and the skeletal-sand banks, but other environments were virtually excluded. The debris included lithified skeletal-fragment grainstone and packstone and uncemented, rounded skeletal fragments, mainly caprinids. Rounding of the skeletal fragments probably occurred in the wave-agitated shallow-shelf environment, prior to redeposition.

Mass movements may have been triggered in several ways. Oversteepening by accretion of skeletal-sand banks could produce slow creep or rapid slope failure. Oversteepening by growth of bound reefs could lead to failure which would produce massive, coherent blocks. Blocks would perhaps not be recognized from cores even if deposited discordantly with the surrounding sediment. Poorly coherent reef debris, a more abundant product from rudist reefs, would be disaggregated during transport. Shallow-water skeletal fragments encased in pelagic matrix (Figs. 17, 24B, 26C) may have had such a history.

Wave and organic erosion of reefal boundstone, submarine-lithified sediment, and island rock would produce lithoclasts. Lithoclasts which may reflect each of these histories have been identified from Tamabra breccia. The presence of islands is suggested by probable supratidal sediments from Golden Lane cores. The Bahama Banks provide modern analogs for muddy supratidal sediments formed leeward of rocky islands of cemented grainstones (Shinn and others, 1969).

The tendency for the steepest part of the shelf edge to be most subject to mass transport may help explain why some wells encounter backreef facies at the present topographic crest of the Golden Lane escarpment. Shelf-edge reefs and sand shoals which would have formed the depositional crest may now be contributing to the more than 300 km^3 of reef-derived material in the Tamabra Limestone (Viniegra and Castillo-Tejero, 1970).

The exact processes which moved material down the submarine slope and 10 to 15 km across the low-relief sea floor will remain the least understood part of the Tamabra depositional history. At least two distinct mechanisms appear to have been active.

One mechanism moved cemented clasts of shelf limestone, some skeletal debris, basinal clasts which were very cohesive or possibly cemented, and a matrix of basinal mud. The resulting deposit is breccia, probably of limited areal extent. Some breccia exhibits graded intervals, clasts floating in pelagic muddy matrix, and clasts oriented at high angles to bedding. The basinal clasts and the pelagic mud of the matrix are lithologies unknown from the Golden Lane escarpment. They must have been incorporated from the lower slope of the escarpment or from the basin floor during transport.

This mechanism is inferred to have been debris flow. The flows did not generally involve turbulent suspension of particles, hence the rarity of strong particle orientation, graded beds, and other sedimentary structures indicating current traction and particle-by-particle sedimentation. Properties of the transported mass must have changed considerably from the onset of transport where it presumably contained skeletal fragments and angular lithoclasts with little matrix. The original motion may have been as a poorly lubricated slump, perhaps of talus accumulated by wave erosion. The final product moved up to 15 km from the foot of the Golden Lane escarpment.

Transport of the rudist wackestone is inferred to have been by a similar mechanism. Relatively little basinal material was incorporated and more mud was present in the original sediment. Some sliding of large blocks of reef material may have been involved, although no blocks have been identified from the subsurface. Some large exotic blocks have been recognized in outcrops flanking the Valles-San Luis Potosi platform (Carrasco V., 1973 and this volume).

Different mechanisms may have transported the rounded skeletal fragments to produce skeletal fragment grainstone and packstone. Matrix was sparse or lacking in these deposits, or its nature was obscured completely by recrystallization. There is little evidence of cementation prior to erosion and transport. No basinal material was incorporated during transport. The accumulated deposits form a fairly homogeneous and extensive blanket of sand. Long axis orientation and imbrication of grains indicate deposition in a regimen dominated by currents, either persistent or spasmodic. The sparsity of matrix could be attributed to winnowing by bottom currents, but adjacent beds of breccia and pelagic wackestone with no indication of winnowing argue against this.

The inception of movement of the sorted and rounded skeletal material is envisioned as occurring through oversteepening or undercutting of shelf-margin sand banks interspersed among reefs along the shelf edge, but the rate and mechanics of movement are purely speculative. It could range from creep down submarine canyons (Shepard and Dill, 1966) or sand chutes (Enos, in press), to turbidity-current flow. Slow creep might be expected to incorporate basinal material from concurrent pelagic sedimentation, although recrystallization, and possibly winnowing could have obscured this effect. Turbidity-current flow would be expected to produce more distinct layering and graded beds.

The grainstone and packstone deposits are analogous to the enigmatic group of clean, well-sorted, massive-bedded, structureless, terrigenous sand bodies which accumulate in wedges or prisms along the margins of some turbidite basins. The unfortunate name "fluxoturbidite" has been applied to these bodies and covers a considerable range of ignorance. It must be confessed that the skeletal-fragment grainstone and packstone belongs in this company.

Application to oil exploration.—The conclusion that the Tamabra Limestone is a basinal deposit rather than a faulted shelf margin or a subsidiary shelf deposit has important consequences for oil exploration. If the Tamabra were a shelf-margin deposit, exploration for this type of reservoir would be confined to the positive side of the shelf edge, where most current exploration is concentrated. The localization of another Poza Rica might depend on major fluctuations of sea level, on faulting of a shelf margin, or on antecedent structure. If, however, the Tamabra is a basinal deposit, then exploration should be directed to the margins of basins adjacent to carbonate shelves, a large and relatively unexplored domain.

The question of which basins should be explored can be approached from a conceptual model. To predict where favorable conditions exist for development of porous basin-margin carbonate debris, it is necessary to isolate the controls. If deposits are formed by gravity-driven mass movements, depositional slope at the edge of the basin should be a prime factor. The length of the slope or, with a given slope angle, the relief between shelf and basin could be an important consideration in the size of body to be expected, both in the distance debris would be distributed from the shelf edge and in the thickness.

Availability of debris is an equally important factor. This is partially dependent upon the skeletal productivity of the shelf margin. In a geologic time framework, the best measure would be the rate of vertical accretion of the shelf edge. For example, isopachs (Fig. 7) indicate that as much as 1500 m of El Abra Limestone accumulated during the Albian-Cenomanian, a time span of perhaps 15 m.y. This gives an accumulation rate of 10 cm/1000 years, a respectable but not startling rate (Holocene sediments of the south Florida shelf margin accumulated at rates of 17 to 100 cm/1000 years; Enos, in press). Rates of cementation and type of reef-building organisms should also influence the availability of erodable material.

The most important controls (or at least the most readily quantified) on the volume or thickness of shelf debris transported into the basin would appear to be slope and relief at the shelf edge. It may be possible to derive an empirical relationship between these two parameters, which can be determined seismically in many cases, that would aid in evaluating the potential for reservoir facies early in an exploration effort. Figure 37 is a conceptual diagram of how this relationship might appear. Data are probably available in petroleum industry files to test the validity of this diagram. If successful, it could serve as a guide for future exploration.

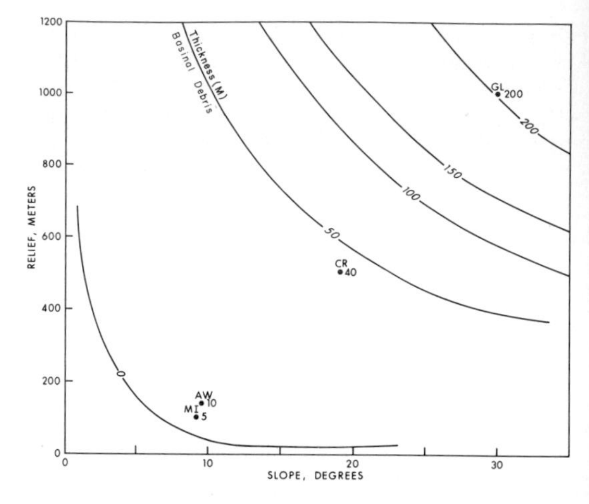

FIG. 37.—Conceptual relationship of shelf-edge relief, slope and thickness of basinal debris, contoured in meters. Four readily available values are plotted to suggest the quantitative relationship. The bare minimum of data presented here (4 points for a 3-dimensional surface) hint that a strong correlation may exist between slope and relief. If the relationship is strong enough, the problem would reduce to 2 dimensions; either parameter could be plotted against thickness. Data are: MI—Miette Reef (Mountjoy, 1967); AW—Ancient Wall (Mountjoy, 1967); CR—Capitan Reef (Newell and others, 1953; personal observations); GL—Golden Lane (this study).

Exploration potential of carbonate basins.—Carbonate basin margins offer new objectives for oil exploration (Cook and others, 1972, p. 467). Past exploration in carbonate provinces has been predicated on the shelf-margin model with its many possibilities for reservoir development. The shelf-margin model is attractive because the shelf edge may be readily recognizable from seismic data or limited well control. Just as this convenience is an impetus to shelf-margin exploration, it can also aid exploration on the "bad" side of the shelf edge, in the basin margin.

A carbonate basin bounded by a high, steep shelf edge has good potential for reservoir development and for exploitable petroleum accumulations at the basin margin. The following considerations converge to optimize potential:

1. Depositional facies—The relief and steepness of the shelf slope are probably the primary controls on the volume of redeposited shelf debris at the basin margin. Primary porosity may be high.

2. Predictability—Relief and steepness of the slope enhance the discovery and definition of the basin margin; hence, those controls producing the best target also give the best chance of prediction.

3. Favorable diagenesis—The conditions which produced extensive leaching in the Tamabra are not known, but an alternative to subaerial leaching is suggested above (ms. in preparation) which would be favored by high relief on the edge of the basin.

4. Source rocks—Fine-grained basinal sediment, whether carbonate or terrigenous, which encases the basin-margin detrital carbonate may juxtapose source rocks with the reservoir. The source-rock potential depends on organic richness and temperature history.

5. Seals—Fine-grained basinal sediment enclosing the detrital carbonates may very effectively seal the reservoir.

Only one Poza Rica has thus far been discovered. With the many miles of steep carbonate shelf edges which ringed the Lower Cretaceous Gulf of Mexico, as well as other carbonate basins, it is unlikely that only one Poza Rica exists.[9]

[9] Fragmentary reports on the Reforma trend in the Isthmus of Tehuantepec suggest that another Poza Rica may have already been discovered (Franco, 1975, 1976). It is not clear whether the geologic references to a Poza Rica analog are based on the shelf-margin model of Barnetche and Illing (1956) and Coogan and others (1972) or on the basin-margin model as presented here. Reference is made to at least two basinal fields within the trend (Franco, 1976).

ACKNOWLEDGEMENTS

This study was conducted for Shell Development Company in cooperation with Petroleós Mexicanos in 1969-70. Cores and work space were made available at the Poza Rica office of Petróleos Mexicanos, through the courtesy of Ing. Francisco Viniegra O. and Ing. Octavio Mejia D. Selected cores were released for detailed study at Shell Development Company. Mechanical logs and well-file data were provided for the cores studied. Two unpublished seismic lines were made available for cursory examination by Petróleos Mexicanos. Cores used by Coogan, Bebout, and Maggio in their study of the Golden Lane and Poza Rica trends were generously provided for study by Esso Production Research, Howard Gould, Manager.

This paper or earlier versions were critically reviewed by A. H. Coogan, H. E. Cook, J. F. Hubert, W. J. Koch, and Juergen Reinherdt, who made many helpful suggestions. Among the many colleagues at Shell and Petróleos Mexicanos who provided stimulation and assistance are M. R. Thomasson, Q. C. Hebrew, E. A. Shinn, W. J. Koch, R. L. Waite, R. M. Lloyd, E. López Ramos, R. Gonzáles G., A. Govela S., and R. Cabrera.

REFERENCES

AGUAYO-C., J. E., 1975, Sedimentary environments and diagenetic implications of the El Abra Limestone at its type locality, east Mexico: Ph.D. Dissert., Univ. Texas, Dallas, Texas, 165 p.

BACK, WILLIAM, 1963, Preliminary results of a study of calcium carbonate saturation of ground water in central Florida: Internat. Assoc. Sci. Hydrology Bull., v. 8, no. 3, p. 43-51.

BARNETCHE, ALFONSO, AND ILLING, L. V., 1956, The Tamabra Limestone of the Poza Rica oilfield, Veracruz, Mexico: 20th Internat. Geol. Congress, Mexico, D. F., 38 p.

BATHURST, R. G. C., 1971, Carbonate sediments and their diagenesis: Elsevier, Amsterdam, The Netherlands, 620 p.

BEBOUT, D. G., COOGAN, A. H., AND MAGGIO, C. M., 1969, Golden Lane-Poza Rica trends, Mexico—An alternate interpretation [abs.]: Am. Assoc. Petroleum Geologists Bull., v. 53, p. 706.

BECERRA-H., A., 1970, Estudio bioestratigráfico de la Formación Tamabra del Cretácico en el distrito de Poza Rica: Inst. Mexicano Petróleo Rev., v. 2, p. 21-25.

BLOXSOM, W. E., 1972, A Lower Cretaceous (Comanchean) prograding shelf and associated environments of deposition, northern Coahuila, Mexico: M.A. Thesis, Univ. Texas, Austin, Texas, 207 p.

BONET-M., FÉDERICO, F., 1952, La facies Urgoniana del Cretácico Medio de la región de Tampico: Asoc. Mexicana Geólogos Petroleros Bol., v. 4, p. 153-262.

——, 1963, Biostratigraphic notes on the Cretaceous of eastern Mexico: *In* Geology of Peregrina Canyon and Sierra de El Abra, Mexico: Corpus Christi Geol. Soc., Guidebook Ann. Field Trip, p. 36-48.

BOYD, D. R., 1963, Geology of the Golden Lane trend and related fields of the Tampico Embayment: *In* Geology of Peregrina Canyon and Sierra de El Abra, Mexico: Corpus Christi Geol. Soc., Guidebook Ann. Field Trip, p. 49-56.

CARRASCO-V., B., 1973, Exotic blocks of forereef slope, Cretaceous Valles-San Luis Potosi platform, Mexico [abs.]: Am. Assoc. Petroleum Geologists Bull., v. 57, p. 772.

CARRILLO-B., J., 1971, La plataforma Valles-San Luis Potosi: Asoc. Mexicana Geólogos Petroleros Bol., v. 23, p. 1-101.

COOGAN, A. H., BEBOUT, D. G., AND MAGGIO, C. M., 1972, Depositional environments and geologic history of Golden Lane and Poza Rica trend, an alternative view: Am. Assoc. Petroleum Geologists Bull., v. 56, p. 1419-1447.

COOK, H. E., MCDANIEL, P. N., MOUNTJOY, E. W., AND PRAY, L. C., 1972, Allochthonous carbonate debris flows at Devonian bank ('reef') margins, Alberta, Canada: Bull. Canadian Petroleum Geology, v. 20, p. 439-497.

CORPUS CHRISTI GEOLOGICAL SOCIETY, 1963, Geology of Peregrina Canyon and Sierra de El Abra, Mexico: Annual Field Trip Guidebook, 96 p.

DUNHAM, R. J., 1972, Capitan Reef, New Mexico and Texas: Facts and questions to aid interpretation and group discussion: Soc. Econ. Paleontologists and Mineralogists Permain Basin Sec., Pub. 72-14, 270 p.

ENOS, PAUL, 1974a, Carbonate sedimentation patterns of the south Florida shelf margin [abs.]: Am. Assoc. Petroleum Geologists, Ann. Meeting Abs., v. 1, p. 21.

——, 1974b, Reefs, platforms, and basins in middle Cretaceous in northeast Mexico: Am. Assoc. Petroleum Geologists Bull., v. 58, p. 800-809.

FRANCO, ALVARO, 1975, Mexico's crude-exporting role may be short-lived: Oil and Gas Jour. v. 73, no. 21, p. 25-27.

——, 1976, New Reforma finds push Mexico to new oil heights: Oil and Gas Jour., v. 74, no. 20, p. 71-74.

FRIEDMAN, G. M., 1965, Occurrence and stability relationships of aragonite, high-magnesium calcite, and low-magnesium calcite under deep-sea conditions: Geol. Soc. America Bull., v. 76, p. 1191-1196.

GONZÁLEZ-G., R., 1969, Areas con possibilidades de producción en sedimentos del Jurásico Superior (Caloviano-Titoniano): Inst. Mexicano Petroleo, Sem. Exploración Petrolera, Mesa Redonda no. 3(3), 19 p.

GRIFFITH, L. S., PITCHER, M. G., AND RICE, G. W., 1969, Quantitative environmental analysis of a Lower Cretaceous reef complex: *In* G. M. Friedman (ed.), Depositional environments in carbonate rocks: Soc. Econ. Paleontologists and Mineralogists, Spec. Pub. 14, p. 120-138.

Guzmán, E. J., 1967, Reef type stratigraphic traps in Mexico: Proc. 7th World Petroleum Congress, v. 2, p. 461-470.

Heim, Arnold, 1940, The front ranges of the Sierra Madre Oriental, Mexico, from Ciudad Victoria to Tamazunchale: Eclogae Geol. Helvetiae, v. 33, p. 313-352.

Islas-J., L., and Equia-A., H., 1961, El comportamiento sismológico del atolón conocido con el nombre de "Faja de Oro," Estado de Veracruz, México: Asoc. Mexicana Geofísicos Exploración Bol., v. 2, p. 19-41.

Joides, 1965, Ocean drilling on the continental margin: Science, v. 150, p. 709-716.

Kauffman, E. G., and Sohl, N. F., 1974, Structure and evolution of Antillean Cretaceous rudist frameworks: Naturf. Gesell. Basel Verh., v. 84, p. 399-467.

Kuenen, P. H., 1956, Experimental abrasion of pebbles. 2. Rolling by current: Jour. Geology, v. 64, p. 336-368.

López Ramos, Ernesto, 1950, Geología del subsuelo de tres pozos de exploración al sureste de Poza Rica, Veracruz: Asoc. Mexicana Geólogos Petroleros Bol., v. 2, p. 381-395.

Middleton, G. V., and Hampton, M. A., 1973, Sediment gravity flows: Mechanics of flow and deposition: *In* G. V. Middleton and A. H. Bouma (eds.), Turbidites and deep-water sedimentation: Soc. Econ. Paleontologists and Mineralogists Pacific Sec., Los Angeles, California, p. 1-38.

Mountjoy, E. W., 1967, Factors governing the development of the Frasnian Miette and Ancient Wall reef complexes (banks and biostromes): *In* D. H. Oswald (ed.), International symposium on the Devonian System, v. 2: Alberta Soc. Petroleum Geologists, Calgary, Alberta, p. 387-408.

Muir, J. M., 1936, Geology of the Tampico region, Mexico: Am. Assoc. Petroleum Geologists, Tulsa, Oklahoma, 280 p.

Murray, G. E., 1961, Geology of the Atlantic and Gulf Coastal Province of North America: Harper, New York, New York, 692 p.

——and Krutak, P. R., 1963, Regional geology of northeastern Mexico: *In* Geology of Pergrina Canyon and Sierra de El Abra, Mexico: Corpus Christi Geol. Soc., Guidebook Ann. Field Trip, p. 1-10.

Newell, N. D., Rigby, J. K., Fischer, A. G., Whiteman, A. J., Hickox, J. E., and Bradley, J. S., 1953, The Permian reef complex of the Guadalupe Mountains region, Texas and New Mexico: Freeman, San Francisco, California, 236 p.

Perkins, B. F., 1974, Paleoecology of a rudist reef complex in Comanche (Cretaceous) Glen Rose Limestone of central Texas: Geoscience and Man, v. 8, p. 131-173.

Rockwell, D. W., and Garcia Rojas, A., 1953, Coordination of seismic and geologic data in Poza Rica-Golden Lane area, Mexico: Am. Assoc. Petroleum Geologists Bull, v. 37, p. 2551-2565.

Salas-G., P., 1949, Geology and development of Poza Rica oil field, Vera Cruz, Mexico: Am. Association Petroleum Geologists Bull., v. 33, p. 1385-1409.

Shepard, F. P., and Dill, R. F., 1966, Submarine canyons and other sea valleys: Rand McNally, Chicago, Illinois, 381 p.

Shinn, E. A., Lloyd, R. M., and Ginsburg, R. N., 1969, Anatomy of a modern carbonate tidal-flat, Andros Island, Bahamas: Jour. Sed. Petrology, v. 39, p. 1202-1228.

Sotomayor-C., A., 1954, Distribución y causas de la porosidad en las calizas del Cretácico Medio en la región de Tampico, Poza Rica: Asoc. Mexicana Geólogos Petroleros Bol., v. 6, p. 157-206.

Stabler, C. L., and Mejia-D., O., 1975, Upper Jurassic of Gulf Coast of Mexico [abs.]: Am. Assoc. Petroleum Geologists Ann. Meeting Abs., v. 2, p. 71.

Viniegra-O., F., and Castillo-Tejero, C., 1970, Golden Lane fields, Veracruz, Mexico: *In* M. T. Halbouty (ed.), Geology of giant petroleum fields: Am. Assoc. Petroleum Geologists, Mem. 14, p. 309-325.

SEPM Special Publication No. 25, p. 315–317, November 1977

ABOUT THE AUTHORS

MICHAEL A. ARTHUR

Michael Arthur was born in Sacramento, California and received his B.A. degree (1971) and M.S. degree (1974) in geology from the University of California at Riverside. He is presently completing his Ph.D. degree at Princeton University. Michael's major interests include sedimentology, sedimentary geochemistry and stable isotope geochemistry.

HELEN K. BARKER

Helen Barker was born in Cleveland, Ohio, and obtained her A.B. degree from the College of Wooster, Wooster, Ohio, in 1973 and her M.S. degree from Brigham Young Univeristy, Provo, Utah in 1976. Her master's research focused on the petrography of carbonates from the oil producing Sunniland Trend of Florida. Currently, she is a geologist with Exxon Company USA, Houston, Texas.

HAROLD J. BISSELL

Harold Bissell was born in Springville, Utah. He received the B.S. degree in geology at Brigham Young University, and the M.S. and Ph.D. degrees in geology at the State University of Iowa. In addition, he pursued graduate studies in geology at Louisiana State University and sampled Holocene sediments in the Mississippi River Delta—Gulf of Mexico area. Harold has been visiting professor of geology at the University of Washington (Seattle), University of North Carolina at Chapel Hill, University of Florida (Gainesville), and the University of California, Riverside. His field studies in the Great Basin, Rocky Mountains, and Colorado Plateau have been, in large measure, in connection with various petroleum companies. Accordingly, his special interests have been centered around Paleozoic and Mesozoic stratigraphy, fusulinid biostratigraphy, and carbonate petrology and petrography. His publications reflect this research. He has taught at Brigham Young University since 1938, where he is currently Professor of Geology.

CHARLES W. BYERS

Charles Byers was born in Philadelphia. He received his B.S. from Marietta College and his M.Phil. and Ph.D degrees (1973) from Yale University. His thesis research involved a comparison of black shale paleoenvironments in rocks of Devonian and Cretaceous age. Charles has continued research on the topics of paleoenvironments and shale fabrics, with studies of Precambrian, Devonian, and Lower Cretaceous formations. In addition, he has initiated paleoecologic studies on the Cambrian and Ordovician rocks of Wisconsin, where he is Assistant Professor of Geology at the University of Wisconsin in Madison.

RICHARD K. CALLAHAN

Richard Callahan was born in Massachusetts and earned his B.S. degree at the University of Massachusetts. After serving three years in the U.S. Navy, he received the M.S. degree at the University of Massachusetts in 1974. Richard is now an exploration geologist with Sun Oil Company in Houston, Texas.

BALDOMERO CARRASCO-V

Baldomero Carrasco was born in Pachuca, Hgo., Mexico. He attended the Universidad Nacional Autonoma de Mexico from which he obtained his Ingeniero Geologo Degree (1962) and subsequently received an M.S. degree (1968) from the University of Texas at Austin. He worked for Petróleos Mexicanos from 1962 to 1967 in field geology, Tertiary micropaleontology, Mesozoic stratigraphy (mainly studies on Cretaceous ammonoids). Baldomero joined the Instituto Mexicano del Petroleo in 1968 where he has served as a Research Scientist in problems on paleontology, stratigraphy, sedimentation petrology and diagenesis of carbonate rocks mainly those of Cretaceous age. Currently, he is Head of the Department of Stratigraphy.

HARRY E. COOK

Harry was born in Fresno, California, where he was reared on Gallo wine and dried apricots. He received his B.A. at the University of California at Santa Barbara (1961) and his Ph.D. at the University of California at Berkeley (1966). While employed at Marathon Oil Company's Denver Research Center (1965–70) he conducted research on shoal-water and basinal carbonate rocks with emphasis on submarine sediment-gravity flow deposits. From 1970–74 he was an Associate Professor at the University of California at Riverside and a member of the JOIDES Deep Sea Drilling Project. Here his interests shifted toward deeper marine environments focusing on the geologic history of the equatorial Pacific and North American stratigraphic principles as applied to deep-sea sediments. Since 1974 he has been with the U.S. Geological Survey in Menlo Park, California where his current research includes stable isotope geochemistry of deep-water carbonates, comparative studies on ancient and modern continental slopes, sediment-gravity flow processes, and energy resources in deep-water marine environments.

GRAHAM R. DAVIES

Graham Davies was born in Perth, Western Australia. He received his B.Sc. in 1964 and Ph.D. in 1969 from the University of Western Australia. His doctoral research on Quaternary carbonate sedimentation in Shark Bay was published in AAPG Memoir 13. Graham was a Postdoctoral Associate at Rice University in 1968 and an Instructor in Geology at Rice in 1969. In 1970–71 he was an NRC Fellow at the Institute of Sedimentary and Petroleum Geology of the Geological Survey of Canada in Calgary, Alberta, and in 1971 he became a permanent staff member of that Institute. Graham has conducted studies in Australia, New Mexico, West Texas, Alberta, Yukon Territory and the Canadian Arctic Archipelago. These studies have included the stratigraphy, sedimentology, and diagenesis of carbonates and evaporites in both shallow and deep-water settings. He presently is Vice-President of Applied Geoscience And Technology (AGAT) Consultants Ltd., in Calgary.

PAUL ENOS

Paul was born and reared on Pennsylvanian limestones near Topeka, Kansas, and received a B.S. degree from the University of Kansas in 1956. A year as a Fulbright scholar at Universität Tübingen and two years as a psychological warrior in the Army were followed by graduate school at Stanford University (M.S. degree, 1961) and Yale University (Ph.D. degree, 1965). A dissertation on Ordovician flysch led naturally to research on recent carbonates in Florida when he was a research geologist with Shell Development Company from 1964–70. These two threads were united in the study of mass-transported basinal carbonates reported in this volume. As an Associate Professor of Geology at the State University of New York at Binghamton since 1970, he has extended his work on Mexican Cretaceous carbonates through outcrop study, core study of the Atlantic shelf-edge in the Early Cretaceous cored by the Deep Sea Drilling Project, and diagenetic study of the Poza Rica rocks while on leave at the University of Liverpool.

IAN EVANS

Ian Evans was born in Llanelli, Wales, and obtained his B.A. degree from the University College, Swansea (1966), his M.S. degree from the University of South Carolina (1968) and his Ph.D. degree from Texas A&M University (1971). Since 1972 he has been associated with the Earth Sciences and Resources Institute of the University of South Carolina which has sponsored his research activities in North Africa. Principal research interests lie in the areas of regional stratigraphy/sedimentation and paleoecology. Ian is currently Associate Professor of Geology at the University of Houston, Houston, Texas.

ALFRED G. FISCHER

Alfred Fischer was born in Germany and received his higher education in the United States (M.A. University of Wisconsin, Ph.D. Columbia University, 1950). He has been affiliated with the Virginia Polytechnic Institute, the Stanolind Oil and Gas Company, the University of Rochester, the University of Kansas, the International Petroleum Co. Ltd. (Esso), and Princeton University, where he holds the Blair Chair of Geology. He has been an NSF Senior Fellow, a Guggenheim Fellow, and holds the Lepold von Buch medal. His main field of interest is the synthesis of paleontological, sedimentological, and tectonic approaches toward an understanding of the behavior and the history of the biosphere. His work has been mainly with marine invertebrates, marine biogenic sediments, and the stratigraphy of marine deposits. He and his students have worked mainly in North America, Peru, the Austrian and German Alps, the Appennines, and the deep sea floor (Atlantic, Pacific). Al is currently directing a team-study of pelagic sediments that utilizes a combination of petrographical, geochemical, paleontological and magnetic approaches.

JOHN C. HOPKINS

John Hopkins was born in Auckland, New Zealand. He obtained his B.Sc. and M.Sc. degrees in geology at the University of Auckland. After a short period of employment with the Ministry of Works in New Zealand, he emigrated to Canada where he received his Ph.D degree at McGill University in 1972. His research interests have centered around petrography of both carbonate and clastic rocks. He headed Texaco's Geological Laboratory in Calgary, Alberta, Canada for several years. Currently John is on the geology department faculty at the University of Calgary, Alberta, Canada.

JOHN F. HUBERT

John Hubert was born in Massachusetts. He received his undergraduate training at Harvard College, followed by the M.S. degree at the University of Colorado, and the Ph.D. degree at Pennsylvania State University. From 1958–70, he taught at the University of Missouri-Columbia and since then at the University of Massachusetts. John's research has included studies on fluvial red beds, deltaic sequences, deep-sea sands, lacustrine rocks, flysch sequences, grain-size textural analyses, interpretation of heavy-mineral assemblages, carbonate petrology, and statistical geology.

CHRISTOPHER G. ST. C. KENDALL

Besides having the longest name of any living geologist, Chris arranged to have his birth in New Dehli, India. He received his B.S. degree at Trinity College in Dublin and his Ph.D. at Imperial College in London. This was followed by post-doctorates at the University of Texas at Austin and the University of Sydney in Australia. Chris came back to the United States where he was an Assistant Professor of Geology at Ohio State University for several years. Later he was a research geologist at Exxon Production Research in Houston, Texas, where he carried out studies on various aspects of carbonate rocks including petrology, environments of deposition and diagensis. Chris is currently Associate Director of the Earth Sciences and Resources Institute of the University of South Carolina.

IAN A. MCILREATH

Ian McIlreath was born in Ontario, Canada, and obtained his B.Sc. and M.Sc. degrees in engineering geology at Queens University. He recently received his Ph.D. from the University of Calgary. His research interest lies in determining the nature of Cambrian carbonate shelf to basin transitions exposed in Western Canada. Ian is now with Shell Canada Resources in Calgary, Alberta, Canada.

JUERGEN REINHARDT

Juergen Reinhardt was born in Entingen, West Germany, and emigrated to the United States in 1949. He received his A.B. degree from Brown University in 1968. Juergen began his studies of carbonate rocks in the Central Appalachians in 1970 at the John Hopkins University where he received his Ph.D. in 1973. Most of his recent research has involved fluvial and nearshore sedimentary rocks, first in Lower Cretaceous rocks in Maryland with the Maryland Geological Survey and since 1975 in Upper Cretaceous rocks in Georgia and Alabama with the U.S. Geological Survey in Reston, Virginia. In both his Cambro-Ordovician and Cretaceous studies Juergen has been most interested in the nature of the clastic-carbonate rock transition.

DONALD L. SMITH

Donald Smith was born in Laramie, Wyoming, and obtained his B.A. degree there at the University of Wyoming in 1965. He received his M.S. degree from the University of Wisconsin (1967) and his Ph.D. from the University of Montana (1972). Since 1972, he has served on the faculties of Idaho State and Montana State universities. Don's major research interests include Mississippian limestones of the Northern Rockies and Holocene carbonates of Bermuda.

ROBERT K. SUCHECKI

Robert Suchecki was born in Massachusetts. After serving two years in Korea with the U.S. Army, he enrolled at the University of Massachusetts, receiving the B.S. degree in 1973, followed by the M.S. degree in 1975. Robert is now studying for his Ph.D. degree at the University of Texas at Austin.

MICHAEL E. TAYLOR

Michael Taylor was born in Salt Lake City, Utah, and received his B.A. and M.S. degrees in geology from Utah State University. He received his Ph.D. in paleontology at the Universtiy of California at Berkeley in 1971. Since 1969, Michael has been employed by the U.S. Geological Survey in Washington, D.C., where he specializes in the paleontology of trilobites and lower Paleozoic stratigraphy. His research has dealt primarily with integration of paleontological and sedimentological data in the reconstruction of early Paleozoic environments and geography. This research deals mainly with the Cambrian and Early Ordovician of the Great Basin and upper New York state. Currently Michael is with the U.S. Geological Survey in Denver, Colorado.

DONALD A. YUREWICZ

Donald Yurewicz was born in Patuxent River, Maryland, and received his B.A. degree from Rutgers University in 1970. In 1973 and 1976 he received his M.S. and Ph.D. degrees at the University of Wisconsin at Madison where his research involved an integrated study of the sedimentology, paleoecology, and diagenesis of the Capitan Limestone. Don recently joined Exxon Production Research in Houston, Texas where he is working on a variety of research dealing with carbonate rocks.

AUTHOR INDEX

SUBJECT INDEX[1]

[1]Where reference is made to specific figures the text page is the number following the hyphen.